Genetically Modified Crops in Asia Pacific

Dedicated to our parents and wives for their love

Genetically Modified Crops in Asia Pacific

Editors: GT Gujar, Y Andi Trisyono and Mao Chen

CSIRO PUBLISHING

CABI

A catalogue record for this book is available from the National Library of Australia.

ISBN: 9781486310906 (hbk)
ISBN: 9781486310913 (epdf)
ISBN: 9781486310920 (epub)

Published in print in Australia and New Zealand, and in all other formats throughout the world, by CSIRO Publishing.

CSIRO Publishing
Locked Bag 10
Clayton South VIC 3169
Australia

Telephone: +61 3 9545 8400
Email: publishing.sales@csiro.au
Website: www.publish.csiro.au

A catalogue record for this book is available from the British Library, London, UK.

Published in print only, throughout the world (except in Australia and New Zealand), by CABI.

ISBN 9781789248418

CABI
Nosworthy Way
Wallingford
Oxfordshire OX10 8DE
UK

Tel: +44 (0)1491 832111
Fax: +44 (0)1491 833508
Email: info@cabi.org
Website: www.cabi.org

CABI
We Work
One Lincoln Street, 24th Floor
Boston, MA 02111
USA

Tel: +1 (617)682-9015
E-mail: cabi-nao@cabi.org

Front cover: (top) Corn field in Son La, Vietnam (photo by Mao Chen); (bottom) Canola field in Australia (photo by Pragna Prathap and Aleena Joy).

Edited by Adrienne de Kretser, Righting Writing
Cover design by James Kelly
Typeset by Envisage
Printed in Singapore by Markono Print Media Pte Ltd

Sep20_01

Foreword

Most of the Asia Pacific region comprises developing countries. Rural population accounts for ~50%, and more than 20% of the population live in poverty. Nearly half of the world's food-insecure people and two-thirds of its undernourished people live in this region. Agriculture is overwhelmed with severe limitations like poor agri-infrastructure, technological development and management of natural resources. The Asia Pacific region has emerged as a major importer of food grains, despite being agriculture-intensive. Therefore, it is of great significance to improve agricultural production capacity through scientific and technological innovations to solve the food scarcity in Asia.

Transgenic technology offers great promise for breeding crops selectively with genes of interest isolated from different sources. This is not possible with conventional crop improvement tactics. In 2018, biotech/genetically modified (GM) crops were grown across 191.7 million ha – the largest area since their introduction in 1996. These crops have helped to enhance productivity and farm income and to improve environmental quality.

The Asia Pacific region in general lags in the use of biotech/GM technologies, despite the success of GM crops in some of its member nations. India, China, Australia and the Philippines are the major countries growing these crops, while many have approved them for food and feed use. Cultivation of GM crops can help alleviate hunger and malnutrition in the Asia Pacific region, as the UN's sustainable development goal of achieving zero hunger by 2030 remains a challenge.

This book *Genetically Modified Crops in Asia Pacific* comprehensively reviews the development and commercialisation of GM crops in Asia Pacific countries along with other issues like sustainability, environmental safety, biotechnology policies and regulations. I believe that its publication will help all stakeholders in Asia Pacific countries in the promotion of agricultural production and economy. I congratulate the editors as well as authors for their contributions in this excellent book.

Kongming Wu
Vice President and Professor
Chinese Academy of Agricultural Sciences
Beijing
20 January 2020

Contents

Preface

This book *Genetically Modified Crops in Asia Pacific* is our attempt to contribute to the food security of the region where most of the world's poor live, among 60% of the global population on 30% of the global land mass. It also represents an immense diversity in terms of food habits, living standards, education, religion, geography and politics. GM crops and other agricultural technologies have emerged as forerunners to protect crops against biotic as well as abiotic stresses. Other traits related to specific nutrients and aesthetic values are also added to these crops through GM technologies. Time and again, elaborate assessments have shown the safety of GM crops (NASEM Report 2016 at http://nas-sites.org/ge-crops/). In 2018, as per ISAAA, biotech crops were planted on 191.7 million hectares in 26 countries, a record increase in area of more than 100-fold, from 1.7 million hectares in 1996. Biotech crops increased crop production by 658 million tons valued at US$186 billion in 1996–2016, and reduced environmental impact by (i) saving 671 million kg of active ingredients of pesticides in 1996–2016, (ii) reducing Environmental Impact Quotient by 18.4% in 1996–2016 and (iii) reducing CO_2 emissions in 2016 by 27 billion kg. Biotech crops also helped alleviate poverty through uplifting the economic conditions of 16–17 million small farmers and their families, totalling >65 million people (http://www.isaaa.org/resources/publications/briefs/54/).

The developed member nations of the Asia Pacific region have readily accepted GM crops for food, feed and environmental release, with Australia in the lead. The developing member states have cautiously followed this path of crop improvement. No member state has given up on the immense potential of biotechnology for the benefit of its people. Almost every country has faced opposition to GM technology due to its allegedly ill effects on human health and environment, and the possibility of misuse of genetic diversity by technology developers, mostly in the developed nations. Over the years, the international community has played a positive role in mitigating this opposition, with 196 nations becoming party to the Convention on Biological Diversity, 171 nations to the Cartegena Protocol on biosafety and 122 to the Nagoya Protocol on access and benefit sharing. Many member nations of Asia Pacific have accordingly enacted necessary legislation for approval of GM crops, which involved collaboration with the international community to solve locally perceived constraints to achieving higher crop productivity. Development of insect-protective Bt brinjal (eggplant), drought-tolerant sugarcane, insect-resistant rice, late blight resistant GM potato and golden rice are pointers in this direction.

This book deals with various aspects of GM crops like research, development, adoption and sustainability. As these crops offer protection against pest insects and disease-causing pathogens, which are likely to evolve resistance to toxins over time, resistance management is prerequisite for their sustainability. The book also discusses the regulatory efforts of member states to ensure evaluation of GM crops for safety for humans and the environment, allaying fears of their ill-effects. It probes into some aspects of gene editing and nanobiotechnology as these hold prospects of crop improvement and protection. We are grateful to our 60+ authors, most of them from the Asia Pacific region, for their excellent contributions that shaped this book. We hope that the book will be useful to stakeholders, especially in the Asia Pacific, give impetus to the development of agricultural technologies in general and biotechnology in particular, and serve the cause of food security for the region in the near future.

We acknowledge with gratitude our reviewers including Drs GT Behere, J Bentur, TVK Singh, MP Moharil, S Subramanian, VK Dilawari, Rupa Nair, VK Baranwal, RN Pudake, Chinnababu Naik, VT Gajbhiye, AU Solanke, VCB Naik, VS Nagrare, M Asim, W Akbar, AR Alam and Nguyen Thi Hanh, along with the contributors.

Thanks are also due to Lauren Webb, Nicki Dennis and Tracey Kudis of CSIRO Publishing for consistent support and guidance throughout this project.

GT Gujar
Y Andi Trisyono
Mao Chen
26 January 2020

About the editors

GT Gujar, currently associated with the AGRIBIOSYS, New Delhi, retired as Head of Division of Entomology from ICAR-Indian Agricultural Research Institute (IARI), New Delhi, India in 2015 after 37 years of academic service and as Visiting Research Professor at DST-Institute of Advanced Study in Science and Technology, Guwahati in 2016. He has wide experience in teaching and research especially related to insect resistance management including Bt toxins, integrated pest management (IPM) and insect-resistant transgenic crops, in collaboration with national and international researchers including post-graduate students. He has an MSc and a PhD in Entomology from the IARI and post-doctoral experience as Marie Curie EU Post-doctoral fellow in the UK and Germany. He has served as an expert on GM crops risk assessment and management and on IPM-related panels in government and educational institutions.

Y Andi Trisyono is a Professor of Entomology in the faculty of Agriculture at the University of Gadjah Mada in Indonesia, where he teaches insecticide toxicology, insect physiology, transgenic crops and integrated pest management. He earned an MS and a PhD in Entomology from Michigan State University and University of Missouri, respectively. His research is focused on risk assessment and risk management of crop protection technologies. He has served as a consultant for national and international organisations. He has been a member of the technical team for environmental risk assessment of GM crops and has served as co-chair for the advisory committee in crop protection for the Ministry of Agriculture in Indonesia.

Mao Chen, currently working as Head of Market Development Field Testing at APAC Bayer Crop Science in Singapore, has an MS and a PhD in Entomology from Zhejiang University, Hangzhou, China and post-doctoral experience at NY State Agricultural Station (Cornell University), Geneva. He has specialised in insect resistance management in Bt crops and published extensively in insect resistance management, integrated pest management and invasive insect species management. He has experience with biotech crop regulations. Mao joined Monsanto in 2010 working on insecticidal gene discovery, crop evaluation and resistance management at Monsanto Research Center, St Louis (USA) then worked as Lead, Biotech Regulatory Affairs for Bayer Crop Science (earlier Monsanto) for the Asia Pacific in Singapore until 2019.

Contributors

Joel L Adorada, Bureau of Plant Industry, Department of Agriculture, Malate, Manila, Philippines.

Edwin P Alcantara, National Institute of Molecular Biology and Biotechnology, University of the Philippines Los Baños College, Laguna 4031, Philippines.

Alberta Dinar Ambarwati, Indonesian Center for Agricultural Biotechnology and Genetic Resources Research and Development, Indonesian Agency of Agricultural Research and Development, Ministry of Agriculture; Jl.Tentara Pelajar No. 3A, Bogor 16111, Indonesia.

Sagar S Arya, TERI-Deakin Nano Biotechnology Centre, The Energy and Resources Institute, Gurgaon 122001, India.

Abul K Azad, Bangladesh Agricultural Research Institute, Gazipur 1701, Bangladesh.

Prem L Bhalla, Plant Molecular Biology and Biotechnology Laboratory, Faculty of Veterinary and Agricultural Sciences, University of Melbourne, Parkville VIC 3010, Australia.

Graham Brookes, PG Economics Ltd, Dorchester, UK.

Merdelyn Caasi-Lit, Entomology Laboratory, Institute of Plant Breeding, College of Agriculture and Food Science, University of the Philippines Los Baños College, Laguna 4031, Philippines.

Barbara L Caoili, Institute of Weed Science, Entomology and Plant Pathology, College of Agriculture and Food Science, University of the Philippines Los Baños College, Laguna 4031, Philippines.

Sally Ceeney, Cotton Australia, 247 Coward St, Mascot NSW 2020, Australia.

Bharat Char, Maharashtra Hybrid Seeds Co. Pvt Ltd, Mahyco Research Centre, Dawalwadi, Jalna-Aurangabad Rd, Jalna 431203, India.

Hafiza Masooma Naseer Cheema, Department of Plant Breeding and Genetics, University of Agriculture, Faisalabad, Pakistan.

Mao Chen, Bayer (South-east Asia) Pte Ltd, OCBC Center East, 63 Chulia St Level 14, 049514 Singapore.

Ratul K Das, TERI-Deakin Nano Biotechnology Centre, The Energy and Resources Institute, Gurgaon 122001, India.

Mukul K Dubey, TERI-Deakin Nano Biotechnology Centre, The Energy and Resources Institute, Gurgaon 122001, India.

1

Biotech/genetically modified crops in Asia Pacific: a way forward

GT Gujar, Y Andi Trisyono and Mao Chen

Introduction

Asia Pacific is generally considered a region in or near the western Pacific Ocean, and includes much of East, South and South-East Asia and Oceania. It has temperate to tropical climates, mountainous to plateau geography, and landlocked to island localities. Asia Pacific has 41 sovereign countries. It occupies ~29.23 million sq km area (22.5% of the world's land area of 130 million sq km), supports 3.86 billion people (more than 50% of the world's population) and had about $25.3 trillion gross domestic product (GDP) (~36% of the world's wealth) in 2017. It also presents a large variation in agriculture. About 20–50% of land in the region is devoted to farming. More than 40% of farm area is cultivated by small farmers owning less than 2 ha. Yet, these farmers have a much larger share in production of staple crops. Asia Pacific produces ~50% of the world's cereals, 45% of roots and tubers, 48% of pulses, 49% of fruits and 66% of vegetables (FAO-RAP 2012). It is the largest producer of rice and spices in the world. Besides crops, it has more vibrant capture and inland fish farming than any other region of the world. Asia Pacific consumes 10–30% or even more of its animal proteins from fish, while most of the developed world uses less than 10% of fish-derived animal proteins. Asia Pacific thus utilises its abundance of natural resources to supplement its nutritional needs.

Agriculture contributes ~18% of the overall GDP in South Asia, 8% in East Asia and 12% each in the South-East Asia and Oceania regions.[1] Agricultural production has increased significantly in the last five decades in this region, contributing to food and nutritional security for its people.

Asia Pacific also represents numerically a large number of hungry people (~491 million, more than 50% of the world's hungry in 2016) as well as stunted, wasted and obese children (136 million, ~56.7% of the world's children in 2017). Changes in dietary habits in recent years have led to obesity in children and more specifically in adults of South Asia. Unfortunately, the rising population of hungry people since 2015 has suffered from malnutrition and consequently poor health. Even deficiency of macronutrients like protein abounds, including in some countries where vegetarians largely depend upon pulses for their needs.

The world population is projected to increase from 7.5 billion in 2018 to ~9.7 billion in 2050, with the greatest increase being in Asia. The region's limited land and other natural resources will act as constraints to achieving higher production. At present, per capita availability of arable land has shrunken from 0.16 ha to 0.09 ha over the last 50 years in the East Asia Pacific region (excluding high-income countries).[2] Similarly, water availability has declined with increasing cropping intensity despite a deepening water table. Untreated waste water is polluting the environment. According to Asian Development Bank (2016), Asia Pacific is fast becoming a hotspot for water insecurity.[3] Furthermore, most of its population is migrating to cities and large townships. Changes in food habits, a rising middle class, and awareness of hygiene and health are increasingly pressing for alternatives to the current food needs. The diversification of food basket to meet nutritional security is slowly taking place, with many countries increasing their fruit and vegetable production; China and India together are the largest producers of fruits and vegetables in the world. Despite increases in agricultural production, the challenge of feeding the ever-increasing population is formidable. China regularly imports food products and is predicted to continue to do so in the range of 3–5% of its domestic production up to 2025 (Huang *et al.* 2017). India imports a huge quantity of edible oil (~15–16 million tons annually) as it faces huge shortfall of edible oilseeds. A survey conducted by Rabobank showed that Asia imports 11% more food products than it exports.[4] It is estimated that as India becomes the most populous nation by 2030, it may be able to meet its food grain production needs but not its fruit, vegetable, edible oil, poultry and bovine meat needs, if productivity and post-harvest constraints are not more effectively addressed (Kumar *et al.* 2016). Similarly, ~60% of India's population may face severe shortfall in terms of three macronutrients like calories (especially from carbohydrates), digestible protein and fat, during 2030–2050 (Ritchie *et al.* 2018). Thus, this region represents a challenge wherein developmental goals of eradication of hunger and all forms of malnutrition by 2030 need to be achieved with the available land and other non-renewable resources in a sustainable manner.

Unfortunately, most countries of the region spend less than 1% of their GDP on agricultural research and development, although it is well known that returns on agricultural research and development are much more than investments. Agriculture will have to be more technology-oriented in the context of socio-economics and the environment.[5]

GM crops in Asia Pacific

Traditionally, crop improvement involves selection of best varieties based on the traits of high productivity and adaptability in the local environment, followed by breeding to combine the traits to the furthest extent possible. However, these efforts often meet with the limited success as traits are not well quantifiable and often linked with other less desirable traits from the donor parents. Moreover, the breeding is restricted to a great extent within a genus as interspecific breeding poses incompatibility. However, genetic modification (GM) makes it possible to introgress traits from any biota with precision and express its full potential, within a reasonable timeframe. GM offers traits like disease resistance (DR), insect resistance (IR), herbicide tolerance (HT), abiotic stress tolerance (AST) especially drought tolerance, pollination control, modified product quality (MPQ) and overall yield improvement.

The first large-scale planting of insect-resistant GM crops was started in 1996 in the USA. It included Bt cotton, Bt corn and Bt potato with IR trait derived from *Bacillus thuringiensis* (Bt) to control key pests. Australia too commercialised Bt cotton in 1996 thus leading a group in the Asia Pacific region, followed by China in 1997. After the Americas (both North and South), the Asia Pacific region has major stakes in GM crops, involving ~10.5% (~19.5 million ha) of its area. However, the development of GM crops varied a lot in this region. China developed GM tobacco as early as 1992 to control viral infection and cultivated it until 1996. India and the Philippines approved their first GM crops (Bt cotton and Bt corn, respectively) as late as 2002. Indonesia planted Bt cotton from 2000–2003. At present, the Asia

Pacific region has many events approved for food, feed and cultivation purposes (Table 1.1). Most of these traits are related to soybean, maize, canola and cotton.

As many as 14 countries of Asia Pacific have approved various traits for 19 GM crops including alfalfa, argentine canola, carnation, cotton, eggplant, maize, papaya, petunia, poplar, potato, rice, rose, safflower, soybean, sugar beet, sugarcane, sweet pepper, tomato and wheat, totalling 1125.[6] Most of these approvals of various events are however in the developed countries such as Australia (135), Japan (317), New Zealand (108), South Korea (166) and Taiwan (141). Only few countries like Australia, China, India, Pakistan, the Philippines and Vietnam are growing GM crops on a large scale. India, China and Pakistan cultivate mostly GM cotton with IR trait to control bollworms, while Australia has GM cotton for IR and HT traits. With Bt cotton, India has emerged as the largest producer (6.2 million tons of lint cotton, 24.8% of the world's production) of cotton and is a significant exporter along with Pakistan and Australia. Australia has also developed GM canola (improved version of rapeseed, *Brassica napus*) which is tolerant to herbicides like glyphosate and glufosinate. The GM canola was first commercialised in 2008 and since then has been planted annually over 440 000 ha (~17.7% of canola area), contributing significantly to the production of ~3.3 million tons of oilseeds. The Philippines first planted GM maize in 2003. GM maize (predominantly with IR and HT traits, constituting ~95% of GM maize area) was planted over ~642 000 ha (46% of total maize area) by ~470 500 farmers in 2017, helping to increase productivity to ~3.31 ton/ha and achieve self-sufficiency. Studies on the impact of these GM crops have shown increased productivity and farm income (Brookes and Barfoot 2018a), decline in pesticide use, reduction in land and environmental footprints, and increase in environmental quality (Brookes and Barfoot 2018b; Zilberman *et al.* 2018). Chapter 2 in this book specifically discusses the impact of GM crops in Asia Pacific.

Rice is the most important crop of this region. It is grown over ~144 million ha in Asia (~88% of the world's 163.3 million ha of rice area). China grows over 30.6 million ha and India 43.4 million ha, and together produce ~358 million tons annually. Asia produces ~87% of the world production of rice. It is predicted that Asia will have to increase its production of rice by 39% by 2050 to meet its current consumption pattern in view of projected population increase (Milovanovic and Smutka 2017). Being a staple crop of the region, rice meets most of the calorie needs and can also supply essential micronutrients if it is fortified. More than 200 million children suffer from vitamin A deficiency (VAD) and a large number of them (2.5–3.0 million) die due to compromised immune systems, succumbing to various infectious diseases arising out of VAD. This deficiency can be met by eating green vegetables, carrots, milk, butter and other dietary sources. However, VAD continues to plague children and adults particularly in South-East Asia, as people ignore the need for or cannot access a balanced nutritious diet. Professors Ingo Portrykus and Peter Beyer envisaged an approach to counter VAD through expressing carotenoid pathway to produce β-carotene, a precursor of vitamin A, in rice seed in 1999.[7] As a result, the first GM event producing 1.6 μg/g of β-carotene in rice grains (golden rice) was developed in 2002. The new experimental lines (GR2) of Asiatic/Japonica GM rice expressed as much as 36–37 μg/g of β-carotene in their grains. Golden rice has proven to be an efficient way of meeting the vitamin A requirement, as a bowl of 100–150 g of cooked rice is enough to meet 60% of a child's vitamin A needs. Golden rice is being bred in to local cultivars in the Philippines, Bangladesh, India and many other countries to adapt them to the local conditions, yield as much as other rice cultivars do, and suit the local taste. In 2018, the US Environment Protection Authority, Health Canada and Food Standards Australia and New Zealand approved golden rice for human consumption in their respective countries.[8] In 2019, the Philippines approved it for human consumption.

Iron deficiency affects nearly 30% of the region's population, leading to anaemia. Johnson and colleagues from the Australian universities of Adelaide and Melbourne have overexpressed

Table 1.1. Current status of GM events of crops approved in the Asia Pacific region

Crop	Country	Events	Traits approved	Cultivation
Alfalfa	Australia	3	HT, MPQ	None
	Japan	5	HT, MPQ	None
	New Zealand	3	HT, MPQ	
	Philippines	2	HT	
	Singapore	3	HT, MPQ	
	South Korea	5	HT, MPQ	
Argentine canola	Australia	23	HT	Yes, HT
	China	12	HT, pollen control	
	Japan	21	HT, pollen control	
	New Zealand	16	HT, pollen control	
	Philippines	2	HT	
	Singapore	2	HT	
	South Korea	14	HT, pollen control	
	Taiwan	9	HT, pollen control	
Carnation	Australia	12	HT, MPQ	
	Japan	6	MPQ	
Cotton	Australia	28	HT, IR	Yes, HT, IR
	China	11	HT, IR	Yes, IR
	India	6	IR	Yes, IR
	Japan	38	HT, IR	
	Myanmar	1	IR	Yes, IR
	New Zealand	23	HT, IR	
	Pakistan	1	IR	Yes, IR
	Philippines	8	HT, IR	
	Singapore	4	HT, IR	
	South Korea	30	HT, IR	
	Taiwan	25	HT, IR	
Eggplant	Bangladesh	1	IR	Yes, IR
Maize	Australia	27	AST, DR, HT, IR, MPQ	
	China	18	AST, HT, IR	
	Indonesia	9	AST, HT, IR	
	Japan	199	HT, IR, MPQ	
	New Zealand	27	HT, IR, MPQ	
	Philippines	52	AST, HT, IR, MPQ	Yes, HT, IR
	Singapore	15	AST, HT, IR	
	South Korea	81	AST, HT, IR, MPQ	
	Taiwan	79	AST, HT, IR, MPQ	
	Thailand	12	AST, HT, IR	
	Vietnam	14	AST, HT, IR	Yes in 2016
Papaya	China	1	DR	Yes in 2015
	Japan	1	DR	
Petunia	China	1	MPQ	
Poplar	China	2	IR	
Potato	Australia	17	DR, IR, MPQ	
	Japan	9	DR, IR, MPQ	
	New Zealand	17	DR, IR, MPQ	
	Philippines	8	DR, IR	
	South Korea	8	DR, IR	

Crop	Country	Events	Traits approved	Cultivation
Rice	Australia	2	HT, MPQ	
	China	2	IR	
	Japan	1	MPQ	
	New Zealand	2	HT, MPQ	
	Philippines	1	HT	
Rose	Australia	1	MPQ	
	Japan	2	MPQ	
Safflower	Australia	2	MPQ	
Soybean	Australia	17	HT, MPQ	
	China	12	HT, MPQ	
	India	5	HT	
	Indonesia	6	HT, MPQ	
	Japan	32	HT, MPQ	
	New Zealand	17	HT, MPQ	
	Philippines	16	HT, MPQ	
	Singapore	10	HT, MPQ	
	South Korea	27	HT, MPQ	
	Taiwan	27	HT, MPQ	
	Thailand	3	HT	
	Vietnam	8	HT, MPQ	
Sugarcane	Indonesia	3	AST	
Sugar beet	Australia	2	HT	
	China	1	HT	
	Japan	3	HT	
	New Zealand	2	HT	
	Philippines	1	HT	
	Singapore	1	HT	
	South Korea	1	HT	
	Taiwan	1	HT	
Sweet pepper	China	1	DR	
Tomato	China	3	DR, MPQ	
Wheat	Australia	1	HT	
	New Zealand	1	HT	

AST = abiotic stress tolerance, DR = disease resistance, HT = herbicide tolerance, IR = insect resistance, MPQ = modified product quality, pollen control for producing hybrids.

OsNAS1, *OsNAS2* and *OsNAS3* genes and reported significant increase in iron and zinc content of rice grains with *OsNAS2* to meet daily dietary needs (Johnson *et al.* 2011). Rice is infested by many insect pests which reduce yields by ~25% (Oerke 2006). Bt rice was developed in China (Huang *et al.* 2005) and subsequently got approval in 2009.[9] However, Bt rice was never allowed to be grown due to social pressure, caused by inadequate understanding and acceptance by consumers (Li *et al.* 2016). Similarly, Bt rice was grown on 4000 ha for seed purpose in 2005, but not commercially grown in Iran. As this region faces water deficit, rice may have to be grown during aerobic conditions with alternate drying and wetting of fields, or in localities where droughts are frequent. Weeds are known to be key constraints for such rice cultivation. Hence, herbicide tolerant rice may have a prospect. The USA approved glufosinate HT rice in 2007. Stacking of these genes could alleviate nutritional deficiencies as well as help to increase productivity through management of resources and cost reduction.

Another important GM crop is Bt brinjal developed by India's Maharashtra Hybrid Seed Co. (Mahyco) in 2000s. Brinjal is widely grown in South-East Asia and West Asia but suffers a lot from insect pests, especially brinjal fruit and shoot borer (*Leucinodes orbonalis*). Farmers resort to repeated insecticidal sprays, thus contaminating the produce with residues, polluting the environment and often incurring losses due to inadequate control. Bangladesh Agricultural Research Institute (BARI) developed four local Bt varieties from Mahyco's EE1 event, that showed efficacy against brinjal fruit and shoot borer. Bt brinjal was approved in Bangladesh in 2013. Since then, stewardship of Bt brinjal development for its sustainability has been carried out under the USAID's PBS and Feed the Future- South Asia Eggplant Improvement Partnership. Bt brinjal has been grown by ~27 000 farmers to date. The success of Bt brinjal is influencing the development of other GM crops in Bangladesh (Shelton *et al.* 2018). Shelton and his colleagues in this book explain the role of Bt brinjal in detail in Chapter 9.

Sugarcane is an important crop of the Asia Pacific region.[10] Asia Pacific produces as much sugarcane as Brazil, the largest producer at 800 million tons annually. India produces ~348.5 million tons while China produces ~122.7 million tons annually. Sugarcane requires huge amounts of water for its growth. The Asia Pacific region is facing huge water deficit with sinking water table and contamination with heavy metals.[11] In 2013 Indonesia approved its first GM crop, drought-resistant sugarcane (NXI-1T, NXI-4T, NXI-6T), for commercial cultivation.[12]

Sugar is an important food commodity traded worldwide. Brazil is the key player, exporting cane sugar to as many as 150 countries, as it produces nearly 41% of the world's sugarcane annually. In 2017, Brazil approved Bt sugarcane (CTB141175/01-A) to control sugarcane borer (*Diatraea saccharalis*) which is reported to cause annual loss of about US$1.4 billion. Cane sugar from Bt sugarcane is approved in the USA, Canada, Australia and New Zealand for human consumption. Similarly, Brazil grows GM soybean (33.7 million ha, 97% of total soybean) which is traded worldwide, thus debunking myths about a possible adverse effect on export of GM crop like rice, as concern for the rice-exporting countries of Asia.

While GM technology is being developed for specific needs, there is a great reluctance to accelerate the development of conventional GM crops like cotton, corn, canola and soybean in this region. China as well as India are not growing new GM crops other than Bt cotton and GM papaya and Bt cotton, respectively. Moreover, there are no updates of Bt cotton incorporating new traits in these two countries. The Philippines, another important grower of GM crops, especially corn, is hesitating to approve other technologies or crops, while Myanmar has only recently approved Bt cotton. At the same time, illegal technologies are spreading among farmers; for example the use of HT cotton in India, HT corn in the Philippines, GM/Bt corn in China, and legal or illegal imports of virus-resistant papaya seeds. India reportedly grew an unapproved HT cotton in as many as eight states in 2017 and 2018. In the state of Maharashtra alone, the government seized ~18 million HT seed packets in 2018 and 13 million HT seed packets in 2017.[13] There seems to be mismatch between farmers' interests and consumers' perceptions as reflected in the government action.

Paradoxically, this region thus offers a contrast in adoption of GM technology that may be explained by its social, political, religious and economical diversity. Impact studies of Bt crops in many countries showed the economic and social benefits of GM technology (Fischer *et al.* 2015; Sadashivappa 2015). Furthermore, GM technology has helped in reducing pesticide use and improving environmental quality. Yet, this argument has not gone well with stakeholders (Kolady 2011). For instance, Bt brinjal is not approved in India, despite its recommendation for environmental release by the regulatory body Genetic Engineering Approval Committee of the Ministry of Environment and Forests as early as 2009. Another important instance is that of pollen control technology, which is significant in developing hybrids which have higher productivity. Despite large-scale investment from the public sector, GM mustard (hybrid) is not approved in India, although this technology is likely to increase oilseed productivity. Similarly,

China's reluctance to allow commercial cultivation of Bt rice and other crops, and upgrades to Bt cotton, is proving costly to its food security. Thus, access to GM crops depends upon political and regulatory settings (Fischer *et al.* 2015). This in turn depends upon the perceptions of consumers in terms of risks and benefits of GM crops, knowledge and trust, and personal values (Lucht 2015). Redundancy or indifference in stewardship of GM technology has cost dearly as the sustainability of GM crops is under environmental stress. Bt cotton approved in India is becoming less economic as the pink bollworm has evolved field-level resistance to Bt toxins and no new updates to Bt technology are made available (Dhurua and Gujar 2011; Naik *et al.* 2018). Resistance to GM crops in the target pest species is spreading widely in many countries and threatening to have worldwide effects similar to antibiotics for gram-negative bacteria, greenhouse gases and carbon dioxide emission (Jørgensen *et al.* 2018). The complexity of GM technology may explain the slow progress in some countries with varying levels of education in science and technology, in developing legal frameworks for crop biosafety, protecting biodiversity and ensuring intellectual property (IP) rights. For example, some native seed companies in India have refused to honour agreements with the technology provider of Bt cotton in view of differences in understanding of IP rights, non-recognition of IP for live material like seed, creation of monopoly and misuse of dominance in the marketplace under Competition Commission of India laws and the *Essential Commodities Act*. The government-determined pricing of Bt cotton seed has further discouraged investments for updating Bt cotton technology in India. A weak implementation framework allows illegal development of GE technology and its adoption by the farmers, as in case of HT cotton (Gulati and Hussain 2017). At the same time, the governmental control over seed pricing irrespective of IP rights claims has helped farmers to reduce input costs, and the seed companies to increase competitiveness in the production of quality Bt cotton seed in India. Bt cotton, which is essentially a cash crop, can have an adverse effect on food security at local levels, as cash from such crops is not eventually utilised for buying food grains or other essentials on a long-term basis. Thus, stewardship of GM technology has social, regulatory, economic, environmental and technological implications impacting human societies in different ways. Clearing these cobwebs will help us to forge ahead in achieving the ultimate goal of food and nutritional security.

Acknowledgements

We are grateful to Drs TM Manjunath and GT Behere for their useful comments.

Endnotes

1. http://www.fao.org/3/mw252en/mw252en.pdf
2. https://data.worldbank.org/indicator/%20AG.LND.ARBL.HA.PC?end=2015&start=1965
3. https://www.adb.org/sites/default/files/publication/189411/awdo-2016.pdf
4. https://economics.rabobank.com/publications/2016/february/asia-pacific-agricultural-perspectives/
5. http://www.fao.org/3/mw252en/mw252en.pdf
6. http://www.isaaa.org/gmapprovaldatabase/default.asp
7. http://goldenrice.org/
8. https://allianceforscience.cornell.edu/blog/2018/05/us-fda-approves-golden-rice/
9. http://www.isaaa.org/kc/cropbiotechupdate/article/default.asp?ID=5112
10. https://www.statista.com/statistics/681930/asia-pacific-sugarcane-production-by-country/
11. https://economics.rabobank.com/publications/2016/february/asia-pacific-agricultural-perspectives/
12. https://www.eurofins.de/food-analysis/food-news/food-testing-news/genetically-modified-sugarcane/
13. https://timesofindia.indiatimes.com/city/mumbai/80000ha-of-cotton-crop-using-illegal-ht-seeds-agri-officials/articleshow/65813266.cms

References

Brookes G, Barfoot P (2018a) Farm income and production impacts of using GM crop technology 1996–2016. *GM Crops and Food: Biotechnology in Agriculture and the Food Chain* **9**, 59–89. doi:10.1080/21645698.2018.1464866

Brookes G, Barfoot P (2018b) Environmental impacts of genetically modified (GM) crop use 1996–2016: Impacts on pesticide use and carbon emissions. *GM Crops and Food: Biotechnology in Agriculture and the Food Chain* **9**, 109–139. doi:10.1080/21645698.2018.1476792

Dhurua S, Gujar GT (2011) Field-evolved resistance to Bt toxin Cry1Ac in pink bollworm, *Pectinophora gossypiella* (Saunders) (Lepidoptera: Gelechiidae) from India. *Pest Management Science* **67**, 898–903. doi:10.1002/ps.2127

FAO-RAP (2012) *Regional Conference for Asia and the Pacific. Selected Indicators of Food and Agricultural Development in the Asia-Pacific Region 2001–2011.* Food and Agriculture Organization of the United Nations, Bangkok.

Fischer K, Ekener-Petersen E, Rydhmer L, Björnberg KE (2015) Social impacts of GM crops in agriculture: a systematic literature review. *Sustainability* **7**, 8598–8620. doi:10.3390/su7078598

Gulati A, Hussain S (2017) Innovators, not pirates. *The Indian Express*, 23 October.

Huang J, Hu R, Rozelle S, Pray C (2005) Insect-resistant GM rice in farmers' fields: assessing productivity and health effects in China. *Science* **308**(5722), 688–690. doi:10.1126/science.1108972

Huang J-K, Wei W, Cui Q, Xie W (2017) The prospects for China's food security and imports: will China starve the world via imports? *Journal of Integrative Agriculture* **16**, 2933–2944. doi:10.1016/S2095-3119(17)61756-8

Johnson AAT, Kyriacou B, Callahan DL, Carruthers L, Stangoulis J, Lambi E, Tester M (2011) Constitutive overexpression of the OsNAS gene family reveals single gene strategies for effective iron- and zinc-biofortification of rice endosperm. *PLoS One* **6**(9), e24476. doi:10.1371/journal.pone.0024476

Jørgensen PS, Aktipis A, Brown Z, Carrière Y, Downes S, Dunn RR, Epstein G, Frisvold GB, Hawthorne D, Gröhn YT, Gujar GT, Jasovský D, Klein EY, Klein F, Lhermie G, Mota-Sanchez D, Omoto C, Schlüter M, Scott HM, Wernli D, Carroll SP (2018) Antibiotic and pesticide susceptibility and the Anthropocene operating space. *Nature Sustainability* **1**, 632–641. doi:10.1038/s41893-018-0164-3

Kolady DE (2011) Are we missing something in the debate on Bt eggplant in India? *Asian Biotechnology and Development Review* **13**, 39–51.

Kumar P, Joshi PK, Mittal S (2016) Demand vs supply of food in India: futuristic projection. *Proceedings of the Indian National Science Academy* **82**, 1579–1586. doi:10.16943/ptinsa/2016/48889

Li Y, Hallerman EM, Liu Q, Wu KM, Peng Y (2016) The development and status of Bt rice in China. *Plant Biotechnology Journal* **14**, 839–848. doi:10.1111/pbi.12464

Lucht JM (2015) Public acceptance of plant biotechnology and GM crops. *Viruses* **7**, 4254–4281. doi:10.3390/v7082819

Milovanovic V, Smutka L (2017) Asian countries in the global rice market. *Acta Universitatis Agriculturae et Silviculturae Medelianae Brunensis* **65**, 679–688. doi:10.11118/actaun201765020679

Naik VCB, Kumbhare S, Kranthi S, Satuja U, Kranthi KR (2018) Field-evolved resistance of pink bollworm, *Pectinophora gossypiella* (Saunders) (Lepidoptera: Gelechiidae) to transgenic Bt cotton expressing Cry1Ac and Cry2Ab in India. *Pest Management Science* **74**, 2544–2554. doi:10.1002/ps.5038

Oerke E-C (2006) Crop losses to pests. *Journal of Agricultural Science* **144**, 31–43. doi:10.1017/S0021859605005708

Ritchie H, Reay D, Higgins P (2018) Sustainable food security in India: domestic production and macronutrient availability. *PLoS ONE* **13**(3), e0193766. doi:10.1371/journal.pone.0193766

Sadashivappa P (2015) Socio-economic impacts of Bt cotton adoption in India: evidence from panel data. *AgBioForum* **18**, 193–208.

Shelton AM, Hossain MJ, Paranjape V, Azad AK, Rahman ML, Khan ASMMR, Prodhan MZH, Rashid MA, Majumdar R, Hossain MA, Hussain SS, Huesing JE, McCandless L (2018) Bt eggplant project in Bangladesh: history, present status and future direction. *Frontiers in Bioengineering and Biotechnology* **6**, 106. doi:10.3389/fbioe.2018.00106

Zilberman D, Holland TG, Trilnick I (2018) Agricultural GMOs: what we know and where the scientists disagree. *Sustainability* **10**, 1514. doi:10.3390/su10051514

2

Economic and environmental impact of genetically modified crops in Asia Pacific

Graham Brookes

Introduction

Genetically modified (GM) crop technology has been widely used in some countries of the Asia Pacific region for more than 20 years, having been first used in Australian cotton crops in 1996. Since then, its use in cotton has been extended to China (1997), India (2002), Myanmar (2007) and Pakistan (2009). The technology has been adopted in corn crops in the Philippines (2003) and Vietnam (2015), canola in Australia (2008) and brinjal in Bangladesh (2014).

The main traits so far commercialised have essentially been derived from bacteria and convey:

- tolerance to specific herbicides (notably glyphosate and glufosinate) in corn, cotton and canola (spring oilseed rape). The herbicide tolerance (HT) technology allows for 'over the top' spraying of crops with these specific broad-spectrum herbicides, that target both grass and broad-leaved weeds;
- resistance to specific insect pests of corn, cotton and brinjal. This insect resistance (IR) or Bt technology involving plant incorporation of toxin genes from the soil bacterium *Bacillus thuringiensis* (Bt) offers resistance in plants to major pests such as corn borers (e.g. Asian corn borer (ACB) – *Ostrinia furnacalis*), corn earworm (CEW) – *Helicoverpa zea*, bollworm and budworm (heliothine species) in cotton and brinjal shoot and fruit borer (*Leucinodes orbonalis*).

This chapter presents an assessment of some of the key economic and environmental impacts associated with the adoption of GM crop technology in the Asia Pacific region. The analysis focuses on:

- gross farm income effects on costs of production, yield, production and farm income;
- changes in the amount of insecticides and herbicides applied to the GM crops relative to conventionally grown alternatives;
- the contribution towards reducing global greenhouse gas (GHG) emissions.

It draws on the material presented in the findings of analysis into the global impact of GM crops by the author (Brookes and Barfoot 2018a,b) that covers the period 1996 to 2016.

Adoption

Although the first commercial GM crops (tomatoes) were planted in 1994, 1996 was the first year in which a significant area of crops containing GM traits, 1.66 million ha, was planted. Since then there has been a dramatic increase in plantings. By 2016 the global planted area was 177.6 million ha.

In terms of the share of the main crops in which GM traits have been commercialised (soybeans, corn (maize), cotton and canola), GM traits accounted for 48% of the global plantings to these four crops in 2016 (Figure 2.1).

Almost all of the global GM crop area derives from soybeans, corn, cotton and canola. In 2016, GM soybeans accounted for the largest share (52%), followed by corn (30%), cotton (13%) and canola (5%).

In terms of the share of total global plantings to these four crops, GM traits accounted for the majority of soybean plantings (75%) in 2016. For the other three main crops, the GM shares were 29% for corn, 72% for cotton and 26% for canola (Figure 2.2).

Figure 2.2 summarises the breakdown of the main GM traits planted globally in 2016. HT soybeans dominated, accounting for 36% of the total, followed by insect-resistant (IR, largely Bt) and HT corn each with 20.6% share, IR soybeans (also containing HT technology) 8.9%, IR cotton 8% and HT cotton 2%. The total plantings by trait produces a higher global planted area (251 million ha) than the global area by crop (177.6 million ha) because of planting of some crops containing the stacked traits of HT and IR.

The USA had the largest share of global GM crop plantings in 2016 (38%), followed by Brazil (28%). The other main countries planting GM crops were Argentina, India, Canada and China (Figure 2.3).

In terms of the GM traits used and the share of crop production in the Asia-adopting countries, in 2016 the technology accounted for important shares of total production of three main crops, in some countries of this region. More specifically:

- in Australia, GM traits now account for almost all cotton production, with IR and HT traited seed present in 95% and 98% respectively of the total crop. HT canola was used on 19% of the canola crop;
- in China, IR cotton, first adopted in 1997, accounted for 95% of the crop in 2016. GM virus-resistant papaya has also been used in China since 2008;
- in India, IR cotton was first adopted in 2002 and its use increased rapidly in subsequent years, so that by 2016 this technology dominated cotton production (96% of the total);
- IR cotton is also grown in Pakistan and Myanmar and dominates production in each country;
- in the Philippines, IR corn was first used commercially in 2003, with HT corn adopted from 2006. Seed containing both traits were planted in 24% of the total crop in 2016;
- Vietnam adopted IR and HT stacked corn in 2015 and 3% of the crop used seed containing this technology in 2016;
- IR brinjal has been grown in Bangladesh since 2014 and in 2017 was planted on about 5% of the crop (2400 ha).

In the context of global plantings of GM crops, the Asia Pacific region accounted for 5%, 1%, 78% and 0% respectively of the global area devoted to GM corn, canola, cotton and soybean crops in 2016.

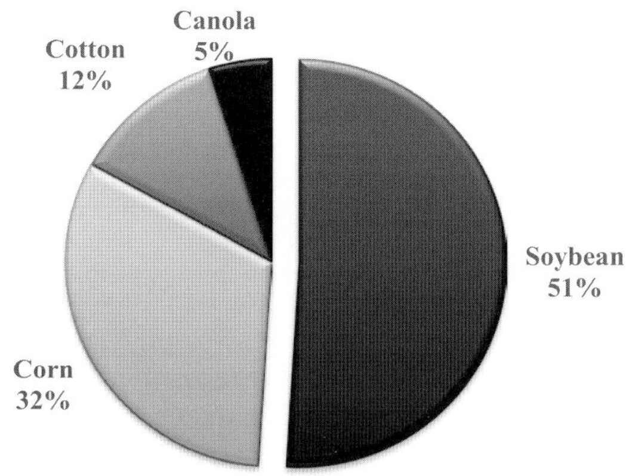

Figure 2.1 GM crop plantings 2016 by crop (base area of the four GM crops: 177.6 million ha). Sources: Various including ISAAA, Canola Council of Canada, CropLife Canada, USDA, CSIRO, ArgenBio, National Ministries of Agriculture (Mexico, the Philippines, Spain, Vietnam), Grains South Africa.

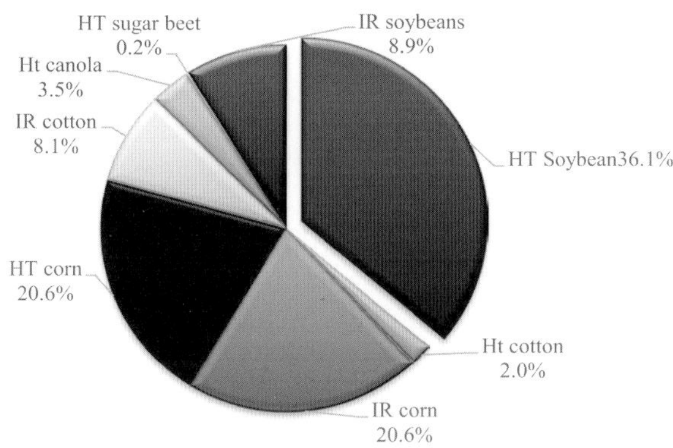

Figure 2.2 Global GM crop plantings by main trait and crop in 2016. Sources: Various including ISAAA, Canola Council of Canada, CropLife Canada, USDA, CSIRO, ArgenBio, National Ministries of Agriculture (Mexico, the Philippines, Spain, Vietnam), Grains South Africa.

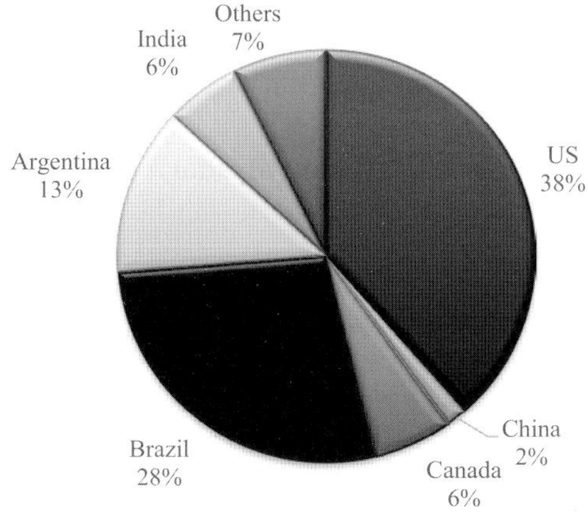

Figure 2.3 Global GM crop plantings 2016 by country. Sources: Various including ISAAA, Canola Council of Canada, CropLife Canada, USDA, CSIRO, ArgenBio, National Ministries of Agriculture (Mexico, the Philippines, Spain, Vietnam), Grains South Africa.

Socio-economic impacts

Impact on yields and farm income

IR crops

The main way in which IR seed technology has impacted on crops and farms has been through lowering the levels of pest damage and hence delivering higher yields. In case of IR corn-growing countries, an average yield increase of 18.2% and 7.2% was reported in the Philippines and Vietnam, respectively. For IR cotton, yield increases ranged widely – China 10%, Pakistan 21%, India 30% and Myanmar 30.7% – on the basis of adoption during 1996–2016.

The greatest improvement in yields has occurred in developing countries, where conventional methods of pest control have been least effective; for example, due to poorly developed extension and advisory services, and lack of access to finance to fund use of crop protection application equipment and products. In developed countries, the IR crops are mainly associated with cost savings due to reduced insecticide use. These effects can be seen in the level of farm income gains from the adoption of these technologies, as shown in Table 2.1.

The substantial benefit to Australian cotton farmers from using IR cotton has arisen from highly effective pest control that has enabled farmers to significantly reduce the use of insecticides. It is interesting to note that higher yields from the use of IR cotton technology have occurred in many user countries except in Australia, where levels of heliothine species pest control with intensive insecticide use and other tactics were good. The main benefits and reasons for adoption of this technology in Australia are the significant cost savings and the associated environmental gains from reduced insecticide use when compared to conventional (non-GM) cotton.

Cotton is a crop that has traditionally been subjected to numerous insecticide treatments in order to control budworm (*Helicoverpa punctigera*) and bollworm (*Helicoverpa armigera*) pests which can devastate cotton crops. Before the availability of IR cotton technology, most Australian cotton crops were typically sprayed with insecticides 11 times per season (range 5–19) in order to deliver effective control of these pests, especially *H. armigera*. The availability

Table 2.1. IR crops: average gross farm income benefit 1996–2016 (US$/ha)

Country	IR corn: cost of technology ($/ha)	IR corn (income benefit after deduction of cost of technology) ($/ha)	Aggregate income benefit IR corn ($ million)	IR cotton: cost of technology ($/ha)	IR cotton (income benefit after deduction of cost of technology) ($/ha)	Aggregate income benefit IR cotton ($ million)
Philippines	30–47	100	553.0	n/a	n/a	n/a
China	n/a	n/a	n/a	38–60	349	19 644.9
Australia	n/a	n/a	n/a	85–299	211	953.7
India	n/a	n/a	n/a	12–54	207	21 121.7
Myanmar	n/a	n/a	n/a	17–20	160	358.4
Pakistan	n/a	n/a	n/a	4–15	235	4794.3
Vietnam	38–42	105	4.0	n/a	n/a	n/a

1. The range in values for cost of technology relates to annual changes in the average cost paid by farmers. It varies for reasons such as the price of the technology set by seed companies, the nature and effectiveness of the technology (e.g. in Australia, second-generation 'Bollgard' cotton offered protection against a wider range of pests than the earlier first generation of 'Bollgard' technology), exchange rates, average seed rates and values identified in different studies.
2. Average across all countries is a weighted average based on areas planted in each user country.
3. n/a = not applicable (not grown).

of IR cotton technology has enabled cotton farmers to substantially reduce the number of insecticide treatments to those necessary to control the cotton pests that IR technology does not target. Cotton crops are now typically subjected to no more than two to four insecticide treatments per crop. This significantly reduced need to spray insecticides has resulted in substantial savings in expenditure for insecticides and their application.

At the aggregate level, the gross farm income gains from using IR corn and cotton in 2016 across the region were US$71.6 million and US$2.83 billion, respectively. Cumulatively since 1996, the gains were US$557 million for IR corn use in the Philippines and Vietnam and US$46.9 billion for IR cotton use in the five countries of Australia, China, India, Myanmar and Pakistan. These aggregate farm income gains equate to an average farm income gain of US$100/ha for the use of IR corn and US$253/hectare for IR cotton.

In the context of global farm income gains, the Asia Pacific region accounted for 2% and 85% of the global farm income gains derived from the use of IR corn and IR cotton, respectively, in 2016.

HT crops

In the Asia Pacific region, the use of HT crop technology has been more limited than IR crop technology, with HT corn used in the Philippines and Vietnam, and HT cotton and HT canola used in Australia.

The main impact of HT (largely tolerant to glyphosate) technology has been to provide more cost-effective (less expensive) and easier weed control for farmers. Some users have also derived higher yields from better weed control (relative to weed control obtained from conventional technology: Brookes and Barfoot 2018a). The magnitude of these impacts varies by country and year, and is mainly due to prevailing costs of different herbicides used in HT systems versus conventional alternatives, the mix and amounts of herbicides applied, the cost of accessing the HT technology and the levels of weed problems.

HT corn

The adoption of HT corn has resulted in yield gains from improved weed control in both the Philippines and Vietnam (Table 2.2) and led to more cost-effective and cheaper weed control. In 2016, the total farm income gain from using this technology in these two countries was US$15.7 million with the cumulative gain over the period 2006–2016 being US$172.4 million. Within this, 65% was due to yield gains and the rest derived from lower costs of production. The average gain across the two countries was US$31/ha.

HT cotton

The use of HT cotton in Australia delivered a gross farm income gain of about $7.6 million in 2016 and the total gross farm income benefit was $99.1 million in the 1997–2016 period. This equates to an average gain of $28/ha, derived entirely from cost savings.

Table 2.2. HT corn: summary of average gross farm income impacts 1996–2016 (US$/ha)

Country	Cost of technology	Average gross farm income benefit (after deduction of cost of technology)	Aggregate income benefit ($ million)	Type of benefit
Philippines	24–47	31	171.0	Cost savings plus yield gains of +5%
Vietnam	26–28	37	1.43	Cost savings plus yield gains of +5%

HT canola

HT canola (tolerant to glyphosate) has been grown in Australia since 2008. The gross farm income impact associated with the adoption of this technology has been $89.9 million, with 78% of the gain arising from higher yields and the balance due to lower costs of weed control. In 2016, the income gain from the adoption of HT technology was $15.5 million. Since 2008, the average farm income gain was $45/ha.

In the context of global farm income gains, the Asia Pacific region accounted for 0.7%, 3% and 11% of the global farm income gains derived from the use of HT corn, canola and cotton, respectively, in 2016.

Crop production effects

Based on the yield impacts in the direct farm income benefit calculations above, GM crops have added important volumes to regional production of corn, cotton and canola (Table 2.3).

The IR traits used in corn and cotton have accounted for 77% of the additional corn production and all of the additional cotton production. Positive yield impacts from the use of this technology have occurred in all user countries, except for IR cotton in Australia where, as indicated above, the levels of heliothine (boll and budworm pests) pest control previously obtained with intensive insecticide use were already very good. The average yield impact across the total area planted to these traits has been +18% for corn and +17% for cotton.

Other impacts

The adoption of biotech crops in Asia Pacific countries has also had several secondary economic impacts.

Impact on labour use

One of the important cost savings arising from the use of IR technology, especially in cotton, comes from reduced frequency of insecticide applications, and, where HT technology is used, reduced need for hand weeding. This has had a knock-on effect in terms of reducing the demand for, and use of, farm labour for spraying and weeding. However, the higher yields associated with biotech IR technology, especially cotton, have increased harvesting costs. For example, in South Africa, Gouse *et al.* (2005, 2006b) reported no significant change in labour costs from the adoption of IR cotton as the higher harvest costs offset the effects of reduced labour for insecticide applications. Gouse *et al.* (2006a) examined the impact of using IR and HT corn in Kwazulu-Natal (South Africa) and found that the net labour requirements fell relative to conventional production systems because of reduced needs for insecticide applications and hand weeding. This affected all types of labour – men, women and children – with women the most affected (having been the main source of labour for hand weeding). In this regional context, the labour changes were viewed positively because there was a general agricultural sector labour shortage, any reduction in labour requirements for children resulted in improved school attendance, it allowed women more

Table 2.3. Additional crop production arising from positive yield effects of GM crops in Asia Pacific countries

	Additional production (tonnes)	
	1996–2016	2016
Corn	3 513 800	453 300
Cotton	22 255 200	1 860 000
Canola	283 000	5600

time for child care and allowed men to take up (better-paid) employment opportunities outside agriculture.

Impact on the wider rural economy

Subramanian and Qaim (2009) examined this issue in India via the monitoring of household expenditure patterns and activities. While this was only a small piece of research, it provided a useful insight into wider economic impacts and was representative of regions in central and southern India. Its key findings were that biotech IR cotton had delivered a net creation of rural employment, with the additional harvest labour requirements being greater than the reductions associated with pest scouting and spraying. This had gender implications given that it has been mostly females who gained, relative to males who lost out. This analysis did, however identify that on average the saved male family labour was re-employed efficiently in alternative agricultural and non-agricultural activities so that the overall returns to male labour in the local economy increased. In some cases, where the labour changes were intra-family, the reduced male labour requirement included redeployment into household chores, thus improving the quality of life for females two-fold via higher income and a reduction in domestic chores. In addition, higher levels of farm income associated with the use of biotech cotton (e.g. wage income for female workers rose by an average of $40/ha) have increased rural household incomes, resulting in a knock-on benefit to the local economy in terms of increased expenditure on household goods and services (e.g. household amenities, repairs and renovation) and social commitments such as the education of children (Kathage and Qaim 2012). This category of benefit was also identified in relation to the adoption of stacked (IR and HT) corn in the Philippines. Here the additional income associated with higher yields enabled farming families to better feed and clothe their families, to undertake property repairs and improvements and to invest in better education for children (Torres *et al.* 2013).

Reduction of risk

Biotech IR technology has been cited in several studies as delivering improved risk management. Essentially, the technology takes away much of the worry of significant pest damage occurring and is, therefore, highly valued by farmers. These studies include Brookes (2008) in relation to biotech IR corn (Spain) and Crost and Shankar (2008) in relation to IR cotton (India). This may be especially beneficial for smaller farmers who tend to have fewer options than larger farmers to reduce their vulnerability to risk (Zilberman *et al.* 2007).

Higher quality of crop

There is a growing body of research evidence relating to the superior quality of biotech IR corn relative to conventional and organic corn from the reduction in borer pest damage (Bakan *et al.* 2002; Poeydemenge 2006a,b). Reduced pest damage results in a reduction in the amount of fumonisin building up in the crop and, as a result, a significant fall in the presence of cancer-causing mycotoxins in corn. This then results in less corn being rejected by users in both the food and feed-using sectors.

Environmental impacts

GM traits in corn, cotton and canola have contributed to a significant reduction in the environmental impact associated with insecticide and herbicide use on the areas devoted to GM crops in the Asia Pacific region (Table 2.4). Since 1996, the use of GM technology has directly resulted in insecticide and herbicide use on the GM crop area in the Asia Pacific region falling by 268 million kg of active ingredient (a 29% reduction).

Table 2.4. Impact of changes in the use of herbicides and insecticides from growing GM crops in the Asia Pacific region, 1996–2016

Trait	Change in volume of active ingredient used (million kg)	Change in active ingredient use on GM crops (%)	% change in environmental impact associated with herbicide or insecticide use on GM crops	Area GM trait 2016 (ha)
HT canola	–1.00	–3.9	–3.4	Australia: 448 000
HT cotton	–4.19	–17.5	–23.0	Australia: 568 000
IR cotton	–260.5	–30.9	–34.0	16 273 000
IR corn	n/a	n/a	n/a	Philippines: 688 000
HT corn	–2.5	–15.6	–32	Philippines: 690 000
Totals	–268.19	–29	–34	

IR corn – no data available. Adoption of this technology mainly applies to the Philippines during this period, where use of insecticides for conventional control of Asian corn boring pests is not available but reported by industry to have been limited.

While changes in volume of pesticides applied to crops can be a useful indicator of environmental impact, it is an imperfect measure because it does not account for differences in the specific pest or weed control programs used in GM and conventional cropping systems. A better measure of the environmental impact associated with pesticide use, known as the environmental impact quotient (EIQ), shows that the environmental impact associated with herbicide and insecticide use on the area planted to GM crops in the Asia Pacific region between 1996 and 2015 fell by 34%. The EIQ (Kovach *et al.* 1992) distils the various environmental and health impacts of individual pesticides in different GM and conventional production systems into a single field value per hectare and draws on key toxicity and environmental exposure data related to individual products. It therefore provides a better measure to contrast and compare the impact of various pesticides on the environment and human health than weight of active ingredient alone. However, it should be noted that the EIQ is an indicator only (primarily of toxicity) and does not take into account all environmental issues and impacts, as elaborated by Brookes and Barfoot (2018b).

In both absolute and per hectare terms, the largest environmental gain has been associated with the adoption of IR cotton. Since 1996, cotton farmers have used 260.5 million kg less insecticide active ingredient in IR cotton crops (a 30.9% reduction) and thus reduced the associated environmental impact of insecticide use on this crop by 34%, compared to the levels of use that would have otherwise occurred if this crop area had used conventional (non-GM) seed. In absolute terms, the use of insecticides (per ha) on the largely GM IR cotton crop has fallen by nearly 90% compared to levels of use on conventional cotton crops in the mid-1990s. GM IR seed technology has been a significant factor in this reduction, together with new insecticides and improved methods of pest monitoring and management.

In recent years, where over 90% of the cotton crop has used GM IR technology, the reduction in insecticide use has annually been equal to ~60% of what would have been used if conventional (non-GM) cotton had been grown.

The significant reduction in insecticide use through adoption of GM IR cotton, coupled with better pest management, has also made an important contribution to improving water quality (e.g. in the north-east rivers of Australia: Kennedy *et al.* 2013).

GHG emission cuts

GM crops have also delivered savings in GHG emissions. This derives from two principal sources:

- reduced fuel use from less frequent herbicide or insecticide applications and/or a reduction in the energy use in soil cultivation. The fuel savings associated with making fewer spray runs (relative to conventional crops) and the switch to conservation, reduced and no-tillage farming systems have resulted in permanent savings in CO_2 emissions;
- the use of no-till and reduced-till farming systems. No-till farming means that the ground is not ploughed at all, while reduced tillage means that the ground is disturbed less than it would be with traditional tillage systems. For example, under a no-till farming system, soybean seeds are planted through the organic material that is left over from a previous crop such as corn, cotton or wheat, or wheat/barley is planted through the organic material of a previous canola crop.

These production systems have increased significantly with the adoption of HT crops because the HT technology has improved farmers' ability to control competing weeds, reducing the need to partly rely on soil cultivation and seedbed preparation as means to getting good levels of weed control. As a result, tractor fuel use for tillage is reduced, soil quality is enhanced and levels of soil erosion cut, leading to lower GHG emissions from soil. These soil-based GHG emission savings have occurred mostly in North and South America and are mainly associated with corn and soybean crop production systems.

In the Asia Pacific region, the main GHG emission savings from the adoption of GM crops have been associated with reduced insecticide spraying on Australian cotton. While soil-based GHG emission savings associated with no/reduced tillage production systems occur in Australia, these production systems were widely used before the availability of GM HT technology in canola and cotton. Therefore, the GHG savings are not directly attributable to the widespread adoption of GM HT technology in Australia even though the availability of this technology has probably helped many Australian canola farmers to continue to use no/reduced tillage production systems. Also, while the adoption of IR cotton has resulted in a large cut in the number of insecticide applications on cotton in China, India, Myanmar and Pakistan, those applications were typically made by individuals with backpack sprayers rather than fuel-using mechanical methods.

Between 1996 and 2016, the widespread adoption of IR cotton has resulted in 36.2 million fewer spray runs on cotton crops, a saving of 30.3 million litres of fuel and a reduction in GHG emissions of 81.1 million kg of CO_2.

Conclusion

The adoption of crop biotechnology by many farmers in the Asia Pacific region over the last 20 years has delivered important economic and environmental benefits. The biotech IR traits in corn and cotton have mostly delivered higher incomes through improved yields, and environmental gains mostly from decreased use of insecticides, while the gains from biotech HT traits in corn, cotton and canola have come largely from reduced costs of production and increased use of more environmentally benign herbicides.

In the last 20 years, crop biotechnology has helped many farmers in the Asia Pacific region grow more food and fibre using fewer resources, by reducing the damage caused by pests. The highest yield increases have occurred in developing countries, such as the Philippines, India and Pakistan. This has contributed to a more reliable and secure food and fibre supply base in these countries.

With higher yields and less time and money spent managing pests and weeds, farmers have earned higher incomes. This has proved to be especially valuable for cotton farmers in the Asia Pacific region where, in 2016, an average $9 was received for each extra dollar invested in biotech IR cotton seeds. This helped them not only invest in their farms and support their communities, but better feed, clothe and provide good medical care, education and housing for their families.

Crop biotechnology has also helped farmers to be more efficient with their application of crop protection products, which not only reduces their environmental impact, but saves time and money. It is changing agriculture's land footprint by allowing farmers to grow more without needing to use additional land. To maintain production levels at 2016 levels without biotech IR technology the region's cotton farmers would, for example, have had to plant an additional 2.4 million ha of cotton – an area equivalent to 17% of the land area of Bangladesh.

References

Bakan B, Melcion D, Richard-Molard D, Canagnier B (2002) Fungal growth and *Fusarium* mycotoxin content in isogenic traditional maize and GM maize grown in France and Spain. *Journal of Agricultural and Food Chemistry* **50**, 728–731. doi:10.1021/jf0108258

Brookes G (2008) The benefits of adopting GM insect resistant (Bt) maize in the EU: first results from 1998–2006. *International Journal of Biotechnology* **10**, 148–166. doi:10.1504/IJBT.2008.018351

Brookes G, Barfoot P (2018a) Farm income and production impacts of using GM crop technology 1996–2016. *GM Crops and Food: Biotechnology in Agriculture and the Food Chain* **9**, 59–89. doi:10.1080/21645698.2018.1464866

Brookes G, Barfoot P (2018b) Environmental impacts of genetically modified (GM) crop use 1996–2016: impacts on pesticide use and carbon emissions. *GM Crops and Food: Biotechnology in Agriculture and the Food Chain* **9**, 109–139. doi:10.1080/21645698.2018.1476792

Crost B, Shankar B (2008) Bt cotton and production risk: panel data evidence. *International Journal of Biotechnology* **10**, 122–130. doi:10.1504/IJBT.2008.018349

Gouse M, Pray C, Kirsten J, Schimmelpfennig D (2005) A GM subsistence crop in Africa: the case of Bt white maize in S Africa. *International Journal of Biotechnology* **7**, 84–94. doi:10.1504/IJBT.2005.006447

Gouse M, Piesse J, Thirtle C (2006a) Output and labour effect of GM maize and minimum tillage in a communal area of Kwazulu-Natal. *Journal of Development Perspectives* **2**, 71–86.

Gouse M, Pray C, Schimmelpfennig D, Kirsten J (2006b) Three seasons of insect resistant maize in South Africa: have small farmers benefited? *AgBioForum* **9**, 15–22.

Kathage J, Qaim M (2012) Economic impacts and impact dynamics of Bt cotton in India. *Proceedings of the National Academy of Sciences of the United States of America* **109**, 11652–11656. doi:10.1073/pnas.1203647109

Kennedy IR, Rose MT, Crossan A, Burns M (2013) Research and practice: environmental action for improving water quality in cotton catchments since 1990. *Crop and Pasture Science* **64**, 1095–1111. doi:10.1071/CP13091

Kovach J, Petzoldt C, Degni J, Tette J (1992) *A Method to Measure the Environmental Impact of Pesticides.* New York's Food and Life Sciences Bulletin. New York State Agricultural Experimentation Station, Cornell University, Geneva, NY, 139. <http://www.nysipm.cornell.edu/publications/EIQ.html>

Poeydemenge C (2006a) *Les bénéfices du maïs Bt observés en France.* Presentation to the SEPROMA General Assembly, 21 September 2006.

Poeydemenge C (2006b) *Bt Maize Benefits observed in France.* Presentation to 10th ICABR conference on agricultural biotechnology, Ravello, Italy, July 2006.

Subramanian A, Qaim M (2009) Village-wide effects of agricultural biotechnology: the case of Bt cotton in India. *World Development* **37**, 256–267. doi:10.1016/j.worlddev.2008.03.010

Torres CS, Daya RA, Osalla MTB, Gopela JN (2013) Adoption and uptake pathway of GM/biotech crops by small-scale, resource-poor Filipino farmers. ISAAA, CDC-UPLB and SEARCA, Los Banos, Laguna, Philippines. <www.isaaa.org>.

Zilberman D, Ameden H, Qaim M (2007) The impact of agricultural biotechnology on yields, risks and biodiversity in low-income countries. *Journal of Development Studies* **43**, 63–78. doi:10.1080/00220380601055544

3

Development and commercialisation of Bt cotton in India

TM Manjunath

Introduction

Prior to the introduction of Bt cotton in India in 2002, chemical control was the most widely adopted practice as, when successful, its results were quick and visible. The number of sprays ranged from five to 20 with an all-India average of seven sprays for a cotton crop. Despite such repeated sprays, bollworms continued to cause yield losses of up to ~60%. Among the factors responsible for such unsatisfactory control were the use of unapproved and spurious insecticides, faulty applications and emergence of insect pest resistance to most of the recommended insecticides including the synthetic pyrethroids. Therefore, farmers, scientists and policy-makers were frustrated and desperately looking for alternatives for bollworm control. Meanwhile, farmers continued to use insecticides as they were left with no option but to 'spray and pray'. In such a challenging scenario, Bt cotton, based on its widespread adoption and demonstrated success in other countries, brought a ray of hope to cotton farmers following its first regulatory approval by the Indian government in March 2002 (Manjunath and Mohan 2002; Barwale *et al.* 2004). Since then, Bt cotton became the most dominant approach for bollworm control, with drastic reduction in insecticidal sprays. However, while Bt cotton kills bollworms and other secondary lepidopteran pests, sucking pests continue to be managed by other suitable methods as required. Since Bt cotton is compatible with all other control measures, it is a major component of integrated pest management (IPM) strategy for cotton pest complex.

Various issues related to the challenges posed by bollworms, the need for and development of Bt cotton in India, its adoption and impact, its regulation pertaining to safety assessment and risk management, opposition to this technology, and its prospects are discussed in this chapter.

Cotton bollworms: a formidable challenge

Cotton bollworms, which belong to the insect order Lepidoptera, are the most destructive pests of cotton crop in India. These include the old world bollworm *Helicoverpa armigera* (Hübner) (Noctuidae), pink bollworm *Pectinophora gossypiella* (Saunders) (Gelechiidae), spotted bollworm *Earias vittella* (Fabricius) (Noctuidae) and spiny bollworm *Earias insulana* (Boisduval). The term 'American bollworm' often used in literature for *H. armigera* is misleading as this species does not occur in the USA (the new world), but is originally distributed and dominant only in Asia,

Africa and Australia. Therefore, the term 'old world bollworm' or simply 'cotton bollworm' is more appropriate. The tobacco caterpillar *Spodoptera litura* (Fabricius) (Noctuidae), also a lepidopteran, is a sporadic pest of cotton. It is highly polyphagous and predominantly a defoliator of tobacco, castor, cotton, maize and many vegetable crops, but can cause serious damage to cotton bolls and other fruiting structures during an outbreak.

Among the bollworms, *H. armigera* was the most dominant and destructive, especially in the pre-Bt era, and it was very difficult to control due to its widespread insecticide resistance (including pyrethroids), multivoltine as well as prolific breeding, and high polyphagy with more than 180 plant species recorded as hosts. Of these, about a dozen crops can be seriously damaged (Manjunath *et al.* 1989).

It was estimated that, before the introduction of Bt cotton, insecticides worth about INR30 billion were used annually in Indian agriculture. Of these, ~50% was spent for the control of cotton pests, including INR12 billion against bollworms alone. In terms of volume, ~50% of the total insecticides used in Indian agriculture were applied on cotton crops, underscoring the economic importance of bollworms in general and *H. armigera* in particular. Cotton area was on the decline in India because of frequent bollworm infestation and outbreaks. The area declined from an average of 8.7 million ha up to 2001 to a meagre 7.6 million ha in 2002 and 2003.

There are several individual approaches for pest control but IPM, where several options are judiciously integrated, has been considered a wise approach. IPM aims at sequential or simultaneous use of non-chemical methods like cultural practices, sex pheromone traps, biological control agents (parasitoids, predators and pathogens including Bt), botanical insecticides and other factors prior to requiring need-based application of chemical insecticides. There are only few reports of successful implementation of IPM (Sharma *et al.* 2001). In general, in a cotton ecosystem where regular spraying had become a habit with farmers, IPM did not infuse enough confidence as it could not match the efficacy of pesticides in providing quick and visible results. In this regard, Bt cotton offered an excellent alternative. Besides being safe to non-target organisms and the environment, it can match the temporal efficacy of insecticides in controlling bollworms. Moreover, Bt cotton is compatible with all other control measures and is a major component in the IPM strategy for cotton pest complex (Fitt 2000; Manjunath 2005, 2011).

Development and commercialisation of Bt cotton

Bacillus thuringiensis Berliner (Bt) is a spore-forming soil-borne bacterium with ubiquitous distribution. It is highly pathogenic to insects. The toxicity of Bt against insect pests is largely due to para-sporal secretion of crystalline (Cry) proteinaceous toxins which are highly specific in their action. A large number of Cry toxins have been characterised (Crickmore *et al.* 1998). Cryl toxins are highly toxic to Lepidoptera, Cry2 to Lepidoptera and Diptera, Cry3 to Coleoptera and Cry4 to Diptera. The required genes of these toxins are introgressed into the host's genome to be heritable traits for development of transgenic crops, also called Bt crops. More than one *cry* gene are introgressed to enhance the spectrum of toxicity against a pest complex; for example, Bt cotton is incorporated with the lepidopteron-specific *cry* gene(s) as it is designed to control bollworms and other Lepidoptera. Similarly, Bt corn, Bt potato, Bt brinjal (eggplant), Bt rice, Bt cowpea etc. use encoded Bt proteins to provide insect control.

Development of transgenic Bt plants

The first genes encoding the insecticidal Bt proteins were cloned in the early 1980s. This paved the way for constructing recombinant bacterial insecticides containing novel combinations of these proteins, and the development of Bt plants. The first Bt cotton plants were developed by the Chinese Academy of Agricultural Sciences in China (but not commercialised) and by Monsanto Co. in the USA in the early 1990s (Perlak *et al.* 1990, 1991). Regulatory approval and large-scale commercial cultivation of Bt crops, which included Bt cotton along with Bt corn and Bt potato developed by Monsanto, took place in the USA in 1996 (Perlak *et al.* 2001). Thus, 1996 marked the beginning of commercialisation of transgenic crops.

Development of Bt cotton in India

After obtaining the necessary permit from the Indian government, Maharashtra Hybrid Seed Co. (Mahyco) imported a small quantity (100 g) of seeds of Coker 312 variety of cotton containing the Bt gene, *cry1Ac* (event MON 531), along with the technology (Bollgard®) under a licence from Monsanto USA, to develop Bt cotton in India. The Bt gene was transferred from transgenic Coker 312 to parental plants of Mahyco's proprietary hybrids through backcrossing, converting the native parental plants into Bt plants. Transgenic parental plants were strongly stabilised for the presence of Bt trait through multiple backcrossings before producing the Bt hybrids, using the same general procedure as for developing cotton hybrids. In fact, cotton hybrids with Indian germplasm were already under commercial cultivation in India even before the incorporation of Bt gene in them. The Bt gene has enabled such hybrids to express its protein that is toxic to bollworm larvae. Such gene(s) can be introduced into any desired cotton cultivars, as shown by Mahyco and subsequently by several other seed companies.

It took approximately seven years for Mahyco to complete all the regulatory trials related to biosafety and agronomic impact and to get official approval for the first batch of three Bt cotton hybrids in India, in March 2002. The chronology of events is summarised in Table 3.1.

It should be clearly understood that only the Bt gene(s) along with the genetic engineering technology had been obtained by Mahyco from Monsanto – Bt cotton plants, as such, were not

Table 3.1. Development and approval of Bt cotton in India

1995	Mahyco applied to Dept of Biotechnology (DBT) for permission to import a small stock of Bollgard® (Bt cotton) seeds from Monsanto Co., USA. DBT gave permission.
1996	A nucleus stock of 100 g of cotton seeds of the American variety Coker 312 containing the Bollgard® Bt gene, *cry1Ac,* was received by Mahyco from Monsanto USA. Mahyco initiated crossing Coker 312 with the Indian cotton breeding lines to introgress *cry1Ac* gene. 40 elite Indian parental lines were converted for Bt trait.
1996–98	Risk assessment studies were conducted, using Indian Bt cotton hybrids, in laboratories and fields designated by Review Committee for Genetic Manipulation (RCGM)/Genetic Engineering Appraisal Committee (GEAC). These included pollen escape, aggressiveness and persistence, biochemical analysis, toxicity and allergenicity.
1998–99	Multi-location field trials at 40 locations in nine states to assess agronomic benefits and safety. Data submitted to RCGM.
1999–2000	Field trials repeated at 10 locations in six states. Data submitted to RCGM.
2000	July 2000: Based on the recommendations of RCGM, the GEAC gave permission for large-scale field trials in 85 ha and seed production in 150 ha.
2001	Kharif 2001: Large-scale field trials covering 100 ha. Field trials were also conducted by All India Coordinated Cotton Improvement Project of the Indian Council of Agricultural Research (ICAR).
2002	On 26 March 2002, GEAC approved Mahyco's three Bt cotton hybrids, i.e. MECH 12 Bt, MECH 162 Bt and MECH 184 Bt, for commercial cultivation in India. This approval, with certain conditions, was initially valid for three years.

introduced into India. Further, when incorporated in a desired plant, the Bt gene imparts only an insecticidal trait. Thus, Bt is only a trait, not an independent cultivar by itself.

Further progress on Bt cotton

The first version of Bt cotton, Bollgard®, which was approved by GEAC in 2002, carried only a single gene, *cry1Ac* (event MON-531). Later, Mahyco developed an advanced version of Bt cotton, Bollgard II®, also obtained from Monsanto. It was stacked with two Bt genes, *cry1Ac* and *cry2Ab2* (event MON-15985) and was approved in 2006.

Later, other technology developers like JK Agri Genetics, Nath Seeds and Metahelix also developed Bt cottons, incorporating Bt genes derived indigenously or obtained from other sources. These were also approved by GEAC after the prescribed regulatory trials. The Central Institute of Cotton Research (CICR) in Nagpur, in collaboration with University of Agricultural Sciences in Dharwad, claimed to have developed its own Bt cotton technology, incorporating the Bt gene into a popular variety, not a hybrid. It was approved by GEAC in 2008. However, following a controversy pertaining to Bt construct and subsequent probe by ICAR, this Bt cotton was withdrawn. The status of various Bt cottons approved in India so far is depicted in Table 3.2.

Thus, four Bt genes in various modified versions (*cry1Ac, cry2Ab2, cry1Ab* and *cry1C)* and altogether five valid events (event MON-531, event MON-15985, event 1, event GFM and event MLS-9124) (excluding event BNLA-601 of CICR) are involved in the Bt cottons that have been so far approved by GEAC. Bollgard II® is the only one that carries two stacked genes, expressing two Bt proteins.

Commercialisation

Starting with Mahyco's three Bt cotton hybrids (MECH 12 Bt, MECH 162 Bt and MECH 184 Bt) in March 2002 that contained *cry1Ac* gene (event MON-531), GEAC has so far approved more than 1000 Bt cotton hybrids featuring five events developed by four private companies for commercialisation. The approvals with details of the hybrids are available on the Indian GMO Research Information System (IGMORIS) website (http://igmoris.nic.in/commercial_approved.asp).

The hybrids are meant for different agro-climatic regions in North Zone (Haryana, Punjab, Rajasthan), Central Zone (Gujarat, Madhya Pradesh, Maharashtra) and South Zone (Andhra Pradesh, Telangana, Karnataka, Tamil Nadu). Some hybrids are approved for more

Table 3.2. Status of Bt cotton approved in India

Six versions of Bt cotton developed and approved in India							
	2002	2006			2008	2009	Total
Developer	Mahyco	Mahyco	Nath Seeds	JK Agri Genetics	CICR and UAS(D) (Public)	Metahelix	5
Gene(s)	*cry1Ac*	*cry1Ac*+ *cry2Ab* (stacked)	*cry1Ac*+ *cry1Ab* (fused)	*cry1Ac*	*cry1Ac*	*cry1C*	4
Events	MON-531 (Bollgard®)	MON-15983 (Bollgard II®)	GFM	JK Event 1	BNLA 106 Event In varieties (withdrawn)	MLS 9124	6
Bt genes in >1000 cotton hybrids for different agro-climatic regions marketed by over 45 Indian companies.							

Adapted from Manjunath (2011)

than one zone. They are marketed by ~50 companies with licensing arrangements (Manjunath 2011).

More than 1000 cotton hybrids developed by various seed companies to suit different agro-climatic regions are commercially available. Mahyco's Bollgard II® is the most preferred Bt cotton by cotton farmers all over India, with a share of 90–95% of total Bt cotton area.

Adoption and impact

Globally, genetic modification (GM) technology is mainly available for three important traits: insect resistance (IR), herbicide tolerance (HT) and virus resistance (VR). The IR and HT traits are deployed in four principal crops – soybean, maize, cotton and canola – either with a single trait (IR or HT) or with two traits (IR+HT) stacked in the same plant. Almost all countries where these four are cultivated as major crops have significantly adopted this technology. As of 2018–19, altogether 191.7 million ha (increased from 189.8 million ha in 2017–18) of GM crops, including 11.6 million ha of Bt cotton in India, were cultivated in 26 (21 developing, five industrial) countries by ~17 million farmers. Crop-wise, 77% of 95.9 million ha or 78% of the global area of 122.9 million ha of soybean included HT, 24.9 million ha or 76% of the total 32.8 million ha of cotton included IR or IR+HT, 58.9 million ha or 30% of 196.3 million ha of maize included IR or IR+HT and 10.1 million ha or 29% of 34.8 million ha of canola included HT. Bangladesh joined the GM countries, giving regulatory approval for commercial cultivation of IR Bt brinjal (eggplant) for control of the fruit-and-shoot borer *Leucinodes orbonalis* in October 2013. The area planted with Bt brinjal increased from 12 ha in 2014, the first year, to 2400 ha planted by ~27 000 farmers in 2017. This comprised 5% of the total brinjal area of 50 000 ha in the country (BARC 2013; ISAAA 2017). Brazil commercialised Bt sugarcane in 2018 and Nigeria commercialised Bt cowpea in 2019.

Biotech crops have expanded beyond the big four (maize, soybeans, cotton, and canola) to give more choices for many of the world's consumers and food producers. Biotech crops of alfalfa, sugar beets, papaya, squash, potatoes, brinjal, apples and sugarcane are already in the market. Additionally, biotech crop research is being conducted by public sector institutions into rice, banana, potato, wheat, chickpea, pigeon pea and mustard to develop various economically important and nutritional quality traits beneficial to food producers and consumers in developing countries (ISAAA 2017, 2018).

Adoption of Bt cotton in India

The area planted with Bt cotton in India was ~30 000 ha in 2002, the first year of adoption. It increased consistently from year to year and reached 11.6 million ha in 2018, equivalent to 95% of the total cotton area of 12.24 million ha grown in the country (Manjunath 2011; ISAAA 2018). Figure 3.1 shows the adoption of Bt cotton in India.

The adoption of Bt cotton has been overwhelming since its approval in 2002. The number of farmers adopting GM technology increased from ~20 000 in 2002 to over 7.5 million by 2018, reflecting their confidence in this technology.

Advantages of Bt cotton

Transgenic Bt cotton, as with other Bt crops, offers several advantages (Manjunath 2011).

- Bt technology is made available in the seed itself. Farmers just sow the Bt seeds as they do conventional seeds. The resulting plants have the in-built ability to produce Bt toxins within their body – leaf, stem, buds, flower, bolls – and defend themselves against the target pests, i.e. bollworms. No extra efforts or equipment are needed to use this technology.

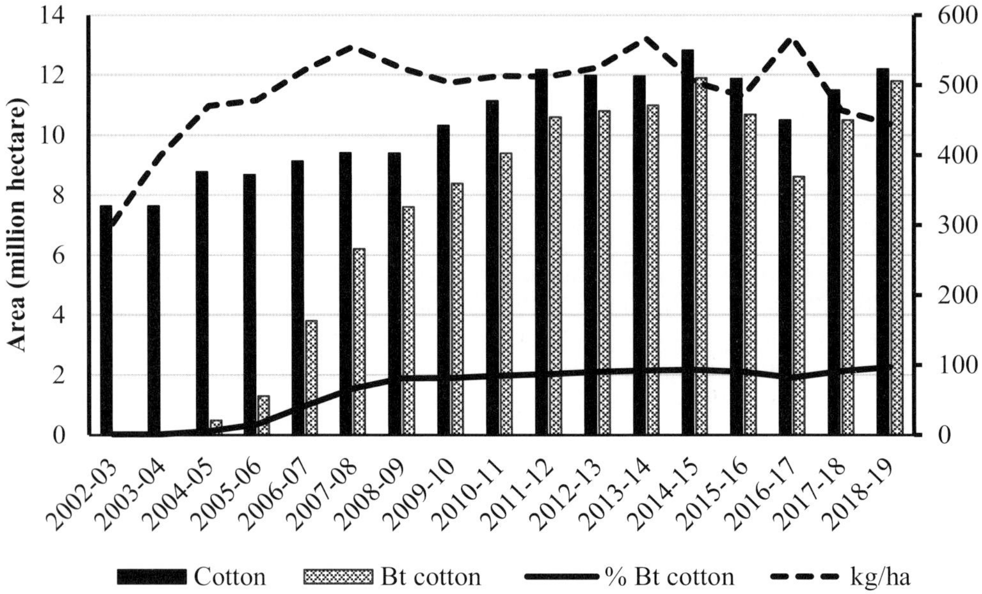

Figure 3.1 Adoption of Bt cotton in India (million ha).

- Bt protein is expressed in all parts of the plant. The newly hatched larvae feeding on any part of the plant – leaves, stem, flowers, squares, bolls – will ingest Bt protein and die within one or two days, thereby preventing any potential serious damage to the crop.
- Bt is present within the plant almost throughout its life, providing pest control day and night.
- Bt proteins, being host-specific, affect only the target pests and are safe for non-target arthropods like predators, parasitoids and other beneficials such as bees and other pollinators.
- Transgenic Bt technology is compatible with other control measures such as biological control, pheromones, botanical insecticides and chemical insecticides that are recommended for IPM in the cotton ecosystem.
- Bt crops help to avoid or minimise insecticide sprays, thus contributing to a cleaner environment, better health, and conservation of biological control agents and biodiversity.
- Bt crops offer protection from target pests right from the early days of the crop, leading to a healthy crop, greater harvest and more profit.

In other words, Bt crops provide social, economic and environment benefits leading to conservation of biological control agents and biodiversity.

Impact of Bt cotton

Various studies and surveys carried out in India by both national and international agricultural scientists and economists have clearly demonstrated effective control of bollworms due to introduction of Bt cotton. As a result, the yield of lint cotton increased from 31% to 67%, from ~300 kg/ha before 2002 to over 500 kg/ha from 2006 onward, mainly due to effective control of bollworms. Other factors such as improved breeding, availability of better hybrids, better insecticides and overall improved farm management also contributed to the success, but Bt technology, as a single input, played a major role. This turned India from an importer of cotton to an exporter. The insecticide sprays for bollworm control have been drastically reduced from 10–15 times during a cropping season to just one or two, only when direly needed. In most years, spraying has been nil. However, farmers may be required to undertake need-based sprays (two or three sprays) against sap-sucking insects and other non-lepidopteran pests that are not controlled by Bt technology. Overall, there has been a reduction in insecticide application by ~55%. The farmers are able to get a profit that ranged from INR6400/ha to INR27 000/ha, thereby improving their economic and social conditions (Manjunath 2011; Kathage and Qaim 2012; Mayee and Choudhary 2013; Mohan and Oufattole 2015). The fact that as of 2018, 11.6 million ha which constitute around 95% of the total cotton acreage in India is covered by Bt cotton and that 7.5 million farmers have adopted this technology, clearly reflects its benefits and popularity (ISAAA 2017, 2018).

As a result of growing Bt cotton on large areas for over 15 years, there is a general decline in the population of the key cotton bollworm *H. armigera,* owing to its effective control. Consequently, farmers growing other crops such as pigeon pea, chickpea, sunflower, maize, tomato and chillies which also serve as important natural hosts of *H. armigera* are experiencing a respite from this notorious pest and reduced spraying on such host crops. Thus, Bt cotton has also helped in conservation of biological control agents and biodiversity. Overall, Bt cotton farmers have experienced tangible benefits such as higher production, greater profits and substantial reduction in insecticide usage for bollworm control along with the attendant environmental benefits.

Over the last few years, however, bollworms – more particularly pink bollworm *P. gossypiella* – have shown resistance to Bt cotton. One of the reasons for this is non-compliance of IR management practices like refuge planting as recommended by the regulatory authorities. Nevertheless, IR to toxins is a natural evolutionary phenomenon that can be retarded proactively with new products to replace older ones when need arises. The development of Bollgard II® (two toxins), Bollgard 3® (three toxins) to counter resistance to single Bt toxins is a progressive step in this direction.

Safety assessment and risk management

Biosafety data of approved genes/events of GM cotton are available at IGMORIS (2013).

Safety to non-target organisms

It is suspected by some that the health of humans and animals exposed to Bt Cry proteins may be affected in some way. To dispel any such doubts, a large number of experiments, prescribed by the regulatory authorities in each country, have been carried out to examine the safety of Cry proteins. Experimental animals including mice, rats, fish, poultry, rabbits and sheep fed with unusually high doses (500, 1000 and 4300 mg/kg bodyweight) of Cry protein showed no acute toxic effect on their health. These animals were found to be equivalent to those not fed with Cry proteins with respect to bodyweight, food consumption, pathology of internal organs and other health aspects. Based on such research data, the US Environmental Protection Agency concluded 'toxicity and infectivity risks of Cry proteins to non-target organisms like birds, freshwater fish, freshwater aquatic invertebrates, estuarine and marine animals, arthropod predators/parasitoids, honey bees, annelids and mammalian wildlife will be minimal to non-existent at the recommended rates of registered *B. thuringiensis* active ingredients' (EPA 2015). Therefore, there is no need for undue fear. The allegations regarding safety were based on speculation.

Biosafety studies carried out in India

Biosafety data on comparative toxicity, allergenicity and feeding studies in experimental animals were generated by product developers before seeking approval from GEAC as per the

Rules for the Manufacture, Use, Import, Export & Storage of Hazardous Micro-organisms, Genetically Engineered Organisms or Cells 1989 and *Revised Guidelines for Research in Transgenic Plants & Guidelines for Toxicity and Allergenicity Evaluation of Transgenic Seeds, Plants and Plant Parts 1998* (GEAC n.d.; IGMORIS 2013).

Feed-safety studies using Bt cotton seed-meal were conducted on goats, buffaloes, cows, rabbits, birds and fish by competent scientists at various institutions including the Central Toxicological Research Institute, Lucknow; National Dairy Research Institute, Karnal; Central Institute of Fisheries Education, Mumbai; Central Avian Research Institute, Izatnagar; National Institute of Nutrition, Hyderabad; and G.B. Pant University of Agriculture and Technology, Pantnagar. The results revealed that the animals fed with Bt cotton seed-meal showed no ill-effects and were comparable to control animals across various tests. In other words, Bt cotton seed-meal was 'substantially equivalent' to its non-Bt counterpart. Based on these scientific data, the regulatory authorities determined Bt cotton as safe.

Fate of Bt protein in the soil

Studies were conducted to determine the amount of Bt protein leached by roots and other plant parts incorporated in the soil and its effect on soil rhizosphere and non-rhizosphere microflora, soil collembola, earthworms, etc. It was found that there was no adverse effect. In fact, there was no difference between the soils obtained from the Bt and non-Bt plots in this respect. It was also found that the Bt insecticidal proteins were readily susceptible to metabolic, microbial and abiotic degradation once they were ingested or excreted into soil. The half-life of Cry1Ac protein in plant tissues has been found to be short (Clark *et al.* 2005) and therefore it cannot accumulate, causing delayed effect.

Bt transgenic crops and biodiversity

There have been allegations that Bt transgenic crops will affect biodiversity. It is not true. It should be realised that most of the cultivars, for example cotton hybrids, before they were chosen for incorporating the Bt gene(s), were already under commercial cultivation and a part of on-farm biodiversity.

Except for the presence of the Bt gene which was deliberately introduced into them for the specific purpose of controlling the target pests, the Bt plants are 'substantially equivalent' to their non-Bt counterparts in all respects. In other words, there is no significant difference in their morphological features or their chemical constituents such as protein, carbohydrate, moisture and acidic contents in the foliage and other parts. Therefore, the adverse impact of Bt crops, if any, on other crops, non-target organisms or biodiversity is no different from that of its traditional counterpart except for affecting the targeted cotton pests. In fact, due to the reduced application of chemical pesticides, the populations of non-target beneficial insects like ladybird beetles, green lacewings, parasitoids and other biological control agents as well as honey bees were reported to be greater in Bt fields, thereby enhancing biodiversity in the crop ecosystem. The allegation that Bt crops adversely affect biodiversity is baseless (Manjunath 2011).

Pollen flow, super weeds, gene pollution etc.

There is an apprehension that the pollen from Bt cotton plants may get transferred to weeds, resulting in super weeds. There is also a fear that the Bt gene(s) incorporated in cotton plants may escape and cause gene pollution or genetic contamination. These are unfounded, as clarified below.

Weeds

Cotton has only one close weed relative, *Gossypium stocksii*, which is found in northern Gujarat where cotton is not cultivated. There is no record of bollworms feeding on *G. stocksii* and there is no other major lepidopteran common to cotton and this weed. Further, cotton pollen is heavy and cannot move beyond a few metres away from cotton fields. Therefore, the possibility of gene transfer and the development of super weed is remote. In other countries and other Bt crops, there is no evidence that super weeds have ever developed since GM technology was introduced in 1996.

Gene flow

In India there are two types of cotton – American cotton (*Gossypium hirsutum* and *G. barbadense*) and Desi (local) cotton (*Gossypium herbaceum* and *G. arboreum*). All the Bt cotton hybrids developed involved *G. hirsutum*, by crossing *G. hirsutum* x *G. hirsutum* or *G. hirsutum* x *G. barbadense*. The American cotton is tetraploid in genetic makeup whereas the Desi cotton is diploid. There is no reproductive compatibility between the two. Even if cross-pollination occurs between the tetraploid and diploid cotton plants, the zygotic embryo will not develop. Terms such as gene pollution and genetic contamination are not scientific; they are buzzwords intended to intimidate lay people (IGMORIS 2013; Manjunath 2011).

Opposition to Bt cotton

Bt cotton faced opposition from a few non-government organisations (NGOs), also called civil society organisations (CSOs), and from certain individuals from 1998 when it was still undergoing the regulatory trials in India. The opposition was premature – at that point, the performance of Bt cotton was hardly known as it was not yet field-tested. The allegations about its safety and benefits are still continuing despite 16 years since its regulatory approval by the Indian government in March 2002. Opponents do not seem to recognise the fact that Bt cotton has consistently made remarkable progress since its introduction and that by 2018 it was grown by ~7.5 million farmers on 11.6 million ha in all the cotton-growing states – with no negative effect. The safety concerns are related to a mixture of political, ideological, social, economic, ethical, emotional and legal issues. Most of these are based on perceptions without scientific rationale, primarily aimed at sensationalism. Their allegations seem to suggest that Bt cotton kills almost everything – except the bollworms for which the technology was specifically developed. Almost all the questions raised by opponents have been taken seriously and proved wrong by conducting experiments (Manjunath 2011).

Safety of Bt cotton and other GM crops

GM crops have been commercially cultivated since 1996 and in 2018–19 were cultivated across 191.7 million ha in 26 countries (21 developing, five industrial), including India with Bt cotton since 2002 (ISAAA 2018). They have proven to be safe and beneficial. So far there has been no credible scientific evidence to show that Bt cotton or other GM crops or products have caused any harm to humans, animals, other non-target organisms or the environment anywhere in the world.

The approval for commercialisation of GM crops is highly regulated. Every GM crop must undergo all the safety tests prescribed by the regulators in each country, mostly in public laboratories. Food, feed and environmental safety data generated during such studies are thoroughly examined at various stages by several technical committees constituted by the regulatory authorities. Only after these bodies are satisfied with the scientific data that the crop is safe, is approval given for its commercialisation. No other crop or food is subjected to such stringent scrutiny as GM crops/products before they are openly cultivated or marketed.

The International Centre for Genetic Engineering and Biotechnology (ICGEB) has listed several websites that provide an exhaustive and comprehensive array of biosafety-related information (ICGEB n.d.). In India, the biosafety data pertaining to Bt cotton and other GM crops can be viewed on the IGMORIS website created by the Department of Biotechnology.

Globally, over 2000 studies dealing with the safety of GMOs have been published in scientific journals. Almost every major scientific body and regulatory agency in the world has reviewed such research data and strongly vouched that the food and feed derived from GM crops are safe. These include the American Medical Association, American Association for the Advancement of Science, National Academy of Sciences, Food Standards Australia & New Zealand, French Academy of Science, UK Royal Society of Medicine, European Commission, Union of German Academies of Sciences and Humanities, six major science academies in India (Indian Academy of Sciences, Indian National Academy of Engineering, Indian National Science Academy, National Academy of Agricultural Sciences) and six other world academies of sciences (Brazil, China, Mexico, Third World Academy of Sciences, Royal Society and National Academy of Sciences of the USA) and the World Health Organization. GM products are a result of thorough research and assessment, and are 'substantially equivalent' to their non-GM counterparts. Therefore, the safety of GM foods is on par with that of conventionally produced food (Entine 2013; Wendel 2013; Manjunath and Mohan 2015). The overwhelming scientific support for GM crops should convince those who are concerned by opponents' allegations regarding their safety and benefits.

Conclusion

Bt cotton has undoubtedly provided relief in India from cotton bollworms, which were previously the most destructive and difficult pests to control. The fact that in India the area cultivated with Bt cotton increased from ~30 000 ha in 2002, the first year, to 11.6 million ha by 2018–19, comprising 95% of the total cotton area, clearly reflects its overwhelming acceptance by the 7.5 million farmers who use it. The cultivation of Bt cotton has resulted in yield increase by 31% to 67% owing to effective control of bollworms, reduction in insecticide sprays from 25% to 55% and increase in profit to farmers from INR6400/ha to INR27 000/ha, contributing to overall social, economic and environmental benefits.

It should be realised that Bt cotton, being lepidopteran-specific, offers control of only bollworms and a few other caterpillar pests. Even for lepidopteran pests, the control paradigm offered by Bt cotton has been on the wane due to various reasons such as non-compliance of refuge planting, spurious seeds, reduced efficacy due to field-evolved resistance and failure of quality check. Bt does not control sucking insects (mealybugs, whiteflies, aphids, jassids, etc.) and non-lepidopteran pests (mirids, weevils, bugs, mites, etc.) or diseases such as blight and wilt. Therefore, to get the full benefit of this technology, it is necessary to be watchful and adopt other methods to tackle such problems as and when required. Bt technology should not be treated as a silver bullet to solve all problems! It is not possible to resolve all problems with any single technology. We should try to integrate and take full advantage of all available technologies, both traditional and modern. Bt cotton is an example in this endeavour. It can serve as a major thrust in the IPM strategy for cotton pest complex as it is compatible with other plant protection methods.

References

BARC (2013) *Bangladesh Agricultural Research Council, Bangladesh Approved BT Brinjal for Commercial Cultivation*. BARC, Ministry of Agriculture, Government of the People's Republic of Bangladesh, 30 October 2013. <http://www.barc.gov.bd/bt_brinjal.php>

Barwale RB, Gadwal VR, Zehr U, Zehr B (2004) Prospects for Bt cotton technology in India. *AgBioForum* 7, 23–26.

pests. The first IRM strategies for Bt cotton and Bt corn were evolved jointly in the 1990s by the USDA-Environment Protection Agency (EPA) in collaboration with other stakeholders such as the Insecticide Resistance Action Committee, academics and technology generators.

The basic principle of IRM is based upon the expression of high dose of toxin in the transgenic crop to kill most of the susceptible target insects, leaving only rare resistant homozygous ones in the presence of adequate refuges that support abundant production of Bt-susceptible individuals which will mate with the surviving resistant homozygous individuals to produce the Bt-susceptible progeny, as resistance is often a recessive trait (Tabashnik 1994; Gould 1998). The IRM strategies are fine-tuned to adapt to the local agroecosystem depending upon the behaviour, biology and population dynamics of the target pests, the baseline sensitivity and the variability among target pest species to the toxins expressed by Bt crops, genetic diversity of pest species, expression of transgene traits in the crop plant parts, intensity of selection pressure, genetics of inheritance of resistance trait in the target species, presence and effectiveness of alternate natural hosts to support the susceptible insect pests and agronomic practices influencing pest management (Gould 1998). Using the biological, ecological and genetic data, modelling studies have helped to estimate the nature and size of refuge required and their mandatory needs in an agroecosystem (Head and Greenplate 2012; Sumerford *et al.* 2013). As more countries commercialised Bt crops, IRM strategies were suitably modified based on the region-specific IRM features. In retrospect, IRM strategies for Bt crops succeeded to varying extents depending upon the nature of Bt resistance and their implementation in various countries (Huang *et al.* 2011; Head and Greenplate 2012; Carrière *et al.* 2016). Field-level resistance in five lepidopterans and one coleopteran pest to Bt toxins has been reported from some countries (Tabashnik *et al.* 2009, 2013; Tabashnik and Carrière 2017). As far as cotton is concerned, cotton bollworm *Helicoverpa zea* is reported to have evolved practical resistance to both Cry1Ac and Cry2Ab2 in the USA (Tabashnik and Carrière 2017) and the pink bollworm to these toxins in India (Dhurua and Gujar 2011; Dennehy *et al.* 2011; Mohan *et al.* 2016; Naik *et al.* 2018; Shahanaz 2019).

IRM for Bt cotton in India: its evolution and current status

The key bollworms targeted by Bt cotton in India are the cotton bollworm (synonyms being American bollworm, old-world bollworm) *H. armigera*, tobacco caterpillar *S. litura*, pink bollworm *Pectinophora gossypiella* and spotted and spiny bollworms *Earias vittella* and *E. insulana*. Certain sporadic lepidopteran pests such as cotton leaf perforator *Anomis flava*, leaf roller *Sylepta derogata* and soybean caterpillar *S. exigua* are also controlled by Bt cotton. Among the key bollworms, *H. armigera* was the most feared in view of its multi-voltine nature, prolific breeding leading to frequent outbreaks and a history of widespread resistance to synthetic pyrethroids and other insecticides in the pre-Bt cotton era of the 1980s and 1990s (Armes *et al.* 1992; Kranthi *et al.* 2002a, b; Ramasubramanian 2004; Ramasubramanian and Regupathy 2004). Cotton cultivation during those decades had become less sustainable and eco-friendly and bollworm control was grossly ineffective, despite spraying with different insecticides and mixtures eight to 10 times or more in a crop season. Bt cotton was introduced in 2002 in view of the desperate situation caused by cotton bollworm and reports of effective control of cotton bollworms in the USA where Bt cotton was commercialised in 1996. A structured non-Bt cotton refuge (in five perimeter rows or amounting to 20% of Bt cotton area, whichever is more) was an essential requirement of resistance management for growing approved Bt transgenic cotton hybrids in India. A brief description of various IRM measures which were initiated for Bt cotton in India, many involving public–private partnerships, is given in Table 4.1.

Table 4.1. Evolution of IRM strategy for Bt cotton in India

#	IRM aspect	Institutions and study
1	(i) Pre-commercialisation baseline-sensitivity of Indian cotton bollworms to Bt proteins in assays (ii) in-commercialisation efficacy of Bt cotton in managing cotton bollworms (iii) toxicity profile of Bt toxins to cotton bollworms	National Bureau of Agricultural Insect Resources (NBAIR), Bangalore (Jalali *et al.* 2004); Indian Agricultural Research Institute (IARI), New Delhi (Chandrashekar *et al.* 2005; Gujar *et al.* 2000, 2007, 2008); University of Agricultural Sciences (UAS), Dharwad (Fakruddin *et al.* 2003); UAS, Raichur (Ranjith *et al.* 2010); Central Institute for Cotton Research (CICR), Nagpur (Kranthi *et al.* 1999, 2001, 2009).
2	Structured refuge for Bt cotton (Bollgard®) – 2002	Genetic Engineering Approval Committee (GEAC), Ministry of Environment and Forest, stipulated structured refuge while approving commercial cultivation of Bt cotton in 2002; Gujar *et al.* (2010).
3	Post-commercialisation annual resistance monitoring in cotton bollworms	CICR and network of public institutions conducted annual monitoring of shift in susceptibility to Bt proteins in cotton bollworm populations from 2002 (Kranthi and Kranthi 2003) to 2014. Data available at Ministry website (http://moef.gov.in/project-approvals/geac-clearances/othinfo/).
4	Bt cotton stewardship on trait quality management to ensure high-dose expression of Bt protein(s) in target tissues and required trait purity in the seed	Event identity, Bt gene(s) expression and bollworm control efficacy were audited by Mahyco Monsanto Biotech Ltd (MMB; Bt technology licensor) before approval by Review Committee for Genetic Manipulation of all new Bt cotton hybrids (Bollgard® and Bollgard II®) produced by Bt technology licensees until 2017. Since then, CICR has been authorised to audit and approve new Bt cotton hybrids based on fulfilment of the criteria on trait quality and bollworm efficacy. MMB independently monitored trait quality and resistance development continuously, post-approval of Bt cotton hybrids. CICR and other laboratories characterised Cry1Ac expression profiles in Bt cotton (Kranthi *et al.* 2005); Cry1Ac resistance evolution (Kranthi *et al.* 2000); inheritance of Cry1Ac resistance (Chandrashekar and Gujar 2004; Kranthi *et al.* 2006; Nair *et al.* 2010) and modelled evolution of resistance to Cry1Ac (Kranthi and Kranthi 2004). The stability of expression of Cry1Ac and Cry2Ab2 proteins in target tissues was assessed in all the Bollgard II® hybrids which were approved for commercialisation between 2004 and 2015 (Dhanaraj *et al.* 2019).
5	Structured refuge – education and extension	Field teams of Mahyco, Monsanto and the Bt technology licensees created awareness among farmers and seed dealers on the importance of planting refuge through pre-season campaigns and farmer demonstration plots.
6	Studies on size and role of natural refuge	Assessed the relative abundance of *H. armigera* on alternate host crops (natural refuge) (Ravi *et al.* 2005). Estimated the relative abundance of alternate host crops of cotton bollworms through satellite imagery (Verma *et al.* 2019).
7	Participation of public institutions in collaborative and independent IRM studies	Several institutions like NBAIR, IARI, UAS, Dharwad and Raichur; Punjab Agricultural University (PAU), Ludhiana; Tamil Nadu Agricultural University (TNAU), Coimbatore; Navsari Agricultural University (NAU), Surat conducted collaborative projects on IRM (resistance monitoring) with Mahyco.

#	IRM aspect	Institutions and study
8	Approval of stacked Bt gene cotton with independent modes of action (Bollgard II®) in 2006	Bt cotton with pyramided Bt genes (Bollgard II®) having improved bollworm control efficacy and a better IRM tool.
9	Experimentation on alternatives to structured refuge	CICR and network of state agricultural universities like PAU, TNAU, NAU, UASRaichur (Chowdary *et al.* 2014) conducted field evaluation of 'Refuge-in-bag' (RIB) approach for Bollgard II® cotton hybrids during kharif seasons of 2011 and 2012.
10	Demonstration trials of RIB for Bollgard II® in 233 and 220 farmer fields in kharif 2013 and 2014 seasons, respectively	Bollgard II® seeds, blended with non-Bt seeds of the corresponding cotton hybrid as refuge so as to constitute a minimum of 5% of refuge seeds in the final seed-blend, were sown in several pan-India 0.4 acre plots to study and demonstrate the suitability of RIB Bollgard II® (Monsanto, unpublished).
11	2016 – approval of RIB for Bollgard II® hybrids by the Ministry of Agriculture and Farmers Welfare	Ministry of Agriculture approved RIB which involved blending of non-Bt refuge cotton seeds at a minimum of 5% level in the blend. The gazette notification also provided guidelines on seed and trait quality to the Bt cotton seed producers. RIB implementation was initiated from the 2017 season and a complete switch-over from structured refuge to RIB would occur by the end of 2019.

Baseline-susceptibility and post-commercialisation shift in susceptibility to Bt toxins expressed by Bt cotton

Prior to the regulatory approval of single Bt-gene cotton in India in March 2002, the technology provider Monsanto and the technology partner and licensor in India, Mahyco, had initiated the mandated India-specific IRM studies in collaboration with academic and governmental institutions. In addition, several resourceful public laboratories contributed data on the pre-commercialisation baseline susceptibility of cotton bollworms to Cry1Ac and later Cry2Ab2, and helped in assessment of the efficacy of Bt cotton in managing bollworms in the initial years of commercialisation. Multiple populations of the *H. armigera* from major cotton localities were screened with Cry1Ac (and later with Cry2Ab2) (Table 4.2). This baseline susceptibility data of *H. armigera* to Cry1Ac was benchmarked for tracking resistance development during the post-commercialisation years. In the subsequent years, many Bt cotton events containing Cry1Ac constructs (JK AgriGenetics's Event 1, Nath Seeds' GFM event and ICAR-CICR-UAS Dharwad's BNLA-601) or Cry1Ca1 (Metahelix Life Sciences' MLS-9124 event) received regulatory approval for commercialisation in India, but published data on the baseline susceptibility and post-release resistance monitoring are available in the public domain only for the Bt toxins expressed by Monsanto events MON 531 (Cry1Ac) and MON 15985 (Cry1Ac, Cry2Ab2).

While approving various Bt cotton events, the Indian regulator (the GEAC), entrusted the CICR with the resistance monitoring of cotton bollworms to various Bt toxins expressed by the approved transgenic cotton events. Toxicity data on Cry1Ac and Cry2Ab2 against the multiple populations of field-sourced *H. armigera* from different locations during the 2002–14 cotton growing seasons are available at the Ministry and CICR websites (http://moef.gov.in/project-approvals/geac-clearances/othinfo/; http://www.cicr.org.in/Evaluation.html) along with field-efficacy reports of the various Bt cotton hybrids under the All India Cotton Improvement Project. Analysis of Bt cotton productivity for these supervised field trials during 2002–10 showed 12–40% increase in central India, 31–40% increase in south India and 32–37% increase in north India over non-Bt cotton (unpublished results). In addition to CICR, several other

Table 4.2. Median lethal concentrations (LC_{50})* of Bt toxins Cry1Ac and Cry2Ab to cotton bollworms

Bt protein	*Helicoverpa armigera*	*Earias vittella*	*Pectinophora gossypiella*	*Spodoptera litura*
Cry1Ac	0.01–0.67[a] 0.02–0.54[b] 0.11–0.71[c] 0.15–1.10[d]	0.01–0.11[e]	0.05[f]; 0.028–0.045[j]	Not sensitive
Cry2Ab2	6.00–28.60[g]		0.05[f]; 0.16–0.5[k]	1.5–18.0[h]; 6.96[i]

* Expressed as μg toxin/mL of diet.
[a] Kranthi *et al.* (2001); [b] Kranthi and Kranthi (2003); [c] Jalali *et al.* (2004); [d] Fakruddin *et al.* (2003); [e] Kranthi *et al.* (1999); [f] Muralimohan *et al.* (2009); [g] Kranthi *et al.* (2009); [h] unpublished data; [i] 96 h neonate bioassays (unpublished); [j] Mittal *et al.* (2016); [k] Malthankar and Gujar (2016).

laboratories of Mahyco, IARI and state agricultural universities have been involved in Bt resistance monitoring efforts. An overview of the CICR post-commercialisation resistance monitoring data is shown in Figure 4.1, which summarises the Cry1Ac-sensitivity profiles of multiple populations of *H. armigera* and contrasts with the pre-commercialisation baseline data (Jalali *et al.* 2004). Bt resistance monitoring by CICR was predominantly restricted to *H. armigera* populations collected from non-cotton crops. It represented all the major cotton growing areas in eight states (four zones) in India. Resistance monitoring was carried out from 2002 to 2004 on the spotted and spiny bollworms and from 2008 on the pink bollworm. It intensified from 2012 in view of reports of high incidence of pink bollworm on dual-toxin Bt cotton.

The mean sensitivity of *H. armigera* to Cry1Ac declined over a period of ~15 years, as reported in terms of increase in the mean LC_{50} value from the pre-commercialisation baseline (1999, 2001) (Jalali *et al.* 2004) of 0.34 μg/mL to the post-commercialisation mean (2002–14) of 0.82 μg/mL (= ppm, parts per million) of diet (a range of means, 0.13–1.99 ppm). The post-commercialisation period of 2002–14 can be divided into two phases, (i) when the mean LC_{50} values from 2002 to 2009 steadied around a mean of 0.35 and (ii) from 2010 to 2014 which saw a steady increase in mean LC_{50} to 1.57 ppm (highest being 1.99 in 2013). Specifically, till 2007, Cry1Ac sensitivity of all populations was below 1.0 ppm but from 2009 onwards, populations of *H. armigera* sourced from the western states of Gujarat (Bharuch, Junagarh, Surat, Ahmedabad, Surendranagar, Vadodara, Amreli) registered LC_{50} values above 2.0 ppm. Some of the populations with exceptionally high tolerance to Cry1Ac were from Bharuch (LC_{50} 5.11 ppm), Ahmedabad (8.69), Hanumangarh (9.07) and Parbhani (13.15) from 2009 to 2014. The resistance ratio (RR) of Cry1Ac, computed with reference to LC_{50} value of 0.1 ppm of the Bt-susceptible reference strain of *H. armigera*, was 51-fold in 2012, 87-fold in 2013 and 132-fold in 2014. Susceptibility of the cotton bollworm was tested with several Bt toxins like Cry1Aa, Cry1Ab, Cry1Ac, spore-crystal complex of Bt subsp. *kurstaki* HD-1 and Bt subsp. *kurstaki* HD-73 and these studies showed a trend of increasing tolerance (Chandrashekar *et al.* 2005). The differences in toxicity of Cry1Ac derived from different sources used in transgenic Bt cotton in India (Gujar *et al.* 2007) could have influenced the increasing tolerance of the bollworm. Ranjith *et al.* (2010) reported resistance in the cotton bollworm *H. armigera* collected from fields in Raichur that survived for two generations on leaves of dual-toxin Bt cotton under laboratory conditions. As supporting evidence for increasing tolerance, Kukanur *et al.* (2018) reported an increase in Cry1Ac resistance allele frequencies of 0.085 and 0.035, assessed in 2013 and 2014, respectively, which were 65- and 29-fold higher than the baseline allele frequency of 2004. Ningthoujam *et al.* (2017) estimated the Cry1Ac resistance allele frequency

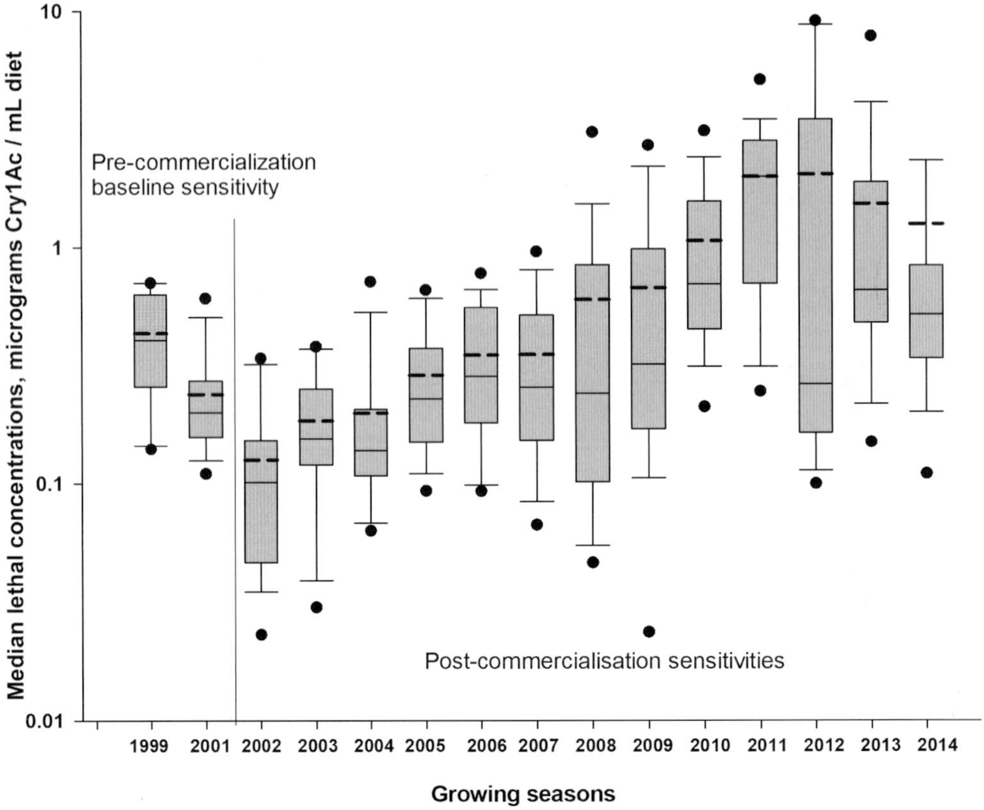

Figure 4.1 Shift in sensitivity profiles of sampled multiple populations of *H. armigera* to Cry1Ac during pre- and post-commercialisation periods, 1998–2014. Bold broken line in box = mean; thin continuous line = median; lower and upper whisker caps indicate 5th and 95th percentiles; lower and upper box margins indicate 25th and 75th percentiles; bold dots are outliers. Source: Resistance monitoring data of ICAR Central Institute of Cotton Research, Nagpur.

of 0.085 for the *H. armigera* population collected from Karnataka in 2013. All the foregoing reports of increased tolerance to Cry1Ac did not, however, translate into high-level crop damage in Bt cotton in these areas, unlike in case of the pink bollworm, wherein high resistance to Bt toxins was positively associated with crop (boll) damage at different locations during crop season of 2009, indicative of field-resistance.

Dual Bt-gene cotton hybrids producing Cry1Ac and Cry2Ab2 were approved for commercial cultivation in 2006 in India as a proactive measure to delay evolution of Bt resistance in cotton bollworms. This IRM strategy, in line with several studies including field experience with stacked Bt cotton in the USA and Australia, showed that co-expression of two or more toxins with independent modes of action in the same plant enhanced efficacy against the target pests by substantially reducing the proportion of resistant bollworms that could otherwise survive on a single-gene Bt cotton (Roush 1998; Zhao *et al.* 2003; Sivasupramaniam *et al.* 2008).

Cry1Ac was highly toxic to all the key Indian cotton lepidopterans (excepting *S. litura*), whereas Cry2Ab2 was moderately toxic to *H. armigera* and *S. litura* and highly toxic, as Cry1Ac, to *P. gossypiella* and *Earias* spp. Similar toxicity profiles were recorded with bollworms in the USA, where Cry2Ab2 was 3-fold and 40-fold less sensitive, relative to Cry1Ac, to the pink bollworm and tobacco budworm *Heliothis virescens*, respectively (Sivasupramaniam *et al.* 2008).

The variation in the susceptibility of multiple populations of *H. armigera* to Cry2Ab2 toxin was studied in 2004–05 and 2007–08, a year prior to and just after the commercialisation of the dual-Bt gene cotton (Kranthi *et al.* 2009). Resistance to Cry2Ab2 in the cotton bollworm was monitored during 2009 and then from 2012 to 2014.[1] Cry2Ab2 caused moderate to severe inhibition of growth and development and a high dose of 40 ppm did not evoke appreciable mortality in several populations of *H. armigera*, implying that Cry2Ab2 had moderate toxicity and at best could complement the high toxicity of Cry1Ac with a different mode of insecticidal action. High variability in the median moult inhibitory concentrations (MIC_{50}) of Cry2Ab2 against cotton bollworm appeared to be its distinctive feature, as studied in 27 populations in 2009 (0.05–69.26 ppm); in nine populations in 2012 (0.01–4.29 ppm); in 15 populations in 2013 (0.09–29.19) and in six populations in 2014 (0.50–4.60 ppm). Toxicity of Cry2Ab2 to the cotton bollworm varied from LC_{50} of 0.12 (the most susceptible population from Parbhani) to that of 101.33 ppm (the most resistant population from Nanded) during 2014–15, marking a trend towards increasing resistance.[2]

The high toxicity of Cry1Ac and Cry2Ab2 toxins in Bollgard II effectively controlled infestations of *E. vittella* (prevalent in central and southern cotton fields) and *E. insulana* (predominant in north India). The median lethal concentration (LC_{50}) of Cry1Ac to field populations of *E. vittella* ranged from 0.28 to 1.22 μg Cry1Ac/cm^2 of treated leaf (Kranthi *et al.* 1999). Resistance monitoring studies in 2002–04 established both species of *Earias* to be highly susceptible to Cry1Ac in diet-incorporation bioassays. The LC_{50} of Cry1Ac ranged from 0.001 to 0.105 ppm for *E. vittella* and from 0.022 to 0.269 ppm for *E. insulana*.[3] The high sensitivity of *E. vittella* to Cry1Ac provided an opportunity to develop an assay to estimate Cry1Ac content in transgenic Bt cotton tissues in India (Farah *et al.* 2003) before the advent of an enzyme-linked immunosorbent assay for specific detection and quantification of Bt proteins, akin to an highly sensitive *H. virescens* bioassay developed in the USA (Greenplate 1999).

Studies on susceptibility of pink bollworm to both Cry1Ac and Cry2Ab were initiated with standardisation of mass rearing methodology as late as 2007. Development of the necessary protocols enabled its inclusion into the annual resistance monitoring efforts. The mean LC_{50} of Cry1Ac for the pink bollworm was 0.040 ppm (range 0.023–0.069) and of Cry2Ab2 was 0.051 ppm (range 0.036–0.073 ppm) in diet incorporation studies (Muralimohan *et al.* 2009).

Cry toxin expression and resistance evolution

Delivering a high dose of Bt toxins is primarily a function of the transformation event. Studies carried out on expression of Cry1Ac in Bt cotton cultivars highlighted the importance of the genetic background of donor parents (Adamczyk and Sumerford 2001). The events MON 531 and MON 15985 in cotton varieties (open pollinated) in the USA had delivered a high dose of toxins for the control of sensitive bollworms like *P. gossypiella* and *H. virescens*, in many varietal backgrounds across the cotton belt. The activity spectrum was enhanced to include *S. frugiperda* and *S. exigua* with the introduction of Bollgard II® (Sivasupramaniam *et al.* 2008). The expression profiles of MON 531 and MON 15985 in diverse hybrid backgrounds across many growing seasons have been extensively studied in India. Kranthi *et al.* (2005) reported higher Cry1Ac content in the upper and lower canopy leaves compared to mid canopy, and found decrease in Cry1Ac content from 5.51 μg/g to 0.05 μg/g in the upper canopy leaves of Bt cotton as the crop age progressed from 30 to 148 days. A similar trend was observed for middle and lower canopy leaves. They also reported that high *H. armigera* larval mortality on leaves was observed until the crop age of 85 days, after which larval mortality declined rapidly as Cry1Ac concentration decreased below a critical concentration of 1.8 μg/g (= ppm) in upper canopy leaves (feeding site of early instars). Cry1Ac toxin expression was highest in the leaves, followed by squares, bolls and flowers; the flowers did not have enough toxin to control cotton bollworm. Saini and Dhawan (2010) reported significant negative correlation

between tannin and Cry toxin contents in Bt cotton hybrids. Toxicity to *H. armigera* and *S. litura* was positively correlated with toxin content (Cry1Ac and Cry2Ab2) and total phenols in the Bollgard II® hybrids and in the Cry1Ac Bt cotton hybrids, suggesting synergistic action among Bt protein and total phenols. Similarly, Singh *et al.* (2016b) reported maximal mortality of cotton bollworm on the terminal leaves of 112-day old plants, followed by 30-day, 88-day and 65-day old JKCH1947 Bt cotton, suggesting possible interaction of phenols and Cry1Ac at different stages of plant growth. Dhanaraj *et al.* (2019) published a comprehensive analysis of season-long expression profiles of Cry1Ac and Cry2Ab2 toxins in 4214 Bollgard II® hybrid season combinations, tested across 12 growing seasons (2004–15) in the four cotton growing zones. Cry1Ac and Cry2Ab2 content (calculated on wet tissue weight basis) in leaves (mean moisture content of 90.9%; moisture estimates for plant parts provided in personal communication) were 0.55–1.05 μg/g and 23.25–48.90 μg/g, respectively; Cry1Ac and Cry2Ab2 contents in pre-candle squares (mean moisture content of 76.9%) were 0.88–1.86 μg/g and 60.86–117.92 μg/g respectively; and Cry1Ac and Cry2Ab2 contents in boll epicarp (mean moisture content of 66.5%) were 1.13–2.21 μg/g and 80.77–154.02 μg/g respectively in multiple Bt cotton hybrids under different agronomic conditions. Many other studies (Knox *et al.* 2008; Li *et al.* 2011; Badigar *et al.* 2013; Manjunatha *et al.* 2017; Nagappa and Khadi 2018) reported similar observations on lower Cry1Ac and relatively higher Cry2Ab2 contents in plant parts on wet weight basis. However, both Cry toxin contents in bolls were lower at least two- to 10-fold than in the leaves (Cry1Ac 1.3–1.5 μg/g; Cry2Ab2 16.8–22.7 μg/g) as reported by Li *et al.* (2011). Similarly, Badigar *et al.* (2013) reported lower Cry toxin contents in the boll rind (0.15–0.95 μg/g Cry1Ac and 9.5–34.6 μg/g Cry2Ab2) and raw seeds (0.15–1.74 μg/g Cry1Ac and 12.34–43.4 μg/g Cry2Ab2) than in leaves (0.42–2.94 μg/g Cry1Ac and 18.28–52.95 μg/g Cry2Ab2). The combined toxicity in flowers and bolls of dual-toxin Bt cotton induced higher mortality to pink bollworm larvae and that in leaves to tobacco caterpillar, relative to the single-toxin Bt cotton. Manjunatha *et al.* (2017) reported ~15–20-fold more Cry2Ab2 (30–63 μg/g) than Cry1Ac (1.6–3.8 μg/g) in the leaves of 60-, 90- and 120-day old dual-toxin Bt cotton crop. Cry toxin content declined more than half over this period. The decline in toxin content over time could hasten Bt resistance evolution, as reported for *H. zea* (Carrière *et al.* 2019b), and change the mode of inheritance from functionally recessive to partially dominant, as reported for *H. armigera* (Bird and Akhurst 2005).

Most cases of field-level resistance in lepidopterans are associated with the lack of high dose of Cry toxins in the Bt crops (Tabashnik and Carrière 2017). It is likely that cotton bollworm *H armigera* might evolve field-resistance to Cry1Ac in Bt cotton. Further, cadherin-based resistant mutants r9 and r10 of *H. armigera* were found to be present in the wild population at a frequency of <0.1 in India (unpublished). Similar early warning of field-resistance is reported for *H. armigera* in China (Jin *et al.* 2015; Tabashnik and Carrière 2017).

However, in the case of pink bollworm populations in India, both Cry proteins are almost at par in their toxicity to the neonates (Muralimohan *et al.* 2009), but over time this pest evolved resistance to both Cry1Ac and Cry2Ab2. Molecular characterisation of Cry1Ac resistance in pink bollworm in India revealed mutations in cadherin gene due to alternate splicing with high variability in cadherin transcripts compared to pink bollworm cadherin mutations in the USA (Fabrick *et al.* 2014). Cry2Ab2 resistance was attributed to mis-splicing of the ATP-binding cassette transporter gene (*PgABCA2*) from field-selected populations in India (Mathew *et al.* 2018).

Natural refuge has delayed Bt resistance in *H. armigera*

H. armigera is polyphagous and multiplies productively on several host crops like pigeon pea, chickpea, sorghum, maize, chillies, sunflower and tomato in India, though ovipositional

preference varies. A mosaic of these host crops, in small land-holdings (~2–5 acres), is a typical cotton cropping landscape in western, central and south India and supplements IRM for Bt cotton by contributing to the pool of Bt-susceptible population of *H. armigera*. Pigeon pea is the preferred host for *H. armigera* over cotton (Ravi *et al.* 2005) and is grown extensively in central India cotton fields, often as an intercrop. An overlap in bloom period with Bt cotton preferentially attracts oviposition on pigeon pea and contributes to the Bt-susceptible *H. armigera* population pool. Considering this aspect, Indian regulators approved both non-Bt cotton and pigeon pea as refuge crops for Bt cotton. While pigeon pea is a good host crop for *H. armigera*, it is a poor host for the pink, spotted and spiny bollworms. Thus, only non-Bt cotton can serve as a refuge crop for all four major target bollworms in India.

Satellite mapping of the relative proportions of cotton and other host crops of *H. armigera* revealed a substantial area (up to 40%) of alternate host crops as natural refuge in the major cotton belts of south, central and western India, but only a sparse area in north India (Verma *et al.* 2019). This is in line with the estimated refuge requirements of 42%, 19% and 0% for north, central and south India based on assessment of risk of Bt resistance evolution and its impact on the sustainability of Bt cotton productivity in India (Singla *et al.* 2012). Thus, the fortuitous availability of natural refuge as part of the cotton cropping landscape of central and south India could be the most probable cause for the observed delay (>18 seasons) in Bt resistance evolution in a polyphagous pest like *H. armigera*. *S. litura* is also a polyphagous pest and can multiply on castor, green gram and several vegetables. However, the host range of the spotted bollworm, *E. vittella*, is restricted to cotton and okra, and the pink bollworm to cotton.

If natural refuge is adequately large and diverse, it could substitute for structured refuge, to an extent. For example, China has not mandated planting of structured refuge for Cry1Ac cotton and relies solely on the presence of natural refuge for *H. armigera* in crops like chickpea, pigeon pea, corn, soybean, peanuts and sesame (Wu *et al.* 2002; Lu *et al.* 2013). Field studies have indicated that this size and diversity of natural refuge adequately delayed resistance evolution to Cry1Ac. However, modelling studies have indicated natural refuge to be inferior to structured refuge (Jin *et al.* 2015). The productivity of heliothine spp. (*H. armigera* and *H. punctifera*) on pigeon pea, maize and sorghum has been studied in determining the size of structured refuge options in Australia for Bt cotton, Ingard®, Bollgard II® and Bollgard 3®.[4] Likewise, the US EPA has waived the need for structured refuge in stacked-Bt cotton-growing areas of the USA where pink bollworm is not endemic because of the presence of adequate natural refuge for *H. zea* and tobacco budworm *H. virescens* in the form of maize and weeds.[5]

Natural refuge cannot be relied upon as a stand-alone IRM measure in India for two main reasons. First, natural refuge of adequate size and diversity for *H. armigera* is available in central and south India but is sparse in the northern Indian states which constitute ~18% of the total Bt cotton acreage. Second, natural hosts of the pink bollworm (conventional cotton hybrids of *G. hirsutum* and indigenous varieties of *G. herbaceum and G. arboreum*) occupy a minuscule area (<5%) and may not be meaningful from an IRM perspective.

Planting non-Bt cotton refuge critical for delaying development of Bt resistance

Since planting non-Bt cotton refuge was integral to Bt cotton stewardship and to sustaining the field-efficacy of Bt cotton, the GEAC, when approving commercial cultivation of Bt cotton (Bollgard®) in March 2002, also stipulated a requirement to plant structured refuge crop of non-Bt cotton in a perimeter belt of five rows or 20% of total sown area, whichever was more. Therefore, the Bt-technology licensee seed companies provided 120 g of non-Bt hybrid cotton seeds to be planted as refuge with every packet of Bt cotton hybrid seeds (450 g) for an acre. The refuge non-Bt cotton hybrid stipulated in 2002 was the non-transgenic counterpart of the

Bt hybrid, later changed and approved in 2006 to be any non-Bt hybrid of same species, maturity and fibre length as that of the Bt hybrid or a non-Bt popular hybrid.[6] In 2009, the GEAC also permitted cultivation of pigeon pea as refuge border crop around BNBt (Bikaneri Narma Bt) variety. However, the size of refuge requirement was unchanged despite approval of stacked Bt gene cotton in 2006. It was realised over the years that compliance with structured refuge planting for Bt cotton remained very low – Indian cotton farmers were reluctant to sow the refuge non-Bt seeds, driven by economic compulsion to maximise returns from a 100% Bt cotton crop rather than set aside 20% area for refuge.

Mallet and Porter (1992) had foreseen a compliance problem with planting of structured refuge, years before Bt crops were commercialised in the USA. Indian Bt cotton farmers are not bound by any sowing agreement to plant refuge, whereas in the USA and Australia planting refuge is legally enforceable. However, despite legal obligation, 25% of US farmers did not comply with refuge planting (Jaffe 2009; Head and Greenplate 2012).

Other reasons which deterred cotton farmers from planting refuge were lack of awareness, intricacy in refuge-sowing and gap-filling operations and often poor quality of refuge seed (Global AgriSystems 2016; Kranthi *et al.* 2017). Thus, lack of refuge planting put Bt cotton technology at great risk of bollworms evolving resistance to it. Bt cotton stakeholders in India had to look at other refuge options to ensure greater compliance.

Alternative to structured refuge

Realising the importance of increasing compliance on refuge planting, Bt technology generators and the US EPA optimised a Refuge-in-bag (RIB) approach for Bt maize in which refuge seeds are delivered through blending Bt (95%) and non-Bt (5%) seeds in a single bag. RIB is applicable only to Bt crops with a stack of two or more Bt genes with better insect-control efficacy and hence a smaller refuge of 5%. RIB was approved for Bt maize stacked with HT in 2011 in the USA and Canada, and in Vietnam in 2014 (Mohan 2018). Taking a lead from the US experience, Indian Bt cotton stakeholders (MMB – Bt technology licensor; CICR and several agricultural universities) experimented with the suitability of RIB for Bollgard II® cotton hybrids as a substitute for structured refuge, beginning in 2011. Over 450 demonstration trials on RIB at 5% level in a seed blend in 2013 and 2014 validated its suitability for Bollgard II® cotton (unpublished). In 2017, the Ministry of Agriculture and Farmers' Welfare approved RIB for Bollgard II® to be implemented from 2017 onwards and a complete switch-over to RIB was mandated from the growing season of 2020.[7] This marked the first approval of RIB for stacked Bt cotton among the countries which have adopted GM crops. The notification specified that Bt cotton with RIB (Bt-RIB) seed pack (475 g) must contain a minimum of 90% and a maximum of 95% cotton seeds positive for each Bt transgene, implying that non-Bt refuge seeds will constitute 5–10% of the blend. The refuge seeds must be of a non-Bt hybrid, an isogenic of the corresponding Bt hybrid. The key features of an RIB for Bollgard II® in India are:

- assured high compliance of refuge on small farms in developing countries where planting structured refuge is ignored;
- uniformity in crop agronomy, especially for bloom period of isogenic non-Bt cotton refuge plants with those of the Bt counterpart to synchronise moth emergence;
- close proximity of susceptible moths from non-Bt refuge plants with the rare resistant moths from Bt plants for mating;
- most suited for the pink bollworm as it completes its larval development within the boll of refuge plant thus avoiding exposure to Bt plant and the attendant risk of evolving resistance.

A major shortcoming of RIB with Bt cotton could arise from the movement of cotton bollworm larvae between the refuge and Bt cotton plant which could hasten Bt resistance

development (Mallet and Porter 1992; Onstad *et al.* 2011; Head and Greenplate 2012; Carroll *et al.* 2012). Among the four major lepidopteran pests of cotton in India, *H. armigera* and *S. litura* move from plant to plant and large larvae of *S. litura* can move across fields, whereas *E. vittella* is not known to move much between plants and *P. gossypiella* completes its larval development within the boll it enters. Larval movement increases with the bushy architecture of cotton hybrids as foliage overlap. In an RIB format, this can mean possible migration of large (>3rd instar) *H. armigera* and *S. litura* larvae from a non-Bt refuge plant onto the adjacent Bollgard II® plants. The migrant larvae will succumb to Bt toxicity at late larval instar or pupal stage, but not before damaging a few squares or bolls. The toxic effect will be quicker to manifest in Bt stacks. However, such movements decrease the efficacy of refuge by killing the bollworm larvae which would have matured into Bt-susceptible moths, had they not moved from the refuge plant. Conversely, larvae moving from Bt plant to non-Bt refuge plant would escape receiving the lethal dose of Bt protein(s) and increase the survivorship of heterozygotes for resistance alleles, which could hasten resistance development. The frequency of crossing over to refuge plants would be low due to high mortality, if the Bt cotton plants express two or more Bt toxins at high dose. The impact of larval movement on Bt resistance evolution was modelled by Carroll *et al.* (2012) who concluded that two- or three-Bt gene pyramids were as durable as the 5% structured refuge counterpart for Bt pyramids.

The spin-off effect of Bt-resistant insects feeding on other host crops may increase fitness costs that will eventually reduce their population, as reported for bollworms (Gassmann *et al.* 2009). In case of pink bollworm, both Cry1Ac- and Cry2Ab2-resistant insects showed fitness costs in terms of prolonged larval period when fed on the untreated diet (Malthankar and Gujar 2016; Mittal *et al.* 2016), but not the Cry1Ac-resistant cotton bollworm *H. armigera* (Nair *et al.* 2010).

As Indian Bt cotton seed producers were preparing for a complete switch from structured refuge to RIB for Bollgard II® cotton from June 2020, quality stewardship in the entire process of producing refuge seeds, which includes maintaining 5–10% non-Bt refuge seeds along with achieving trait and quality standards as per seed Act, assumed an important role. It would be the single most important factor deciding the success or failure of RIB (Mohan and Sadananda 2019; Muralimohan and Mahesh 2020). Primarily, non-Bt refuge seed should be of high quality so that refuge plants synchronise with Bt cotton plants in all respects, and support random mating between susceptible and resistant adults to produce susceptible progeny. Maintaining isogenic lines of non-Bt parents of the current popular Bt cotton hybrids and using them for refuge seed production, however challenging, will ensure the success of RIB in the long run. Outreach on RIB to ~7.5 million farmers at various resource levels will ensure that the farmers retain the refuge plants even if they are phenotypically distinguishable or because of bollworm incidence are likely to reduce yield expectation. Notwithstanding these challenges with RIB stewardship, the assured compliance with refuge planting will help in sustaining Bt cotton technology.

Pink bollworm: first and only bollworm to evolve field-resistance to Bt cotton

Pink bollworm has always been an enigmatic pest and a destructive one in many cotton-growing areas of the world (Naranjo *et al.* 2001). Bt cotton provided the much-needed respite from this pest in the USA and India for many years, until the evolution of Bt field-resistance in India. Pink bollworm, first described in 1843, was then a serious pest impacting the quality and quantity of cotton produce and ultimately the cotton trade, necessitating the British-administered Madras government to promulgate the *Pest Act 1919* for its management (Ramachandra Rao 1921). It remained one of three major pests of cotton along with *Earias fabia* and *E. insulana* (Ayyar 1932; Maheshwariah and Puttarudriah 1956); *Heliothis armigera* was a minor pest. Pink bollworm is believed to have originated in the India–Pakistan region or

eastern Indian Ocean and to have reached Mexico and south-western USA via Egypt through imported infested seed cotton, to become a key pest. As an introduced pest, it is now spread all over the world.[8]

Pink bollworm is practically monophagous on cotton and is reported to infest more than a dozen plant species, including a few weeds (Dastur *et al.* 1960). It is typically a late-season pest, mainly infesting mature green cotton bolls from 100 days after sowing (DAS) onwards with peak infestation around 140 DAS. Loss of seed-cotton yield was reported to be 37–50% in Punjab and ranged widely in other localities (Dhawan *et al.* 1986). Damage to bolls and the fibre is particularly severe in long-duration hybrids and wherever the cropping period has been extended beyond 140 days, up to 220 days. Insecticidal sprays have limited efficacy on egg and larval stages which are largely inaccessible; insecticide resistance was reported in early 2000s (Kranthi *et al.* 2002a).

Pink bollworm is unique among the bollworms as being equally sensitive to Cry1Ac and Cry2Ab2. It was effectively controlled by the single- and dual-gene Bt cotton in India before resistance evolution (Bambawale *et al.* 2004; Naik *et al.* 2014).

Bt resistance in pink bollworm

Cotton bollworms in India have been exposed to Bt cotton for ~20 seasons but, from a selection pressure perspective, a substantial area of Bt cotton became available only from 2006. Field-resistance to Bt cotton evolved in a sequential manner following exposure to Bt proteins, first to Cry1Ac from 2002 and then also to Cry2Ab2 from 2006 via Bollgard II®. Field-resistance to Cry1Ac was first observed in 2008 in a population sourced from non-Bt cotton fields in Amreli (Dhurua and Gujar 2011) and in a single field of single-toxin Bt cotton (100-day old) in Amreli district of Gujarat state with unusually high incidence of pink bollworm larvae in >90% of the sampled bolls (Mohan *et al.* 2016). The subsequent seasons of 2009–10 witnessed its inadequate control in several districts of Gujarat state (Sharma 2010) and in 2010–11 in the adjoining states of Maharashtra and Madhya Pradesh, coupled with high tolerance (>2000-fold less susceptible to Cry1Ac) to a diagnostic concentration of 10 ppm Cry1Ac (Mohan *et al.* 2016). Adjoining Bollgard II® fields were well protected against all bollworms, as the Cry1Ac-resistant pink bollworm larvae were also susceptible to Cry2Ab2 in assays. Mittal *et al.* (2016) also reported high tolerance to Cry1Ac in populations collected from Bharuch in Gujarat and Jalgaon in Maharashtra in 2011–12. This unequivocally indicated that resistance had evolved to Cry1Ac expressed by Bt cotton. Resistance to Cry2Ab2 in the pink bollworm populations steadily evolved. Naik *et al.* (2018) presented evidence of resistance to both Cry1Ac and Cry2Ab2 toxins in pink bollworm larvae sourced from infested Bollgard II® fields from 2013 to 2017 in central and south India. Their studies reported 29–73% larval recovery from Bollgard II® bolls and significant increase in tolerance to both Cry1Ac and Cry2Ab2. The mean LC_{50} of Cry1Ac for populations in central and south India increased from mean of 0.33 μg/mL of diet in 2013 to a mean of 6.94 μg/mL of diet, and the resistance ratio from 47 in 2013 to 1387 in 2017. The mean LC_{50} of Cry2Ab2 across multiple populations in central and south India was 0.014 μg/mL of diet in 2013, which increased to a mean of 12.51 μg/mL. The resultant resistant ratio rose from a mean of 5.4 in 2013 to 4196 in 2017. However, resistance ratios for pink bollworm populations in north India were relatively low (26–262 for Cry1Ac and 1–108 for Cry2Ab2) and there was little incidence on Bt cotton during the same period. During 2016–18, high incidence of *P. gossypiella* (74–100% infestation) in central and south India in concurrence with high LC_{50} values to Cry1Ac (0.22–3.13 μg/mL of diet), Cry2Ab2 (0.13–1.69 μg/mL of diet) and to the mixture of Cry1Ac and Cry2Ab2 expressed in Bollgard II® (MRC7031 seed containing 21 ng Cry1Ac and 4830 ng Cry2Ab/mg powder) (0.18–3.64 μg/mL of diet) compared to laboratory-susceptible population, i.e. Cry1Ac (0.058 μg/mL of diet), Cry2Ab (0.038 μg/mL

of diet) and Cry1Ac + Cry2Ab2 mixture (0.045 µg/mL of diet) provided evidence of continued field-evolved resistance in the pink bollworm to Bollgard II® cotton (Shahanaz 2019).

In hindsight, resistance evolved as selection pressure built up with Bt cotton (single and dual-Bt cotton, together) occupying 66% of total cotton area in 2007–08, 81% in 2008–09 then plateauing around 95% from 2013–14 onwards, while dual-toxin Bt cotton alone formed ~27% of total Bt cotton in 2008 and 57% in 2009. It increased to ~90% and above from 2012 onwards.

A second tipping point in resistance evolution was the local practice of extended cropping of Bt cotton beyond 160 DAS to the last pick of cotton around 230 DAS in irrigated tracts of Gujarat and the central Indian states of Maharashtra and Madhya Pradesh along the Narmada river belt. In such areas, Bt-resistant pink bollworm populations gain the selective advantage of multiplying on Bt cotton for three to four additional generations, resulting in a large residual population in the bolls and in the soil at harvest. If the pest cycle is not broken by destroying unopened/partially opened bolls and deep ploughing, it invariably results in a large carryover to the next cropping season with a strong likelihood of early appearance in open flowers in the new season (June sowings).

A third major contributing factor to Bt resistance in the pink bollworm was low compliance on IPM practices like post-harvest summer deep-ploughing to destroy pupae and hibernating larvae, destruction of unopened bolls on cotton stalks and of insect-infested seeds in cotton gins, and mass trapping of adult moths with pheromone traps. Regular adoption of these practices could have broken the pest cycle and minimised the carryover of Bt-resistant pink bollworm to the next cotton season. Transport of pink bollworm-infested cotton seeds helped spread infestation and resistance to neighbouring states, which primarily bought them for extraction of cotton seed oil.

A fourth factor which could have hastened Bt resistance evolution in India appears to be planting of illegal and substandard Bt cotton in Gujarat, a few years before the official approval of Bt cotton in 2002 (Menon 2001) and thereafter in many other localities (Jayaraman 2004; Showalter *et al.* 2009; Ramaswami *et al.* 2012; Singh *et al.* 2016b).[9] Possible poor expression of Cry1Ac in the target tissues of these illegal varieties could have favoured increase in population of individuals with semi-dominance of Bt resistance, as dominance is reported to be concentration-dependent in the pink bollworm (Liu *et al.* 2001; Tabashnik *et al.* 2002). Inheritance of Cry1Ac resistance was found to be partially recessive (Nair *et al.* 2016) as well as semi-dominant (Mittal *et al.* 2016). Cry2Ab2 resistance was semi-dominant (Malthankar and Gujar 2016). Such resistance may help its spatial spread, even if susceptible insects are available. Compounding these issues, the Cry1Ac and Cry2Ab2 traits in Bt cotton are in hemizygous status in the hybrid and segregation of Bt traits in seeds developing seeds may support growth and development of pink bollworm larvae which thus acquire an ability to tolerate toxins and evolve resistance. However, there is limited knowledge on the probability of survival under these circumstances and their relative contribution to resistance evolution. Using a seven-day bioassay to evaluate the toxicity distribution in the seeds of single-toxin and dual-toxin cotton, Singh *et al.* (2016b) reported that only 46% of seeds of single-toxin Bt cotton and 22% of seeds of dual-toxin Bt cotton caused 100% of the neonate mortality of the cotton bollworm. This suggests the possibility of larval survival on many seeds with Bt trait.

So far, cotton farmers in the three northern states of Rajasthan, Haryana and Punjab (constituting ~20% of Indian cotton production) have not encountered Bt resistance in the pink bollworm, though these areas are equally deficient in refuge sowing. Cotton crop is terminated early (within 150 DAS) in these regions to facilitate sowing of wheat and other winter crops. Short- to medium-duration cotton hybrids harvested by December appear to have delayed the evolution of Bt resistance. In fact, these measures are the key recommendations to contain Bt resistance in the central Indian states of Gujarat, Maharashtra and Madhya Pradesh. Field-resistance to Bollgard II®, first detected in 2014, spread to the

southern states of Telangana, Andhra Pradesh and Tamil Nadu in subsequent years. In those states, the Bt-resistant populations also multiplied on summer cotton which is grown in irrigated tracts, thus providing year-long continuity for the pest.

The other key bollworms (*H. armigera* and *S. litura*), though exposed to Bt cotton for the same number of kharif seasons, have not shown indications of field-resistance to Bt cotton, primarily because of their polyphagous nature and the availability of a wide range of alternate host crops as refuge which are co-cultivated alongside Bt cotton in the growing season. However, the spotted and spiny bollworms (*E. vittella* and *E. insulana*) need to be closely watched for Bt resistance evolution as they too depend upon cotton as the main host – a secondary host is okra, which occupies a relatively smaller area (0.5 million ha in all three seasons of a year).

Can the current Bt resistance in pink bollworm be managed?

Given the fact that pink bollworm has evolved field-resistance to two Bt proteins in India due to grossly inadequate refuge planting and neglect of IPM practices over many years, the immediate question is whether this situation can be remedied, so that Indian farmers continue to derive benefits from the Bt cotton technology. The next-generation Bt cotton technology Bollgard 3® that produces three toxins (Cry1Ac, Cry2Ab2 and Vip3a) could be effective against the Bollgard II®-resistant pink bollworm. The upgrade was approved in Australia in 2016 to ensure the sustainability of GM technology. Until an effective technology is available, there is no choice but to manage Bt resistance in pink bollworm through conventional approaches, involving (i) wide-area adoption of cultural practices aimed to break the pink bollworm pest cycle in combination with (ii) popularising early-maturing Bt cotton hybrids and/or implementing early termination of medium-/long-duration Bt cotton hybrids in pink bollworm endemic areas, and (iii) rigorous implementation of area-wide community based IPM methodologies including need-based insecticide application (Kranthi 2015; Mohan 2017). In short, extension advisories should strongly advocate cultivation of Bt cotton as a component in the local IPM package because this approach also ensures management of sap-sucking insect pests of cotton not targeted by Bt technology.

In China, resistance to Cry1Ac in the pink bollworm in Yangtze valley was reversed by using F2 seeds which provided 25% refuges that suppressed 96% of population and required 69% fewer insecticidal sprays over 11 years (Wan *et al.* 2017). Successful control of pink bollworm in Arizona was based upon a proactive area-wide pink bollworm suppression/eradication program based upon (i) cultural practices, (ii) transgenic Bt cotton, (iii) mating disruption using pheromones, (iv) release of radiation sterilised moths and (v) need-based insecticide application between 2001 and 2008 (Grefenstette *et al.* 2009). India, faced with Bollgard II®-resistant pink bollworm, could initiate a similar pink bollworm management program once the pheromone formulation PB Rope L and 'Splat'-based mating disruption is approved. The success of PB Rope L in combination with need-based use of insecticide on Bt cotton was demonstrated in 20 ha trials during 2005–06 and 2006–07.[10]

Conclusion

Indian cotton farmers in particular, and the cotton trade-chain as a whole, have benefited immensely from Bt cotton since its introduction in 2002. Bt cotton adoption has steadied around 12 million ha constituting >95% of total cotton area – an open endorsement of the value GM technology has provided across nearly 20 growing seasons. However, Bt cotton has suffered loss of field-efficacy since 2010 due to sequential evolution of resistance in the pink bollworm to Cry1Ac and Cry2Ab2, probably due to selection from pre-existing resistance alleles and/or *de novo*

mutation (Hawkins *et al.* 2019). Resistance management is both technical and involves socio-ecological complexities, as reportedly found from the comparison of resistance governance in four Bt cotton-growing countries (Australia, Brazil, India and the USA) (Carrière *et al.* 2019a). But for the setback in India, the Bt technology has been effective in managing all other lepidopteran pests including *H. armigera* populations which are constantly monitored for evolution of resistance. Since there are no prospects of upgrade in Bt cotton technology such as use of better GM constructs (Singh *et al.* 2016a), stacking with additional genes as in Bollgard 3® or pyramiding of Bt genes with RNAi (Ni *et al.* 2017), IRM and IPM measures need to be implemented to sustain it. Fortunately, the Indian government has taken proactive measures for assuring Bt trait-cum-seed purity and a compliance-assured refuge delivery system in the form of RIB for Bt cotton. It is imperative that India maintains a long-term strategy involving new technologies to sustain its use of Bt cotton.

Acknowledgements

The authors are grateful to Dr J Bentur and Dr R Nair for their useful comments.

Endnotes

1. http://moef.gov.in/project-approvals/geac-clearances/othinfo/
2. http://moef.gov.in/project-approvals/geac-clearances/othinfo/
3. http://moef.gov.in/project-approvals/geac-clearances/othinfo/
4. https://www.monsantoglobal.com/global/au/products/Documents/Bollgard-3-Resistance-Management-Plan.pdf
5. https://www.epa.gov/regulation-biotechnology-under-tsca-and-fifra/insect-resistance-management-bt-plant-incorporated
6. http://www.envfor.nic.in/divisions/csurv/geac/geac-71.pdf
7. http://www.indianemployees.com/uploads/documents/22016/1483101466-3087.pdf
8. https://www.cabi.org/isc/datasheet/39417#todistributionTable
9. http://moef.gov.in/wp-content/uploads/2018/03/vrguj.pdf
10. https://wcrc.confex.com/wcrc/2007/techprogram/P1712.HTM

References

Adamczyk JJ, Sumerford DV (2001) Potential factors impacting season long expression of Cry1Ac in 13 commercial varieties of Bollgard® cotton. *Journal of Insect Science* **1**, 1–6. doi:10.1673/031.001.1301

Armes NJ, Jadhav DR, Bond GS, King ABS (1992) Insecticide resistance in *Helicoverpa armigera* in South India. *Pest Management Science* **34**, 355–364. doi:10.1002/ps.2780340409

Ayyar TVR (1932) Insects affecting the cotton plant in India. *Madras Agriculture Department Bulletin* **28**, 1–28.

Badigar HK, Patil SB, Udikari SS (2013) Expressional dynamics of Cry proteins in first and second generation interspecific Bt cotton hybrids and their comparative efficacy against pink bollworm, *Pectinophora gossypiella* (Saunders) and tobacco caterpillar, *Spodoptera litura* (Fab.). *Bioinfolet* **10**, 1069–1075.

Bambawale OM, Singh A, Sharma OP, Bhosle BB, Lavekar RC, Dhandapani A, Kanwar V, Tanwar RK, Rathod KS, Patange NR, Pawar VM (2004) Performance of Bt cotton (MECH-162) under integrated pest management in farmers' participatory field trials in Nanded district, Central India. *Current Science* **86**, 1628–1633.

Bird LJ, Akhurst RJ (2005) Fitness of Cry1A resistant and susceptible *Helicoverpa armigera* (Lepidoptera: Noctuidae) on transgenic cotton with reduced levels of Cry1Ac. *Journal of Economic Entomology* **98**, 1311–1319. doi:10.1603/0022-0493-98.4.1311

Brookes G, Barfoot P (2018a) Farm income and production impacts of using GM crop technology 1996–2016. *GM Crops and Food* **9**, 59–89. doi:10.1080/21645698.2018.1464866

Brookes G, Barfoot P (2018b) Environmental impacts of genetically modified (GM) crop use 1996–2016: impacts on pesticide use and carbon emissions. *GM Crops and Food: Biotechnology in Agriculture and the Food Chain* **9**(3), 109–139 doi:10.1080/21645698.2018.1476792.

Businessline (2019) <https://www.thehindubusinessline.com/economy/agri-business/india-to-produce-record-cotton-crop-for-2019-20-predicts-usda/article29735164.ece>

CAI (2019) Cotton Association of India. <http://www.caionline.in/site/archives?article_type=Press+Releases>

Carrière Y, Fabrick JA, Tabashnik BE (2016) Can pyramids and seed mixtures delay resistance to Bt crops? *Trends in Biotechnology* **34**, 291–302. doi:10.1016/j.tibtech.2015.12.011

Carrière Y, Brown ZS, Downes SJ, Gujar G, Epstein G, Omoto C, Storer NP, Mota-Sanchez D, Jørgensen PS, Carroll SP (2019a) Governing evolution: a socioecological comparison of resistance management for insecticidal transgenic Bt crops among four countries. *Ambio* **49**, 1–16. doi:10.1007/s13280-019-01167-0

Carrière Y, DeGain B, Unnithan GC, Harpold VS, Li X, Tabashnik BE (2019b) Seasonal declines in Cry1Ac and Cry2Ab concentration in maturing cotton favour faster evolution of resistance to pyramided Bt cotton in *Helicoverpa zea* (Lepidoptera: Noctuidae). *Journal of Economic Entomology* **112**, 2907–2914. doi:10.1093/jee/toz236

Carroll MW, Head G, Caprio M (2012) When and where a seed mix refuge makes sense for managing insect resistance to Bt plants. *Crop Protection* **38**, 74–79. doi:10.1016/j.cropro.2012.02.015

Chandrashekar K, Gujar GT (2004) Development and mechanisms of resistance to *Bacillus thuringiensis* endotoxin Cry1Ac in the American bollworm, *Helicoverpa armigera* (Hübner). *Indian Journal of Experimental Biology* **42**, 164–173.

Chandrashekar K, Archana K, Kalia V, Gujar GT (2005) Baseline susceptibility of the American bollworm, *Helicoverpa armigera* (Hübner) to *Bacillus thuringiensis* Berl. var. *kurstaki* and its endotoxins in India. *Current Science* **88**, 167–175.

Choudhary B, Gaur K (2015) *Biotech Cotton in India, 2002 to 2014*. ISAAA Series of Biotech Crop Profiles. ISAAA, Ithaca, NY.

Chowdary LR, Bheemanna M, Hosamani AC, Prabhuraj A, Naik MK, Nidagundi JM (2014) Built in refuge for the management of pink bollworm, *Pectinophora gossypiella* Saunders (Gelichidae: Lepidoptera) in Bt cotton. *Journal of Applied and Natural Science* **6**, 202–206. doi:10.31018/jans.v6i1.401

Dastur RH, Asana RD, Sawhney K, Sikka SM, Vasudeva RS, Khan Q, Rao VP, Sethi BL (1960) Cotton in India: a monograph. In *Insect and Mite Pests*. Chapter IV. Indian Central Cotton Committee, Mumbai.

Dennehy TJ, Head GP, Moar W, Greenplate JT, Mohan KS, Ravi KC (2011) Field failure of first generation Bt cotton documented with pink bollworm in Gujarat state India. In *Proceedings of the Beltwide Cotton Production Research Conference*, Atlanta, USA, 4–7 January, 1063.

Dhanaraj AL, Willse AR, Kamath SP (2019) Stability of expression of Cry1Ac and Cry2Ab2 proteins in Bollgard II hybrids at different stages of crop growth in different genotypes across cropping seasons and multiple geographies. *Transgenic Research* **28**, 33–50. doi:10.1007/s11248-018-0102-1

Dhawan AK, Simwat GS, Sidhu AS, Madan VK (1986) Pink bollworm damage in cottons at different levels of loculi infestation. *Indian Journal of Agricultural Sciences* **56**, 738–742.

Dhurua S, Gujar GT (2011) Field-evolved resistance to Bt toxin Cry1Ac in the pink bollworm, *Pectinophora gossypiella* (Saunders) (Lepidoptera: Gelechiidae), from India. *Pest Management Science* **67**, 898–903. doi:10.1002/ps.2127

DNFI (2019) *Top 10 Cotton-producing Countries in the World*. Discover Natural Fibres Initiative. <https://dnfi.org/cotton/top-10-cotton-producing-countries-in-the-world_4785/>

Fabrick JA, Ponnuraj J, Singh A, Tanwar RK, Unnithan GC, Yelich AJ, Li X, Carrière Y, Tabashnik BE (2014) Alternate splicing and highly variable cadherin transcripts associated with field-evolved resistance of pink bollworm to Bt cotton in India. *PLoS One* **9**(5), e97900. doi:10.1371/journal.pone.0097900

Fakruddin B, Prasad PRB, Prakash SH, Krishnareddy KB, Patil BV, Kuruvinashetty MS (2003) Baseline resistance to Cry1Ac toxin in cotton bollworm *Helicoverpa armigera* (Hubner) in south Indian cotton ecosystem. *Current Science* **84**, 1304–1307.

Farah D, Nandi JN, Anuradha K, Ravi KC, Mohan KS, Manjunath TM (2003) An insect-based assay to quantify *Bacillus thuringiensis* insecticidal protein Cry1Ac expressed in planta. *Entomon* **28**, 27–33.

Gassmann AJ, Carrière Y, Tabashnik BE (2009) Fitness costs of insect resistance to *Bacillus thuringiensis*. *Annual Review of Entomology* **54**, 147–163. doi:10.1146/annurev.ento.54.110807.090518

Global AgriSystems (2016) *Final Report Impact Evaluation & Socio-Economic Study of Bt Cotton*. Submitted to Department of Agriculture & Cooperation, Ministry of Agriculture, New Delhi. <http://re.indiaenvironmentportal.org.in/files/file/Final%20Report%20%20-%20BT%20Cotton.pdf>

Gould F (1998) Sustainability of transgenic insecticidal cultivars: integrating pest genetics and ecology. *Annual Review of Entomology* **43**, 701–726. doi:10.1146/annurev.ento.43.1.701

Greenplate JT (1999) Quantification of *Bacillus thuringiensis* insect control protein Cry1Ac over time in Bollgard cotton fruit and terminals. *Journal of Economic Entomology* **92**, 1377–1383. doi:10.1093/jee/92.6.1377

Grefenstette B, El-Lissy O, Staten RT (2009) *Pink Bollworm Eradication Program in the USA*. <https://www.aphis.usda.gov/plant_health/plant_pest_info/cotton_pests/downloads/pbw-erad-plan2-09.pdf>

Gujar GT, Kumari A, Kalia V, Chandrashekar K (2000) Spatial and temporal variation in susceptibility of the American bollworm, *Helicoverpa armigera* (Hubner) to *Bacillus thuringiensis* var. *kurstaki* in India. *Current Science* **78**, 995–1001.

Gujar GT, Kalia V, Kumar A, Singh BP, Mittal A, Nair R, Mohan M (2007) *Helicoverpa armigera* baseline susceptibility to *Bacillus thuringiensis* (Bt) Cry toxins and resistance management for Bt cotton in India. *Journal of Invertebrate Pathology* **95**, 214–219. doi:10.1016/j.jip.2007.03.011

Gujar GT, Nair R, Singh BP, Kumari A, Kalia V (2008) Toxicity of the cotton bollworm, *Helicoverpa armigera*, of some Cry1Ac toxins expressed in cotton in India. *Crop Protection* **27**, 537–544. doi:10.1016/j.cropro.2007.08.008

Gujar GT, Kalia V, Bunker GK, Dhurua S (2010) Impact of different levels of non-Bt cotton refuges on pest populations, bollworm damage, and Bt cotton production. *Journal of Asia-Pacific Entomology* **13**, 249–253. doi:10.1016/j.aspen.2010.06.004

Hawkins NJ, Bass C, Dixon A, Neve P (2019) The evolutionary origins of pesticide resistance. *Biological Reviews of the Cambridge Philosophical Society* **94**, 135–155. doi:10.1111/brv.12440

Head GP, Greenplate J (2012) The design and implementation of insect resistance management programs for Bt crops. *GM Crops and Food: Biotechnology in Agriculture and the Food Chain* **3**(3), 144–153. doi:10.4161/gmcr.20743

Huang F, Andow DA, Buschman LL (2011) Success of the high-dose/refuge resistance management strategy after 15 years of Bt crop use in North America. *Entomologia Experimentalis et Applicata* **140**, 1–16. doi:10.1111/j.1570-7458.2011.01138.x

ISAAA (2018) *Global Status of Commercialized Biotech/GM Crops in 2017: Biotech Crops Continue to Help Meet the Challenges of Increased Population and Climate Change*. ISAAA Brief No. 54. ISAAA, Ithaca, NY. <https://www.isaaa.org/resources/publications/briefs/54/executivesummary/default.asp>

Jaffe G (2009) *Complacency on the Farm: Significant Non-compliance with EPA's Refuge Requirements Threatens the Future Effectiveness of Genetically Engineered Pest Protected Corn*. Center for Science in the Public Interest, 9 November. <http://cspinet.org/new/pdf/complacencyonthefarm.pdf>

Jalali SK, Mohan KS, Singh SP, Manjunath TM, Lalitha Y (2004) Baseline susceptibility of the old-world bollworm, *Helicoverpa armigera* (Lepidoptera: Noctuidae) populations from India to *Bacillus thuringiensis* Cry1Ac insecticidal protein. *Crop Protection* **23**, 53–59. doi:10.1016/S0261-2194(03)00170-4

Jayaraman KS (2004) Illegal seeds overtake India's cotton fields. *Nature Biotechnology* **22**, 1333–1334. doi:10.1038/nbt1104-1333

Jin L, Zhang H, Lu Y, Yang Y, Wu K, Tabashnik BE, Wu Y (2015) Large-scale test of the natural refuge strategy for delaying insect resistance to transgenic Bt crops. *Nature* **33**(2), 169–174.

Kathage J, Qaim M (2012) Economic impacts and impact dynamics of Bt (*Bacillus thuringiensis*) cotton in India. *Proceedings of the National Academy of Sciences of the United States of America* **109**, 11652–11656. doi:10.1073/pnas.1203647109

Kavi PI, Amarnath JS, Sivasankari B (2018) Economic and environmental impact of pesticide use in conventional cotton and Bt cotton. *Economic Affairs* **63**, 425–431.

Knox OGG, Gupta VVSR, Roberts GN, Downes SJ (2008) Improving environmental loading assessments of Cry protein from GM plants based on experimentation in cotton. *Open Agriculture Journal* **2**, 105–112. doi:10.2174/1874331500802010105

Kranthi KR (2012) *Bt Cotton Questions and Answers*. Indian Society for Cotton Improvement, Mumbai.
Kranthi KR (2015) Pink bollworm strikes Bt cotton. *Cotton Statistics and News* **35**, 1–6.
Kranthi KR, Kranthi S (2003) Monitoring for shifts in baseline susceptibility (development of tolerance/resistance) in the cotton bollworms *Helicoverpa armigera*, and *Earias vittella* against Cry1A(c) toxin in various cotton growing regions of the country. *Annual Report 2002–2003*. Central Institute for Cotton Research. <http://164.100.154.103/divisions/csurv/geac/cicr0203.htm>
Kranthi KR, Kranthi NR (2004) Modelling adaptability of cotton bollworm, *Helicoverpa armigera* (Hübner) to Bt cotton in India. *Current Science* **87**, 1096–1107.
Kranthi S, Kranthi KR, Lavhe NV (1999) Baseline toxicity of Cry1A toxins to the spotted bollworm, *Earias vittella* F. *Crop Protection* **18**, 551–555. doi:10.1016/S0261-2194(99)00058-7
Kranthi KR, Kranthi S, Ali S, Banerjee SK (2000) Resistance to Cry1Ac δ-endotoxin of *Bacillus thuringiensis* in a laboratory selected strain of *Helicoverpa armigera* (Hübner). *Current Science* **78**, 1001–1004.
Kranthi KR, Kranthi S, Wanjari RR (2001) Baseline toxicity of Cry1A toxins to *Helicoverpa armigera*. *International Journal of Pest Management* **47**, 141–145. doi:10.1080/09670870151130651
Kranthi KR, Jadhav DR, Kranthi S, Wanjari RR, Ali S, Russell D (2002a) Insecticide resistance in five major insect pests of cotton in India. *Crop Protection* **21**, 449–460. doi:10.1016/S0261-2194(01)00131-4
Kranthi KR, Russell D, Wanjari R, Kherde M, Munje S, Lavhe N, Armes N (2002b) In-season changes in resistance to insecticides in *Helicoverpa armigera* (Lepidoptera: Noctuidae) in India. *Journal of Economic Entomology* **95**, 134–142. doi:10.1603/0022-0493-95.1.134
Kranthi KR, Naidu S, Dhawad CS, Tatwawadi A, Mate K, Patil E, Bharose AA, Behere GT, Wadaskar RM, Kranthi S (2005) Temporal and intra-plant variability of Cry1Ac expression in Bt cotton and its influence on the survival of the cotton bollworm, *Helicoverpa armigera* (Hübner) (Noctuidae: Lepidoptera). *Current Science* **89**, 291–298.
Kranthi KR, Dhawad CS, Naidu SR, Mate K, Behere GT, Wadaskar RM, Kranthi S (2006) Inheritance of resistance in Indian *Helicoverpa armigera* (Hübner) to Cry1Ac toxin of *Bacillus thuringiensis*. *Crop Protection* **25**, 119–124. doi:10.1016/j.cropro.2005.03.011
Kranthi S, Dhawad CS, Naidu S, Bharose A, Chaudhary A, Sangode V, Nehare SK, Bajaj SR, Kranthi KR (2009) Susceptibility of the cotton bollworm, *Helicoverpa armigera* (Hubner) (Lepidoptera: Noctuidae), to the *Bacillus thuringiensis* toxin Cry2Ab before and after the introduction of Bollgard II. *Crop Protection* **28**, 371–375. doi:10.1016/j.cropro.2008.12.001
Kranthi S, Satija U, Pusadkar P, Kumar R, Shastri CS, Ansari S, Santosh HB, Monga D, Kranthi KR (2017) Non-Bt seeds provided by seed companies in India: are they suitable as refuge for Bt cotton? *Current Science* **112**, 1992–1993.
Krishna VV, Qaim M (2012) Bt cotton and sustainability of pesticide reductions in India. *Agricultural Systems* **107**, 47–55. doi:10.1016/j.agsy.2011.11.005
Kukanur VS, Singh TVK, Kranthi KR, Andow D (2018) Cry1Ac resistance allele frequency in the field populations of *Helicoverpa armigera* (Hübner) collected in Telangana and Andhra Pradesh, India. *Crop Protection* **107**, 34–40. doi:10.1016/j.cropro.2018.01.008
Li Y, Romeis J, Wang P, Peng Y, Shelton AM (2011) A comprehensive analysis of the effects of Bt cotton on *Coleomegilla maculata* demonstrates no detrimental effects by Cry1Ac and Cry2Ab. *PLoS One* **6**(7), e22185. doi:10.1371/journal.pone.0022185
Liu Y-B, Tabashnik BE, Meyer SK, Carrière Y, Bartlett AC (2001) Genetics of pink bollworm resistance to *Bacillus thuringiensis* toxin Cry1Ac. *Journal of Economic Entomology* **94**, 248–252. doi:10.1603/0022-0493-94.1.248
Lu Z, Zalucki MP, Perkins LE, Wang D, Wu L (2013) Towards a resistance management strategy for *Helicoverpa armigera* in Bt cotton in north-western China: an assessment of potential refuge crops. *Journal of Pesticide Science* **86**, 695–703. doi:10.1007/s10340-013-0517-7
Maheshwariah BM, Puttarudriah M (1956) Insect pests of cotton in Mysore. *Mysore Agriculture Journal* **31**, 220–226.
Mallet J, Porter P (1992) Preventing insect adaptation to insect-resistant crops: are seed mixtures or refugia the best strategy? *Proceedings of the Royal Society of London. Series B, Biological Sciences* **250**, 165–169. doi:10.1098/rspb.1992.0145

Malthankar PA, Gujar GT (2016) Toxicity of *Bacillus thuringiensis* Cry2Ab and the inheritance of Cry2Ab resistance in the pink bollworm, *Pectinophora gossypiella* (Saunders). *Indian Journal of Experimental Biology* **54**, 586–596.

Manjunatha SB, Biradar DP, Aladakatti YR (2017) Cry protein expression in Bt cotton hybrids as influenced by nutrient management based on target yield. *Biochemical and Cellular Archives* **17**, 141–145.

Mathew LG, Ponnuraj J, Mallappa B, Chowdary LR, Zhang J, Tay WT, Walsh TK, Gordon KHJ, Heckel DG, Downes S, Carrière Y, Li X, Tabashnik BE, Fabrick JA (2018) ABC transporter mis-splicing associated with resistance to Bt toxin Cry2Ab in laboratory- and field-selected pink bollworm. *Scientific Reports* **8**, 13531. doi:10.1038/s41598-018-31840-5

Menon P (2001) Waking up to GM cotton. *Frontline* **18**(23). <http://www.thehindu.com/fline/fl1823/18230440.htm>

Mittal A, Ningthoukhongjam J, Malthankar PA, Ningthoujam K, Kalia V, Gujar GT (2016) Toxicity of Cry1Ac, inheritance of Cry1Ac resistance and fitness cost association in the pink bollworm, *Pectinophora gossypiella* (Lepidotera: Gelechiidae). *Biopesticides International* **12**, 127–138.

Mohan KS (2017) An area-wide approach to pink bollworm management on Bt cotton in India: a dire necessity with community participation. *Current Science* **112**, 1988–1989.

Mohan KS (2018) Refuge-in-bag for Bt cotton. *Current Science* **114**, 726–727.

Mohan KS, Sadananda AR (2019) Success of refuge-in-bag for Bt cotton hinges on good stewardship. *Current Science* **117**, 739–740.

Mohan KS, Ravi KC, Suresh PJ, Sumerford D, Head GP (2016) Field resistance to the *Bacillus thuringiensis* protein Cry1Ac expressed in Bollgard® hybrid cotton in pink bollworm, *Pectinophora gossypiella* (Saunders), populations in India. *Pest Management Science* **72**, 738–746. doi:10.1002/ps.4047

Muralimohan K, Kamath SP, Mohan KS, Ravi KC, Deeba G, Sivasupramanian S, Head GP (2009) Mass rearing diet for the pink bollworm *Pectinophora gossypiella* (Lepidoptera: Gelechiidae) and its susceptibility to insecticidal Bt proteins. *International Journal of Tropical Insect Science* **29**, 102–107. doi:10.1017/S1742758409990129

Muralimohan K, Mahesh HM (2020) Cry toxin expression in *Bt*-cotton hybrid seeds: impact on 'refuge-in-bag' strategy for managing resistance in bollworms. *Current Science* **118**, 1493–1494.

Nagappa BH, Khadi BM (2018) Expression of Bt gene (*Cry1Ac*) on different plant parts at different stages in Bt cotton genotypes. *International Journal of Current Microbiology and Applied Sciences* **7**(1), 3339–3353. doi:10.20546/ijcmas.2018.701.396

Naik VC, Jothi DB, Dabhade PL, Kranthi S (2014) Pink bollworm *Pectinophora gossypiella* (Saunders) infestation on Bt and non Bt hybrids in India in 2011–2012. *Cotton Research Journal* **6**, 37–40.

Naik VC, Kumbhare S, Kranthi S, Satuja U, Kranthi KR (2018) Field-evolved resistance of pink bollworm, *Pectinophora gossypiella* (Saunders) (Lepidoptera: Gelechiidae) to transgenic Bt cotton expressing Cry1Ac and Cry2Ab in India. *Pest Management Science* **74**, 2544–2554. doi:10.1002/ps.5038

Nair R, Kalia V, Aggarwal KK, Gujar GT (2010) Inheritance of Cry1Ac resistance and associated biological traits in the cotton bollworm, *Helicoverpa armigera* (Lepidoptera: Noctuidae). *Journal of Invertebrate Pathology* **104**, 31–38. doi:10.1016/j.jip.2010.01.004

Nair R, Kamath SP, Mohan KS, Head G, Summerford DV (2016) Inheritance of field-resistance to the *Bacillus thuringiensis* protein Cry1Ac in *Pectinophora gossypiella* (Lepidoptera: Gelachiidae) collected from India. *Pest Management Science* **72**, 558–565. doi:10.1002/ps.4023

Naranjo SE, Butler GD Jr, Henneberry TJ (2001) *A Bibliography of the Pink Bollworm, Pectinophora gossypiella (Saunders).* US Department of Agriculture, Agricultural Research Service, Bibliographies and Literature of Agriculture No. 136.

Ni M, Ma W, Wang X, Gao M, Dai Y, Wei X, Zhang L, Peng Y, Chen S, Ding L, Tian Y, Li J, Wang H, Wang X, Xu G, Guo W, Yang Y, Wu Y, Heuberger S, Tabashnik BE, Zhang T, Zhu Z (2017) Next generation transgenic cotton: pyramiding RNAi and Bt counters insect resistance. *Plant Biotechnology Journal* **15**, 1204–1213. doi:10.1111/pbi.12709

Ningthoujam K, Sharma K, Gujar GT (2017) Frequency of alleles conferring resistance to Cry1Ac toxin of *Bacillus thuringiensis* in population of *Helicoverpa armigera* (Hubner) (Lepidoptera: Noctuidae) from India. *Indian Journal of Experimental Biology* **55**, 728–733.

Onstad DW, Mitchell PD, Hurley TM, Lundgren JG, Porter RP, Krupke CH, Spencer JL, Difonozo CD, Baute TS, Hellmich RL, Buschman LL, Hutchison WD, Tooker JF (2011) Seeds of change: corn seed

mixtures for resistance management and integrated pest management. *Journal of Economic Entomology* **104**, 343–352. doi:10.1603/EC10388
Ramachandra Rao Y (1921) The pink bollworm and the pest act. *Madras Agricultural Journal* **9**(5), 121–125.
Ramasubramanian T (2004) Magnitude, mechanism and management of pyrethroid resistance in *Helicoverpa armigera* Hübner in India. *Journal of Entomology* **1**, 6–11. doi:10.3923/je.2004.6.11
Ramasubramanian T, Regupathy A (2004) Pattern of cross-resistance in pyrethroid-selected populations of *Helicoverpa armigera* Hübner (Lep., Noctuidae) from India. *Journal of Applied Entomology* **128**, 583–587. doi:10.1111/j.1439-0418.2004.00897.x
Ramaswami B, Pray CE, Lalitha N (2012) The spread of illegal transgenic cotton varieties in India: biosafety regulations, monopoly and enforcement. *World Development* **40**, 177–188. doi:10.1016/j.worlddev.2011.04.007
Ranganathan T, Gaurav S, Halder II (2018) Pesticide usage by cotton farmers in India: changes over a decade. *Economic and Political Weekly* **53**(19), 43–51.
Ranjith MT, Prabhuraj A, Srinivasa YB (2010) Survival and reproduction of natural populations of *Helicoverpa armigera* on Bt cotton hybrids in Raichur, India. *Current Science* **99**, 1602–1606.
Ravi KC, Mohan KS, Manjunath TM, Head G, Patil BV, Angeline Greba DP, Premalatha K, Peter J, Rao NGV (2005) Relative abundance of *Helicoverpa armigera* (Lepidoptera: Noctuidae) on different host crops in India and the role of these crops as natural refuge for *Bacillus thuringiensis* cotton. *Environmental Entomology* **34**, 59–69. doi:10.1603/0046-225X-34.1.59
Roush RT (1998) Two-toxin strategies for management of insecticidal transgenic crops: can pyramiding succeed where pesticide mixtures have not? *Philosophical Transactions of the Royal Society of London. Series B, Biological Sciences* **353**, 1777–1786. doi:10.1098/rstb.1998.0330
Saini MK, Dhawan AK (2010) Expression of biochemical contents in transgenic cotton with single and stacked genes. *Journal of Insect Science* **23**, 247–256.
Shahanaz (2019) Characterization of BG II resistance in field-collected strains of pink bollworm *Pectinophora gossypiella* (Saunders) (Lepidoptera: Gelechiidae). PhD thesis, ICAR-Indian Agricultural Research Institute, New Delhi.
Sharma DC (2010) *Bt Cotton has Failed, Admits Monsanto.* <https://www.indiatoday.in/india/north/story/Bt cotton-has-failed-admits-monsanto-68749-2010-03-06)>
Showalter AM, Heuberger S, Tabashnik BE, Carrière Y (2009) A primer for using transgenic insecticidal cotton in developing countries. *Journal of Insect Science* **9**, 22. doi:10.1673/031.009.2201
Singh AK, Paritosh K, Kant U, Burma PK, Pental D (2016a) High expression of Cry1Ac protein in cotton (*Gossypium hirsutum*) by combining independent transgenic events that target the protein to cytoplasm and plastids. *PLoS One* **11**(7), e0158603. doi:10.1371/journal.pone.0158603
Singh BP, Sandhu SS, Kalia VK, Gujar GT, Dhillon MK (2016b) Assessment of Bt trait purity in different generations of transgenic cottons. *Indian Journal of Experimental Biology* **54**, 237–244.
Singla R, Johnson PN, Misra SK (2012) Examination of regional-level efficient refuge requirements for Bt cotton in India. *Agriculture Biological Forum* **15**, 303–314.
Sivasupramaniam S, Moar WJ, Ruschke LG, Osborn JG, Jiang C, Sebaugh JL, Brown GR, Shappley ZW, Oppenhuizen ME, Mullins JW, Greenplate JT (2008) Toxicity and characterization of cotton expressing *Bacillus thuringiensis* Cry1Ac and Cry2Ab2 proteins for control of lepidopterous pests. *Journal of Economic Entomology* **101**, 546–554. doi:10.1603/0022-0493(2008)101[546:TACOCE]2.0.CO;2
Sumerford DV, Head GP, Shelton A, Greenplate J, Moar W (2013) Field evolved resistance: assessing the problem and ways to move forward. *Journal of Economic Entomology* **106**, 1525–1534. doi:10.1603/EC13103
Tabashnik BE (1994) Evolution of resistance to *Bacillus thuringiensis*. *Annual Review of Entomology* **39**, 47–79. doi:10.1146/annurev.en.39.010194.000403
Tabashnik BE, Carrière Y (2017) Surge in insect resistance to transgenic crops and prospects for sustainability. *Nature Biotechnology* **35**, 926–935. doi:10.1038/nbt.3974
Tabashnik BE, Liu YB, Dennehy T, Sims MA, Sisterson MS, Biggs RW, Carrière Y (2002) Inheritance of resistance to Bt toxin Cry1Ac in a field-derived strain of pink bollworm (Lepidoptera: Gelechiidae). *Journal of Economic Entomology* **95**, 1018–1026. doi:10.1093/jee/95.5.1018
Tabashnik BE, Van Rensburg JBJ, Carrière Y (2009) Field-evolved insect resistance to Bt crops: definition, theory and data. *Journal of Economic Entomology* **102**(6), 2011–2025. doi:10.1603/029.102.0601

Tabashnik BE, Brévault T, Carrière Y (2013) Insect resistance to Bt crops: lessons from the first billion acres. *Nature Biotechnology* **31**, 510–521. doi:10.1038/nbt.2597

Verma D, Kalubarme MH, Saroha GP, Mohan KS, Ravi KC, Brajendra, Singh AN (2019) Monitoring changes in cotton acreage and alternate host crops of cotton bollworm using remote sensing and GIS in major cotton growing regions of India. *International Archives of the Photogrammetry, Remote Sensing and Spatial Information Sciences* **XLII-3/W6**, 525–533. doi:10.5194/isprs-archives-XLII-3-W6-525-2019

Wan P, Xu D, Cong S, Jiang Y, Huang Y, Wang J, Wu H, Wang L, Wu KM, Carrière Y, Mathias A, Li X, Tabashnik BE (2017) Hybridizing transgenic Bt cotton with non-Bt cotton counters resistance in pink bollworm. *Proceedings of the National Academy of Sciences of the United States of America* **114**, 5413–5418. doi:10.1073/pnas.1700396114

Wu K, Guo Y, Gao S (2002) Evaluation of the natural refuge function for *Helicoverpa armigera* (Lepidoptera: Noctuidae) within *Bacillus thuringiensis* transgenic cotton growing areas in north China. *Journal of Economic Entomology* **95**, 832–837. doi:10.1603/0022-0493-95.4.832

Zhao JZ, Cao J, Li Y, Collins HL, Roush RT, Earle ED, Shelton AM (2003) Transgenic plants expressing two *Bacillus thuringiensis* toxins delay insect resistance evolution. *Nature Biotechnology* **21**, 1493–1497. doi:10.1038/nbt907

5

Twenty years of successful Bt cotton production in Australia

Kristen Knight, Paul Grundy, Greg Kauter and Sally Ceeney

Introduction

Bt cotton (cotton transformed with insecticidal gene(s), *cry* isolated from soil bacterium *Bacillus thuringiensis*) has been commercially grown in Australia since 1996. The deployment of Bt cotton in Australia has been within a regulatory framework that requires technology users to adopt practices that aim to reduce the likelihood of pest resistance. This framework has included the implementation of refuge requirements that are among the most stringent in the world. The ratio of structured refuge to Bt cotton as well as naturally occurring non-Bt host crops and native vegetation that contribute additional refugia has likely been a significant factor in the success of resistance management for *Helicoverpa* budworm and bollworm species in Australia (Tabashnik *et al.* 2013).

In this chapter we describe how a unique process of genetically modified (GM) crop regulation and industry-led stewardship has underpinned the successful deployment of Bt cotton and contributed to its continuing high level of pest control efficacy against one of the world's most difficult caterpillar pests – *Helicoverpa armigera* (cotton bollworm). We also examine how this technology has delivered a range of benefits for the Australian cotton industry that extend beyond improved pest management, to include agronomic, industry development, environmental and social rewards.

History of deployment of Bt cotton technology

A diverse range of stakeholders representing the industry including the technology provider, growers, scientists and the Australian community through government regulatory bodies have been involved in the successful commercialisation of three different Bt traits. These stakeholders have also been involved in the development of ecologically based stewardship strategies that aimed to delay the evolution of resistance. In Australia, Bt cotton was first released as a single gene (*cry1Ac*) Ingard® technology, followed by the twin stack (*cry1Ac* and *cry2Ab*) Bollgard II® then the triple stack (*cry1Ac*, *cry2Ab* and *vip3A*) Bollgard 3®, which produced respective protein toxins. The chronology for the release of these Bt technologies and key resistance events are detailed in Figure 5.1.

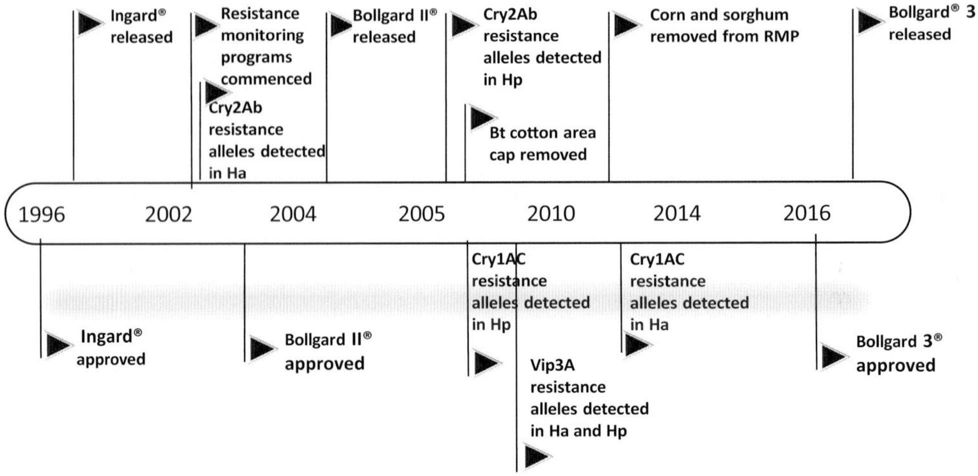

Figure 5.1 Chronology for the approval and commercial release of Bt cotton technologies in Australia and key resistance monitoring events. Ha = *H. armigera*, Hp = *H. punctigera*, RMP = resistance management plan.

How a history of pesticide resistance management informed stewardship for resistance management for Bt cotton

Prior to the commercialisation of the first Bt cotton varieties, the Australian cotton industry had already developed and undertaken a range of measures aimed at limiting the severity of resistance to pyrethroids and carbamates by *H. armigera* that had been first detected during the early 1980s. The first resistance management strategy, employed during the 1983–84 (Australian summer) growing season, entailed restricting synthetic pyrethroid use with the intention of confining exposure to a single generation of *H. armigera* (Forrester *et al.* 1993). This initial industry intervention was followed by a program of resistance monitoring for several target pests to a range of insecticides that continues to the current day (Bird 2015, 2017). Growers and field agronomists were consulted annually about the resistance monitoring program results and recommendations were consequently made by this group regarding how insecticide usage might be altered by the industry in response to any changes. This program became known as the Insecticide Resistance Management Strategy (IRMS) for the cotton industry (Forrester *et al.* 1993). IRMS limitations on the use of insecticides during defined time periods are voluntary, but industry compliance has been historically high due to a strong understanding of, and support for, the need to ensure sustainable production practices for the future (GHD 2003; Rossiter *et al.* 2008). The IRMS has expanded to recommend usage patterns for insecticides and acaricides against all major pests of cotton. It includes consideration of relevant non-chemical tactics that enhance resistance management, such as planting windows, utilising naturally occurring biological control agents and post-harvest pupae destruction (Wilson *et al.* 2018).

As a result of the IRMS, cultural acceptance for the need to pre-emptively manage resistance was well established in Australia before the 1996 release of commercial cotton varieties containing the Bt Cry1Ac protein (Ingard®).

In 1993, a group of agricultural research scientists and representative cotton and grain growers formed a Bt Management Working Group in recognition of the need to develop a stewardship plan that would pre-emptively mitigate the potential of *Helicoverpa* spp. to develop resistance to Bt proteins. Its role was to develop guidelines for the release of transgenic varieties of any crop type expressing the Bt endotoxin proteins and to communicate these to the relevant Australian government regulatory authorities. As part of this effort, several cotton research projects were undertaken to develop data that might inform a future Resistance Management Plan (RMP). Laboratory selection experiments designed to evaluate the ability of pests to develop resistance, studies that considered *Helicoverpa* spp. diapause ecology and the potential for using tillage techniques to target end-of-season surviving pupae, and the potential impact of Bt cotton on non-target invertebrates were studied (CRCSCP 1997). Working with the Bt Management Working Group and incorporating information derived from local research, Monsanto developed plans for the commercialisation of Ingard® cotton in Australia including the regulatory approval process, product performance and commercial licence provisions in 1994.

The cotton industry grower representative body at the time, known as the Australian Cotton Growers Research Association (ACGRA[1]) commenced discussions with Monsanto regarding the process for commercial release of Ingard® and potential tactics for managing resistance. As a result, the Transgenic and Insecticides Management Strategies (TIMS) Committee was formed. It was tasked with preserving the future interest of Australian cotton producers by providing a forum to decide the best resistance strategies for transgenic crops, insecticides, miticides and herbicides (ACGRA 1994). Committee members represent cotton growers, research organisations, crop consultants and grains industries. The TIMS Committee review process has evolved to include several subcommittee panels that provide specific advice for each technology group – insecticides, herbicides (including herbicide-tolerant crops) and Bt crops (TIMS 2014).

The Bt Technical Panel membership consists of research organisation stakeholders and Cotton Australia. This panel examines the results arising from scientific research that pertains to resistance monitoring, pest ecology, farming system changes and population genetics as well RMP industry compliance data collected by product registrants (e.g. Monsanto) from a scientific perspective. An annual statement of scientific advice is provided to the main TIMS Committee, which assesses it alongside any other practical industry considerations before providing responses or advice to product registrants and statutory government agencies regarding resistance management measures for Bt technologies. Through this process the TIMS Committee has direct involvement in the design of resistance management measures for transgenic technologies and a key role in ratifying the RMP proposals put forward by potential product registrants as part of the federal government's registration approval process (Figure 5.2) (TIMS 2014). Key milestones for the Committee have been the ratification of the RMP for both Bollgard II® in 2002 and Bollgard 3® in 2014.

Australian government regulatory environment for GM technologies

Australia has a comprehensive, transparent and science-based evaluation regulatory framework for genetically modified organisms (GMOs) with discrete agencies that evaluate environmental and human health risk assessment, food safety and product efficacy and use.

Environment and human health risk regulation

The Office of the Gene Technology Regulator (OGTR) within the Australian government Department of Health is responsible for the regulation of all GMOs within an integrated legislative framework across all regulatory agencies (Polya 2008). In Australia, the OGTR provides oversight for the protection of the health and safety of people and the environment, by identifying risks posed by or as a result of gene technology including transgenic crops. The

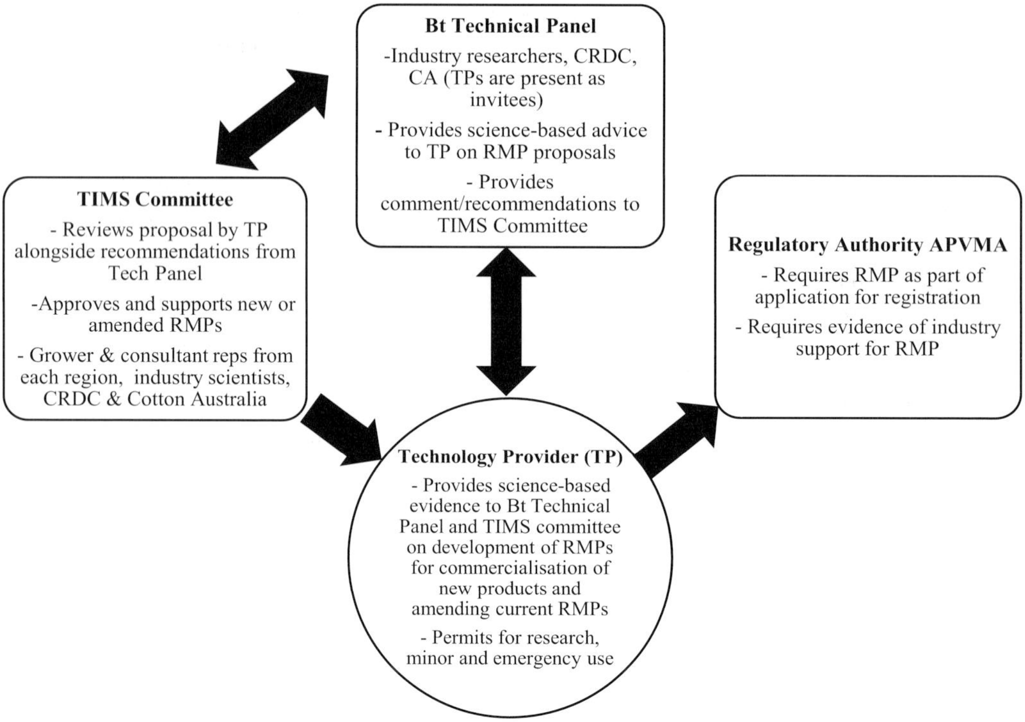

Figure 5.2 Process for developing RMPs as required by the regulatory authority in Australia for Bt cotton.

OGTR seeks to manage these risks through regulating dealings with GMOs and providing a licensing system based on the assessment of scientific risk and community consultation. The OGTR's Risk Assessment and Risk Management Plan has legislated a process of consultation with a broad range of stakeholders (OGTR 2018). The OGTR licensed and authorised the commercial release of Bollgard II® (DIR 012) in 2002 and Bollgard 3® (DIR 124) in 2014 (OGTR 2019).

Prior to the establishment of the OGTR, a non-statutory body known as the Genetic Manipulation Advisory Committee (GMAC) (formed in 1987 within the Australian government Department of Industry, Science and Resources) was responsible for overseeing the development and use of GM techniques. The GMAC assessed potential hazards to the community and environment and made recommendations on appropriate safety and containment procedures to be used by organisations undertaking research for commercialising GMOs (GMAC 1999).

The single-gene Ingard® cotton trait was one of the first GMOs to be considered for development and commercial release in Australia. Due to a range of concerns with a very new technology, GMAC took a precautionary approach to regulating the development of this product with a system of permits and approvals for highly regulated and contained field research sites that enabled the collection of data on a range of efficacy and environmental factors (CRCSCP 1995, 1997; Constable *et al.* 1998). These permits had several regulatory conditions, including adverse incident reporting, the use of pollen traps in the form of conventional non-GM cotton, cleaning of planting, harvest and transport equipment, specific crop residue destruction and post-harvest trial site monitoring requirements. The first commercial release was approved by GMAC in 1996 with an initial 30 000 ha approved to be grown in cotton regions located south of the Tropic of Capricorn. As subsequent research better defined several risks (e.g. out-crossing to native *Gossypium*) this area was expanded, allowing growers to sow up to 30% GMO in their sown total cotton area by 1999.

Food safety regulation

Harvested cotton is processed to separate the lint from seed. The seed by-product may be crushed to produce edible oil utilised in food products including frying and cooking oil, dressings, mayonnaise and margarine. Lint fibres or linters are a cellulose by-product of seed processing and may be used as a base in packaging, in high-fibre products and as a viscosity enhancer in certain foods. The remaining seed oil crushing by-products – seed meal and hulls – are primarily used in livestock feed rations (FSANZ 2019).

The Australia and New Zealand Food Authority (ANZFA), an Australian government Department of Health statutory body, is tasked with safety assessment of foods and food ingredients derived from gene technology, and providing information to consumers, including product labelling. The issue of regulation for food products derived from GM crops was clarified by the introduction of Food Standard Code (18A) which stated that foods derived from gene technology cannot be sold unless they have undergone a safety assessment and are approved for sale (ANZFA 1997). In Australia, regulatory assessment requirements for food safety and labelling are based on a 'composition of final food' framework, differing from the 'method of production' labelling for GM foods in the European Union (FSANZ 2003).

Approval for commercial Bt cotton determined that oil and linters can be regarded as 'substantially equivalent to unmodified cottonseed in respect of composition, safety, wholesomeness and end use' (ANZFA 1999). No product labelling is required for products derived from cottonseed oil or linters from Bt cotton in Australia as these food products contain no GM protein and are deemed equivalent to food from other commercially available cotton varieties in terms of safety for human consumption and nutritional adequacy (FSANZ 2004).

Product efficacy and stated purpose regulation

The Australian Pesticides and Veterinary Medicines Authority (APVMA) (established in 1993 as the National Registration Authority until the name change in 2003) is a statutory body responsible for the administration of the *Agricultural and Veterinary Chemicals Code Act 1994* (APVMA 2018). This agency ensures that crops containing GM traits fulfil any claims of product efficacy and that the product registrant has strategies in place to maintain efficacy and prevent pest resistance over the longer term. In the case of Ingard® cotton, the APVMA required the registrant (Monsanto) to develop an RMP for the technology and have it ratified by the TIMS Committee on behalf of the cotton industry. The National Registration Authority granted Monsanto a licence for Ingard® cotton in August 1998 after two seasons of limited commercial production under permit.

State government regulation of GM crops

Although there is a coordinated and cooperative national regulatory system for GMOs in Australia, individual state governments may choose to limit or ban GMOs based on trade or market considerations even if all federal approvals or licences have been obtained, under the Gene Technology Recognition of Designated Areas Principle 2003.[2] Bt cotton is currently grown in the moratoria-free jurisdictions of Queensland, Victoria, Western Australia (moratoria Act repealed 2016) and the Northern Territory, and under exemption from the GM crop moratoria in New South Wales (*Gene Technology Act 2003* NSW).[3] South Australia, where cotton is not grown, still has a moratorium on all GM crops. The political, and to some degree community, imprimatur for the unrestricted production of approved commercial GM cotton has enabled a clear market development pathway for GM cotton trait providers and an incentive for continued technology development which has arguably kept the Australian cotton industry internationally competitive (GHD 2003; Innovact-Consulting 2012; Wilson *et al.* 2018).

Resistance management stewardship program in Australia

RMPs are developed to be pre-emptive and to delay the evolution of resistance. They are based on knowledge regarding the product efficacy, the genetics of insect resistance and insect behaviour (Shelton *et al.* 2000; Tabashnik *et al.* 2003; Head and Dennehy 2010).

In Australia, an RMP has been developed for each iteration of Bt cotton released to industry from the technology provider, Monsanto, in consultation with the Cotton Australia Bt Technical Panel. Since the first Australian RMP was developed in 1996 for the introduction of Ingard® (Fitt 2000), each version has been based on three key assumptions and principles.

Assumptions for the Australian environment:

1. resistance to the Cry proteins expressed in Bt cotton was functionally recessive at plant-expressed concentrations;
2. resistance genes occurred at low frequencies in field populations;
3. random mating occurred between individuals from the refuges and Bt cotton (Roush 1998).

The principles are as follows:

1. limiting exposure to the Bt insecticidal proteins through calendar date-based planting to limit the number of *Helicoverpa* spp. generations exposed to the Bt proteins. This is also supported through specified end-of-season crop destruction practices and control of any re-emergent Bt cotton;
2. providing the opportunity for diluting resistance through the production of additional susceptible target pests within purposely planted structured non-Bt host crop refuges. Refuge crops are required to produce populations of moths that have not been exposed to selection with Bt proteins. Mating of these moths with survivors from Bt cotton crops can help dilute selection for resistance and slow the rate at which resistance evolves;
3. limiting the escape of pests carrying resistance alleles at the end of the season through mandatory cultivation (pupae busting) in cool climates or trap cropping in warm environments. *Helicoverpa* spp. larvae can form pupae that will diapause over winter in cool regions. Such pupae pose a risk as any individuals carrying resistance alleles at the end of the season (having survived exposure to toxins in Bt cotton) will enter diapause. If left to emerge as adults, those that are carrying resistance alleles may mate and resistance could begin to develop. When larvae pupate, they burrow and form open tunnels to the pupation site. These tunnels must be clear for the adults to emerge in the spring. By destroying the emergence tunnels the adults are unable to surface, and die without the need for chemical control. Tillage also directly exposes a proportion of pupae to unfavourable weather and predation by birds. In warm climates, *Helicoverpa* diapause is less frequent, and trap crops of pigeon pea (*Cajanus cajan*) are grown to attract ovipositing *Helicoverpa* into an area where they can be destroyed with targeted trap crop destruction and tillage.

The RMPs for all the Bt cotton technologies grown in Australia since 1996 can be seen in Table 5.1.

Table 5.1. RMPs for each Bt cotton technology – Ingard®, Bollgard II® & Bollgard 3®

RMP requirement	Ingard®/Bollgard RMP	Bollgard II® RMP	Bollgard 3® RMP
Years	1996–2004	2004–16 or even later	2016 to date
Planting window	All transgenic cotton crops to be planted by 15 November	All transgenic cotton crops to be planted by, or between, specified dates within each production region	All transgenic cotton crops to be planted in southern regions between 1 August and 31 December Window 1: all transgenic cotton crops to be planted between 1 August and 31 October Window 2: all transgenic cotton crops to be planted between 1 November and 31 December
Refuge requirements (% of Bt cotton grown)	Unsprayed irrigated conventional cotton 10% Sprayed irrigated conventional cotton 50% Maize or sorghum irrigated and managed to flower between 15 January and 28 February and with no Helicoverpa control 20% Ingard® was also capped to 30% of the cotton grown	Sprayed irrigated conventional cotton 100% Unsprayed irrigated conventional cotton 10% Pigeon pea irrigated unsprayed 5% Sorghum* irrigated unsprayed (with special conditions, including at least 3 planting dates) 15% Maize* irrigated unsprayed (with special conditions, including at least 3 planting dates) 20%	Sprayed irrigated conventional cotton 100% Unsprayed irrigated conventional cotton 5% Pigeon pea irrigated unsprayed 2.5% Central Queensland planted after 31 October Sprayed irrigated conventional cotton 100% Unsprayed irrigated conventional cotton 10% Pigeon pea irrigated unsprayed 5%
End-of-season management	As soon as possible after Ingard® crop harvest, pupae busting is to occur in Ingard® fields but refuges left uncultivated	In southern areas, pupae busting is to occur in Bollgard II® fields within 4 weeks of crop harvest, but refuges left uncultivated preferably until October. In central Queensland a late summer pigeon pea trap crop is grown to attract any late season *Helicoverpa* emerging from Bollgard II® and then destroyed	In southern areas, pupae busting not mandatory if first defoliation applied by 31 March Mandatory pupae busting required on fields not defoliated by 31 March to a depth of 10 cm and a width of 30 cm either side of the plant line In central Queensland a late summer pigeon pea trap crop is grown to attract any late season *Helicoverpa* emerging from Bollgard 3® and then destroyed

* Sorghum and maize refuge options were removed during 2010/11 season after detection of resistance alleles in *H. punctigera*. These crops are not a suitable host for *H. punctigera*.

The applicability and efficacy of these three core principles of Bt cotton RMP in Australia is regularly reviewed through continuing research and data collection. Priorities for research are assessed collaboratively between Monsanto and the Bt Technical Panel based on identified information gaps that either support the ongoing resistance management stewardship of existing Bt technologies or where a new trait is under consideration for commercialisation. For instance, the research requirements identified through this process before the release of Bollgard 3® consisted of seven general criteria, similar to those described by Head and Dennehy (2010). These are:

1. the genes used in a stack/pyramid (GM plant with more than one transgene) must be sufficiently different to make it unlikely that a single form of resistance will confer cross-resistances to all insecticidal proteins;
2. a stack/pyramid works maximally and that all insecticidal proteins kill susceptible and heterozygote individuals at a high level with their concentrations expressed in the plant;
3. there is no evidence for cross-resistance with existing deployed events (an event is the incorporation of a particular package of genetic material in a defined place in the plant genome);
4. existing levels of resistance to specific insecticidal proteins in the target populations are considered;
5. resistance characteristics are defined, i.e. recessive, dominant, number of genes/alleles involved, fitness costs (Bt resistance in *Helicoverpa* spp. is recessive, meaning that susceptible and heterozygote individuals are killed by the insecticidal proteins);
6. predictive resistance modelling is undertaken utilising relevant data and assumptions for key factors to test the robustness of RMP assumptions;
7. any other factors that may affect product performance and aspects, such as changing pest complexes or population dynamics, are considered.

As a general principle, knowledge gaps relating to specific matters of new product efficacy and modelling of how a new product might fit with existing or modified RMP tactics were the responsibility of the product registrant. Broader questions pertaining to the efficacy of the tactics deployed by industry for Bt resistance management were supported partly by grower levy-funded research managed by the Cotton Research and Development Corporation (CRDC) in collaboration with partnering university and government research institutions (e.g. CSIRO) and the technology provider.

In the case of Bollgard 3®, field and laboratory research was undertaken by Monsanto and CSIRO to develop new data on survival of susceptible, resistant and heterozygote *H. armigera* and *H. punctigera* to each of the modes of action of insecticidal proteins *in planta* over a two-year period. The subsequent data and a case for how they might contribute to a future RMP for Bollgard 3® was provided to the TIMS Bt Technical Panel for independent review. The Panel then provided a statement of advice to the TIMS Committee regarding technical aspects of the RMP proposal for consideration. The registrant also had the opportunity to provide advice to the TIMS Committee. The process is robust and requires participants to consider a range of factors and make predictions of likely outcomes. The role of the TIMS Committee is to balance the Technical Panel and technology provider's advice on the suitability of various aspects of the proposed RMP from a scientific perspective to arrive at an endorsed RMP version for submission to the APVMA for registration.

Commercial enactment and stewardship of the RMP

APVMA approves the active constituents of pesticides in Australia, including the insecticidal proteins in Bt cotton. The label and conditions of registration for each Bt cotton product demands both the registrant (Monsanto) and the technology user (grower) fulfil several responsibilities to ensure both the efficacy of Bt cotton and the management of resistance. These requirements are detailed in Table 5.2.

To ensure a consistent and robust approach to managing instances where technology users are in a situation of RMP non-compliance, Monsanto developed a Resistance Risk Management Plan (RRMP) template that details additional measures a technology user might put in place to contain or ameliorate any resistance risks (Table 5.2). Any technology user found to be non-compliant at any stage of the crop production process will receive an RRMP developed to mitigate the risk according to the nature of the non-compliance. The technology user signs the RRMP documentation to acknowledge its receipt and their agreement to make every effort to implement the applicable risk mitigation strategies. All non-compliant fields are followed up by Monsanto. Each season, qualitative and quantitative data are collected on technology user compliance with the RMP from throughout the industry and are presented to APVMA and the TIMS Committee annually. In Australia, growers are required to follow due diligence as

Table 5.2. Commercial compliance as required by APVMA via the Bollgard 3® label

Label and conditions of registration	Compliance assessments: 3 times through the season	Resistance risk management plans	Monsanto's reporting obligations
All growers must sign a Technology User Agreement (TUA)	After planting, to assess areas of Bollgard cotton and refuge planted and whether there are any refuge non-compliances	Loss or non-establishment of refuges	Hectares of total cotton containing Bollgard technology per valley Grower compliance with the TUA and the RMP The adequacy of the RMP to manage resistance via, but not limited to, resistance bioassays with target pests
		Incorrect refuge width	
		Refuge greater than 2 km from the associated Bollgard cotton	
TUA: legal and binding document binds Monsanto and the grower to undertake certain activities including abiding by the RMP and Monsanto auditing the grower's property to assess compliance with the RMP		Bt cotton volunteers or ratoons present in the refuge Non-compliant pupae busting	
All growers intending to grow Bt cotton must undertake an educational accreditation program	Mid-season, to assess refuge health and attractiveness and measure compliance	No trap crop (for areas where *Helicoverpa* spp. do not go into diapause)	
Monsanto must take remedial action in instances of non-compliance, to mitigate resistance risk	Post crop, to ensure end-of-season strategies are undertaken and are compliant	No trap crop (for areas where *Helicoverpa* spp. do not go into diapause) Trap crop not removed at the appropriate time Bollgard cotton not destroyed in a timely manner Pupae busting unable to be undertaken due to climatic constraints	

described above, and compliance with the RMP has been exceptional since the introduction of Bt cotton (Innovact Consulting 2012). On average, over 96% of all TUAs have been fully compliant with the RMP (Monsanto, unpublished data).

Measuring the effectiveness of the RMP

The rate at which insect resistance may evolve to Bt cotton is determined by the same factors that affect resistance evolution to conventional insecticides (Head and Dennehy 2010).

Resistance monitoring and surveying crop consultants about any changes in survival are the primary mechanisms that the Australian industry uses to measure the effectiveness of the RMP and to detect relatively low incidences of *Helicoverpa* spp. individuals carrying field-evolved resistance in order to allow sufficient lead-time for proactive management to prevent further escalation and field failure (Tabashnik *et al.* 2013).

Resistance testing for Bt, resourced through the CRDC, was first conducted in the mid-1990s by the New South Wales Department of Primary Industries using the F0 screening method to test field-collected insects at a discriminating dose of Cry1Ac (Downes *et al.* 2007; Mahon *et al.* 2007). In 2002 the program shifted to the CSIRO, and in 2003 Monsanto also commenced an annual resistance monitoring program that continues to the current day.

An F2 screening process developed by Andow and Alstad (1998) was explored in 2002–03 and subsequently deployed as part of the monitoring program with the full scale release of Bollgard II® varieties. This screening process improved the sensitivity of the estimates of resistant allele frequencies as well as the ability to identify and recover resistance alleles from field populations (Downes *et al.* 2007; Mahon *et al.* 2007).

The inclusion of the F2 screen enabled the industry to establish the Cry1Ac resistant allele frequency and the baseline Cry2Ab frequencies before the introduction of Bollgard II® (Mahon *et al.* 2007).

Surviving *Helicoverpa* larvae from the F2 screen bioassay technique have homozygous-resistant alleles to the insecticidal protein and can be used to generate a resistant colony for ongoing study purposes. Moths from a homozygous colony can be used to conduct F1 bioassay tests to more efficiently screen for resistance alleles among field-collected insects to the insecticidal protein. The F1 test involves mating a field insect with a moth from the known resistant colony. The F1 generations are then screened using the discriminating dose. The F1 screen only takes six weeks, compared to 12 weeks for the F2 test.

Monsanto began incorporating the F1 screen with the CSIRO-developed *H. armigera* Cry2Ab-resistant colony SP15 in 2004–05. It was followed by CSIRO several years later when F2 testing had shown that field-collected material presented with the one type of resistance. Cry2Ab- and Vip3A-resistant lines suggest a single resistance allele for each toxin that is unique in *H. armigera* and *H. punctigera* (Downes *et al.* 2016). As more resistant colonies were developed for *H. armigera* and *H. punctigera,* and the single resistance allele confirmed, the program's focus switched from F2 to F1 screens.

While *H. armigera* rapidly developed resistance to insecticide sprays, there was only one such case for *H. punctigera* (Downes *et al.* 2007). Because of this, the RMP and the resistance monitoring programs were originally designed to delay the evolution of and to detect resistant alleles in *H. armigera*. In 2004–05 CSIRO included *H. punctigera* in the monitoring program and identified resistance alleles to Cry2Ab; Monsanto added the species in 2007–08. The initial data from the CSIRO monitoring program indicated that the resistance allele frequency to Cry2Ab in *H. punctigera* was higher than expected (Downes *et al.* 2007). The original RMP focused on the perceived threat of resistance developing in *H. armigera* and included sorghum and corn/maize as refuge options. As sorghum and corn/maize do not act as hosts for *H. punctigera*, the RMP was changed in 2010 to remove them as refuge options.

While resistant alleles to all the proteins Cy1Ac, Cry2Ab and Vip3A have been identified in both *H. armigera* and *H. punctigera*, there have been no field failures or field-evolved resistance to Bt cotton in Australia. Both monitoring programs have shown that there has been no statistically significant increase in resistance allele frequencies over time (Downes and Mahon 2012; Downes *et al.* 2016). Resistance alleles remain rare in both species for Cry1Ac (Downes and Mahon 2012).

Bt cotton benefits for the environment and community

The Australian cotton industry has at times had a difficult relationship with other agricultural industries and the broader community. Resistance of several key pests to insecticides became rampant during the 1990s, resulting in significant increases in the application of pesticides in an attempt to minimise crop damage. This increase in pesticide use led to increased environmental contamination as growers experienced difficulties minimising off-target spray-drift onto neighbouring farms, populated areas and riparian areas (Baker 1998; GHD 2003; Roth 2011). This in turn led to community-based calls for the cessation of cotton production in Australia (Downey 1996; Roth 2011). In response to growing community and political concern, the industry implemented programs to improve its pesticide application practices and management of storm water runoff. It also funded independent programs that measured pesticide movement within the environment (Baker 1998). The commercial release of Ingard® Bt cotton coincided with implementation of these programs.

The efficacy of *Helicoverpa* spp. control by the first-generation Ingard® crops provided an immediate 68% reduction of insecticide active constituent use on a per hectare basis compared to non-Bt cotton in the first commercial season (Pyke 1997).

Data on insecticide use on cotton fields have been recorded since the 1993–94 season. These data are collated by the CRDC via an annual quantitative survey of farm advisors. Before the availability of Bt technology, most Australian cotton crops were typically sprayed 10–14 times per season with insecticides for *Helicoverpa* spp. management (Wilson *et al.* 2013) in order to deliver effective control. Over the six years of Bt cotton Ingard® (which was limited to 30% of the area), the insecticide active constituent per hectare was 50% of that used in non-Bt cotton (GHD 2003). The introduction of Bollgard II® after 2002 accelerated this process, resulting in a 94% reduction per hectare over a 15-year period since Bollgard II® and Bollgard 3® varieties have dominated the Australian marketplace. Insecticide active constituent has averaged ~360 g/ha for the last decade, compared to an average of 6380 g/ha on non-Bt cotton during the 10 years before the introduction of Bollgard II®. No other technology, technique or process has had such a significant impact on insecticide use in cotton production in Australia as Bt cotton (Pyke 1997).

This reduction in overall insecticide use associated with the widespread adoption of Bt cotton represents a clear health and safety benefit as exposure to pesticides by workers, bystanders and the community has been significantly reduced (Frager and Temperley 2008). Complaints about cotton growing to the New South Wales state government Environmental Protection Agency fell from up to 60 per year in the late 1990s to under six per year throughout the 2000s (CA and CRDC 2014).

The reduction in insecticide use since 1996 on Bt cotton has favourable environmental impacts, compared to the levels of use that would have continued if the crop was grown conventionally (non-GM) as measured using the environmental impact quotient devised by Kovach *et al.* (1992) (Brookes 2016). For example, detections of endosulfan in north-western

New South Wales rivers declined from ~50% of samples in the 1990s to negligible levels throughout the 2000s (Mawhinney 2011, 2012). The significant decline of Environmental Impact Quotient values for insecticides applied to cotton during the introduction of Bt cotton indicates reduced hazard and consequential improvements in on-farm water quality and worker safety with increasing adoption of GM technologies (Kennedy *et al.* 2013; Brookes and Barfoot 2018).

The adoption of Bollgard II® on 80% of the cotton area within 10 years of its introduction, rising to 99% over 20 years, has been due in large part to these environmental benefits accruing to individual growers and the industry as a whole (Brookes and Barfoot 2014).

In 2012, the national representative organisation for Australian cotton growers, Cotton Australia, became a producing country member of the Better Cotton Initiative (BCI). Established in 2009, BCI is a collaboration of environmental non-government organisations and cotton supply chain organisations which have defined a set of environmental and social standards for the production of sustainable cotton. The BCI now has 5 million licensed farmers generating ~30% of world production (BCI 2019). The Australian cotton industry has developed its own set of criteria for the sustainable production of cotton, known as myBMP. Under this program, growers achieve accreditation for complying with a broad range of measures that pertain to the use of best practices for cotton production and environmental management. In 2014, myBMP was recognised as fulfilling the requirements of the international BCI initiative; as a result, myBMP-accredited growers qualified for monetary reward through the Better Cotton Claim Units (per kilogram lint) on traded bales of lint (CA 2019). The BCI standard for crop protection is 'BCI Farmers Minimise the Harmful Impact of Crop Protection Practices' and this global identity standard for environmentally sustainably grown cotton is specifically technology-neutral in regard to GM cotton (BCI 2018). This reflects the reality and value of Bt cotton in enabling more sustainable production systems around the world, including Australia.

Pest management revolutionised with Bt cotton

In Australia, Bt cotton has provided strong support for integrated pest management (IPM) (Wilson *et al.* 2018). Prior to Bollgard II® cotton, the management of noctuid pests was largely reliant on the use of stringent sampling and threshold protocols whereby crops were routinely scouted for eggs and small larvae and insecticides rapidly applied when numbers were found to breach pre-determined damage thresholds (Fitt 1994). The abilities of noctuid moths to migrate into crops from both nearby source crops (Sequeira 2001) and more distant regions (Gregg *et al.* 1995) and to deposit significant numbers of eggs within cotton crops over several days were primary factors for the use of a reactionary sample, spray and re-sample management approach.

The advent of widespread insecticide resistance by *H. armigera* to a broad range of insecticides (Gunning *et al.* 1984; Forrester *et al.* 1993) demonstrated the deficiency of this approach. In an effort to reduce the cotton industry's dependence on insecticides for *Helicoverpa* spp. management, significant resources were devoted to developing alternate control tactics before the Australian market release of Bollgard II® in 2003. This research included examining how to better exploit natural mortality by predators and parasitoids (Dillon *et al.* 1994; Mensah 2002), the development and commercial use of nucleo-polyhedrovirus biopesticides (Buerger *et al.* 2007), deployment of cultural control tactics such as pupae busting (Duffield 2004; Lloyd *et al.* 2008) and trap cropping (Sequeira 2001; Grundy *et al.* 2004), and the development of an IRMS (Wilson *et al.* 2013). The latter tactics evolved into an area-wide management (AWM) program that aimed to better coordinate the management of *Helicoverpa* spp. across the entire farming system (both cotton and grains)

(Downes *et al.* 2017). The AWM approach recognised that *Helicoverpa* spp. pest activity (Sequeira and Playford 2001) and resistance levels had resulted from the actions of individual farming operations collectively across the industry.

Despite the widespread adoption of the IRMS and deployment of AWM tactics, *Helicoverpa* spp. continued to threaten the viability of cotton production in Australia. The introduction, during the late 1990s, of new insecticides that had more selective modes of action saw rapid and widespread usage driven both by a greater consciousness to adopt products that better conserved non-target species and by the reduced efficacy of existing insecticides. However, despite the IRMS guidelines, even insecticides with a novel mode of action such as Spinosad faltered within three years of coming to market (Rossiter *et al.* 2008). Further complicating pest management was the increased prevalence of sap-sucking pests – mites, *Tetranychus* spp., cotton aphid, *Aphis gossypii* and silverleaf whitefly, *Bemisia tabaci*. In particular, *Tetranychus urticae* became a frequent damaging pest of cotton (Wilson 1993), flaring up through the use of non-selective insecticides for early season pest control (Wilson *et al.* 1998). Both mite and aphid management were further challenged by the emergence of strains that became resistant to a range of insecticides (Herron *et al.* 2001a, b). The silverleaf whitefly *B. tabaci* Meam type 1 was first detected in Australian coastal regions during 1994 (Gunning *et al.* 1995) and is thought to have arrived from North America (De Barro *et al.* 2000). The first outbreak in cotton occurred during 2001 in central Queensland (Moore *et al.* 2004; Sequeira *et al.* 2009); it subsequently spread south (Liu *et al.* 2007) throughout the industry. Management of these sucking pests in conventionally grown cotton presents significant challenges due to the disruption to naturally occurring biological control agents (Trapero *et al.* 2016) caused by insecticide use for *Helicoverpa* spp. control.

Improved IPM for sucking pests

The release and subsequent removal of restrictions on how much Bollgard II® could be grown in 2004 proved to be a critical circuit-breaker for pest management in Australian cotton. Within a single season, the need to spray for *Helicoverpa* spp. was virtually eliminated. Consequently, Bollgard II® provided an ideal platform for the implementation of IPM that better exploited natural mortality agents for sucking pests that had become problematic. In the instance of cotton aphids, a combination of industry strategies to manage resistance and a general decline in the frequency of outbreaks of this pest since the introduction of Bollgard II® has been reflected in a decrease in recorded levels of resistance over the past decade (Herron and Wilson 2017).

The exotic silverleaf whitefly has been an ongoing management challenge as it has spread throughout all regions of the Australian cotton industry. The relevant IPM strategy seeks to minimise the use of disruptive insecticides to better conserve a complex of biological control agents that enact natural pest suppression (CRDC 2018). Insect growth regulators have provided effective and selective control, supported by a complex of biological control agents (Ellsworth and Martinez-Carrillo 2001). These natural enemies are more easily conserved in the low insecticide usage environment enabled by Bollgard II®.

The detection of the mealy bug *Solenopsis solani*, in central Queensland and Burdekin region cotton crops during 2009 (Wilson *et al.* 2013) and later in other regions, has reinforced the need to implement pest management practices that conserve naturally occurring predators and parasitoids (CRDC 2018). The primary control measures for this pest have been the implementation of non-host fallow or crop rotations that reduce in-field mealybug densities between cotton crops (Khan *et al.* 2012) and the preservation of naturally occurring predators and parasitoids that provide biological control. Again, the low insecticide usage environment that Bollgard II® offers (Wilson *et al.* 2013) has enabled conservation of naturally occurring biological agents for mealybug control.

Pest status of *Helicoverpa* for broadacre agriculture changes due to Bt cotton

Prior to the advent of Bt cotton varieties, conventionally grown cotton had been a major host crop and driver of *Helicoverpa* spp. numbers within the farming system (Sequeira and Playford 2001). The lifting of restrictions on Bollgard II® acreage in 2005 had the potential to change cotton from being a major host crop that enabled significant recruitment, to being a sink crop that might suppress *Helicoverpa* spp. recruitment at the landscape scale.

Data from a long-term, broad-scale pheromone trapping program for *H. armigera* conducted throughout the cropping regions of eastern Australia were analysed to determine the potential impact of Bt cotton on the suppression of this pest at a landscape level (Baker and Tann 2017). Suppression of *H. armigera* was not evident during the initial phase of single-gene Bt cotton (Ingard®) from 1996 to 2004. This response was attributed, at least in part, to a precautionary cap (30% of total cotton grown, by area) that was applied to Ingard®. However, with the replacement of Ingard® with Bollgard II® cotton that was uncapped in 2005, the number of *H. armigera* trapped no longer reflected the trend of cotton acreage grown, and moth numbers decreased (Figure 5.3). The efficacy of Bollgard II® varieties, the removal of the acreage cap and a strong industry uptake of improved IPM programs were suggested as the mechanisms most likely to be responsible for the suppression of *H. armigera* (Baker and Tann 2017).

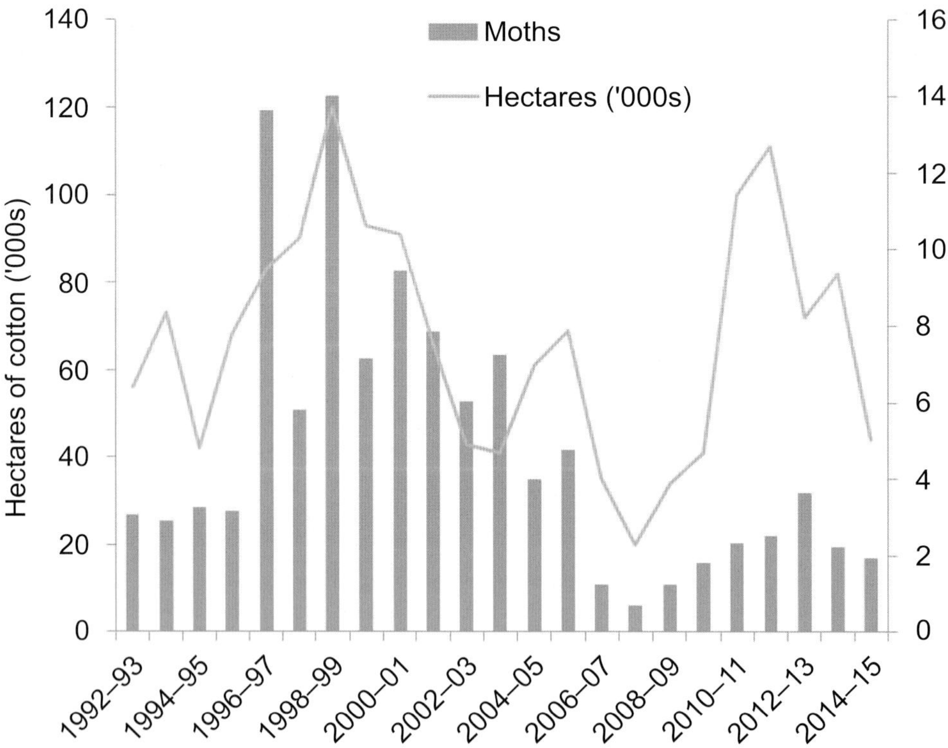

Figure 5.3 Numbers of 3rd + generation *H. armigera* moths trapped each year near ACRI (Narrabri, NSW) and hectares of cotton sown in the Namoi Valley from 1992 to 2015. Source: Cotton data provided by Cotton Australia (Baker and Tann 2017).

H. armigera resistance to conventional insecticides changes after the full-scale commercial release of Bollgard II®

Further evidence about the impact of the full release of Bollgard II® on the pest status of *Helicoverpa* is demonstrated by significant changes in the level of resistance to various insecticides that were commonly used on non-Bt cotton. Prior to Bollgard II®, heavy reliance on pyrethroids and carbamates had selected for resistance to these compounds (Gunning *et al.* 1992; Forrester and Bird 1996). The uptake of Bollgard II® varieties, and consequent reduction in spraying for *Helicoverpa* spp., resulted in a decline in resistance to all chemical classes of insecticides by *H. armigera* and in some instances reversion to completely susceptible baseline levels.

Longer-term insecticide resistance testing demonstrated that bifenthrin resistance increased from 30% to over 60% of individuals screened from 1996 to 2002 and then declined steadily until 2007, at which point resistance could no longer be detected (Bird 2018).

During the same period, fenvalerate resistance increased from 71% in 2008 to 91% in 2011–12, remaining above 90% until 2015 (Bird 2018). Methomyl resistance in *H. armigera* was 33% in 2008, and remained at moderate levels of 21–40% until 2015 (Bird 2018). The increase in fenvalerate and more static methomyl resistance levels are likely to be the result of increased chickpea and mungbean pulse production in Australia during this period; pest management in these crops still depends on those products. Cotton recorded a reduction of 60% for pyrethroid and 80% carbamate use during this period, that overlaps broad-scale cropping with Bollgard II® (Bird 2018).

More recently, *Helicoverpa* spp. management has become reliant on newer selective insecticides across several crop commodities including the small acreages of non-Bt cotton that continue to be grown in some regions. These insecticides include emamectin benzoate, chlorantraniliprole and indoxacarb, which have become popular due to their reduced impact on beneficial insects, making them more suited for use in IPM programs (CRDC 2018). Resistance monitoring programs for these new insecticides has been in place since 2013 using F2 screens. Prior to this, from 2002, field resistance to these insecticides in Australian populations of *H. armigera* was determined by the use of topical application of a discriminating dose of insecticide to larvae in the F0 generation, to measure the phenotypic frequency of resistance. Resistance to emamectin benzoate was not detected using F2 screens. There were low but detectable levels of survival at discriminating concentrations of chlorantraniliprole and indoxacarb (Bird 2015). The amount of these products used in the Australian cotton industry is low, compared to usage in the grains and horticulture industry in eastern Australia. Prior to the introduction of Bollgard II®, the cotton industry was a key driver for resistance to insecticides used to control *Helicoverpa* spp. However, the introduction of Bollgard II® has not only significantly reduced the amount of insecticide used, but also had a pest suppressive impact on *H. armigera* in other crops (Baker and Tann 2017). It is likely that product use in the grains and horticulture industry is the key driver for resistance frequencies.

Other industry benefits from Bt cotton technology

The benefits gained by growers and the cotton industry have extended well beyond more efficient pest management, reduced environmental contamination and improved community acceptance. Bt cotton varieties have contributed to the increase of cotton yields and helped underpin increased production efficiency.

Australian cotton farmers using Bt cotton varieties have seen significant economic gains, with an average increase in farm income between 1996 and 2015 valued at $287/ha from insecticidal-resistant traits and $37/ha for herbicidal-tolerant traits (Brookes 2016).

The substantial benefit to Australian cotton farmers from using this technology has arisen from highly effective pest control with significantly reduced use of insecticides. Cotton crops are now typically subject to zero to three insecticide treatments per crop (Wilson *et al.* 2013). The reduction of insecticide use has resulted in substantial savings in expenditure for both product and application costs. During this period, average farm yields increased by 38% (Brookes 2016).

Productivity gains assisted by Bt cotton technology

For the first time since the advent of insecticide resistance in *H. armigera* during the 1980s, the uncapping of Bolgard II® varieties in 2005 enabled the primary focus of crop managers to shift from the prevention of insect-related crop loss to the optimisation of agronomic inputs that could maximise yield potential.

Increased retention of buds and bolls during canopy development due to the high level of *Helicoverpa* spp. control efficacy afforded by Bollgard II® presented both challenges and opportunities for crop managers. Changes in the timing of reproductive demand (boll number and size) can impact both the length of the flowering period and, subsequently, crop maturity and yield (Bange and Milroy 2004). An industry concern at the time was that the increased fruit retention of Bollgard II® may reduce the within-plant resources for continued growth and flowering (premature cutout), thus resulting in smaller plants, shortened flowering and reduced yield (Kelly *et al.* 2004).

Investigating the implications of higher early season fruit retention of Bollgard II® varieties for agronomic management became a significant focus for crop managers and agronomic research. Aspects such as how changes to planting date (Bange *et al.* 2008) or irrigation practices (Yeates *et al.* 2006) might be used to prevent premature cutout and increase yield were a primary focus. Sowing date was found to have no significant influence on the yield potential of Bollgard II® compared to non-Bt cultivars, and yield potential was maintained despite delayed sowing due to the higher retention of early fruit and shorter fruiting cycle (Bange *et al.* 2008). The ability of Bollgard II® varieties to maintain yield potential over a much broader sowing time window has provided a range of farming system and yield advantages for cotton growers in Australia. Being able to sow cotton later in the spring and early summer period has allowed many growers to grow cotton as a dryland crop, to take advantage of rainfall that generally increases during November in northern New South Wales and southern Queensland, a time of year when the sowing of conventional cotton varieties is traditionally riskier. Similarly, the rapid fruiting accumulation cycle of Bollgard II® varieties enables the crop to prosper in shorter season, cooler environments and has been a major factor underpinning significant industry expansion and the construction of new ginneries around the Lachlan, Murrumbidgee and Murray River irrigation regions during the last decade.

Better tailoring irrigation management to suit the altered growth and development of Bollgard II® varieties has also been found to improve crop yield potential. The more rapid accumulation of fruit by Bollgard II® varieties required better targeted use of irrigation, particularly around key growth stages of late flowering and crop cutout, compared to conventional non-Bt varieties (Yeates *et al.* 2006). Bollgard II® varieties have consistently improved yield potential in response to refined irrigation scheduling that aimed to reduce the soil water deficit between irrigations (more frequent timing with less water applied) (Yeates and Roberts 2008). This research led to significant industry-wide changes to how irrigation water is scheduled and applied to cotton crops in Australia. In traditional flood irrigation techniques, there has been a major emphasis on changing field grades and application rates to ensure that water is moved on and off fields quickly in order to minimise deep drainage and waterlogging so that, despite more frequent irrigation, less water is used (Wigginton *et al.* 2012).

The growth and development of cotton cultivars containing Bt traits and how they are grown has changed since the initial release of the first Bollgard II® varieties in 2004. In particular, a new cultivar, marketed in Australia as Sicot 74BRF, combined a longer-cycle growth habit (Stiller 2010) with the insect protection afforded by the Bollgard II® traits, resulting in a plant type variety that would respond to different styles of agronomic management. Where conditions and resources are conducive to continued growth, this variety could be encouraged to flower for longer, setting more bolls and increasing yields via additional bolls retained on an expanded upper canopy. The unique characteristics of this variety have been continued in more recent Bollgard 3® cultivar releases and the interaction between tailored agronomic management and cultivar improvements has been a significant factor for the rise of Australian cotton yields (Liu *et al.* 2013). The increase in yields from newer Bollgard varieties, combined with more efficient water management systems, underpinned a 40% improvement of water-use productivity by Australian cotton growers between 2003 and 2013 (Roth *et al.* 2013).

Industry expansion with Bt cotton in Australia's arid tropics

The failure of commercial cotton production on the Ord River in north-western Australia in 1975 just 10 years after cropping commenced, due to extensive insecticide resistance in *H. armigera* (Michael and Woods 1980), halted the prospects for conventional cotton production in Australia's tropical regions. As a potential solution, the feasibility of growing cotton during the winter dry season as an alternate production strategy to escape insect exposure was assessed during the mid-1990s (Strickland *et al.* 2003; Yeates *et al.* 2013). This strategy sought to grow cotton when key lepidopteran pests were naturally un-abundant compared to the wet season summer months when cotton had been previously grown. This approach was demonstrated to be effective from both pest management (Strickland *et al.* 2003) and agronomic perspectives (Yeates *et al.* 2010). However, the possibility of spring rainfall causing pre-harvest lint losses or quality downgrades in some seasons (Yeates *et al.* 2007), combined with strong public opposition to cotton production in Australia's tropical north, was sufficient to thwart renewed commercial production at the time.

The release of stacked Bt varieties (Bollgard II® and Bollgard 3®) offered renewed potential for a platform from which to develop an IPM program for *H. armigera*, tobacco cutworm *Spodoptera litura* and pink bollworm *Pectinophora gossypiella* on cotton in northern Australia (Yeates *et al.* 2013). The release of Bollgard 3® in particular has coincided with a more favourable political environment and renewed interest in identifying opportunities for sustainable development of irrigated cropping precincts in northern Australia (Hart *et al.* 2019). Testing with Bollgard 3® cultivars has found that it provides field control of the primary northern Australia noctuid pest *S. litura* under most conditions (Grundy 2018, unpublished data). This offers a basis for developing a sustainable management strategy for this pest, just as Bollgard II® has done for the related noctuid *Helicoverpa* spp. in southern Australia.

Detailed climate, soil and water studies have identified a range of sites in tropical Australia that may be well suited for irrigated cropping (Petheram *et al.* 2013, 2018). Agronomic research examining crop responses at several of these sites has shown considerable promise for reliable high-yield cotton production using Bollgard II® and 3® varieties and production practices tailored to the local climate and pest ecology (Grundy *et al.* 2012; Yeates 2018, unpublished data). The learnings that emerged after the failure of tropical cotton production (Michael and Woods 1980), combined with the development of Bt cotton-based IPM practices, enabled commercial test farming to recommence in tropical Australia and provided confidence for the construction of a new commercial ginning facility, scheduled for completion in 2021. The ginnery is located near the town of Kununurra, the very same location where the industry failed due to insecticide resistance 45 years ago. It is unlikely that the current endeavour would be possible without Bt cotton technology.

Endnotes

1. The ACGRA [now Cotton Australia] was a grower lead body representing some 15 regional Cotton Grower Associations, as well as cotton processing and planting seed companies. It was established in 1972 with a charter to encourage, advance and promote the interests of cotton growers in Australia through research (ACGRA 1972).
2. https://www.legislation.gov.au/Details/F2007B00679
3. https://www.legislation.nsw.gov.au/#/view/act/2003/11

References

ACGRA (1972) ACGRA Constitution. Australian Cotton Growers Research Association.

ACGRA (1994) ACGRA Meeting Minutes, 14 September 1994. Australian Cotton Growers Research Association.

Andow DA, Alstad DN (1998) F2 screen for rare resistance alleles. *Journal of Economic Entomology* **91**, 572–578. doi:10.1093/jee/91.3.572

ANZFA (1997) *Assessment Guidelines for Foods and Food Ingredients to be Included in Standard A18: Food Derived from Gene Technology*. Australia and New Zealand Food Authority, Canberra.

ANZFA (1999) *Final Assessment Report A341 INGARD ANZFA 1991*. Australia and New Zealand Food Authority, Canberra.

APVMA (2018) *Australian Pesticides and Veterinary Medicines 2018*. Australian Pesticides and Veterinary Medicines Authority, Canberra. <https://apvma.gov.au/node/1063>

Baker H (1998) Best management practice: why it must work. In *9th Australian Cotton Conference*. Broadbeach, QLD, pp. 221–227. <http://www.insidecotton.com/jspui/handle/1/690>

Baker GH, Tann CR (2017) Broad-scale suppression of cotton bollworm, *Helicoverpa armigera* (Lepidoptera: Noctuidae), associated with Bt cotton crops in northern New South Wales, Australia. *Bulletin of Entomological Research* **107**, 188–199. doi:10.1017/S0007485316000912

Bange MP, Milroy SP (2004) Growth and dry matter partitioning of diverse cotton genotypes. *Field Crops Research* **87**, 73–87. doi:10.1016/j.fcr.2003.09.007

Bange MP, Caton SJ, Milroy SP (2008) Managing yields of high fruit retention in transgenic cotton (*Gossypium hirsutum* L.) using sowing date. *Australian Journal of Agricultural Research* **59**, 733–741. doi:10.1071/AR07423

BCI (2018) *Annual Report 2018*. Better Cotton Initiative. <https:// Stories.bettercotton.com/2018-AnnualReport/index.html>

BCI (2019) *About BCI: History*. Better Cotton Initiative. <https://bettercotton.org/about-bci/bci-history/>

Bird LJ (2015) Baseline susceptibility of *Helicoverpa armigera* (Lepidoptera: Noctuidae) to indoxacarb, emamectin benzoate, and chlorantraniliprole in Australia. *Journal of Economic Entomology* **108**, 294–300. doi:10.1093/jee/tou042

Bird LJ (2017) Genetics, cross-resistance and synergism of indoxacarb resistance in *Helicoverpa armigera* (Lepidoptera: Noctuidae). *Pest Management Science* **73**, 575–581. doi:10.1002/ps.4334

Bird LJ (2018) Pyrethroid and carbamate resistance in Australian *Helicoverpa armigera* (Lepidoptera: Noctuidae) from 2008 to 2015: what has changed since the introduction of Bt cotton? *Bulletin of Entomological Research* **108**, 781–791. doi:10.1017/S0007485317001316

Brookes G (2016) *Adoption and Impact of GM Crops in Australia: 20 Years' Experience*. CropLife Australia, Canberra.

Brookes G, Barfoot P (2014) Economic impact of GM crops. *GM Crops and Food: Biotechnology in Agriculture and the Food Chain* **5**, 65–75. doi:10.4161/gmcr.28098

Brookes G, Barfoot P (2018) Environmental impacts of genetically modified (GM) crop use 1996–2016: impacts on pesticide use and carbon emissions. *GM Crops and Food: Biotechnology in Agriculture and the Food Chain* **9**, 109–139. doi:10.1080/21645698.2018.1476792

Buerger P, Hauxwell C, Murray D (2007) Nucleopolyhedrovirus introduction in Australia. *Virologica Sinica* **22**, 173–179. doi:10.1007/s12250-007-0019-y

CA (2019) *Australian Cotton's myBMP Environmental Program*. Cotton Australia. <https://cottonaustralia.com.au/cotton-library/fact-sheets/cotton-fact-file-cottons-mybmp-environmental-program>

CA, CRDC (2014) *Australian Grown Cotton Sustainability Report*. <https://www.crdc.com.au/publications/australian-grown-cotton-sustainability-report>

Constable GA, Llewellyn DJ, Reid PE (1998) Biotechnology risks and benefits: the Ingard cotton example. In *9th Australian Agronomy Conference*, Wagga Wagga, NSW (Eds DL Michalk, JE Pratley) pp. 92–98. <http://www.regional.org.au/au/asa/1998/plenary/>

CRCSCP (1995) *Annual Report 1994–95*. Cooperative Research Centre for Sustainable Cotton Production, Narrabri, NSW.

CRCSCP (1997) *Annual Report 1996–97*. Cooperative Research Centre for Sustainable Cotton Production, Narrabri, NSW.

CRDC (2018) *Cotton Pest Management Guide 2018–19*. Cotton Research and Development Corporation, Narrabri, NSW.

De Barro PJ, Driver F, Trueman JWH, Curran J (2000) Phylogenetic relationships of world populations of *Bemisia tabaci* (Gennadius) using ribosomal ITS1. *Molecular Phylogenetics and Evolution* **16**, 29–36. doi:10.1006/mpev.1999.0768

Dillon GE, Fitt GP, Forrester NW (1994) Natural mortality of *Helicoverpa* eggs on cotton. In *7th Australian Cotton Conference*, Broadbeach, QLD. <http://www.insidecotton.com/jspui/handle/1/3132>

Downes S, Mahon R (2012) Evolution, ecology and management of resistance in *Helicoverpa* spp. to Bt cotton in Australia. *Journal of Invertebrate Pathology* **110**, 281–286. doi:10.1016/j.jip.2012.04.005

Downes S, Mahon R, Olsen K (2007) Monitoring and adaptive resistance management in Australia for Bt-cotton: current status and future challenges. *Journal of Invertebrate Pathology* **95**, 208–213. doi:10.1016/j.jip.2007.03.010

Downes S, Walsh T, Tek Tay W (2016) Bt resistance in Australian insect species. *Current Opinion in Insect Science* **15**, 78–83. doi:10.1016/j.cois.2016.04.002

Downes S, Kriticos D, Parry H, Paull C, Schellhorn N, Zalucki MP (2017) A perspective on management of *Helicoverpa armigera*: transgenic Bt cotton, IPM, and landscapes. *Pest Management Science* **73**, 485–492. doi:10.1002/ps.4461

Downey J (1996) Mainstream environmental issues. Australian Conservation Foundation. In *8th Australian Cotton Conference*, Broadbeach, QLD, pp. 699–711. <http://www.insidecotton.com/jspui/handle/1/587>

Duffield S (2004) Evaluation of the risk of overwintering *Helicoverpa* spp. pupae under irrigated summer crops in south-eastern Australia and the potential for area-wide management. *Annals of Applied Biology* **144**, 17–26. doi:10.1111/j.1744-7348.2004.tb00312.x

Ellsworth PC, Martinez-Carrillo JL (2001) IPM for *Bemisia tabaci*: a case study from North America. *Crop Protection* **20**, 853–869. doi:10.1016/S0261-2194(01)00116-8

Fitt GP (1994) Cotton pest management: Part 3. An Australian perspective. *Annual Review of Entomology* **39**, 543–562. doi:10.1146/annurev.en.39.010194.002551

Fitt GP (2000) An Australian approach to IPM in cotton: integrating new technologies to minimise insecticide dependence. *Crop Protection* **19**, 793–800. doi:10.1016/S0261-2194(00)00106-X

Forrester NW, Bird LJ (1996) The need for adaptation to change in insecticide resistance management strategies: the Australian experience. In *Molecular Genetics and Evolution of Pesticide Resistance* (Ed. TM Brown.) pp. 160–168. American Chemical Society: Washington, DC.

Forrester NW, Cahill M, Bird LJ, Layland JK (1993) Management of pyrethroid and endosulfan resistance in *Helicoverpa armigera* (Lepidoptera: Noctuidae) in Australia. *Bulletin of Entomological Research: Supplement Series* Number 1 Supplement, 132 pp.

Frager L, Temperley J (2008) The impact of biotechnology and other factors on health and safety in the Australian cotton industry. AgHealth Australia, University of Sydney, Sydney.

FSANZ (2003) *Report on the Review of Labelling of Genetically Modified Foods*. Food Standards Australia and New Zealand, Canberra.

FSANZ (2004) *Final Assessment Report: Food Derived from Insect-protected Cotton Line COT102 A509*. Food Standards Australia and New Zealand, Canberra. <http://www.foodstandards.gov.au/code/applications/pages/applicationa509foodd2222.aspx>

FSANZ (2019) *History of FSANZ, Food Law, Treaties and Agreements*. Food Standards Australia and New Zealand, Canberra. <http://www.foodstandards.gov.au/about/foodlawandtreaties/history/pages/default.aspx>

GHD (2003) *2nd Australian Cotton Industry Environmental Audit Report*. Cotton Research and Development Corporation, Narrabri, NSW.

GMAC (1999) *Annual Report: Genetic Manipulation Advisory Committee (GMAC) 1998–99*. Genetic Manipulation Advisory Committee, Canberra.

Gregg PC, Fitt GP, Zalucki MP, Murray DAH (1995) Insect migration in an arid continent. II. *Helicoverpa* spp. in eastern Australia. In *Insect Migration: Tracking Resources Through Space and Time* (Eds VA Drake and GA Gatehouse), pp. 151–172. Cambridge University Press, Cambridge.

Grundy PR, Sequeira RV, Short KS (2004) Evaluating legume species as alternative trap crops to chickpea for management of *Helicoverpa* spp. (Lepidoptera: Noctuidae) in central Queensland cotton cropping systems. *Bulletin of Entomological Research* **94**, 481–486. doi:10.1079/BER2004327

Grundy P, Yeates S, Grundy T (2012) *NORpak: Cotton Production and Management Guidelines for the Burdekin and North Queensland Coastal Dry Tropics Region 2012*. Cotton Catchments Communities CRC, Narrabri, NSW.

Gunning RV, Easton CS, Greenup LR, Edge VE (1984) Pyrethroid resistance in *Heliothis armigera* (Hübner) (Lepidoptera: Noctuidae) in Australia. *Journal of Economic Entomology* **77**, 1283–1287. doi:10.1093/jee/77.5.1283

Gunning RV, Balfe ME, Easton CS (1992) Carbamate resistance in *Helicoverpa armigera* (Hübner) (Lepidoptera: Noctuidae) in Australia. *Australian Journal of Entomology* **31**, 97–103. doi:10.1111/j.1440-6055.1992.tb00464.x

Gunning RV, Byrne FJ, Conde BD, Connelly MI, Hergstrom K, Devonshire AL (1995) First report of B-biotype *Bemisia tabaci* (Gennadius) (Hemiptera: Aleyrodidae) in Australia. *Australian Journal of Entomology* **34**, 116. doi:10.1111/j.1440-6055.1995.tb01298.x

Hart B, O'Donnell E, Horne A (2019) Sustainable water resources development in northern Australia: the need for coordination, integration and representation. *International Journal of Water Resources Development* doi:10.1080/07900627.2019.1578199

Head G, Dennehy T (2010) Insect resistance management for transgenic Bt cotton. In *Cotton: Biotechnological Advances* (Ed. UB Zehr) pp. 113–125. Springer, Berlin.

Herron GA, Wilson LJ (2017) Can resistance management strategies recover insecticide susceptibility in pests? A case study with cotton aphid *Aphis gossypii* (Aphididae: Hemiptera) in Australian cotton. *Austral Entomology* **56**, 1–13. doi:10.1111/aen.12236

Herron GA, Powis K, Rophail J (2001a) Insecticide resistance in *Aphis gossypii* Glover (Hemiptera: Aphididae): a serious threat to Australian cotton. *Australian Journal of Entomology* **40**, 85–91. doi:10.1046/j.1440-6055.2001.00200.x

Herron GA, Rophail J, Wilson LJ (2001b) The development of bifenthrin resistance in two-spotted spider mite (Acari: Tetranychidae) from Australian cotton. *Experimental & Applied Acarology* **25**, 301–310. doi:10.1023/A:1017967118609

Innovact Consulting (2012) *3rd Environmental Assessment Report*. Prepared for Cotton Research and Development Corporation, Narrabri, NSW. <http://www.insidecotton.com/xmlui/handle/1/4094>

Kelly D, McLennen A, Pyke B, Hickman M, Deutscher S, Kauter G (2004) Industry perceptions on management issues associated with Bollgard II. In *Australian Cotton Conference*, Broadbeach, QLD, pp. 215–220.

Kennedy IR, Rose MT, Crossan A, Burns M (2013) Research and practice: environmental action for improving water quality in cotton catchments since 1990. *Crop and Pasture Science* **64**, 1095–1111. doi:10.1071/CP13091

Khan M, Miles M, Maas S, Byers K, Spargo G (2012) Overwintering strategy of *Solenopsis* mealybug. *Australian Cottongrower* **33**, 22–24.

Kovach J, Petzoldt C, Degni J, Tette J (1992) A method to measure the environmental impact of pesticides. *Food and Life Sciences Bulletin* **139**, 8.

Liu S-S, De Barro PJ, Xu J, Luan J-B, Zang L-S, Ruan Y-M, Wan F-H (2007) Asymmetric mating interactions drive widespread invasion and displacement in a whitefly. *Science* **318**, 1769–1772. doi:10.1126/science.1149887

Liu SM, Constable GA, Reid PE, Stiller WN, Cullis BR (2013) The interaction between breeding and crop management in improved cotton yield. *Field Crops Research* **148**, 49–60. doi:10.1016/j.fcr.2013.04.006

Lloyd RJ, Murray DAH, Hopkinson JE (2008) Abundance and mortality of overwintering pupae of *Helicoverpa armigera* (Hübner) (Lepidoptera: Noctuidae) on the Darling Downs, Queensland, Australia. *Australian Journal of Entomology* **47**, 297–306. doi:10.1111/j.1440-6055.2008.00661.x

Mahon RJ, Olsen KM, Downes S, Addison S (2007) Frequency of alleles conferring resistance to the Bt toxins Cry1Ac and Cry2Ab in Australian populations of *Helicoverpa armigera* (Lepidoptera: Noctuidae). *Journal of Economic Entomology* **100**, 1844–1853. doi:10.1603/0022-0493(2007)100[1844:FOACRT]2.0.CO;2

Mawhinney W (2011) *Namoi Water Quality Project Final Report 2002–2007.* NSW Office of Water, Sydney.

Mawhinney W (2012) Case study, catchment water quality and cotton: northern NSW In *WATERpak: A Guide for Irrigation Management in Cotton and Grain Farming Systems*, pp. 452–459. Cotton Research & Development Corporation, Narrabri, NSW.

Mensah RK (2002) Development of an integrated pest management programme for cotton. Part 1: establishing and utilizing natural enemies. *International Journal of Pest Management* **48**, 87–94. doi:10.1080/09670870110095377

Michael PJ, Woods WM (1980) *An Entomological Review of Cotton Growing in the Ord River Area of Western Australia*. Technical Bulletin 48. WA Department of Agriculture and Food, Perth.

Moore AD, Sequeira RV, Woodger TA (2004) Susceptibility of crop plants to *Bemisia tabaci* (Gennadius) B-biotype (Hemiptera: Aleyrodidae) in central Queensland, Australia. *Australian Entomologist* **31**, 69–74.

OGTR (2018) *How the Legislation is Used to Regulate GMO Dealings*. Office of the Gene Technology Regulator, Canberra. <http://www.ogtr.gov.au/internet/ogtr/publishing.nsf/Content/regs-requirements>

OGTR (2019) *Table of Authorisations for Commercial Releases of GM Plants*. Office of the Gene Technology Regulator, Canberra. <http://www.ogtr.gov.au/internet/ogtr/publishing.nsf/Content/cr-1>

Petheram C, Watson I, Stone P (2013) *Agricultural Resource Assessment for the Gilbert Catchment*. CSIRO, Canberra. <https://doi.org/10.4225/08/584ee5f996e6f>

Petheram C, Watson I, Bruce C, Chilcott C (2018) *Water Resource Assessment for the Mitchell Catchment, Australia*. CSIRO, Canberra. <https://publications.csiro.au/rpr/pub?pid=csiro:EP186909>

Polya R (2008) *Chronology of Genetic Engineering Regulation in Australia: 1953–2008*. Parliamentary Library Research Publications, Canberra. <https://www.aph.gov.au/About_Parliament/Parliamentary_Departments/Parliamentary_Library/pubs/BN/0809/ChronGeneticEngineeringA>

Pyke B (1997) Second INGARD consultants survey highlight variation. *Australian Cottongrower* **18**, 10–11.

Rossiter L, Gunning RV, McKenzie F (2008) Silver anniversary of resistance management in the Australian cotton industry: an overview of the current situation for *Helicovpera armigera*. In *14th Australian Cotton Conference*, Broadbeach, QLD. <http://www.insidecotton.com/xmlui/bitstream/handle/1/2640/L%20Rossiter%20et%20al.pdf?sequence=1&isAllowed=y>

Roth G (2011) Retaining the social licence: the Australian cotton industry case study. In *Defending the Social Licence of Farming: Issues, Challenges and New Directions for Agriculture* (Eds J Williams and P Martin) pp. 69–82. CSIRO Publishing, Melbourne.

Roth G, Harris G, Gillies M, Montgomery J, Wigginton D (2013) Water-use efficiency and productivity trends in Australian irrigated cotton: a review. *Crop and Pasture Science* **64**, 1033–1048. doi:10.1071/CP13315

Roush RT (1998) Two toxin strategies for management of insecticidal transgenic crops: can pyramiding succeed where pesticide mixtures have not? *Philosophical Transactions of the Royal Society of London. Series B, Biological Sciences* **353**, 1777–1786. doi:10.1098/rstb.1998.0330

Sequeira R (2001) Inter-seasonal population dynamics and cultural management of *Helicoverpa* spp. in a central Queensland cropping system. *Australian Journal of Experimental Agriculture* **41**, 249–259. doi:10.1071/EA00051

Sequeira RV, Playford CL (2001) Abundance of *Helicoverpa* (Lepidoptera: Noctuidae) pupae under cotton and other crops in central Queensland: implications for resistance management. *Australian Journal of Entomology* **40**, 264–269. doi:10.1046/j.1440-6055.2001.00234.x

Sequeira RV, Shields A, Moore A, De Barro P (2009) Inter-seasonal population dynamics and pest status of *Bemisia tabaci* (Gennadius) biotype B in an Australian cropping system. *Bulletin of Entomological Research* **99**, 325–335. doi:10.1017/S000748530800638X

Shelton AM, Tang JD, Roush RT, Metz TD, Earle ED (2000) Field tests on managing resistance to Bt-engineered plants. *Nature Biotechnology* **18**, 339–342. doi:10.1038/73804

Stiller W (2010) Sicot 74BRF. *Plant Varieties Journal* **23**, 143–146.

Strickland GR, Annells AJ, Ward AL (2003) Assessing the feasibility for cotton in tropical Australia: research for the development of sustainable pest management systems. In *World Cotton Research Conference*, Cape Town, South Africa (Ed. A Swanepoel) pp. 975–986. International Cotton Advisory Committee, Washington, DC.

Tabashnik BE, Carrière Y, Dennehy TJ, Morin S, Sisterson MS, Roush RT, Shelton AM, Zhao J-Z (2003) Insect resistance to transgenic Bt crops: lessons from the laboratory and field. *Journal of Economic Entomology* **96**, 1031–1038. doi:10.1093/jee/96.4.1031

Tabashnik BE, Brévault T, Carrière Y (2013) Insect resistance to Bt crops: lessons from the first billion acres. *Nature Biotechnology* **31**, 510–521. doi:10.1038/nbt.2597

TIMS (2014) *Transgenic and Insecticide Management Strategies Committee Charter.* Cotton Australia, Sydney.

Trapero C, Wilson IW, Stiller WN, Wilson LJ (2016) Enhancing integrated pest management in GM cotton systems using host plant resistance. *Frontiers of Plant Science* 7, 1–12. doi:10.3389/fpls.2016.00500

Wigginton D, Carter M, Grabham M (2012) *Surface Irrigation Performance and Operation.* Cotton Research and Development Corporation, Narrabri, NSW.

Wilson L, Downes S, Khan M, Whitehouse M, Baker G, Grundy P, Maas S (2013) IPM in the transgenic era: a review of the challenges from emerging pests in Australian cotton systems. *Crop and Pasture Science* **64**, 737–749.

Wilson LJ (1993) Spider mites (Acari: Tetranychidae) affect yield and fiber quality of cotton. *Journal of Economic Entomology* **86**, 566–585. doi:10.1093/jee/86.2.566

Wilson LJ, Bauer LR, Lally DA (1998) Effect of early season insecticide use on predators and outbreaks of spider mites (Acari: Tetranychidae) in cotton. *Bulletin of Entomological Research* **88**, 477–488. doi:10.1017/S000748530004222X

Wilson LJ, Whitehouse MEA, Herron GA (2018) The management of insect pests in Australian cotton: an evolving story. *Annual Review of Entomology* **63**, 215–237. doi:10.1146/annurev-ento-020117-043432

Yeates S, Roberts J (2008) Toward better water management of Bollgard II cotton. In *14th Australian Cotton Conference*, Broadbeach, QLD. Cotton Australia. <http://www.insidecotton.com/jspui/handle/1/2650>

Yeates S, Richards D, Roberts J, Gregory R (2006) Progress in evaluating the moisture stress response of Bollgard II compared with conventional cotton. In *14th Australian Cotton Conference*, Broadbeach, QLD. Cotton Australia. <http://www.insidecotton.com/jspui/handle/1/2567>

Yeates S, Strickland G, Moulden J, Davies A (2007) *NORpak Ord River Irrigation Area Cotton Production and Management Guidelines for the Ord River Irrigation Area (ORIA).* Cotton Catchment Communities Cooperative Research Centre, Narrabri, NSW.

Yeates SJ, Constable GA, McCumstie T (2010) Irrigated cotton in the tropical dry season. I: yield, its components and crop development. *Field Crops Research* **116**, 278–289. doi:10.1016/j.fcr.2010.01.005

Yeates SJ, Strickland GR, Grundy PR (2013) Can sustainable cotton production systems be developed for tropical northern Australia? *Crop and Pasture Science* **64**, 1127–1140. doi:10.1071/CP13220

6

Advances in managing insect resistance to Bt cotton in China

Qingsong Liu, Lanzhi Han and Yunhe Li

Introduction

Genetically engineered (GE) crops have the potential to help meet the increasing demand for food and fibre for the world's growing population. In particular, insect-resistant (IR) GE crops expressing insecticidal genes, mainly crystal (*cry*) and to some extent vegetative insecticidal proteins (*vip*) isolated from the soil bacterium *Bacillus thuringiensis* (Bt), have revolutionised pest control. Since their first commercial planting in the USA in 1996, Bt crops have been rapidly adopted worldwide (ISAAA 2017). In 2017, a total of 101 million ha of Bt crops were planted worldwide. These IRGE crops include mainly Bt soybean, Bt corn, Bt cotton and, to a small extent, Bt eggplant. About 77% of Bt crop area involves stacked traits like IR and herbicide tolerance (HT) (ISAAA 2017).

The successful planting of Bt crops has benefited humans by suppressing pests, decreasing the application of chemical insecticides, conserving natural enemies, and increasing yields, farm profits and environmental quality (Brookes and Barfoot 2018a, b). China was one of the few countries planting Bt cotton as early as 1997 and since then has grown it substantially (ISAAA 2017). In view of the evolution of Bt resistance in target pests threatening Bt cotton sustainability in China and other countries (Wu 2014; Tabashnik and Carrière 2017; Xiao and Wu 2019), in this chapter we review the development and adoption of Bt cotton, elucidate the resistance mechanisms, evaluate potential for the evolution of target pest resistance to Bt crops, and finally discuss the strategies to manage insect resistance to Bt cotton in relation to China more specifically.

Development and use of Bt cotton in China

Cotton is an important cash crop that provides natural fibre, edible oil and other industrial by-products. Cotton production is constrained by many factors, especially herbivorous insects. More than 300 insect species attack cotton crops in China (Wu and Guo 2005). Among cotton insect pests, the following lepidopteran pests have caused enormous yield losses annually: the cotton bollworm (*Helicoverpa armigera*), the pink bollworm (*Pectinophora gossypiella*), the corn borer (*Ostrinia furnacalis*), the beet armyworm (*Spodoptera exigua*), the spiny bollworms (*Earias cupreoviridis*, *E. fabia* and *E. insulana*) and the cotton looper (*Anomis flava*). To manage these pests, China once relied heavily on chemical insecticides (Wu and Guo 2005). As an alternative to insecticide use, IRGE cotton was developed by *Agrobacterium*-mediated transformation with the Bt-modified *cry1A* gene in 1994. The resulting Bt cotton line was found to be highly resistant against cotton bollworm, with more than 80% control. This Bt cotton variety was quickly commercialised in 1997 (Zhang *et al.* 2007). In 1996, a dual-gene IRGE cotton line expressing both *cry1A* and *CpTI* (cowpea trypsin inhibitor) genes was developed, and was found to be effective for controlling cotton bollworm (Zhang *et al.* 2007). Following these initial successes, multiple Bt cotton lines with high resistance to insect pests were developed in China (Guo *et al.* 2015). Recently, Ni *et al.* (2017) developed the first pyramided cotton lines combining Bt and RNA interference (RNAi). These novel IRGE cotton lines stacked with the double-stranded RNA (dsRNA) targeting the juvenile hormone (JH) acid methyl transferase or JH-binding protein of bollworm and Bt Cry proteins showed high resistance against both Bt-resistant and Bt-susceptible bollworms.

Since 1997, the area planted with Bt cotton in China rapidly expanded. It reached 3.75 million ha in 2004, remained stable between 3 and 4.14 million ha from 2004 to 2015 then declined to 2.78 million ha until 2016 (Figure 6.1). Bt cotton area increased from 90% in 2013 to 95.9% of total cotton planting area in 2017 (ISAAA 2017). With the increase in the planting of Bt cotton, cotton bollworm, pink bollworm and other target pests were regionally suppressed, accompanied by a decrease in the application of broad-spectrum chemical insecticides and an enhancement of biological control (Wu *et al.* 2008; Zhang *et al.* 2018). With the large-scale adoption of Bt cotton in China and other countries, however, mirid bugs and other non-target secondary sucking pests became important pests, undermining the benefits of Bt cotton and emphasising the need for integrated pest management (Lu *et al.* 2010; Lu and Wu 2011; Lu *et al.* 2012; Catarino *et al.* 2015).

Development of target pest resistance to Bt cotton and potential mechanisms

Like long-term use of chemical insecticides, long-term large-scale growing of Bt crops will inevitably lead to the evolution of insect resistance in the field. Field-level resistance in corn

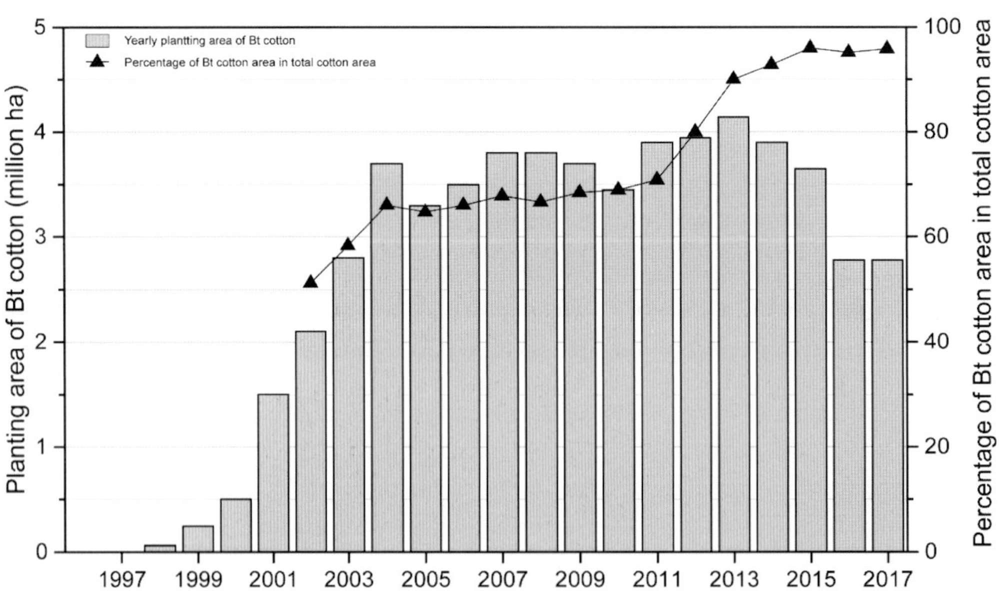

Figure 6.1 Yearly planting of Bt cotton and its percentage of the total area of cotton in China since 1997. Source: ISAAA.

earworm (*Helicoverpa zea*) in the USA and pink bollworm (*P. gossypiella*) in India was reported to Bt cotton expressing Cry1Ac first, and later to both toxins, Cry1Ac and Cry2Ab (Tabashnik and Carrière 2017).

Evolution of resistance in target pests significantly reduces the efficacy of Bt crops. Understanding how Cry toxins work and how insects develop resistance will provide the basis for designing appropriate measures to manage the evolution of Bt resistance (Wu 2014; Han *et al.* 2016; Xiao and Wu 2019). The mode of action of Cry toxins has been extensively studied and reviewed (Pardo-López *et al.* 2013; Wu 2014; Jurat-Fuentes and Crickmore 2017). Although some details of the action of Cry toxins remain unknown, the major steps (crystal solubilisation, proteolytic activation, passage through the peritrophic membrane, receptor binding, membrane insertion, pore formation, and osmotic lysis of midgut cells) are widely accepted (Ferré and Van Rie 2002). Theoretically, resistance to Bt can occur at any of these steps.

Alteration in toxin activation

Proteinase

The first step in the action of Cry toxins is solubilisation of protoxins in the alkaline gastric juice of midgut, followed by activation by proteases to release insecticidal toxins (Pardo-López *et al.* 2013; Jurat-Fuentes and Crickmore 2017). Improper activation of protoxins (insufficient activation or increased degradation) can result in resistance to Bt crops (Oppert 1999). More specifically, changes in the proteases in the insect midgut can reduce the activation of Cry proteins and thus increase Bt resistance (Oppert 1999; Ferré and Van Rie 2002; Jurat-Fuentes and Crickmore 2017; Xiao and Wu 2019). Chandrashekar and Gujar (2004) reported increase in protease activity in the 31-fold Cry1Ac-resistant cotton bollworm that affected stable toxin formation and its availability for binding with midgut receptors. However, the reduced activity of serine protease HaSP2 in the laboratory-selected 72-fold Cry1Ac-resistant larvae of Akola-R strain of cotton bollworm was associated with Cry1Ac protoxin resistance (Rajagopal *et al.* 2009). But the same strain did not show resistance to activated Cry1Ac toxin. Liu *et al.* (2015) reported that mutations in the promoter of a trypsin gene led to high Cry1Ac resistance in cotton bollworm. Combined transcriptome and proteome analyses of the midgut and peritrophic matrix in a Bt-resistant (XJ10) and a Bt-susceptible (XJ) strain of cotton bollworm indicated the down-regulation of mRNA and several trypsins' association with protoxin activation in a Bt-resistant strain (Jin *et al.* 2019). Further, a XJ10 strain was three-fold more resistant to Cry1Ac protoxin than to activated Cry1Ac toxin (Jin *et al.* 2019). The greater resistance to Cry protoxins than to activated toxins in Bt-resistant strains provides additional evidence that changes in proteinases can contribute to Bt resistance (Wu 2014).

Esterase sequestration

A class of serine hydrolases, also called non-specific esterases which could hydrolyze insecticidal esters and sequester xenobiotics, has been implicated in insecticide resistance in many insects. The esterases were found to bind with the Cry1Ac protoxin and activated toxin, and prevent the latter from binding at the midgut receptors of the cotton bollworm (Gunning *et al.* 2005). Likewise, an esterase E4-like gene was up-regulated in the Bt-resistant strain XJ10 of cotton bollworm (Jin *et al.* 2019). These results suggest that esterase sequestration is a potential mechanism of Bt resistance.

Binding site alteration

Altered binding of Cry toxins to target sites on the brush border membrane vesicles (BBMV) of the larval midgut is another mechanism of insect resistance to Bt crops (Ferré and Van Rie

2002). Based on the results of binding competition experiments, Luo *et al.* (2006) proposed a three-site model for the binding of three Cry1A toxins (Cry1Aa, Cry1Ab and Cry1Ac) to BBMV in Bt-susceptible cotton bollworm larvae. According to this model, site 1 binds with all three Cry1A toxins, site 2 with both Cry1Ab and Cry1Ac, and site 3 only with Cry1Ac. In the Cry1Ac-resistant strain, the binding of site 1 with Cry1Ab decreased while site 2 did not bind with Cry1Ac. Further studies showed that binding site alteration was responsible for its resistance to Cry1Ac (Luo *et al.* 2006). In another study, binding site alteration was found to be responsible for field-evolved resistance of cotton bollworm and *Helicoverpa punctigera* (Australian bollworm) to Cry2A (Caccia *et al.* 2010). These results indicate that binding site alteration is a common mechanism for resistance to both Cry1 and Cry2 toxins. Further discussion in this section deals with specific binding sites (enzymatic and non-enzymatic structural proteins) responsible for Cry toxin resistance development in the insects.

Aminopeptidase-N

Aminopeptidase-N (APN) in the insect midgut membrane is an important receptor for Cry toxins (Xiao and Wu 2019). Mutation or altered expression of genes that encode APN confers Bt resistance in cotton bollworm (Wu 2014). The deletion mutation at amino acids 938–1004 sequence of HaAPN1 in the laboratory-selected Bt-R strain of cotton bollworm directly leads to the loss of Cry1Ac binding activity and thereby Cry1Ac resistance (Zhang *et al.* 2009). The expression of HaAPN1 was recently found to be 1.66-fold up-regulated in a Bt-resistant strain relative to a susceptible strain of cotton bollworm (Jin *et al.* 2019). However, the detailed mechanism by which APNs contribute to Bt resistance in cotton bollworm remains unclear (Xiao and Wu 2019).

Cadherin

Cadherins play an important role in many biological processes including cell adhesion, morphogenesis, cytoskeletal organisation and cell sorting/migration, as well as in pathological conditions (Angst *et al.* 2001). Cadherins are well characterised as receptors of Cry toxins (Fabrick and Wu 2015). Insect midgut cadherins are transmembrane proteins composed of four domains – an extracellular domain consisting of nine to 12 cadherin repeats, a membrane proximal region, a single transmembrane domain and an intracellular cytoplasmic domain – all mediating Cry protein toxicity in lepidopteran pests (Wu 2014; Zhang *et al.* 2017; Zhao *et al.* 2017; Xiao and Wu 2019). Cadherin mutations that confer resistance to Bt toxins have been identified in the tobacco budworm (Gahan *et al.* 2001), cotton bollworm (Xu *et al.* 2005; Zhang *et al.* 2012a, b; Nair *et al.* 2013) and pink bollworm (Morin *et al.* 2003).

Altered expression of cadherin genes is a major mechanism for Bt resistance in cotton bollworm and pink bollworm. To date, 18 alleles of the cadherin gene *HaCad* (*r1–r17* and *mHaCad*) responsible for Cry toxin resistance have been identified in cotton bollworm; HaCad r1-r15 and mHaCad in China (Xiao *et al.* 2017) and *r16–17* from cotton bollworm in India (Fabrick and Wu 2015). The *r1* mutation was first identified in a laboratory-selected GYBT strain and was later identified in three resistant strains isolated from the field-selected Anyang population in Henan province of northern China (Xu *et al.* 2005; Zhang *et al.* 2012a). The *r1* allele has a deletion mutation corresponding to exons 8–24 in *HaCad*, which causes a frame shift and introduction of a premature stop codon in the coding sequence (Xu *et al.* 2005). Introgression of the *r1* allele into a Bt-susceptible strain greatly increased its resistance to Cry1Ac toxin (Yang *et al.* 2009). The *r2*, *r3*, and *r5–8* mutations are all generated by insertion of putative transposable DNA elements and encode a truncated HaCad protein (Yang *et al.* 2007; Zhao *et al.* 2010). Alleles *r4* and *r10–14* have mutations causing amino acid substitutions (Zhao *et al.* 2010; Zhang *et al.* 2012a). The *r9* mutation is produced by incorrect mRNA splicing between exons 24 and 25 (Zhang *et al.* 2012a). The *r15* allele has a 55 amino acid deletion in the intracellular cytoplasmic domain of cadherin and has been genetically linked to non-recessive resistance to Cry1Ac toxin in each of three field-selected cotton bollworm populations (Zhang *et al.* 2012b). A DNA-based PCR method to screen the *r15* allele in field populations showed that all of the *r15* alleles had the same deletion in the DNA sequence, but at least four different indels caused loss of exon 32 in the genomic DNA sequences (Zhang *et al.* 2013). Xiao *et al.* (2017) recently identified a new allele (*mHaCad*) in a laboratory-selected 516-fold Bt Cry1Ac-resistant cotton bollworm strain, which affected intracellular protein trafficking in the endoplasmic reticulum. Sequence analysis of *r16* (*r9* in the original study) showed a deletion of 45 base pairs (bp) in the intron region, and mutation in the 5′-splice site of exon 26 resulted in an additional amino acid (Nair *et al.* 2013). In the *r17* (*r10* in the original study) allele, a premature stop codon resulted in the putative translation into two different proteins of 1343 and 335 amino acids relative to a single protein of 1756 amino acids of the susceptible strain (Nair *et al.* 2013).

In the pink bollworm, *PgCad1* encodes a cadherin protein that binds to Cry1Ac in susceptible larvae. Previous studies identified 12 alleles of *PgCad1* associated with resistance to Cry1Ac: *r1*–*r4* from laboratory-selected Arizona strains in the USA and *r5*–*r12* from field-selected populations in India (Morin *et al.* 2003; Fabrick and Tabashnik 2012; Fabrick *et al.* 2014). Alleles *r1*–*r4* were predicted to cause the loss of amino acids upstream of the Cry1Ac-binding region of the cadherin protein (Morin *et al.* 2003; Fabrick and Tabashnik 2012). Analyses of cDNA of *r5*–*r12* revealed a total of 19 transcript isoforms, each of which contains a premature stop codon or a deletion of at least 99 bp, or both (Fabrick *et al.* 2014). Novel alleles of *PgCad1*, including *r15* and *r16*, were recently identified in field populations in China and were associated with resistance to Cry1Ac. The *r15* allele in *PgCad1* has a 3370 bp insertion, which is a miniature inverted repeat transposable element (MITE) that contains two additional transposons and produces two mis-spliced transcript variants (*r15A* and *r15B*). A strain homozygous for *r15* allele with high resistance to Cry1Ac but with no or little cross-resistance to Cry2Ab completed its life-cycle on Cry1Ac-expressing Bt cotton (Wang *et al.* 2019b). The *r16* allele has 1545 bp of a degenerate transposon inserted in exon 20 of *PgCad1*, which generates a mis-spliced transcript containing a premature stop codon (Wang *et al.* 2019a). Like the *r15* homozygous strain, the *r16* homozygous strain had high resistance to Cry1Ac and low cross-resistance to Cry2Ab; the resistant strain could complete its life-cycle on Bt cotton producing Cry1Ac (Wang *et al.* 2019a).

In most cases, resistance to the Bt toxin Cry1Ac is associated with mutations that disrupt the extracellular or intracellular cytoplasmic domains of cadherin proteins (Fabrick and Wu 2015). Wang *et al.* (2018) reported that the transmembrane mutation of *PgCad1* in a field population of pink bollworm in the Yangtze River Valley was associated with Cry1Ac resistance. The mutant allele has a 207 bp deletion and encodes a cadherin protein that lacks the transmembrane domain, resulting in disrupted cellular trafficking of cadherin (Wang *et al.* 2018). A strain homozygous for this allele had high resistance to Cry1Ac and low cross-resistance to Cry2Ab. Feeding on Bt cotton plants producing Cry1Ac, no susceptible larvae survived, but the resistant individuals could complete the life-cycle (Wang *et al.* 2018). Moreover, knockdown of *PgCad1 lncRNA*, a long non-coding RNA encoded by the pink bollworm cadherin gene (*PgCad1*), repressed the transcription of *PgCad1* and significantly decreased the susceptibility to Cry1Ac toxin (Li *et al.* 2019). The authors speculated that *PgCad1 lncRNA* enhanced the expression of its genomically associated cadherin and thereby increased the susceptibility of pink bollworm to the Cry1Ac toxin (Li *et al.* 2019). The latter study suggested that lncRNA-mediated resistance may be a novel mechanism underlying the resistance of target pests to Bt toxins.

Cadherin alleles from laboratory- and field-selected cotton bollworm and pink bollworm populations are currently assigned to five groups:

1. a truncation group: a premature stop codon is present at any site on the extracellular domain of cadherin;
2. a group with a deletion at the extracellular domain: deletions from the Bt toxin-binding region or other site at the extracellular domain affect the normal interaction between cadherin and Bt toxin;
3. a group with a deletion at the intracellular domain: the deletion does not affect toxin binding but affects post-binding events;
4. a group with a deletion at the transmembrane domain: these deletions disrupt cellular trafficking of cadherin
5. a group with single amino acid mutations in the Cry toxin-binding region of the cadherin receptor: these mutations result in cadherin mislocalisation and underpin resistance (Wu 2014; Xiao *et al.* 2017; Wang *et al.* 2018).

Alkaline phosphatase

Reduced levels of midgut membrane-bound alkaline phosphatase (ALP) protein and reduced expression of the corresponding RNA have been associated with Cry1Ac resistance in tobacco budworm and cotton bollworm. In tobacco budworm, the HvALP was proposed to be a Cry1Ac-binding protein, and its levels were significantly lower in the resistant strain YHD2 than in the susceptible strain YDK (Jurat-Fuentes and Adang 2004). Based on proteomic and genomic analyses, Jurat-Fuentes *et al.* (2011) reported that, although reduced levels of ALP were common in tobacco budworm, cotton bollworm and other lepidopterans resistant to Cry toxins, a direct correlation between HvALP levels and resistance was still lacking.

In cotton bollworm, HaALP1 and HaALP2 were speculated to be receptors for Cry1Ac toxin, and N-glycosidase digestion of HaALPs reduced their Cry1Ac binding (Ning *et al.* 2010). It remains to be determined whether reduced expression of ALP in resistant strains represents a direct decrease in functional Cry toxin receptors or is only the consequence of adaptation to the toxin (Jurat-Fuentes *et al.* 2011). Expression of the *HaALP2* gene was recently determined to be 2.24-fold more in the Bt-resistant strain XJ10 than in the susceptible strain XJ of cotton bollworm (Jin *et al.* 2019). Activity of ALP was found to be higher and the isozyme pattern different in the Cry2Ab-resistant pink bollworm larvae when compared with the susceptible insects (Malthankar and Gujar 2014). Zhao *et al.* (2017) reported importance of ALP for toxicity of Cry2Aa in the cotton bollworm.

Adenosine triphosphate-binding cassette transporters

Insect ABC transporters have been implicated as putative receptors in the insecticidal action of Cry toxins (Wu 2014). A high level of resistance to Bt toxin associated with a mutation of the ABC transporter ABCC2 has been found worldwide in many insects including tobacco budworm, cotton bollworm and pink bollworm (Xiao and Wu 2019).

In China, Xiao *et al.* (2015) reported that mis-splicing of the ABCC2 gene involving a 73 bp insertion in the cDNA of a Bt-resistant strain of cotton bollworm resulted in a premature stop codon, which was expected to generate a truncated ABCC2 protein and confer a 1000-fold increase in resistance to Cry1Ac toxin relative to its susceptible parent strain. Interestingly, this resistant strain with a mutation in the ABCC2 gene has increased sensitivity to abamectin, a widely used neurotoxic insecticide derived from the soil bacterium *Streptomyces avermitilis*, which suggests that Bt resistance mediated by mutation in the ABC transporter may cause negative cross-resistance to bacterial insecticides (Xiao *et al.* 2016). Using the clustered regularly interspaced short palindromic repeats (CRISPR)/Cas9 genome editing system, Wang *et al.* (2017) created two strains of *HaABCA2* gene knockout cotton bollworm with high levels of resistance to Cry2Aa (>120-fold increase in resistance) and Cry2Ab (>100-fold increase in resistance) compared with the original strain.

Other receptors

In addition to the receptors mentioned above, other receptors on the cell membrane also have been found to interact with Cry toxins (Xiao and Wu 2019). In the diamondback moth *Plutella xylostella*, for example, glucosinolate sulfatases GSS1 and GSS2 can directly bind to Cry1Bd, and their expression is crucial for Cry1Bd toxicity (Chen *et al.* unpublished). However, no receptor of this kind is reported in the target pests of Bt cotton.

Symbiotic microbes

Symbiotic microbes play an important role in insect growth and development and in insect interactions with the environment (Henry *et al.* 2015; Lu *et al.* 2016). The involvement of symbiotic microbes in insect interactions with Bt toxins is controversial (Caccia *et al.* 2016; Xiao and Wu 2019). Broderick *et al.* (2006) reported that gut microbiota are required for the insecticidal activity of Bt toxins. However, midgut microbes are not required for the efficacy of Cry toxins against diamondback moth and tobacco hornworm *Manduca sexta* (Johnston and Crickmore 2009; Raymond *et al.* 2009). In addition, Jiang *et al.* (2010) found that the midgut bacterial communities were similar in a Bt-resistant and a Bt-sensitive strain of cotton bollworm, and concluded that gut microbes had no direct effect on the Bt resistance of cotton bollworm.

Resistance monitoring

The management of insect resistance to Bt crops depends on early detection via monitoring (Tabashnik and Carrière 2017). Bt resistance monitoring generally includes four steps:

1. determining baseline susceptibility and initial resistance allele frequency of target pests to Bt toxins, before introduction of IRGE crop;
2. detecting occurrence and levels of resistance to Bt toxins over the time-scale of Bt crop adoption to forewarn of a resistance problem;
3. determining the effectiveness of the implemented resistance management strategies;
4. documenting losses of insect control in the field associated with resistance (Wu 2014).

The methods used for these studies include dose–response and diagnostic concentration assays, F1, F2 and DNA screens for the target pests using toxins expressed in IRGE crops. These bioassay methods involve use of insect populations from different locations of Bt crops over a period of time to gauge spatial and temporal changes in response to Bt toxins. The results of target pest response are also expressed in terms of resistance intensity and the frequency of resistant alleles. The dose–response assays are widely used to estimate the toxin concentration that kills 50% of a pest population (LC_{50}). The resistance ratios (RRs) are calculated by comparing the LC_{50} values obtained from field populations exposed to Bt cotton against those of Bt-susceptible strains. The diagnostic concentration assays are commonly used to measure the frequency of resistant individuals as the proportion of field populations surviving at a concentration which kills all or nearly all individuals of the susceptible populations, but few or no resistant individuals. The F1 screen used various crosses between field-collected insects and a laboratory-selected Bt-resistant strain, and the F1 neonates of the crosses were used for bioassay aimed at determining Bt resistance allele frequency. The principle of the F2 screen is that F1 offspring of each mated female collected from the field are sib-mated to produce the F2 generation, and larvae in the F2 generation are screened with a diagnostic concentration of Bt toxin (Andow and Alstad 1998). If both the field-collected female and male insects carry one recessive resistance allele (*r*), the F2 generation will have ~6.25% survival (*rr*) when screened with Bt toxins. The DNA screen uses the genomic DNA of parents and progeny of F1 reciprocal crosses, backcrosses and F2 of reciprocal crosses for estimating resistance data. These two types of detection methods are complementary, and have proved effective for monitoring Bt resistance in cotton bollworm and pink bollworm in China (Wu 2014).

In China, bioassays were used to monitor the susceptibility of cotton bollworm field populations to Cry1Ac protein from 1997 to 2006 (Wu 2007). Cotton bollworm field populations remained susceptible to Cry1Ac protein and did not show a shift towards resistance, leading to the conclusion that there are rare resistance alleles in the field populations. From 2014–15, bioassays with Cry1Ac and Cry2Ab using nine populations of cotton bollworm collected from north-western China (Xinjiang Uygur Autonomous Region) indicated higher susceptibility to Cry1Ac than populations in northern China. However, susceptibility to Cry2Ab did not differ between north-western and northern China (Wang *et al.* 2019c). In 2000, in six provinces of the Yangtze River Valley, millions of small farmers planted Bt cotton that produced Cry1Ac; pink bollworm was one of the primary target pests. In this area, the percentage of cotton area planted with Bt cotton was 9% in 2000, 52% in 2004, 84% in 2006, 92% in 2008 and 94% in 2009 and 2010 (Wan *et al.* 2012). Laboratory diet bioassays with 51 field-derived strains of pink bollworm from 16 sites in this area showed that susceptibility to Cry1Ac was significantly lower in 2008–10 than in 2005–07 (Wan *et al.* 2012). The percentage of populations with one or more larvae surviving at the diagnostic concentration increased from 0% in 2005–07 to 56% in 2008–10, but decreased to 0% in 2011–15 (Wan *et al.* 2017).

Although insect bioassays have been widely used for resistance monitoring, they often fail to detect resistance when the resistance frequency is <0.005. In contrast, F1 and F2 screens can dependably detect low resistance frequencies in target pests. With F1 screen and molecular cloning, 15 cadherin alleles (*r1–r15*) associated with Cry1Ac resistance were discovered in field populations of cotton bollworm from China (Yang *et al.* 2007; Zhao *et al.* 2010; Zhang *et al.* 2012a, b). Using F1 screen, Jin *et al.* (2015) evaluated the resistance to Cry1Ac of cotton bollworm collected from 2010 to 2013 at 17 field sites in six provinces of northern China, and found that that the percentage of resistant insects increased from 0.93% in 2010 to 5.5% in 2013. In China, F2 screening data of 363 field-derived iso-female families of cotton bollworm showed that the frequency of Cry1Ac resistance was three times higher for a field population from northern China (0.052), where there is intensive planting of Bt cotton, than for field populations from north-western China (0.016), where the planting of Bt cotton is limited (Zhang *et al.* 2012a). From 2002 to 2013, F1 and F2 screens were used to monitor Bt resistance in Xiajin county of Shandong province and Anci county of Hebei province in northern China (An *et al.* 2015). The results showed that although the susceptibility of cotton bollworm to Cry1Ac protein significantly decreased between 2002 and 2013, resistance to Bt cotton in field populations slowly increased (An *et al.* 2015).

Bioassay-based detection methods (including F1 and F2 screens) have been used as primary tools to detect and monitor Bt resistance, but these methods require substantial labour and time for sampling and laboratory-rearing of progeny (Zhang *et al.* 2013; Wu 2014). Compared with bioassay-based methods, DNA screening can save labour and time because it can detect resistance genotypes in individual insects at any life-stage. DNA screens are also efficient in detecting recessive resistance alleles at low frequencies (Zhang *et al.* 2013). For example, DNA-based screening has been used to detect Cry1Ac resistance in the field for *r15* allele of *HaCad* in cotton bollworm which has been proved to confer non-recessive resistance to Cry1Ac (Zhang *et al.* 2013). The results showed that, during 2011–12, the *r15* allele frequency was estimated to be 0.0027 in northern China, where Bt cotton is intensively planted, while there was no *r15* allele detected in cotton bollworm collected from the Xinjiang Uygur Autonomous Region, where the planting of Bt cotton is limited.

The use of DNA screening for monitoring Bt resistance is still in its infancy for two main reasons: the diverse mechanisms of resistance in field populations make it difficult to accurately predict which mechanism may be operative in a specific field, and multiple mutations occurring at a single resistance locus complicate the DNA screen (Wu 2014).

Resistance management strategies in China

As discussed above, the evolution of resistance by target pests is the major problem threatening the sustainable use of IRGE crops. To delay and manage resistance outbreaks in the field, multiple strategies are used, including co-expression of multiple *cry* genes in the same plant (pyramided plants), high dose of Bt toxin combined with the application of refuge (high-dose/refuge strategy), and pyramids of cotton lines combining protection from a Bt toxin and RNA interference (RNAi) (Li *et al.* 2017; Ni *et al.* 2017).

Among these IR management strategies, the high-dose/refuge strategy is the most commonly used worldwide (Li *et al.* 2017; Tabashnik and Carrière 2017). The idea is that a certain fraction of the resistant individuals that survive on Bt crops will mate with the relatively abundant susceptible individuals in the refuges and thus produce Bt-susceptible heterozygous offspring. The offspring of such matings will be killed by the high dose of toxin expressed in the Bt crops. Generally, the refuges are either structured and mandated or unstructured and natural. The structured refuges contain the same crop as the Bt crop, but in a non-Bt form, while the unstructured and natural refuges contain other crops as host plants of the target pests, and even contain grasses that are naturally present in the cropping system around the Bt crops (Carrière *et al.* 2015; Li *et al.* 2017). In most cases, structured mandated refuges are most common in developed countries where monocropping is practised, while natural refuges are common in China and other developing countries, where farm size is small, mixed cropping a norm and multiple cropping quite frequent (Li *et al.* 2017).

In northern China, field-evolved resistance in cotton bollworm to Cry1Ac was detected less than eight years after Bt cotton planting exceeded 80% of the total area planted with cotton (Zhang *et al.* 2011). In the Yangtze River Valley, the first evidence of pink bollworm resistance to Cry1Ac was detected in 2008, eight years after Bt cotton was introduced there (Wan *et al.* 2012). Nevertheless, no widespread failure of Bt cotton to control the two insects has been reported. The sustained susceptibility of cotton bollworm in north-western China is attributed to the limited adoption of Bt cotton in this area (Wang *et al.* 2019c; Zhang *et al.* 2011). For instance, the area planted with Bt cotton in 2012 in Xinjiang Uygur Autonomous Region was 53% of cotton area in that region; the area planted with Bt cotton in six provinces of northern China increased from 50% in 2000 to 91% in 2004 and to nearly 100% of total area planted with cotton in 2014 (Wang *et al.* 2019c; Zhang *et al.* 2018). The foremost resistance management strategy in northern China is to rely on natural refuges of other host plants, such as wheat, corn, soybean and peanut (Jin *et al.* 2015). With the future commercialisation of Bt corn, the key natural refuge plants for cotton bollworm will be reduced. Under this scenario, the pest may quickly develop resistance to Bt corn and Bt cotton (Han *et al.* 2016). Moreover, the natural refuge strategy is based on the recessive nature of the resistance trait in the target pest. The monitoring of field populations of cotton bollworm in northern China revealed that non-recessive resistance increased faster than recessive resistance (Jin *et al.* 2015). This suggests that the natural refuge strategy may gradually become less effective in delaying the resistance of cotton bollworm to Bt cotton.

Pink bollworm is not the major pest targeted by Bt Cry1Ac cotton throughout China, except in the Yangtze River Valley (Wan *et al.* 2012). In contrast to cotton bollworm, the pink bollworm is an oligophagous pest feeding almost exclusively on cotton. The natural refuge strategy is therefore not applicable against pink bollworm. With the increased percentage of cotton area planted with Bt cotton in the Yangtze River Valley, it was expected that this species would quickly develop resistance to Bt cotton. In fact, Wan *et al.* (2012) reported significantly higher resistance in the pink bollworm to Cry1Ac in 2008–10 than in 2005–07, and predicted rapid increase in resistance in pink bollworm in the Yangtze River Valley after 2010. Unexpectedly, this prediction did not become a reality. On the contrary, the percentage of pink bollworm populations with one or more larvae surviving at the diagnostic concentration declined from 56% in 2008–10 to 0% in 2011–15 (Wan *et al.* 2017). This unexpected decrease in resistance has been attributed to a serendipitously implemented seed mixture strategy in the area. In practice, production of F1 seeds requires costly hand-pollination; to reduce costs, seed producers crossed Bt cotton with non-Bt cotton to create F1 hybrids that produce Bt toxin. Subsequent self-pollination by F1 hybrids creates F2 seeds for sale that are expected to consist of 25% Bt homozygotes and 50% hemizygotes that produce Bt toxin and 25% non-Bt homozygotes that produce no Bt toxins (Zhang *et al.* 2000). Wan *et al.* (2017) found that F2 seeds had been popularly planted in the Yangtze River Valley since 2010. Thus, a seed mixture produced by F1 hybrids generates a built-in refuge with roughly 25% non-Bt plants that are randomly interspersed in cotton fields. Eleven years of field monitoring data showed that the planting of F2 cotton seeds delayed or even reversed the development of pink bollworm resistance to Bt cotton in the Yangtze River cotton region (Carrière *et al.* 2016; Wan *et al.* 2017).

In the USA and Australia, the use of structured non-Bt crops as refuges is quite feasible and effective (Li *et al.* 2017; Tabashnik and Carrière 2017; Wilson *et al.* 2018). In Arizona, for example, refuges with non-Bt cotton represented an annual mean of 37% of the total cotton area each year during 1997–2005, which was when the first generation of Bt cotton, which produced only Bt Cry1Ac toxin, was planted. These non-Bt cotton refuges sustained the susceptibility of pink bollworm to Bt cotton (Tabashnik *et al.* 2012). From 2006 to 2009 in Arizona, cotton expressing both Cry1Ac and Cry2Ab was planted and the mean percentage of non-Bt cotton decreased to 7%, such that large areas planted with Bt cotton lacked refuges. Monitoring data indicated that there was no increase in resistance. The sustained susceptibility of pink bollworm to Bt cotton is partly attributed to the release of sterile moths during 2006–09, which greatly reduced pink bollworm numbers in Arizona (Tabashnik *et al.* 2010, 2012). In Australia, growers were required to cap Bt cotton cultivation at 30% of total cotton area from 1997 to 2003, when the first generation of Bt cotton was planted for resistance management in view of monocropping over a large scale, thereby leaving remaining 70% as refuge (Wilson *et al.* 2018). The susceptibility of Australian bollworm to Cry1Ac cotton was maintained until Cry1Ac cotton was replaced completely by pyramided cotton that produced both Cry1Ac and Cry2Ab, in 2004. With the introduction of pyramided cotton, the non-Bt cotton refuge requirement was reduced to 10%. The combined use of refuges and pyramided cotton has prevented an increase in Australian bollworm resistance to Bt cotton in Australia (Wilson *et al.* 2018). However, in India, although non-Bt refuges were mandated by the government, the mandate was poorly executed due to the small land holdings. As expected, pink bollworm resistance to Bt cotton expressing Cry1Ac was detected in 2008, six years after the first planting (Dhurua and Gujar 2011).

The monitoring of Bt resistance in cotton bollworm and pink bollworm in China has forewarned the necessity for implementation of proactive resistance management. One option to counter resistance is to switch to Bt cotton that produces two or more Bt toxins (pyramided Bt cotton). Current pyramided Bt crops produce two or more different toxins that belong to either the Cry or to the *Vip* + Cry protein families (Carrière *et al.* 2016; Chapter 5 in this book). Since their first planting in 2003, pyramided crops have become increasingly prevalent in the USA and other countries (Carrière *et al.* 2016; Tabashnik and Carrière 2017). However, they may also have a shortfall (Brévault *et al.* 2013). For example, pyramided Bt cotton producing Cry1Ac and Cry2Ab, first commercialised in 2006, rapidly occupied more than 80% of Bt cotton area in 2008 and then accounted for 90–96% of Bt cotton in India from 2012 – with a mosaic of single-toxin Bt events (Choudhary and Gaur 2015). This may create the opportunity for target pests to evolve cross-resistance or multiple resistance. A laboratory-selected pink bollworm strain with high resistance to Cry2Ab, for example, also showed high cross-resistance to Cry1Ac (Tabashnik *et al.* 2009). In India, pink bollworm resistance to both toxins has spread to many localities in the central and southern cotton-growing regions (Tabashnik and Carrière, 2019). In addition, two field-collected Cry1Ac-resistant cotton bollworm strains from northern China showed low but significant cross-resistance to Cry2Ab (Jin *et al.* 2013). To minimise the chance that mutations in a single gene could confer resistance to both Cry1Ac and Cry2Ab, the best approach is to completely replace existing transformation events with new, preferably unrelated ones like in Australia across a short time-frame, without leaving a mosaic of events over the period. Since 2016, Australia has implemented three-toxin Bt cotton (Bollgard 3®) that produces Cry1Ac, Cry2Ab and Vip3A. In China, scientists have developed next-generation IRGE cotton that produces Bt Cry1Ac and RNAi, which causes 91–93% mortality of cotton bollworm in the field (Ni *et al.* 2017). The current commercial Bt cotton varieties in China produce only Cry1A toxin. To guarantee the future efficacy of Bt cotton against target insects in China, growers must switch to the next-generation Bt cotton (Li *et al.* 2020). Other options for countering resistance to Bt cotton are to increase the planting of non-Bt cotton and to integrate other pest management tactics including sterile moth release in Bt cotton fields (Tabashnik *et al.* 2010; Wan *et al.* 2012).

Conclusion

Long-term growing of Bt crops worldwide has proved efficacious for controlling target pests, and has brought significant benefits to humans and the environment. However, the development of resistance in the target pests threatens these benefits of Bt crops. Hence, resistance management is an essential component of Bt crop development in many countries. The success of management strategies depends upon technology *per se*, the biology of target pests and associated agroecological conditions. Evolving newer plant- as well as insect-based technologies will ensure the sustainability of Bt crops.

Acknowledgements

We are grateful to the editors and three anonymous reviewers for constructive comments on an early draft of the manuscript. This work was supported by the National GMO New Variety Breeding Program of the People's Republic of China (2016ZX08011-001).

References

An J, Gao Y, Lei C, Gould F, Wu K (2015) Monitoring cotton bollworm resistance to Cry1Ac in two counties of northern China during 2009–2013. *Pest Management Science* **71**, 377–382. doi:10.1002/ps.3807

Andow DA, Alstad D (1998) F2 screen for rare resistance alleles. *Journal of Economic Entomology* **91**, 572–578. doi:10.1093/jee/91.3.572

Angst BD, Marcozzi C, Magee AI (2001) The cadherin superfamily: diversity in form and function. *Journal of Cell Science* **114**, 629–641.

Brévault T, Heuberger S, Zhang M, Ellers-Kirk C, Ni X, Masson L, Li X, Tabashnik BE, Carrière Y (2013) Potential shortfall of pyramided transgenic cotton for insect resistance management. *Proceedings of the National Academy of Sciences of the United States of America* **110**, 5806–5811. doi:10.1073/pnas.1216719110

Broderick NA, Raffa KF, Handelsman J (2006) Midgut bacteria required for *Bacillus thuringiensis* insecticidal activity. *Proceedings of the National Academy of Sciences of the United States of America* **103**, 15196–15199. doi:10.1073/pnas.0604865103

Brookes G, Barfoot P (2018a) Farm income and production impacts of using GM crop technology 1996–2016. *GM Crops and Food: Biotechnology in Agriculture and the Food Chain* **9**, 59–89. doi:10.1080/21645698.2018.1464866
Brookes G, Barfoot P (2018b) Environmental impacts of genetically modified (GM) crop use 1996–2016: impacts on pesticide use and carbon emissions. *GM Crops and Food: Biotechnology in Agriculture and the Food Chain* **9**, 109–139. doi:10.1080/21645698.2018.1476792.
Caccia S, Hernández-Rodríguez CS, Mahon RJ, Downes S, James W, Bautsoens N, Van Rie J, Ferré J (2010) Binding site alteration is responsible for field-isolated resistance to *Bacillus thuringiensis* Cry2A insecticidal proteins in two *Helicoverpa* species. *PLoS One* **5**, e9975. doi:10.1371/journal.pone.0009975
Caccia S, Di Lelio I, La Storia A, Marinelli A, Varricchio P, Franzetti E, Banyuls N, Tettamanti G, Casartelli M, Giordana B (2016) Midgut microbiota and host immunocompetence underlie *Bacillus thuringiensis* killing mechanism. *Proceedings of the National Academy of Sciences of the United States of America* **113**, 9486–9491. doi:10.1073/pnas.1521741113
Carrière Y, Crickmore N, Tabashnik BE (2015) Optimizing pyramided transgenic Bt crops for sustainable pest management. *Nature Biotechnology* **33**, 161–168. doi:10.1038/nbt.3099
Carrière Y, Fabrick JA, Tabashnik BE (2016) Can pyramids and seed mixtures delay resistance to Bt crops? *Trends in Biotechnology* **34**, 291–302. doi:10.1016/j.tibtech.2015.12.011
Catarino R, Ceddia G, Areal FJ, Park J (2015) The impact of secondary pests on *Bacillus thuringiensis* (Bt) crops. *Plant Biotechnology Journal* **13**, 601–612. doi:10.1111/pbi.12363
Chandrashekar K, Gujar GT (2004) Development and mechanisms of resistance to *Bacillus thuringiensis* endotoxin Cry1Ac in the American bollworm, *Helicoverpa armigera* (Hübner). *Indian Journal of Experimental Biology* **42**, 164–173.
Choudhary B, Gaur K (2015) *Biotech cotton in India, 2002 to 2014*. ISAAA Series of Biotech Crop Profiles. ISAAA, Ithaca, NY.
Dhurua S, Gujar GT (2011) Field-evolved resistance to Bt toxin Cry1Ac in the pink bollworm, *Pectinophora gossypiella* (Saunders) (Lepidoptera: Gelechiidae), from India. *Pest Management Science* **67**, 898–903. doi:10.1002/ps.2127
Fabrick JA, Tabashnik BE (2012) Similar genetic basis of resistance to Bt toxin Cry1Ac in boll-selected and diet-selected strains of pink bollworm. *PLoS One* **7**, e35658. doi:10.1371/journal.pone.0035658
Fabrick JA, Wu Y (2015) Roles of insect midgut cadherin in Bt intoxication and resistance. In *Bt Resistance: Characterization and Strategies for GM Crops Producing Bacillus thuringiensis Toxins.* (Eds M Soberon, YL Gao and A Bravo) pp.69–86. CABI, Wallingford/Boston.
Fabrick JA, Ponnuraj J, Singh A, Tanwar RK, Unnithan GC, Yelich AJ, Li X, Carrière Y, Tabashnik BE (2014) Alternative splicing and highly variable cadherin transcripts associated with field-evolved resistance of pink bollworm to Bt cotton in India. *PLoS One* **9**, e97900. doi:10.1371/journal.pone.0097900
Ferré J, Van Rie J (2002) Biochemistry and genetics of insect resistance to *Bacillus thuringiensis. Annual Review of Entomology* **47**, 501–533. doi:10.1146/annurev.ento.47.091201.145234
Gahan LJ, Gould F, Heckel DG (2001) Identification of a gene associated with Bt resistance in *Heliothis virescens. Science* **293**, 857–860. doi:10.1126/science.1060949
Gunning RV, Dang HT, Kemp FC, Nicholson IC, Moores GD (2005) New resistance mechanism in *Helicoverpa armigera* threatens transgenic crops expressing *Bacillus thuringiensis* Cry1Ac toxin. *Applied and Environmental Microbiology* **71**, 2558–2563. doi:10.1128/AEM.71.5.2558-2563.2005
Guo S, Wang Y, Sun G, Jin S, Zhou T, Meng Z, Zhang R (2015) Twenty years of research and application of transgenic cotton in China. *Scientia Agricultura Sinica* **48**, 3372–3387.
Han L, Jiang X, Peng Y (2016) Potential resistance management for the sustainable use of insect-resistant genetically modified corn and rice in China. *Current Opinion in Insect Science* **15**, 139–143. doi:10.1016/j.cois.2016.04.004
Henry LM, Maiden MC, Ferrari J, Godfray HCJ (2015) Insect life history and the evolution of bacterial mutualism. *Ecology Letters* **18**, 516–525. doi:10.1111/ele.12425
ISAAA (2017) *Global Status of Commercialized Biotech/GM Crops in 2017: Biotech Crop Adoption Surges as Economic Benefits Accumulate in 22 Years.* ISAAA Briefs 53. ISAAA, Ithaca, NY.
Jiang W, Liang G, Lin Y, Shu C, Song F, Zhang J (2010) Comparison of midgut bacterial community between Bt-resistant and sensitive *Helicoverpa armigera. Acta Microbiologica Sinica* **50**, 828–834.

Jin L, Wei Y, Zhang L, Yang Y, Tabashnik BE, Wu Y (2013) Dominant resistance to Bt cotton and minor cross-resistance to Bt toxin Cry2Ab in cotton bollworm from China. *Evolutionary Applications* **6**, 1222–1235. doi:10.1111/eva.12099

Jin L, Zhang H, Lu Y, Yang Y, Wu K, Tabashnik BE, Wu Y (2015) Large-scale test of the natural refuge strategy for delaying insect resistance to transgenic Bt crops. *Nature Biotechnology* **33**, 169–174. doi:10.1038/nbt.3100

Jin M, Liao C, Chakrabarty S, Wu K, Xiao Y (2019) Comparative proteomics of peritrophic matrix provides an insight into its role in Cry1Ac resistance of cotton bollworm *Helicoverpa armigera*. *Toxins* **11**, 92. doi:10.3390/toxins11020092

Johnston PR, Crickmore N (2009) Gut bacteria are not required for the insecticidal activity of *Bacillus thuringiensis* toward the tobacco hornworm, *Manduca sexta*. *Applied and Environmental Microbiology* **75**, 5094–5099. doi:10.1128/AEM.00966-09

Jurat-Fuentes JL, Adang MJ (2004) Characterization of a Cry1Ac-receptor alkaline phosphatase in susceptible and resistant *Heliothis virescens* larvae. *European Journal of Biochemistry* **271**, 3127–3135. doi:10.1111/j.1432-1033.2004.04238.x

Jurat-Fuentes JL, Crickmore N (2017) Specificity determinants for Cry insecticidal proteins: insights from their mode of action. *Journal of Invertebrate Pathology* **142**, 5–10. doi:10.1016/j.jip.2016.07.018

Jurat-Fuentes JL, Karumbaiah L, Jakka SR, Ning C, Liu C, Wu K, Jackson J, Gould F, Blanco C, Portilla M, Perera O, Adang M (2011) Reduced levels of membrane-bound alkaline phosphatase are common to lepidopteran strains resistant to Cry toxins from *Bacillus thuringiensis*. *PLoS One* **6**, e17606. doi:10.1371/journal.pone.0017606

Li S, Hussain F, Unnithan GC, Dong S, UlAbdin Z, Gu S, Mathew LG, Fabrick JA, Ni X, Carrière Y, Tabashnik BE, Li X (2019) A long non-coding RNA regulates cadherin transcription and susceptibility to Bt toxin Cry1Ac in pink bollworm, *Pectinophora gossypiella*. *Pesticide Biochemistry and Physiology* **158**, 54–60. doi:10.1016/j.pestbp.2019.04.007

Li Y, Gao Y, Wu K (2017) Function and effectiveness of natural refuge in IRM strategies for Bt crops. *Current Opinion in Insect Science* **21**, 1–6. doi:10.1016/j.cois.2017.04.007

Li Y, Hallerman EM, Wu K, Peng Y (2020) Insect-resistant genetically engineered crops in China: development, application, and prospects for use. *Annual Review of Entomology* **65**, 273–292. doi:10.1146/annurev-ento-011019-025039

Liu C, Xiao Y, Li X, Oppert B, Tabashnik BE, Wu K (2015) Cis-mediated down-regulation of a trypsin gene associated with Bt resistance in cotton bollworm. *Scientific Reports* **4**, 7219. doi:10.1038/srep07219

Lu M, Hulcr J, Sun J (2016) The role of symbiotic microbes in insect invasions. *Annual Review of Ecology Evolution and Systematics* **47**, 487–505. doi:10.1146/annurev-ecolsys-121415-032050

Lu Y, Wu K (2011) Mirid bugs in China: pest status and management strategies. *Outlooks on Pest Management* **22**, 248–252. doi:10.1564/22dec02

Lu Y, Wu K, Jiang Y, Xia B, Li P, Feng H, Wyckhuys KA, Guo Y (2010) Mirid bug outbreaks in multiple crops correlated with wide-scale adoption of Bt cotton in China. *Science* **328**, 1151–1154. doi:10.1126/science.1187881

Lu Y, Wu K, Jiang Y, Guo Y, Desneux N (2012) Widespread adoption of Bt cotton and insecticide decrease promotes biocontrol services. *Nature* **487**, 362–365. doi:10.1038/nature11153

Luo S, Wang G, Liang G, Wu KM, Bai L, Ren X, Guo Y (2006) Binding of three Cry1A toxins in resistant and susceptible strains of cotton bollworm (*Helicoverpa armigera*). *Pesticide Biochemistry and Physiology* **85**, 104–109. doi:10.1016/j.pestbp.2005.11.003

Malthankar PA, Gujar GT (2014) Association of Cry2Ab resistance with alkaline phosphatase in the pink bollworm, *Pectinophora gossypiella* (Saunders) (Lepidoptera: Gelechiidae). *Biopesticides International* **10**, 143–151.

Morin S, Biggs RW, Sisterson MS, Shriver L, Ellers-Kirk C, Higginson D, Holley D, Gahan LJ, Heckel DG, Carrière Y, Dennehy TJ, Brown JK, Tabashnik BE (2003) Three cadherin alleles associated with resistance to *Bacillus thuringiensis* in pink bollworm. *Proceedings of the National Academy of Sciences of the United States of America* **100**, 5004–5009. doi:10.1073/pnas.0831036100

Nair R, Kalia V, Aggarwal KK, Gujar GT (2013) Variation in the cadherin gene sequence of Cry1Ac susceptible and resistant *Helicoverpa armigera* (Lepidoptera: Noctuidae) and the identification of mutant alleles in resistant strains. *Current Science* **104**, 215–223.

Ni M, Ma W, Wang X, Gao M, Dai Y, Wei X, Zhang L, Peng Y, Chen S, Ding L, Tian Y, Li J, Wang H, Wang X, Xu G, Guo W, Yang Y, Wu Y, Heuberger S, Tabashnik BE, Zhang T, Zhu Z (2017) Next-generation transgenic cotton: pyramiding RNAi and Bt counters insect resistance. *Plant Biotechnology Journal* **15**, 1204–1213. doi:10.1111/pbi.12709

Ning C, Wu K, Liu C, Gao Y, Jurat-Fuentes JL, Gao X (2010) Characterization of a Cry1Ac toxin-binding alkaline phosphatase in the midgut from *Helicoverpa armigera* (Hubner) larvae. *Journal of Insect Physiology* **56**, 666–672. doi:10.1016/j.jinsphys.2010.02.003

Oppert B (1999) Protease interactions with *Bacillus thuringiensis* insecticidal toxins. *Archives of Insect Biochemistry and Physiology* **42**, 1–12. doi:10.1002/(SICI)1520-6327(199909)42:1<1::AID-ARCH2>3.0.CO;2-%23

Pardo-López L, Soberon M, Bravo A (2013) *Bacillus thuringiensis* insecticidal three-domain Cry toxins: mode of action, insect resistance and consequences for crop protection. *FEMS Microbiology Reviews* **37**, 3–22. doi:10.1111/j.1574-6976.2012.00341.x

Rajagopal R, Arora N, Sivakumar S, Rao NG, Nimbalkar SA, Bhatnagar RK (2009) Resistance of *Helicoverpa armigera* to Cry1Ac toxin from *Bacillus thuringiensis* is due to improper processing of the protoxin. *Biochemical Journal* **419**, 309–316. doi:10.1042/BJ20081152

Raymond B, Johnston PR, Wright DJ, Ellis RJ, Crickmore N, Bonsall MB (2009) A mid-gut microbiota is not required for the pathogenicity of *Bacillus thuringiensis* to diamondback moth larvae. *Environmental Microbiology* **11**, 2556–2563. doi:10.1111/j.1462-2920.2009.01980.x

Tabashnik BE, Carrière Y (2017) Surge in insect resistance to transgenic crops and prospects for sustainability. *Nature Biotechnology* **35**, 926–935. doi:10.1038/nbt.3974

Tabashnik BE, Carrière Y (2019) Global patterns of resistance to Bt crops highlighting pink bollworm in the United States, China, and India. *Journal of Economic Entomology* **112**, 2513–2523. doi:10.1093/jee/toz173

Tabashnik BE, Unnithan GC, Masson L, Crowder DW, Li X, Carrière Y (2009) Asymmetrical cross-resistance between *Bacillus thuringiensis* toxins Cry1Ac and Cry2Ab in pink bollworm. *Proceedings of the National Academy of Sciences of the United States of America* **106**, 11889–11894. doi:10.1073/pnas.0901351106

Tabashnik BE, Sisterson MS, Ellsworth PC, Dennehy TJ, Antilla L, Liesner L, Whitlow M, Staten RT, Fabrick JA, Unnithan GC, Yelich AJ, Ellers-Kirk C, Harpold VS, Li X, Carrière Y (2010) Suppressing resistance to Bt cotton with sterile insect releases. *Nature Biotechnology* **28**, 1304–1307. doi:10.1038/nbt.1704

Tabashnik BE, Morin S, Unnithan GC, Yelich AJ, Ellers-Kirk C, Harpold VS, Sisterson MS, Ellsworth PC, Dennehy TJ, Antilla L, Liesner L, Whitlow M, Staten RT, Fabrick JA, Li X, Carrière Y (2012) Sustained susceptibility of pink bollworm to Bt cotton in the United States. *GM Crops and Food: Biotechnology in Agriculture and the Food Chain* **3**, 194–200. doi:10.4161/gmcr.20329

Wan P, Huang Y, Wu H, Huang M, Cong S, Tabashnik BE, Wu K (2012) Increased frequency of pink bollworm resistance to Bt toxin Cry1Ac in China. *PLoS One* **7**, e29975. doi:10.1371/journal.pone.0029975

Wan P, Xu D, Cong S, Jiang Y, Huang Y, Wang J, Wu H, Wang L, Wu K, Carrière Y, Mathias A, Li X, Tabashnik BE (2017) Hybridizing transgenic Bt cotton with non-Bt cotton counters resistance in pink bollworm. *Proceedings of the National Academy of Sciences of the United States of America* **114**, 5413–5418. doi:10.1073/pnas.1700396114

Wang J, Wang H, Liu S, Liu L, Tay WT, Walsh TK, Yang Y, Wu Y (2017) CRISPR/Cas9 mediated genome editing of *Helicoverpa armigera* with mutations of an ABC transporter gene *HaABCA2* confers resistance to *Bacillus thuringiensis* Cry2A toxins. *Insect Biochemistry and Molecular Biology* **87**, 147–153. doi:10.1016/j.ibmb.2017.07.002

Wang L, Ma Y, Wan P, Liu K, Xiao Y, Wang J, Cong S, Xu D, Wu K, Fabrick JA, Li X, Tabashnik BE (2018) Resistance to *Bacillus thuringiensis* linked with a cadherin transmembrane mutation affecting cellular trafficking in pink bollworm from China. *Insect Biochemistry and Molecular Biology* **94**, 28–35. doi:10.1016/j.ibmb.2018.01.004

Wang L, Ma Y, Guo X, Wan P, Liu K, Cong S, Wang J, Xu D, Xiao Y, Li X, Tabashnik BE, Wu K (2019a) Pink bollworm resistance to Bt toxin Cry1Ac associated with an insertion in cadherin exon 20. *Toxins* **11**, 186. doi:10.3390/toxins11040186

Wang L, Wang J, Ma Y, Wan P, Liu K, Cong S, Xiao Y, Xu D, Wu K, Fabrick JA, Li X, Tabashnik BE (2019b) Transposon insertion causes cadherin mis-splicing and confers resistance to Bt cotton in pink bollworm from China. *Scientific Reports* **9**, 7479. doi:10.1038/s41598-019-43889-x

Wang P, Ma J, Head GP, Xia D, Li J, Wang H, Yang M, Xie Z, Zalucki MP, Lu Z (2019c) Susceptibility of *Helicoverpa armigera* to two Bt toxins, Cry1Ac and Cry2Ab, in north-western China: toward developing an IRM strategy. *Journal of Pest Science* **92**, 923–931. doi:10.1007/s10340-018-1047-0

Wilson LJ, Whitehouse MEA, Herron GA (2018) The management of insect pests in Australian cotton: an evolving story. *Annual Review of Entomology* **63**, 215–237. doi:10.1146/annurev-ento-020117-043432

Wu K (2007) Monitoring and management strategy for *Helicoverpa armigera* resistance to Bt cotton in China. *Journal of Invertebrate Pathology* **95**, 220–223. doi:10.1016/j.jip.2007.03.012

Wu Y (2014) Detection and mechanisms of resistance evolved in insects to Cry toxins from *Bacillus thuringiensis*. In *Advances in Insect Physiology*. (Eds TS Dhadialla and SS Gill), pp. 297–342. Academic Press, Oxford.

Wu KM, Guo YY (2005) The evolution of cotton pest management practices in China. *Annual Review of Entomology* **50**, 31–52. doi:10.1146/annurev.ento.50.071803.130349

Wu KM, Lu YH, Feng HQ, Jiang YY, Zhao JZ (2008) Suppression of cotton bollworm in multiple crops in China in areas with Bt toxin containing cotton. *Science* **321**, 1676–1678. doi:10.1126/science.1160550

Xiao Y, Wu K (2019) Recent progress on the interaction between insects and *Bacillus thuringiensis* crops. *Philosophical Transactions of the Royal Society of London. Series B, Biological Sciences* **374**, 20180316. doi:10.1098/rstb.2018.0316

Xiao Y, Zhang T, Liu C, Heckel DG, Li X, Tabashnik BE, Wu K (2015) Mis-splicing of the *ABCC2* gene linked with Bt toxin resistance in *Helicoverpa armigera*. *Scientific Reports* **4**, 6184. doi:10.1038/srep06184

Xiao Y, Liu K, Zhang D, Gong L, He F, Soberon M, Bravo A, Tabashnik BE, Wu K (2016) Resistance to *Bacillus thuringiensis* mediated by an ABC transporter mutation increases susceptibility to toxins from other bacteria in an invasive insect. *PLoS Pathogens* **12**, e1005450. doi:10.1371/journal.ppat.1005450

Xiao Y, Dai Q, Hu R, Pacheco S, Yang Y, Liang G, Soberon M, Bravo A, Liu K, Wu K (2017) A single point mutation resulting in cadherin mislocalization underpins resistance against *Bacillus thuringiensis* toxin in cotton bollworm. *The Journal of Biological Chemistry* **292**, 2933–2943. doi:10.1074/jbc.M116.768671

Xu X, Yu L, Wu Y (2005) Disruption of a cadherin gene associated with resistance to Cry1Ac δ-endotoxin of *Bacillus thuringiensis* in *Helicoverpa armigera*. *Applied and Environmental Microbiology* **71**, 948–954. doi:10.1128/AEM.71.2.948-954.2005

Yang Y, Chen H, Wu Y, Yang Y, Wu S (2007) Mutated cadherin alleles from a field population of *Helicoverpa armigera* confer resistance to *Bacillus thuringiensis* toxin Cry1Ac. *Applied and Environmental Microbiology* **73**, 6939–6944. doi:10.1128/AEM.01703-07

Yang YH, Yang YJ, Gao WY, Guo JJ, Wu YH, Wu YD (2009) Introgression of a disrupted cadherin gene enables susceptible *Helicoverpa armigera* to obtain resistance to *Bacillus thuringiensis* toxin Cry1Ac. *Bulletin of Entomological Research* **99**, 175–181. doi:10.1017/S0007485308006226

Zhang BH, Guo TL, Wang QL (2000) Inheritance and segregation of exogenous genes in transgenic cotton. *Journal of Genetics* **79**, 71–75. doi:10.1007/BF02728948

Zhang R, Wang Y, Meng Z, Sun G, Guo S (2007) Retrospect and prospect of research on Chinese transgenic insecticidal cotton. *Journal of Agricultural Science and Technology* **9**, 32–42.

Zhang S, Cheng H, Gao Y, Wang G, Liang G, Wu K (2009) Mutation of an aminopeptidase N gene is associated with *Helicoverpa armigera* resistance to *Bacillus thuringiensis* Cry1Ac toxin. *Insect Biochemistry and Molecular Biology* **39**, 421–429. doi:10.1016/j.ibmb.2009.04.003

Zhang H, Yin W, Zhao J, Jin L, Yang Y, Wu S, Tabashnik BE, Wu Y (2011) Early warning of cotton bollworm resistance associated with intensive planting of Bt cotton in China. *PLoS One* **6**, e22874. doi:10.1371/journal.pone.0022874

Zhang H, Tian W, Zhao J, Jin L, Yang J, Liu C, Yang Y, Wu S, Wu K, Cui J, Tabashnik BE, Wu Y (2012a) Diverse genetic basis of field-evolved resistance to Bt cotton in cotton bollworm from China.

Proceedings of the National Academy of Sciences of the United States of America **109**, 10275–10280. doi:10.1073/pnas.1200156109

Zhang H, Wu S, Yang Y, Tabashnik BE, Wu Y (2012b) Non-recessive Bt toxin resistance conferred by an intracellular cadherin mutation in field-selected populations of cotton bollworm. *PLoS One* **7**, e53418. doi:10.1371/journal.pone.0053418

Zhang H, Tang M, Yang F, Yang Y, Wu Y (2013) DNA-based screening for an intracellular cadherin mutation conferring non-recessive Cry1Ac resistance in field populations of *Helicoverpa armigera*. *Pesticide Biochemistry and Physiology* **107**, 148–152. doi:10.1016/j.pestbp.2013.06.007

Zhang H, Yu S, Shi Y, Yang Y, Fabrick JA, Wu Y (2017) Intra- and extracellular domains of the *Helicoverpa armigera* cadherin mediate Cry1Ac cytotoxicity. *Insect Biochemistry and Molecular Biology* **86**, 41–49. doi:10.1016/j.ibmb.2017.05.004

Zhang W, Lu Y, van der Werf W, Huang J, Wu F, Zhou K, Deng X, Jiang Y, Wu K, Rosegrant MW (2018) Multidecadal, county-level analysis of the effects of land use, Bt cotton, and weather on cotton pests in China. *Proceedings of the National Academy of Sciences of the United States of America* **115**, E7700–E7709. doi:10.1073/pnas.1721436115

Zhao J, Jin L, Yang Y, Wu Y (2010) Diverse cadherin mutations conferring resistance to *Bacillus thuringiensis* toxin Cry1Ac in *Helicoverpa armigera*. *Insect Biochemistry and Molecular Biology* **40**, 113–118. doi:10.1016/j.ibmb.2010.01.001

Zhao M, Yuan X, Wei J, Zhang W, Wang B, Khaing MM, Liang G (2017) Functional roles of cadherin, aminopeptidase-N and alkaline phosphatase from *Helicoverpa armigera* (Hübner) in the action mechanism of *Bacillus thuringiensis* Cry2Aa. *Scientific Reports* **7**, 46555. doi:10.1038/srep46555

7

Bt cotton in Pakistan

Hafiza Masooma Naseer Cheema, Asif Ali Khan and Kashif Noor

Introduction

Biotech crops were grown on the largest ever area of 191.7 million ha in 2018 in 26 countries, from the initial area of 1.7 million ha in 1996 when they were first commercialised. Biotech cotton constitutes ~72.6% of total cotton area (~32.9 million ha) in the world and is grown in 15 countries. Pakistan grows biotech cotton on ~2.8 million ha and is ranked eighth in biotech area in the world (ISAAA 2018). It is the fourth largest producer of cotton, contributing ~7.6% of the world's cotton production.

Biotech cotton is essentially insect-protective transgenic cotton expressing *cry1ac* and/or *cry2ab* genes isolated from naturally occurring soil-borne spore-forming bacteria *Bacillus thuringiensis* (Bt) and/or herbicide-tolerant (HT) transgenic cotton with genes sourced from bacteria like *Agrobacterium* or *Streptomyces*. Bt produces Cry toxins which are highly toxic to various insects. Of the toxins known, Cry1 and Cry2 are specific to Lepidoptera with high specificity among different species of lepidopterans. In 1996, a single-toxin Bt cotton producing only Cry1Ac toxin was developed by Monsanto to control bollworms, followed by dual-toxin Bt cotton, called Bollgard II® producing Cry1Ac and Cry2Ab, in 2004 to delay evolution of resistance in the target heliothine pests and enhance the spectrum of activity to include other lepidopterans. Bt cotton is stacked with HT trait to provide for an efficient weed management, especially with glyphosate, and was commercialised in 2012. In 2015, Monsanto developed Bollgard 3® that produces three toxins – Cry1Ac, Cry2Ab and Vip3 – to be more useful in resistance management, with Australia in the lead to grow it. Most developed countries (USA, Australia, Brazil) are using these stacks (Bollgard II®/Roundup Ready Flex) to optimise crop protection against insect pests as well as weeds, and to reduce related costs. On the other hand, India, China and Pakistan, with an area of 52.5% of total world's cotton area (32.9 million ha), grow insect-protective Bt cotton with dual-toxin Bt cotton on only~11.6 million ha in India, and single-toxin Bt cotton on about 2.9 million ha in China and 2.8 million ha in Pakistan. And this is reflected in the fact that of 24.8 million ha of global biotech cotton, ~75% of Bt cotton is insect-resistant (IR), 22% is stacked for IR and HT while 3% is exclusively for HT (ISAAA 2017).

Cotton in Pakistan

Cotton cultivation originated more than 5000 years ago in the Indus Valley civilisation, which covers the present-day Pakistan and north-western India. The soil for cotton cultivation ranges from sandy to clay loam. Climate varies a lot, with temperature exceeding 32°C in some zones and rainfall ranging from 150 mm to 750 mm annually. The crop is quite vulnerable to climate change. Cotton is grown during the rainy season; seeds are sown from April to June and harvested from September to December. Pre-monsoon cotton needs irrigation and is harvested early. Cotton cultivation has many abiotic and biotic constraints to realising its full yield potential. Cotton is grown in small farm holdings as 86% of farm areas are <5 ha, making it quite vulnerable to various factors due to limited investment for increasing productivity. About 145 insects and mites attack cotton throughout its crop season (Huque 1994). These biotic constraints are mainly bollworm complex – American bollworm (*Helicoverpa armigera*), spotted bollworm (*Earias insulana*), pink bollworm (*Pectinophora gossypiella*) – and to some extent defoliator tobacco caterpillar (*Spodoptera litura*). Bollworms feed on squares, flowers and bolls, causing direct yield loss (Abro *et al.* 2003, 2004; Hamed and Nadeem 2010a; Wakil *et al.* 2012), while sucking pests like aphids, whitefly, jassids, mealybugs and mites feed essentially during vegetative growth of crop and spread diseases, causing yield reduction (Nizamani *et al.* 2002; Amjad and Aheer 2007; Shahid *et al.* 2012). It is estimated that pesticides worth about US$300 million are used in Pakistan for crop protection, 80% of them for cotton crops alone. Pakistan grows mostly upland cotton (*Gossypium hirsutum*) in Punjab (76%), followed by Sindh (22.5%) and to some extent Balochistan (1.5%) (Figure 7.1). Over the last two decades, cotton area has remained constant at 2.4–3.2 million ha. Most cotton area is rainfed supplemented with irrigation, with production of ~9.9–14.5 million bales (170 kg/bale) and productivity of 572–815 kg lint/ha (Figure 7.2). Increases in productivity (40%) and production (46%) were higher than land area increase (33%), due to crop improvement and protection technologies. Pakistan is the fourth largest producer and fifth largest consumer of cotton, and the largest producer of cotton yarn. About 1.3–1.6 million farmers cultivate cotton on ~2.7–3 million ha which covers 15% of the cultivated area of country. On the whole, ~30–40% of the cotton is consumed at the domestic level and the remainder is exported as raw material. Cotton production supports the largest industrial sector of Pakistan comprising ~400 textile mills,700 spindles, 2700 looms in the mill sector and 2500 in the non-mill sector, 700 knitwear units, 4000 garments units and 5000 oil expellers. By all measures, it is a vital economic sector of the country and has enormous connections. Cotton contributes ~0.8% of the nation's gross domestic product and ~4.5% of the value addition in agriculture (Govt of Pakistan 2019). That Economic Survey showed that cotton production has marginally declined over the years due to contraction of area as a result of unfavourable prices, climatic factors and pink bollworm, sucking and other pests. The decline in cotton production was as much as 29% in 2015–16 and 17.5% in 2018–19 compared to their respective previous years. In 2019–20, production is likely to be 3.44% lower (~10.2 million bales) despite a 14.2% increase in area (of ~2.78 million ha) compared to the previous year. The decline in yield is attributed to failure to create a conducive environment for the introduction of new seed varieties, following the enactment of the *Plant Breeders' Rights Act 2016* (Rana 2019).

Indiscriminate and extensive use of insecticides during the pre-Bt cotton era has led to extensive resistance evolution in target pests including cotton bollworm, spotted bollworm and pink bollworm (Karim 2000; Ahmad *et al.* 2008; Ahmad and Iqbal 2009; Qayyum *et al.* 2015). Furthermore, regular incidence of polyphagous cotton bollworm has affected other crops (Karim 2000). Extensive pesticide use has also affected environmental quality and led to tragic events such as affecting the health of farmers and others (Jabbar and Mallick 1994; Tariq *et al.* 2007; Ali *et al.* 2010).

Bt cotton in Pakistan

Prior to the official release of Bt cotton in 2010, there were reports of illegal spurious Bt cotton being planted for at least three years in Pakistan. This is possibly due to unofficial access to Bt

Figure 7.1 Major cotton growing districts of Pakistan.

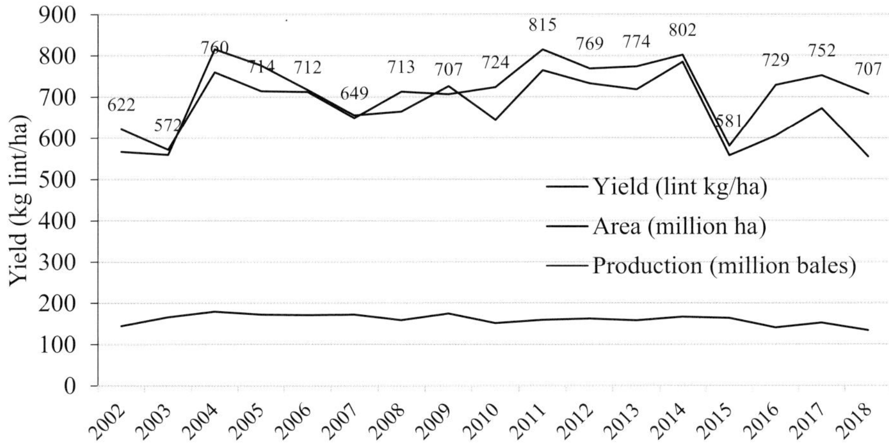

Figure 7.2 Area (million ha), production (million bales) and yield (kg lint/ha) of cotton, 2002–2018. Source: Based on Pakistan Economy Surveys.

cotton varieties by some farmers and the stealthy diffusion of Bt cotton in the cotton seed market, leading to adulteration and poor-quality seed dissemination (Nazli *et al.* 2012; Cheema *et al.* 2016). A Pakistan Economic Survey stated that ~80% of the cotton area in Punjab and Sindh in the 2009 monsoon season was under illegal Bt cotton (Australian Bt) with high susceptibility to cotton leaf curl virus (CLCV) infection (Govt of Pakistan 2010). Despite official release of Bt cotton, a faulty regulatory system and lack of quality seeds led to the proliferation of substandard cotton varieties that did not enhance yields as expected (Rana *et al.* 2013; Spielman *et al.* 2015).

In 2010, IR Bt cotton was officially released in Pakistan by the National Biosafety Committee (NBC), operating under the auspices of the Pakistan Environmental Protection Agency (EPA) as per the provisions of National Biosafety Rules and Guidelines 2005. Eight Bt cotton varieties based upon Monsanto 531 event expressing *cry1Ac* gene – ARS-703, ARS-802 (Ali Akbar Seeds, Multan), IR-3701, IR-1524 (NIBGE, Faisalabad), MG-6, Sitara-008 (Nawab Gurmani Foundation, Multan), FH-113 (CRI, AARI, Faisalabad), Neelam-121 (Neelam Seeds, Multan) – were approved at that time. Two hybrids – HyBt GN-2085 and HyBtGN-31 – were also approved. In 2012, another set of eight Bt cotton varieties – Tarzen-1, MNH-886, NS-141, FH-114, IR-NIBGE-3, IR-NIBGE-901, CIM-598 and VH-259 – were approved for cultivation. Two Bt cotton varieties – NIAB-886 and NIAB-777 – were developed by the Nuclear Institute of Agriculture and Biology, Faisalabad. In view of single-toxin Bt cotton being highly likely to be susceptible to insects evolving resistance, 10 dual-toxin Bt cotton (producing Cry1Ac and Cry2A) varieties – CA-12, CA-904, CA-919, CEMB-02, CEMB-22, CEMB-33, CEMB-44, CEMB-55, CEMB-66 and CR-333 – began development (Azam *et al.* 2013). The Pakistan Central Cotton Committee, which is the national institution that oversees cotton research and development, has also developed some dual-toxin Bt cotton varieties: Bt CIM-595, Bt CIM-599 and Bt CIM-602 in 2013; and Bt CIM-600, Bt Cyto-177 and Bt CRIS-508 in 2017. Giving further impetus to Bt cotton development, 85 Bt cotton varieties (44 from private and 41 from public institutions) were tested for their agronomic suitability in national coordinated trials in 2019.

Bt cotton transgenesis was carried out via *Agrobacterium*-mediated transformation with locally sourced Bt genes like *cry1Ab* and *cry1Ac* using the local variety MNH093 (Rashid *et al.* 2008; Khan *et al.* 2011, 2013). Bakhsh *et al.* (2010, 2011) developed transgenic cotton with two *cry1Ac* and *cry2A* genes. Puspito *et al.* (2015) reported transformation of a local cotton variety VH 289 with two Bt genes (*cry1Ac* and *cry2A*) and a herbicide-resistance gene (*cp4 epsps*) using the *Agrobacterium*-mediated transformation method, and found transformed plants in T0 generation effective against cotton bollworm larvae and with tolerance to glyphosate. Siddiqui *et al.* (2019) used a dual-gene construct containing *cry1Ac* and *cry2Ab* (pGA482-12R) for cotton transformation and found *cry2Ab* expression of ~3.3–7.7 μg/g of fresh leaf tissues, offering high protection against tobacco caterpillar.

Bt cotton varieties were reported to contain ~0.03–1.99 μg/g Cry1Ac toxin in their leaf tissues (Cheema *et al.* 2016) compared to the standard production of ~1.5 μg/g of leaf tissue (Malik and Ahsan 2016). These estimates are low when compared to reported high values of 2–7 μg/g in terminal leaf tissues of cotton (Kranthi *et al.* 2005). An extensive survey was made in 2013 of the cotton belt of Punjab and Sindh to discover Cry1Ac in fresh leaves and boll samples from 593 and 589 farmers' fields at 70 DAS (days after sowing) and 120 DAS. The results showed an average concentration of 0.98 and 0.70 μg/g in leaves, sampled at 70 and 120 DAS respectively; and 0.59 and 0.53 μg/g in bolls sampled at 70 and 120 DAS in Punjab (Spielman *et al.* 2017). Another study was carried out in 2018–19 to estimate levels of Cry1Ac and Cry2A in different plant parts at different growth stages of 10 genotypes (G1–G10) of dual-toxin Bt cotton compared with non-Bt cotton (G11). The dual-gene Bt cotton genotypes (G1 and G2) were provided by CEMB, while G3–G10 were F1 hybrids of G1 and G2 with non-Bt parents (MNH-786 and FH942). Cry1Ac toxin was higher at 70 DAS and gradually decreased with plant age till it became minimum at 110 DAS. But Cry2A toxin levels remained constant with no significant differences between genotypes (Figure 7.3). Cry1Ac concentration was highest in leaves, followed by bolls and then flowers, where it was minimum at 70 DAS. Cry2A concentration was higher in leaves than bolls and lower in flowers at 70 DAS. A similar trend for Cry2A was found in these plant parts at 110 DAS. On average, the concentration of Cry1Ac was approximately four times lower than that of Cry2A. In addition, the expression level of Cry1Ac varied significantly across different genotypes but the level of Cry2A was consistent among the genotypes. The toxin concentration gradually decreased in various plant parts as the plants grew older (Figure 7.3).

The results were in agreement with previous studies conducted in Bollgard I® and Bollgard II® (Greenplate 1999; Kranthi *et al.* 2005), in which the relative level of Cry2Ab protein remained constant throughout plant growth during the whole season (Adamczyk *et al.* 2001).

Concurrent with toxin estimates, the toxicity of Cry1Ac and Cry2Ab alone and together, as expressed in transgenic cotton leaves, was evaluated against the cotton bollworm. The bioassay was conducted in 2018 at seedling stage (35 DAS) of Bt cotton genotypes; namely, G1 and G2 with a single gene (*cry1Ac*), G3 and G4 with dual genes (*cry1Ac+cry2Ab*), and G5 and G6 with a single gene (*cry2Ab*). G7 was non-Bt (control). The efficacy of Bt toxins was tested by feeding the second instar larvae of cotton bollworm with leaves of the above genotypes. Results showed 100% larval mortality of the cotton bollworm when fed with leaves having 1.46 μg/g of Cry1Ac and 6.03 μg/g of Cry2Ab in dual-gene Bt cotton (Figure 7.4). Toxicity of Cry1Ac was reduced two-fold over the period of four years when compared with the recent report of Ahmad *et al.* (2019).

The studies also showed that Cry1Ac and Cry2Ab toxins in the dual-gene Bt cotton were more effective than either of them alone, suggesting synergistic or joint action of differing toxins acting at different sites. Besides, dual-toxin Bt cotton has potential to increase the spectrum of activity against lepidopterans like tobacco caterpillar, which is least susceptible to Cry1Ac alone (Greenplate *et al.* 2000; Stewart *et al.* 2001). It also showed the potential of next-generation Bt cotton with stacked genes to control the target pests effectively.

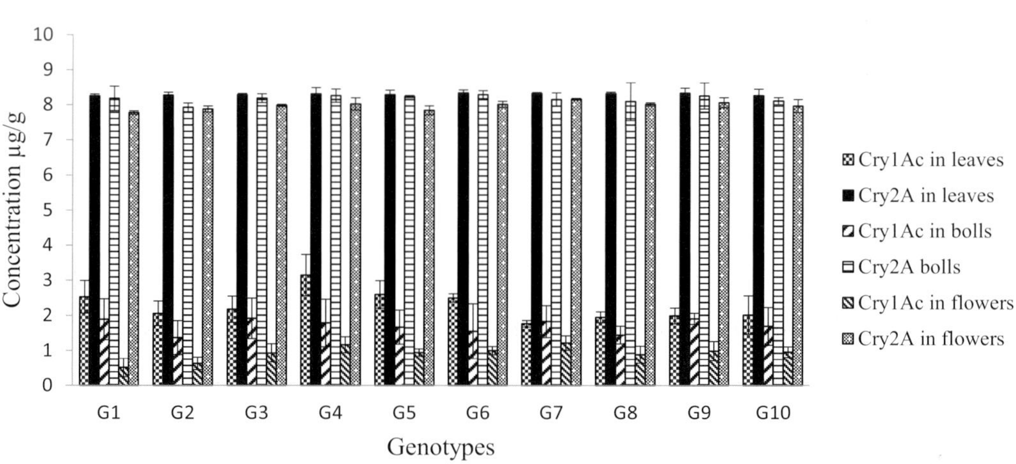

Figure 7.3 Concentration of Cry proteins in different plant parts at 70 DAS. The plants used in this experiment were either of dual-toxin Bt cotton – CKC1 or CKC2 (Cry1Ac and Cry2A), provided by CEMB – or F1 hybrids of these two parents with non-Bt parents (MNH-786 or FH-942). G1 = CKC1, G2 = CKC2, G3 = FH942 × CKC2, G4 = FH942 × CKC1, G5 = MNH786 × CKC1, G6 = MNH786 × CKC2, G7 = CKC2 × FH942, G8 = CKC1 × FH942, G9 = CKC2 × MNH786, G10 = CKC1 × MNH786, G11 = MNH-786 (not shown). The first parent of cross is female; G1, G2 and G11 are self-crossed. Source: Noor *et al.* (unpublished data).

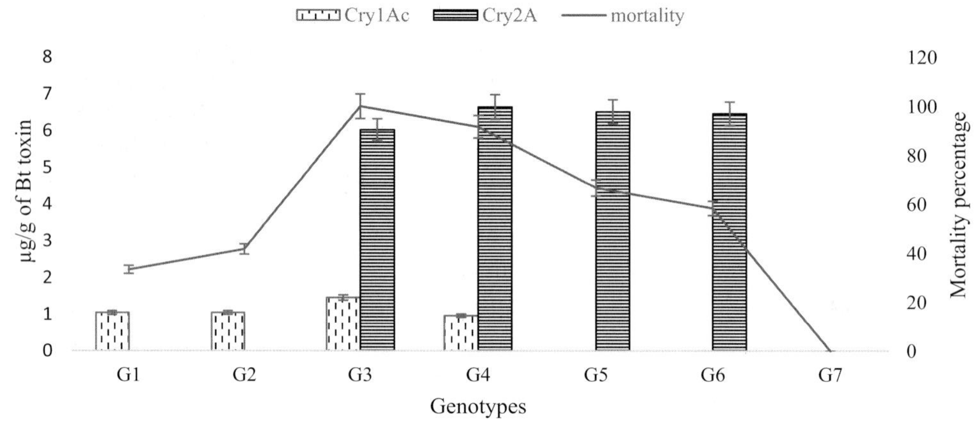

Figure 7.4 Relationship of toxin concentrations in single-toxin and dual-toxin Bt cotton with larval mortality of cotton bollworm. G1 and G2 expressing single gene (*cry1Ac* with event MON 531), G3 and G4 both Bt genes (*cry1Ac+cry2Ab* with event MON 15985) and a non-Bt genotype (G7:MNH786) as a control (Amjad 2019).

Field performance of Bt cotton in Pakistan

Among different *cry* genes, *cry1Ac* gene, a part of MON 531 transformation event of cotton, was effective for insect resistance and was introduced in the local improved germplasm (Spielman *et al.* 2015). Many studies, including those by various provincial seed councils done before the approval of Bt cotton varieties, have shown the effectiveness of Bt cotton against bollworms. Certain documented studies showed that Bt cotton adoption increased yield per hectare up to 10%, reduced pesticide use by 22% per hectare and increased the cropped area for cotton cultivation in Pakistan (Bakhsh 2013; Bakhsh *et al.* 2016). However, Bt cotton did not control sucking pests, which necessitated the use of insecticides in the event of pests crossing economic threshold levels. Another important factor, which is causing huge loss to Bt cotton in Pakistan, is CLCV transmitted by white fly, a sucking pest. Therefore, for sustainable Bt cotton cultivation in the country, breeders need to develop varieties which are genetically resistant to CLCV (Nazir *et al.* 2018). Due to an unregulated seed system, several unapproved cotton lines (the elite breeding lines, without passing through formal evaluation and approval system) are cultivated by farmers. This led to the adoption of genetically similar cotton accessions by farmers, making the genetic base of Bt cotton very narrow (Ashraf *et al.* 2016).

Sustainability of Bt cotton in Pakistan through resistance management

In view of Bt technology's unique ability to offer durable crop protection against bollworms and other lepidopterans, it is mandated to ensure sustainability by implementing resistance management tactics. The foremost factor in this strategy is regular monitoring of targets pests to detect resistance evolution. Baseline susceptibility studies are therefore important forerunners to Bt cotton development, to ensure that target pests possess high-level susceptibility to the toxins at the concentrations expressed in the plant. Similar baseline studies have been conducted on the neonates of cotton bollworms (Gujar *et al.* 2000; Kranthi *et al.* 2001; Jalali *et al.* 2004). They showed high toxicity of Cry1Ac to cotton bollworm neonates. Regular monitoring of target pest resistance levels to Cry1Ac has been carried out since the introduction of Bt cotton in India (Gujar *et al.* 2007; www.cicr.org.in). In Pakistan, there is no regular monitoring of resistance levels in the target bollworms to Bt toxins similar to that done in India, China and Australia through bioassays as a part of institutional commitment. Limited studies on baseline susceptibility of cotton bollworm to Cry1Ac and Cy2Ab toxins have been done. Alvi *et al.* (2012) reported high variation (580-fold) in susceptibility of cotton bollworm (from five locations in 2010 and 2011) to Cry1Ac and could select nearly10-fold more (5440-fold) Cry1Ac-resistant population over five generations of selection pressure in the laboratory. The Cry1Ac selected resistant population, field population and even unselected laboratory population of cotton bollworm were found to survive on several Bt cotton varieties (Bt-121, Bt-866, Bt-456, Bt-802, Bt-703) as compared to the most susceptible population. These studies showed Cry1Ac-resistant trait in cotton bollworm to be semi-dominant, autosomal and controlled by a single gene. They point out the possibility of field-level evolution of resistance in bollworms, instability of resistance in the absence of selection pressure, and the susceptibility of Cry1Ac-resistant insects to conventional insecticides.

In another extensive study on field-level expression of Cry1Ac in different accessions of Bt cotton (MON 531) collected from farmers' fields and its relationship with larval mortality of cotton bollworm, Ahmad *et al.* (2019) reported variation in Cry1Ac content of 0.97–2.73 µg/g in upper leaves, 0.23–0.89 µg/g in middle leaves and 0–0.21 µg/g in lower leaves. Larval survival was ~25–75%, correlating negatively with Cry1Ac concentration. Nearly half of leaf samples (49%) contained on average ~0.5 µg/g Cry1Ac, which was close to LC_{50} of 0.62 µg/g by leaf feeding to the second instar larvae of bollworms.

Production in plants of sufficiently high toxin content to kill most toxin-susceptible homozygous and heterozygous target insects is a requisite for resistance management. It is defined as the concentration, 25 times that of LC_{50} for the susceptible population of target insect species, required in the plant part mostly affected by the pest or *in vivo* bioassay of transgenic tissues over a seven-day period that kills most individuals of target insect pests. Cotton bollworms predominantly oviposit singly on the tender young terminal leaves of the cotton plant. Thus, high Bt toxin production in young terminal leaves will ensure high mortality and subsequent control of bollworms in the following generations. Further, allelochemicals in cotton are known to synergise action of Cry toxins. Aziz *et al.* (2011) evaluated several Bt cotton varieties and found Bt-121 performed best in terms of yield and quality of lint. The Bt toxin level in the recommended varieties appears to be good enough to control spotted bollworm and pink bollworm under field conditions; while significant damage is seen on non-Bt cotton varieties (Arshad *et al.* 2015). Similarly, Khan *et al.* (2013) reported significant mortality of cotton bollworm feeding on transgenic lines of Bt cotton developed using *cry1Ab* gene. Table 7.1 shows the toxin levels in the transgenic Bt cotton lines (MNH-886, FH-142) which have been adopted by farmers. In general, to date, Bt cotton varieties have performed relatively well against cotton bollworm, as there are few reports of bollworm incidence causing substantial field losses.

Coupled with high toxin content, it is necessary to provide non-Bt cotton refuge in which the susceptible insects can multiply. Their mating with the relatively few resistant insects will

Table 7.1. Bt toxin level (µg/g fresh weight) in the leaves of cotton varieties, 2011–15

Varieties / genotypes	2011 (av. 2 laboratories)	2012 (av. 4 laboratories)	2013 (av. 3 laboratories)	2014 (av. 4 laboratories)	2015 (ABRI)
MNH-886	1.49	0.795	3.20	4.41	3.18
FH-142	-	0.965	2.19	1.298	1.83
FH-Lalazar	-	-	2.77	1.361	2.00
FH-Noor	-	-	-	0.842	1.39

Source: Agricultural Biotechnology Research Institute, Faisalabad.

produce susceptible progeny as the larvae inherit recessive resistance trait. Most Cry1Ac-resistant traits are recessive in nature, although some are found to be semi-recessive or semi-dominant (Kranthi *et al.* 2006; Alvi *et al.* 2012) and dominant (Jin *et al.* 2013) in cotton bollworm. Provision of refuges helps delay resistance evolution (Gould 1998). In China, natural refuges are considered enough to delay resistance evolution and hence there is no mandatory structured refuge for Bt cotton (Jin *et al.* 2015). In India, alternate host crops like pigeon pea, chilli, sunflower and corn are grown along with Bt cotton, causing delayed resistance in *H. armigera* even after 16 years of cultivating Bollgard II®, whereas pink bollworm developed resistance against Cry1Ac and Cry2Ab being monophagous in nature (Ravi *et al.* 2005; Verma *et al.* 2019). In Pakistan, structured refuge for Bt cotton is not mandatory as it is believed that there are sufficient natural refuges and other crops growing in the vicinity of Bt cotton or as mixed or intercrops to support susceptible individuals of cotton bollworm. However, the mere presence of alternative natural hosts may not ensure the synchronous emergence of adults of resistant as well as susceptible individuals from their host crops, to produce susceptible progeny. Recent experience with field-level evolution of resistance in the pink bollworm and consequent crop failure in India and Pakistan shows an acute need of structured refuge in the form of non-Bt cotton or refuge-in-bag (non-Bt cotton seed mixed with Bt cotton seed) in Pakistan. In the neighbouring country, India, where structured refuge is mandatory, farmers did not comply. Possible factors include the expectation of reduced yields, the intricacies of non-Bt cotton cultivation and the poor quality of non-Bt cotton seed supplied. An alternative is refuge-in-bag, wherein non-Bt cotton seed is mixed at 5–10% with the Bt cotton seed, ensuring compliance by default; this is now practised in India.

Alternating use of toxin and stacking of toxins are novel tactics to delay and contain resistance in the target pests. However, in practice, stacking or pyramiding of toxins is the most preferred and practical method of overcoming resistance to single toxins. Dual-toxin Bt cotton (Bollgard II®) with different sites of action for the Cry1Ac and Cry2A toxins is more effective (Zhao *et al.* 2003). Australia has successfully used this tactic, replacing single-toxin Bt cotton in 2004 with Bollgard II® cotton to overcome resistance in the target insects. In India, Bollgard II® cotton had replaced single-toxin Bt cotton by 94% by 2013 (Choudhary and Gaur 2015), but the mosaic of single-toxin and dual-toxin Bt cotton created a situation in which pink bollworm could evolve resistance to Cry1Ac (Dhurua and Gujar 2011; Mohan *et al.* 2016) and later to both Cry1Ac and Cry2Ab toxins (Malthankar and Gujar 2016; Mittal *et al.* 2016; Naik *et al.* 2018). It is therefore essential that sustainable Bt cotton production in Pakistan involves robust resistance management with comprehensive use of dual-toxin Bt cotton, with no mosaic of other events. At present, there are practically no dual-toxin Bt cotton varieties grown in the country. A recent socioecological comparison of resistance management of Bt crops showed a strong connection between resistance risk assessment and growers (Carrière *et al.* 2020). This will help in follow-up of mandated and optional resistance management tactics to ensure sustainability of Bt crops, as it is relevant to Pakistan.

Constraints for Bt cotton sustainability

The trait purity of Bt cotton seed is of utmost importance in ensuring optimal and sustained control of bollworms. However, seed quality is a casualty in many developing countries. Ali *et al.* (2010) conducted a survey in 10 districts of Sindh and 11districts in Punjab during the 2007–08 cotton season and found that 10% of samples taken in Punjab and 19% in Sindh were not positive for the presence of Cry1Ac toxin, and only 41% of samples in Sindh and 36% of samples of Bt cotton in Punjab contained high enough Bt protein concentrations to be effective for bollworm control. Another survey was conducted in 2011 to analyse samples of seed sold in markets as Bt cotton. After planting, results showed that 30% of the varieties tested did not show the presence of Cry toxin (Ali *et al.* 2012). A survey of the cotton growing belt of Punjab conducted in 2012 showed that 14% of Bt cotton plant samples were non-Bt, and that only 2% of the positive samples contained optimum transgene expression for bollworm control when compared to the reported lethal level of toxin (Cheema *et al.* 2016). Spielman *et al.* (2017) also reported the sublethal expression of transgene coupled with the variable Cry1Ac in various tissues. The analysis of field-collected leaf and boll samples from the cotton growing belts of Punjab and Sindh showed that samples from 11% of holdings did not contain Bt toxin. The studies concluded that the low level of transgene expression might be due to adulteration, poor breeding methods and improper genetic background in the host. Several other studies further showed that *cry1Ac* gene has a spatio-temporal expression pattern in Pakistani cotton genotypes, leading to a high level of variability of toxin expression in older leaves and during high-temperature regimes (Bakhsh *et al.* 2011; Malik *et al.* 2015; Khan *et al.* 2018). The studies confirmed earlier findings on the major factors influencing the variable toxin levels in Bt cotton like genetic background, climatic conditions, agronomic factors, abiotic and biotic stresses, plant growth and tissue type (Xu *et al.* 2008). Further, the genetic background of unapproved Bt cotton seed in some cases showed high susceptibility to CLCV (Kakakhel 2009). Bakhsh (2013) reported constraints on availability of quality seeds of Bt cotton, while the presence of too many Bt cotton varieties makes it difficult for regulatory authorities to monitor seed quality or compliance to specifications for quality seed (Spielman *et al.* 2017). As Bt technology is specific to bollworm, integrated pest management including resistance management is essential for the sustainability of Bt cotton crops.

Socioeconomic impact of Bt cotton

Bt cotton was adopted illegally by some farmers in the early 2000s (Malik and Ahsan 2016), then spread quite fast. As discussed above, the Pakistan Economic Survey reported 80% adoption of illegal Bt cotton by 2009 (Govt of Pakistan 2010), a year before legal approval of Bt cotton varieties developed by private and public institutions. Abdullah (2010) reported the growing popularity of Bt cotton in Pakistan, which has increased to almost 100% in some localities (Malik and Ahsan 2016). Obviously, the fast rate of adoption of Bt cotton is due to its ability to effectively control the bollworm complex. Pesticide use declined due to Bt cotton (Nazli *et al.* 2010; Nazli 2010). The number of pesticide sprays required during the growing season reduced to three or four on Bt cotton, compared with the 12–13 sprays required on non-Bt cotton. Abedullah and Qaim (2015) estimated gross profit per acre of Bt cotton at US$204, including financial, environmental and health benefits. This translated into US$1.7 billion for the entire cotton area of Pakistan. Bakhsh (2013) reported high net revenue of US$626/ha of Bt cotton relative to non-Bt cotton in Pakistan, including 22% less pesticide use on Bt cotton. Bakhsh *et al.* (2016) reported that Bt cotton farmers harvested more cotton than did non-Bt cotton farmers. Analysis of data on production, yield and area (Figure 7.2) shows that before 2010 the average yield of cotton was 681 kg/ha (average of eight years) which increased by ~8.5% to 739.2 kg/ha (average of next nine years, following official Bt cotton commercialisation). This modest increase must be seen against a background of yield losses during 2015 due to pink bollworm and during following years due to abiotic and biotic stresses, including reduction in crop area caused by low prices.

Resistance evolution in the pink bollworm: threat to Bt cotton sustainability

Bt cotton continues to perform well in the field in Australia, the USA and other countries where no field-level evolution of resistance in bollworms has been reported, possibly due to robust resistance management tactics. However, reports of increased frequency of Bt-resistant bollworms under field conditions have appeared in many countries, including China and India

(Tabashnik *et al.* 2013). The first evidence of field-level evolution of pink bollworm resistance in Bt cotton was reported from Amreli (in Gujarat, India) in 2008, where the insect population showed up to 38-fold resistance to Cry1Ac (Dhurua and Gujar 2011). In 2010, Monsanto reported widespread resistance in pink bollworm to Bt cotton in four Indian districts – Amreli, Rajkot, Junagarh and Bhavnagar of Saurashtra. Later, Mohan *et al.* (2016) and Naik *et al.* (2018) reported extensive pink bollworm resistance in Bt cottons, both single-toxin and dual-toxin, resulting in economic losses to farmers in many Indian states in 2014 and 2015.

Pink bollworm, a key pest of cotton over many decades, completes as many as four generations with moth emergence peaks in August, September and October which coincide with the reproductive stage of cotton in Faisalabad, Pakistan (Hamed and Nadeem 2010b). The outbreak of pink bollworm in Pakistan during 2015 was due to ecological factors like drought and heat and, to some extent, the sublethal expression of Cry toxins in Bt cotton (Cheema *et al.* 2016). Dual-toxin Bt cotton (Bollgard II® cotton event 15985) offers better protection than the single-toxin Bt cotton and delays resistance. However, in view of mosaics of Bt events, even dual-toxin Bt cotton has failed to provide requisite protection against the pink bollworm in India (Naik *et al.* 2018). As the cotton growing area in Pakistan is contiguous with that of north-western India where Bt cotton is grown, and the possibility of pink bollworm resistance remains high, it is likely that wind movement from east to west during the cotton season may expand resistance level to both toxins in Pakistan or that pink bollworm may evolve cross resistance to both toxins. However, to date there is no scientific proof of either event. Another possibility is that importation of Indian cotton might lead to the spread of Bt-resistant populations of pink bollworm.

Integrated pest management and crop management

Stewardship embodies the responsible planning and management of resources for the sustainable impact of transgenic technology in order to harness its maximum benefits. It encompasses the whole life-cycle of the technology, from inception to discontinuation. For Bt cotton technology, it refers to the sensible administration of plant development, regulatory approval, seed production and its dissemination to farmers, implementation of prescribed principles for prolonged insect resistance, crop production, and marketing of the produce worldwide. Following these principles is crucial for the success of Bt cotton at field level, otherwise the target insects will become resistant more speedily than expected.

Bt cotton technology has revolutionised production in developed countries, but its adoption in developing countries like Pakistan did not have similar success. This was largely due to lack of proper stewardship, wherein the system was not supervised and regularised as it was in the developed world, ranging from seed development and marketing to farm education. The stealthy diffusion of single-gene Bt cotton in Pakistan during the first decade of this century was the first indicator of poor monitoring of the spread of this technology. Since the introduction of Bt cotton in Pakistan, biotechnology has become a crucial component of its cotton breeding program for the incorporation of two important traits – IR and HT – but the systematic indifference towards testing and meeting the legalities of the transgenic seed contributed towards the decline in cotton production. The Pakistani economy has suffered a huge loss due to the lack of stewardship of Bt cotton. The lack of regularised seed marketing, the dissemination of poor-quality seed, the absence of integrated pest management and non-compliance with refuge crop cultivation are the main hurdles to making Bt cotton cultivation a success story in Pakistan.

Status of biosafety regulations for transgenic crops

Like other countries, Pakistan has a GM regulatory framework. The Pakistan *Environmental Protection Act 1997* formulated the Pakistan Biosafety Rules and National Biosafety Guidelines in 2005 (Govt of Pakistan 2005). After becoming party to the Cartagena Protocol in 2009, the Pakistan Biosafety Rules formulated regulations for the storage, manufacturing and import of microorganisms, genetically modified organisms (GMOs) and living modified organisms (LMOs) (Gronvall *et al.* 2016). The Rules outlined the composition and functions of the National Biosafety Committee (NBC), Technical Advisory Committee (TAC) and Institutional Biosafety Committees (IBCs) and delineated the duties of a Biosafety Officer. The Pakistan Biosafety Rules ensure that research protocols and biosafety practices comply with the National Biosafety Guidelines. The Guidelines are intended for research with recombinant DNA, transgenic organisms, GMO production, vaccine production and any field trial or intentional environmental release (Govt of Pakistan 2005).

The NBC works under the Ministry of Environment and is chaired by the Secretary of the Ministry of Environment (now Ministry of Climate Change, after the 18th amendment to the constitution). It is responsible for broad-scale risk assessment and management, laboratory regulation, release of GMOs, LMOS and other research products. The NBC and IBC jointly approve and regulate high-level research facilities and Risk Category 3 projects. The TAC functions as an intermediary between the IBCs and the NBC. It advises the NBC about project approval, and reviews safety protocols and research methodologies. It also regulates large-scale GMO projects and their field trials. The IBCs are specific to each institution and report to the NBC through the TAC. They receive research application proposals from the institution, review the biosafety protocol, perform facility inspections and regulate project activity, especially matters related to Risk Category 1 research.

Like in India and China, Pakistan followed an ambiguous path in the development of GM products. After approving Bt cotton in 2010 at federal level, no new Bt crops were released for some time as the approval process was stalled for political reasons (Ashraf *et al.* 2013). This is probably because the federal government devolved matters like the environment and natural ecology from the concurrent federal list to the provinces after the 18th amendment to the constitution and therefore it no longer exercised authority over such subjects. The Pakistan *Environment Protection Act 1997* was passed when the environment was a federal responsibility. The four cotton-growing provinces have now passed their own laws in relation to the environment – Punjab *Environmental Protection (Amendment) Act 2012*, the Balochistan *Environmental Protection Act 2013*, the Khyber Pakhtunkhwa *Environmental Protection Act 2013* and the Sindh *Environment Protection Act 2014*. Punjab also enacted the Punjab Biosafety Rules in 2014. The Pakistan *Environment Protection Act 1997* is not repealed and continues to be operational under the jurisdiction of the federal capital of Islamabad. Bt cotton varieties based upon the MON 531 event continue to be approved after undergoing trials under the state Seed Councils like the Punjab Seed Council and Sindh Seed Council, which operate under the chairmanship of the Agricultural Minister. Every new Bt cotton variety is evaluated for field performance in terms of agronomic benefit as well as ability to resist bollworm attack, through national coordinated varietal trials conducted under the supervision of the Central Cotton Committee. Each variety needs to pass tests on distinctness, uniformity and stability under the Federal Seed Certification and Registration Department, comprising value for cultivation and use tests across ~30–40 locations, as well as biosafety trials. The Punjab Seed Council formally approved 11 Biotech cotton varieties for the Punjab region in 2012. The Sindh Seed Council has also approved Bt cotton varieties for its province. The provincial councils conduct spot examinations with the help of expert subcommittees supervising the field trials. Most of these Bt cotton varieties are based upon the MON 531 event, which is not patentable in Pakistan. Bt cotton MON 531 event is initially approved for three years and is renewable every two years as per National Biosafety Rules. As far as other GM cotton events are concerned, a dual-toxin Bt cotton offering IR and HT, developed by the Centre for Excellence in Molecular Biology, has been approved by the NBC and is registered with the provincial Seed Councils.

In July 2018, the NBC approved as many as 23 single-toxin Bt cotton varieties for cultivation. It is reported to have earlier approved GM corn too. The NBC is still executing its powers as the Pakistan *Environmental Protection Act 1997* remains in force with original executive structure under the new federal Ministry of Climate Change.[1]

Like some developing countries adopting Bt technology, Pakistan is struggling with trait/seed purity as well as seed production as these sectors are highly informal and less regulated (pers. comm. from Dr Muhammad Anjum Ali, DG Agri Ext, Punjab). An ongoing quality issue is that farmers save some seed from harvest for the next year's sowing, thus perpetuating the problem of poor-quality seed. Seed companies may not show high discretion in developing, packaging and marketing their branded seeds; spurious misbranded seeds may occur down the marketing chain. The *Seed Act 1976* was amended via the *Seed Amendment Act 2015* to protect the interests of all stakeholders and provide farmers with quality seeds. Recent requirements of 96% genetic purity, 98% trait purity and >1.5 μg/g expression of Cry1Ac in leaves 90 DAS are likely to ensure quality of Bt cotton varieties. It is likely that in future these issues will be regularly dealt with at state level as Bt cotton is very important to their economies.

Acknowledgements

We are grateful to Dr Muhammad Anjum Ali, Agri Extn, Lahore, Punjab; Mr Muhammad Asim, Bayer Pakistan Crop Science, Lahore; Dr Waseem Akbar of Bayer US Crop Science, Chesterfield, USA; Dr TH Malik of Pakistan Central Cotton Committee, Multan; Dr Abdul Malik Sheikh of Jaffer Brothers Ltd; and Mr Ahmad Rafay Alam of Saleem Alam and Co., for their help.

Endnote

1. http://courtingthelaw.com/2016/04/03/uncategorized/gmo-no-a-constitutional-gridlock/

References

Abdullah A (2010) An analysis of Bt cotton cultivation in Punjab, Pakistan using the Agriculture Decision Support System (ADSS). *Journal of Agrobiotechnology Management and Economics* **13**, 274–287.

Abedullah KS, Qaim M (2015) Bt cotton, pesticide use and environmental efficiency in Pakistan. *Journal of Agricultural Economics* **66**, 66–86. doi:10.1111/1477-9552.12072

Abro GH, Syed TS, Dayo ZA (2003) Varietal resistance of cotton against *Earias* spp. *Pakistan Journal of Biological Sciences* **6**, 1837–1839. doi:10.3923/pjbs.2003.1837.1839

Abro GH, Syed TS, Tunio GM, Khuhro MA (2004) Performance of transgenic Bt cotton against insect pest infestation. *Biotechnology (Faisalabad)* **3**, 75–81. doi:10.3923/biotech.2004.75.81

Adamczyk J, Adams LC, Hardee DD (2001) Field efficacy and seasonal expression profiles for terminal leaves of single and double *Bacillus thuringiensis* toxin cotton genotypes. *Journal of Economic Entomology* **94**, 1589–1593. doi:10.1603/0022-0493-94.6.1589

Ahmad M, Iqbal AM (2009) Resistance of Pakistani field populations of spotted bollworm, *Earias vittella* (Lepidoptera: Noctuidae) to pyrethroid, organophosphorus and new chemical insecticides. *Pest Management Science* **65**, 433–439. doi:10.1002/ps.1702

Ahmad M, Sayyed AH, Ahmad M, Ahmad M (2008) Evidence for field-evolved resistance to newer insecticides in *Spodoptera litura* (Lepidoptera: Noctuidae) from Pakistan. *Crop Protection* **27**, 1367–1372. doi:10.1016/j.cropro.2008.05.003

Ahmad S, Cheema HMN, Khan AA, Khan RSA, Ahmad JN (2019) Resistance status of *Helicoverpa armigera* against Bt cotton in Pakistan. *Transgenic Research* **28**, 199–212. doi:10.1007/s11248-019-00114-9

Ali S, Shahid H, Shahid M, Ali GM, Yusuf Z (2010) Status of Bt cotton cultivation in major growing areas of Pakistan. *Pakistan Journal of Botany* **42**, 1583–1594.

Ali S, Shah SH, Ali GM, Iqbal A, Asad M, Zafar Y (2012) Bt Cry toxin expression profile in selected Pakistani cotton genotypes. *Current Science* **102**, 1632–1636.

Alvi AH, Sayyed AH, Naeem M, Ali M (2012) Field-evolved resistance in *Helicoverpa armigera* (Lepidoptera: Noctuidae) to *Bacillus thuringiensis* toxin *Cry1Ac* in Pakistan. *PLoS ONE* **7**, e47309. doi:10.1371/journal.pone.0047309

Amjad M (2019) Resistance and susceptibility level of army worm (*Spodoptera litura*) and American bollworm (*Helicoverpa armigera*) against double- and single-gene Bt cotton. MSc (Hons) thesis. University of Agriculture, Faisalabad, Pakistan.

Amjad A, Aheer GM (2007) Varietal resistance against sucking insect pests of cotton under Bahawalpur ecological conditions. *Journal of Agricultural Research* **45**, 205–208.

Arshad M, Khan HAA, Rehman MA, Saeed NA (2015) Incidence of insect predators and parasitoids on transgenic Bt cotton in comparison to non-Bt cotton varieties. *Pakistan Journal of Zoology* **47**, 823–829.

Ashraf S, Iftikhar M, Khan GA, Hassan MZY (2013) Worldwide impact of BT cotton: implications for agricultural research in Pakistan. *Greener Journal of Agronomy, Forestry and Horticulture* **1**, 12–20. doi:10.15580/GJAFH.2013.1.082013797

Ashraf J, Malik W, Iqbal MZ, Ali KA, Qayyum A, Noor E, Abid MA, Cheema HMN, Ahmad MQ (2016) Comparative analysis of genetic diversity among Bt cotton genotypes using EST-SSR, ISSR and morphological markers. *Journal of Agricultural Science and Technology* **18**, 517–531.

Azam S, Samiullah TR, Yasmeen A, ud Din S, Iqbal A, Rao AQ, Nasir IA, Rashid B, Shahid AA, Ahmad M, Husnain T (2013) Dissemination of Bt cotton in cotton growing belt of Pakistan. *Advancements in Life Sciences* **1**, 18–25.

Aziz A, Akhtar N, Afzal M, Ashraf M, Tanveer A, Ahmed R, Ehsan MS, Ahmad S (2011) Comparative performance of Bt cotton with some elite conventional cotton cultivars under arid to semi-arid conditions. *African Journal of Agricultural Research* **5**, 1600–1606.

Bakhsh K (2013) *Economic and Environmental Impact of Bt Cotton: Evidence from Pakistan*. Working Paper No. 79–13. South Asian Network for Development and Environmental Economi, Lalitpur, Nepal.

Bakhsh A, Rao AQ, Shahid AA, Husnain T, Riazuddin S (2010) CaMV 35S is a developmental promoter being temporal and spatial in expression pattern of insecticidal genes (*cry1Ac* & *cry2A*) in cotton. *Australian Journal of Basic and Applied Sciences* **4**, 37–44.

Bakhsh A, Shahzad K, Husnain T (2011) Variation in the spatio-temporal expression of insecticidal genes in cotton. *Czech Journal of Genetics and Plant Breeding* **47**, 1–9. doi:10.17221/131/2010-CJGPB

Bakhsh K, Akram W, Jahanzeb A, Khan M (2016) Estimating productivity of Bt cotton and its impact on pesticide use in Punjab (Pakistan). *Pakistan Economic and Social Review* **54**, 15–24.

Carrière Y, Brown Z, Downes S, Gujar GT, Epstein G, Omoto C, Storer N, Mota-Sanchez D, Jørgensen PS, Carroll SP (2020) Governing evolution: a socioecological comparison of resistance management for insecticidal transgenic Bt crops amongst four countries. *Ambio* **49**, 1–16. doi:10.1007/s13280-019-01167-0

Cheema HMN, Khan AA, Khan MI, Aslam U, Rana IA, Khan IA (2016) Assessment of Bt cotton genotypes for the Cry1Ac transgene and its expression. *Journal of Agricultural Science* **154**, 109–117. doi:10.1017/S0021859615000325

Choudhary B, Gaur K (2015) *Biotech Cotton in India, 2002 to 2014*. ISAAA Series of Biotech Crop Profiles. ISAAA, Ithaca, NY.

Dhurua S, Gujar GT (2011) Field-evolved resistance to Bt toxin *Cry1Ac* in the pink bollworm, *Pectinophora gossypiella* (Saunders) (Lepidoptera: Gelechiidae), from India. *Pest Management Science* **67**, 898–903. doi:10.1002/ps.2127

Gould F (1998) Sustainability of transgenic insecticidal cultivars: integrating pest genetics and ecology. *Annual Review of Entomology* **43**, 701–726. doi:10.1146/annurev.ento.43.1.701

Govt of Pakistan (2005) *National Biosafety Guidelines*. Pakistan Environmental Protection Agency, Ministry of Environment, Government of Pakistan, Islamabad.

Govt of Pakistan (2010) *Pakistan Economic Survey (2009–10)*. Economic Advisors Wing, Government of Pakistan, Islamabad.

Govt of Pakistan (2019) *Pakistan Economic Survey (2018–19)*. Economic Advisors Wing, Government of Pakistan, Islamabad.

Greenplate JT (1999) Quantification of *Bacillus thuringiensis* insect control protein *Cry1Ac* over time in Bollgard cotton fruits and terminals. *Journal of Economic Entomology* **92**, 1377–1383. doi:10.1093/jee/92.6.1377

Greenplate JT, Penn SR, Mullins JW, Oppenhuizen M (2000) Seasonal *cry1Ac* levels in DP50B: The 'Bollgard basis' for Bollgard II. In *Proceedings of Beltwide Cotton Conference,* San Antonio, 4–8 January. (Eds CP Dugger and DA Richter) pp. 1039–1040. National Cotton Council, Memphis, TN.

Gronvall GK, Shaerer MP, Collins H (2016) *National Biosafety System: Case Studies to Analyze Current Biosafety Approaches and Regulations for Brazil, China, India, Israel, Pakistan, Kenya, Russia, Singapore, the United Kingdom, and the United States.* UPMC Center for Health Security.

Gujar GT, Kumari A, Kalia V, Chandrashekar K (2000) Spatial and temporal variation in susceptibility of American bollworm, *Helicoverpa armigera* (Hübner) to *Bacillus thuringiensis* var. *kurstaki. Current Science* **78**, 995–1001.

Gujar GT, Kalia V, Kumari A, Singh BP, Mittal A, Nair R, Mohan M (2007) *Helicoverpa armigera* baseline susceptibility to *Bacillus thuringiensis* Cry toxins and resistance management for Bt cotton in India. *Journal of Invertebrate Pathology* **95**, 214–219. doi:10.1016/j.jip.2007.03.011

Hamed M, Nadeem S (2010a) Forecasting of spotted bollworm, *Earias vitella* (Fab.) (Lepidoptera: Noctuidae) occurrence in cotton. *Pakistan Journal of Zoology* **42**, 575–580.

Hamed M, Nadeem S (2010b) Prediction of pink bollworm (*Pectinophora gossypiella*) (Saunder) population cycles in cotton by accumulating thermal units in the agro-climate of Faisalabad. *Pakistan Journal of Zoology* **42**, 431–435.

Huque H (1994) Insect pests of fibre crops. In *Insect Pest Management: Cereal and Cash Crops.* (Ed. AA Hashmi) pp. 193–260. Pakistan Agricultural Research Council, Islamabad.

ISAAA (2017) *Global Status of Commercialized Biotech/GM Crops in 2017: Biotech Crop Adoption Surges as Economic Benefits Accumulate in 22 Years.* ISAAA Brief No. 53. ISAAA, Ithaca, NY.

ISAAA (2018) *Global Status of Commercialized Biotech/GM Crops in 2018: Biotech Crops Continue to Help Meet the Challenges of Increased Population and Climate Change.* ISAAA Brief No. 54. ISAAA, Ithaca, NY.

Jabbar A, Mallick S (1994) *Pesticide and Environmental Situation in Pakistan.* Working Paper Series No. 19. Sustainable Development Policy Institute, Islamabad.

Jalali S, Mohan K, Singh S, Manjunath T, Lalitha Y (2004) Baseline-susceptibility of the old-world bollworm, *Helicoverpa armigera* (Hübner) (Lepidoptera: Noctuidae) populations from India to *Bacillus thuringiensis* Cry1Ac insecticidal protein. *Crop Protection* **23**, 53–59. doi:10.1016/S0261-2194(03)00170-4

Jin L, Wei Y, Zhang L, Yang Y, Tabashnik BE, Wu Y (2013) Dominant resistance to Bt cotton and minor cross-resistance to Bt toxin *Cry2Ab* in cotton bollworm from China. *Evolutionary Applications* **6**, 1222–1235. doi:10.1111/eva.12099

Jin L, Zhang H, Lu Y, Yang Y, Wu K, Tabashnik BE, Wu Y (2015) Large-scale test of the natural refuge strategy for delaying insect resistance to transgenic Bt crops. *Nature Biotechnology* **33**, 169–174. doi:10.1038/nbt.3100

Kakakhel I (2009) Govt decides to use CLCV tolerant Bt cotton varieties, Lahore, Pakistan. *Daily Times.* <http://www.dailytimes.com.pk/default.asp?page=2009%5C09%5C30%5Cstory_30-9-2009_pg5_17>

Karim S (2000) Management of *Helicoverpa armigera*: a review and prospectus for Pakistan. *Pakistan Journal of Biological Sciences* **3**, 1213–1222. doi:10.3923/pjbs.2000.1213.1222

Khan GA, Bakhsh A, Riazuddin S, Husnain T (2011) Introduction of *cry1Ab* gene into cotton *(Gossypium hirsutum)* enhances resistance against lepidopteran pest (*Helicoverpa armigera*). *Spanish Journal of Agricultural Research* **9**, 296–302. doi:10.5424/sjar/20110901-136-10

Khan MH, Ahmad N, Rashdi SMM, Rauf I, Ismail M, Tofique M (2013) Management of sucking pest complex in Bt cotton through the application of different plant products. *Pakistan Journal of Life Science* **1**, 42–48.

Khan MI, Khan AA, Cheema HMN, Khan RSA (2018) Spatio-temporal and intra-plant expression variability of insecticidal gene (*cry1Ac*) in upland cotton. *International Journal of Agriculture and Biology* **20**, 715–722.

Kranthi K, Kranthi S, Wanjari R (2001) Baseline toxicity of *Cry1A* toxins to *Helicoverpa armigera* (Hubner) (Lepidoptera: Noctuidae) in India. *International Journal of Pest Management* **47**, 141–145. doi:10.1080/09670870151130651

Kranthi KR, Naidu S, Dhawad C, Tatwawadi A, Mate K, Patil E, Bharose A, Behere G, Wadaskar R, Kranthi S (2005) Temporal and intra-plant variability of Cry1Ac expression in Bt-cotton and its influence on the survival of the cotton bollworm, *Helicoverpa armigera* (Hübner) (Noctuidae: Lepidoptera). *Current Science* **89**, 291–298.

Kranthi KR, Dhawad CS, Naidu SR, Mate K, Behere GT, Wadaskar RM, Kranthi S (2006) Inheritance of resistance in Indian *Helicoverpa armigera* (Hübner) to *Cry1Ac* toxin of *Bacillus thuringiensis. Crop Protection* **25**, 119–124. doi:10.1016/j.cropro.2005.03.011

Malik TH, Ahsan MZ (2016) Review of the cotton market in Pakistan and its future prospects. *Oilseeds & Fats Crops and Lipids* **23**, D606. doi:10.1051/ocl/2016043

Malik W, Abid MA, Cheema HMN, Khan AA, Iqbal MZ, Qayyum A, Hanif M, Bibi N, Shuna Y, Yasmeen A, Mahmood A, Ashraf J (2015) From Qutn to Bt cotton: development, adoption and prospects. *Cytology and Genetics* **49**, 408–419. doi:10.3103/S0095452715060055

Malthankar PA, Gujar GT (2016) Toxicity of *Bacillus thuringiensis Cry2Ab* and the inheritance of *Cry2Ab* resistance in the pink bollworm, *Pectinophora gossypiella* (Saunders) (Lepidoptera: Gelechiidae). *Indian Journal of Experimental Biology* **54**, 586–596.

Mittal A, Ningthoukhongjam J, Malthankar PA, Ningthoujam K, Kalia V, Gujar GT (2016) Toxicity of Cry1Ac, inheritance of Cry1Ac resistance and fitness cost association in the pink bollworm, *Pectinophora gossypiella* (Lepidoptera: Gelechiidae). *Biopesticides International* **12**, 127–138.

Mohan KS, Ravi KC, Suresh PJ, Sumerford D, Head GP (2016) Field resistance to *Bacillus thuringiensis* protein Cry1Ac expressed in Bollgard hybrid cotton in pink bollworm, *Pectinophora gossypiella* (Saunders) populations in India. *Pest Management Science* **72**, 738–746. doi:10.1002/ps.4047

Naik VCB, Kumbhare S, Kranthi S, Satija U, Kranthi KR (2018) Field evolved-resistance of pink bollworm, *Pectinophora gossypiella* (Saunders) (Lepidoptera: Gelechiidae) to transgenic Bt-cotton expressing Cry1Ac and Cry2Ab in India. *Pest Management Science* **74**, 2544–2554. doi:10.1002/ps.5038

Nazir MS, Khan AA, Khan RSA, Cheema HMN, Shakeel A (2018) Sustainable cotton production under CLCuD threat. *Pakistan Journal of Agricultural Sciences* **55**, 279–285. doi:10.21162/PAKJAS/18.5007

Nazli H (2010) Impact of Bt cotton adoption on farmers' wellbeing in Pakistan. PhD thesis. University of Guelph, Canada.

Nazli H, Sarkar R, Meilke K, Orden D (2010) Economic performance of Bt cotton varieties in Pakistan. Paper presented at the *Agricultural and Applied Economics Association 2010 AAEA, CAES and WAEA Joint Annual Meeting*, 25–27 July, Denver, CO.

Nazli H, Orden D, Sarker R, Meilk K (2012) Bt cotton adoption and well-being of farmers in Pakistan. Paper presented at the *28th Conference of the International Association of Agricultural Economists*, 18–24 August, Foz do Iguaçu, Brazil.

Nizamani IA, Talpur MA, Khuhro DR, Nizamani SM (2002) Relative resistance of cotton cultivars to sucking complex. *Journal of Applied Science* **2**, 686–689.

Puspito AN, Rao AQ, Hafeez MN, Iqbal MS, Bajwa KS, Ali Q, Rashid B, Abbas MA, Latif A, Shahid AA, Nasir IA, Husnain T (2015) Transformation and evaluation of *Cry1Ac+Cry2A* and *GT* gene in *Gossypium hirsutum* L. *Frontiers of Plant Science* **6**, 943. doi:10.3389/fpls.2015.00943.

Qayyum MA, Wakil W, Arif MJ, Sahi ST (2015) *Bacillus thuringiensis* and nuclear polyhedrosis virus for the enhanced biol.-control of *Helicoverpa armigera. International Journal of Agriculture and Biology* **17**, 1043–1048. doi:10.17957/IJAB/15.0025

Rana S (2019) Cotton production likely to miss target by wide range. *The Tribune*, Lahore, 5 October. <https://tribune.com.pk/story/2072442/2-cotton-production-likely-miss-target-wide-margin/>

Rana MA, Khawar H, Gilani AT, Rana HS (2013) *Exploring Dynamics of Cotton Seed Provision in Sindh.* Working Paper. International Growth Centre, London School of Economics and Political Science, London.

Rashid B, Saleem Z, Husnain T, Riazuddin S (2008) Transformation and inheritance of Bt genes in *Gossypium hirsutum. Journal of Plant Biology* **51**, 248–254. doi:10.1007/BF03036123

Ravi KC, Mohan KS, Manjunath TM, Head G, Patil BV, Greba DA, Rao NGV (2005) Relative abundance of *Helicoverpa armigera* (*Lepidoptera: Noctuidae*) on different host crops in India and the role of these crops as natural refuge for *Bacillus thuringiensis* cotton. *Environmental Entomology* **34**, 59–69. doi:10.1603/0046-225X-34.1.59

Shahid MR, Farooq J, Mahmood A, Ilahi F, Riaz M, Shakeel A, Farooq A (2012) Seasonal occurrence of sucking insect pest in cotton ecosystem of Punjab, Pakistan. *Advances in Agriculture & Botanics* **4**, 26–30.

Siddiqui HA, Asif M, Asad S, Naqvi RZ, Ajaz S, Umer N, Anjum N, Rauf I, Sarwar M, Arshad M, Amin I, Saeed M, Mukhtar Z, Bashir A, Mansoor S (2019) Development and evaluation of double gene transgenic cotton lines expressing Cry toxins for protection against chewing insect pests. *Scientific Reports* **9**, 11774. doi:10.1038/s41598-019-48188-z

Spielman DJ, Nazli H, Ma X, Zambrano P, Zaidi F (2015) Technological opportunity, regulatory uncertainty, and the economics of *Bt* cotton in Pakistan. *Journal of Agrobiotechnology Management and Economics* **18**, 98–112.

Spielman DJ, Zaidi F, Zambrano P, Khan AA, Ali S, Cheema HMN, Nazli H, Khan RSA, Iqbal A, Zia MA, Ali GM (2017) What are farmers really planting? Measuring the presence and effectiveness of Bt cotton in Pakistan. *PLoS One* **12**, e0176592. doi:10.1371/journal.pone.0176592

Stewart SD, Adamczyk JJ, Knighten KS, Davis FM (2001) Impact of Bt cottons expressing one or two insecticidal proteins of *Bacillus thuringiensis* Berliner on growth and survival of noctuid (Lepidoptera) larvae. *Journal of Economic Entomology* **94**, 752–760. doi:10.1603/0022-0493-94.3.752

Tabashnik BE, Brévault T, Carrière Y (2013) Insect resistance to Bt crops: lessons from the first billion acres. *Nature Biotechnology* **31**, 510–521. doi:10.1038/nbt.2597

Tariq MI, Afzal S, Hussain I, Sultana N (2007) Pesticides exposure in Pakistan: a review. *Environment International* **33**, 1107–1122. doi:10.1016/j.envint.2007.07.012

Verma D, Kalubarme MH, Saroha GP, Mohan KS, Ravi KC, Singh AN (2019) Monitoring changes in cotton acreage and alternate host crops of cotton bollworm using remote sensing and GIS in major cotton growing regions of India. *International Archives of the Photogrammetry, Remote Sensing and Spatial Information Sciences* **XLII-3/W6**, 525–533. doi:10.5194/isprs-archives-XLII-3-W6-525-2019

Wakil W, Ghazanfar MU, Kwon YJ, Ullah E, Islam S, Ali K (2012) Testing *Paecilomyces lilacinus*, diatomaceous earth and *Azadirachta indica* alone and in combination against cotton aphid *(Aphis gossypii* Glover) (Insecta: Homoptera: Aphididae). *African Journal of Biotechnology* **11**, 821–828. doi:10.5897/AJB11.2446

Xu N, Fok M, Baid L, Zhou Z (2008) Effectiveness and chemical pest control of Bt cotton in the Yangtze River Valley, China. *Crop Protection* **27**, 1269–1276. doi:10.1016/j.cropro.2008.04.003

Zhao JZ, Cao J, Li Y, Collins HL, Roush RT, Earle ED, Shelton AM (2003) Transgenic plants expressing two *Bacillus thuringiensis* toxins delay insect resistance evolution. *Nature Biotechnology* **21**, 1493–1497. doi:10.1038/nbt907

8

Transgenic maize and its potential commercialisation in China

Yueqin Wang, Zhenying Wang, Kanglai He and Jingfei Guo

Introduction

Maize (*Zea mays*) ranks first in planting area and production among all crops in China. Most of the production is devoted to livestock feed, with the rest for starch and sweetcorn (Lu 2016). In 2018, Chinese growers planted 42 million ha of maize and produced 257 million metric tons of grains, at an average yield of 6.1 tons/ha.[1] Maize production is concentrated in the corn belt of China, especially the north spring corn region and the Huang-Huai-Hai summer corn region, which make up 45% and 37% of the total corn planting area, respectively (Li and Wang 2010). Figure 8.1 shows the area of maize in the provinces of China in 2017.

Asian corn borer *Ostrinia furnacalis* is considered the most destructive insect pest of maize throughout China. It causes ~10% yield loss in spring maize and 20–30% in summer maize annually, and much greater in an outbreak year, resulting in considerable economic loss (Wang *et al.* 2000). The young larvae initially feed on the leaves of whorl. As they grow, they move on to feed initially on tassel (male flowers) and then the ear, first on silk and then boring into the cob to feed on kernels, thus causing direct yield loss as well as ear rot contaminated with mycotoxins that lower the grain quality and pose health hazards to livestock (Shen *et al.* 2016). The mature larvae bore into the stalk, pupating inside until emergence. The life-cycle is completed in 30–40 days, and generations are both continuous and overlapping.

Other lepidopteran pests of concern for maize in China include cotton bollworm *Helicoverpa armigera*, yellow peach moth *Conogethes punctiferalis*, beet armyworm *Spodoptera exigua* and common armyworm/oriental armyworm *Mythimna separata*. The incidence of yellow peach moth and cotton bollworm in corn appears to be increasing as cropping systems from the 1990s have changed. For instance, the more frequent use of no-till farming resulted in increased survival of cotton bollworm pupae in the soil during hibernation (Wang *et al.* 2002, 2006).

Fall armyworm *Spodoptera frugiperda*, a native to tropical and subtropical regions of the Americas, invaded Yunnan province in January 2019 and thereafter spread rapidly to the south hilly maize region and south-western mountain maize region in May 2019. It reached Shandong province and Gansu province in June and July 2019, respectively. The invasion of fall armyworm poses a great constraint to maize production in China (Jing *et al.* 2019). It is

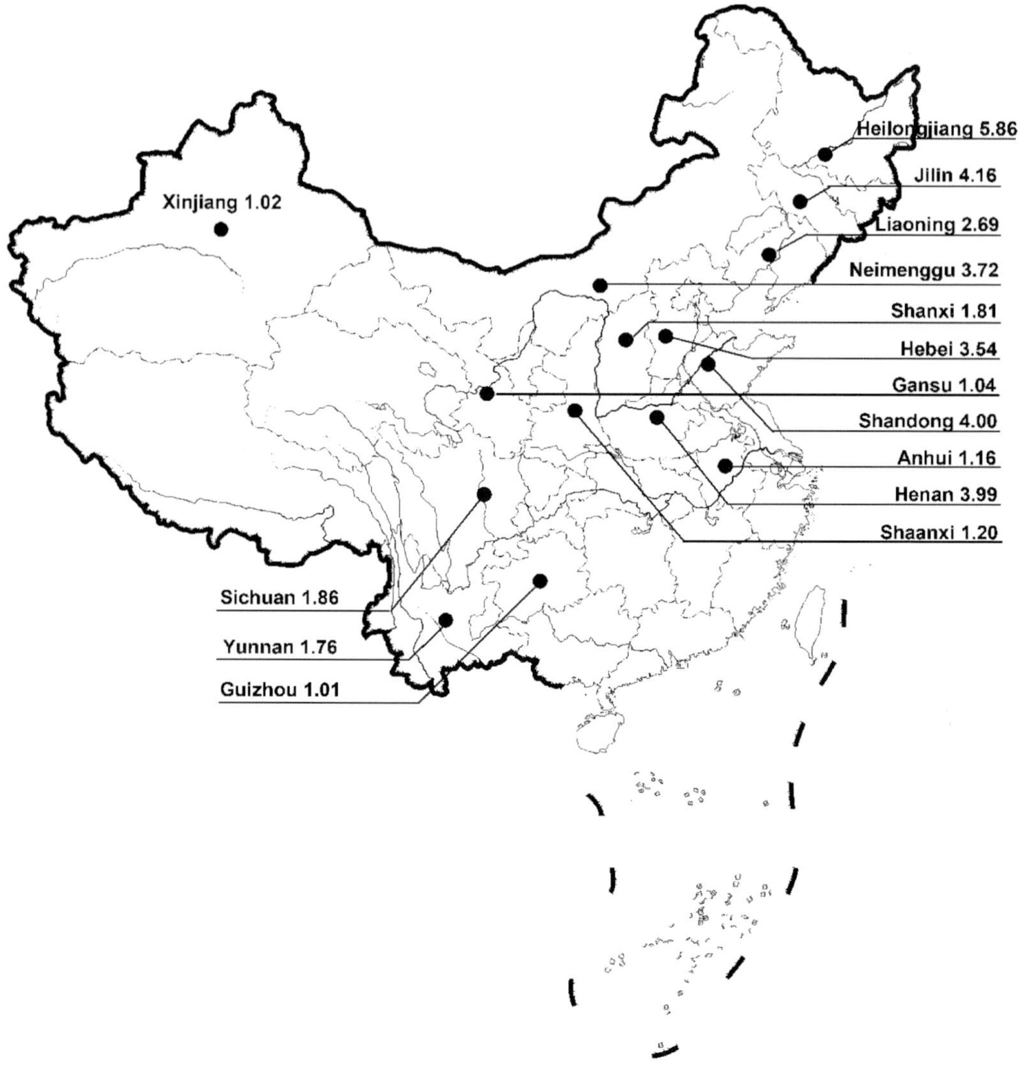

Figure 8.1 Area of maize (>1 million ha) in the provinces of China, 2017.

highly polyphagous with a host range of 353 plant species, preferably feeding on members of Poaceae family, such as maize, sorghum and rice, and on the Fabaceae family such as soybean, peanut and alfalfa (Montezano *et al.* 2018). This poses a broad threat to Chinese food security.

Integrated pest management is broadly practised for various crops including maize. Zhou *et al.* (1995) developed pest management for Asian corn borer in maize that combined various tactics including use of insecticides, host plant resistance, cultural practices (manipulating cropping system, tillage practices) and biological control agents. Fall armyworm is controlled through insect-resistant (IR) transgenic maize combined with other tactics in the Americas (Yu 1992; Bernardi *et al.* 2016; Okuma *et al.* 2018). Recently, Zhang and Wu (2019) reported efficacy of Chinese-developed Bt transgenic maize against larvae of fall armyworm under contained conditions. This calls for adapting novel transgenic approaches compatible with other tactics for maize pest management.

Transgenic maize research history and present status

Transgenic *Bacillus thuringiensis* (Bt) maize that produces Bt toxins efficiently in all crop stages provides a new tool for control of Asian corn borer (Zhang *et al.* 2013; Wang *et al.* 2014b). Transgenic maize has generally been associated with reduced use of chemical insecticides, thereby reducing the risk to the environment and non-target beneficial arthropods (Guo *et al.* 2016; Liu *et al.* 2016b).

China, one of the earliest countries in the world to apply genetic engineering technology, began the research and development of transgenic maize in the 1980s. Under the auspices of the National GMO New Variety Breeding Program, novel genes with valuable applications were isolated and characterised and independent intellectual property rights for these genes were obtained. For insecticidal activity, Bt served as a native source of novel *cry* genes such as *cry1Ah* and *cry1Ie*. The latter was isolated from the Btc007 strain and was very toxic to insects. These genes helped in the development of many transgenic maize events with IR trait, such as Bt799 and IE09S034 (Zhang *et al.* 2013; Wang *et al.* 2014b). In view of the importance of weeds, an important gene, *epsps*, was sourced from Bt by Chinese scientists and was used to develop herbicide-tolerant (HT) maize (Wang *et al.* 2014a, 2016a) and dual-stacked IR-HT maize (Ren *et al.* 2017). Table 8.1 shows the transgenic maize events developed in China.

Although Bt maize is not commercialised in China to date, the central government is paving the way to commercial cultivation of genetically engineered staple crops, especially Bt

Table 8.1. Transgenic maize events developed in China

No.	Event	Trait	Genes	Status	References
1	Bt506 (Bt799)	IR	*cry1Ac*	FT3	Wang *et al.* (2014b)
2	IE09S034	IR	*cry1Ie*	FT2	Zhang *et al.* (2013)
3	CM8101	IR	*cry1Ab-Ma+bar*	FT3	Lu *et al.* (2018)
4	G1F-19	IR	*mcry1F*		Yue (2018)
5	HGK60	IR	*cry1Ah*	Apply for BSC	Song *et al.* (2016)
6	C0030.3.5	IR	*cry1Ab*	Apply for BSC	Zhang and Wu (2019)
7	DBN3601	IR	*cry1Ab+Vip3Aa*	FT1	Zhang and Wu (2019)
8	DBN3608	IR	*cry1Ab+Vip3Aa*		Zhang and Wu (2019)
9	CC-2	HT	*maroACC (EPSPS)*	Apply for BSC	Wang *et al.* (2014a)
10	C0010.1.1	HT	*epsps/pat*	Apply for BSC	He *et al.* (2018)
11	CL38–1	HT	*Epsps*		Wang *et al.* (2016a)
12	GH5112–117C	IR/HT	*cry1Ah+mG2-aroA*		He *et al.* (2012)
13	GAB-3	IR/HT	*cry1Ab/mcry2Ab+G10-epsps*	FT3	Li *et al.* (2015)
14	SK12–5	IR/HT	*cry1Ab/cry2Aj+G10evo-epsps*	Apply for BSC*	Chang *et al.* (2013); Wang *et al.* (2016b)
15	SK12–6	IR/HT	*cry1Ab/cry2Aj+ G10evo-epsps*	FT3	Zhao (2016)
16	DBN9936	IR/HT	*cry1Ab+epsps*	Apply for BSC*	Ren *et al.* (2017)
17	CC-2 × BT799	IR/HT	*cry1Ac+epsps*		Jiang *et al.* (2016)
18	BVLA430101	Phytase	*phyA2*	Got BSC	Guo *et al.* (2014)

IR = insect resistance, HT = herbicide-tolerant, IR/HT = insect resistance/herbicide tolerant, phytase (phosphatase) for solubilising indigestible organic phosphorus found in foodgrains and oilseeds, FT1 = intermediate test, FT2 = environmental release, FT3 = declaration or productive test, BSC = biosafety certificates, * = approved on 2 December 2019.

maize. Phytase transgenic maize event BVLA430101, developed by the Biotechnology Research Institute, CAAS, Beijing, received biosafety certificate in 2009 and renewed later in 2014 which lapsed in 2019. A DBN9858 event (*epsps/pat* gene) was approved on 11 June 2020.

In China, the Ministry of Agriculture and Rural Affairs of the People's Republic of China is the only regulatory agency in charge of biosafety data review, biotech event field trial permits issue, biosafety certificate authorisation and biotech crops commercial approval. However, according to the Regulations on the Safety Control of Genetically Modified Organisms in Agriculture issued by the State Council of the People's Republic of China, other ministries or government agencies such as those dealing with science and technology, environment, public health, foreign trade and economic cooperation will be consulted on major issues.[2]

Efficacy of transgenic maize against major insect pests and tolerance to herbicide

Extensive laboratory bioassays and field trials have been conducted to evaluate the resistance of transgenic maize events to target lepidopteran pests. In laboratory studies, Bt maize was evaluated on the basis of mortality or growth inhibition of Asian corn borer by feeding larvae on whorl leaves, silk, kernels, ear and husk tissues, compared with results on non-Bt maize. The studies showed that the currently developed Bt maize lines were highly toxic to the Asian corn borer, and significantly inhibited growth of the neonates (He *et al.* 2003, 2004; Wang *et al.* 2004). Cotton bollworm neonates did not survive when fed on silk, ear and husk tissues of Bt maize in the laboratory. However, there was some survival of cotton bollworm larvae after artificial infestation at silk stage of maize plants in the field (Chang *et al.* 2006). Chinese domestic Bt-Cry1Ab maize event (C0030.3.5) and Bt stacked (Cry1Ab+Vip3Aa) maize events (DBN3601 and DBN3608) produced target insecticidal proteins efficiently and showed high toxicity to the first to fourth instar larvae of fall armyworm (Zhang and Wu 2019). Laboratory tests also demonstrated that *cry1Ab*-expressing maize such as transgenic events MON 810 and Bt11, the two most commercialised events in the world, were highly effective against the oriental armyworm (Wang *et al.* 2005; Chang *et al.* 2007).

In field trials, Bt and non-Bt control maize plants were artificially infested with neonates of Asian corn borer. A few days later, leaf-feeding ratings (on a 1–9 scale, in which 1 the most resistant and 9 the most susceptible) and borer density were standardised (He *et al.* 2000). The leaf-feeding ratings by Asian corn borer on all Bt maize lines were tested against non-Bt maize and most Bt maize lines rated below 3, which was characterised as 'highly resistant' or 'resistant' (Table 8.2) (Wang *et al.* 2012, 2014b; Chang *et al.* 2013; Dai *et al.* 2014; Song *et al.* 2016).

HT maize lines were also effective in tolerating glyphosate and controlling weeds. Jiang *et al.* (2016) reported that transgenic HT maize, CC-2 expressing *epsps* gene, offered an excellent tolerance to glyphosate even at two times higher recommended rates at the whorl stage. Transgenic IR and HT stacked maize event Shuangkang12–5 survived four weeks after treatment with glyphosate, whereas non-transgenic maize showed phytotoxicity within a week and gradually died (Wang *et al.* 2016b).

Environmental risk assessment of transgenic Bt and HT maize

An IR transgenic maize represents an efficient way for control of insect pests, but may also have potential adverse effects on the environment. Therefore, it must be subjected to a rigorous

Table 8.2. Resistance levels of transgenic maize with artificial infestation in whorl stage

Events	Toxins	LFR	Resistance level	References
BT799	Cry1Ac	1.1	HR	Wang *et al.* (2014b)
BT-X	Cry1Ac	1.2	HR	Wang *et al.* (2012)
HGK60	Cry1Ah	1.3	HR	Song *et al.* (2016)
G186	Cry1Ah	1.3	HR	Dai *et al.* (2014)
IE09S034	Cry1Ie	<3.0	HR	Zhang *et al.* (2013)
SK12–5	Cry1Ab/Cry2Aj	1.0–1.2	HR	Chang *et al.* (2013)
HIF21	Cry1Ah/Cry1Ie	2.1	HR	Yang *et al.* (2012)

LFR = leaf-feeding rating on a scale of 1–9, HR = highly resistant level rating (<3).

risk assessment before the decision whether to cultivate these crops commercially is made (Conner *et al.* 2003; Nap *et al.* 2003).

As maize is highly open-pollinated, it is capable of outcrossing, with its gene flowing from Bt transgenic maize to other cultivars of the same species or their wild and weedy relatives. However, Christou (2002) did not find any evidence of transgenic DNA introgressing into maize landraces in Mexico. Many ecological and agronomic factors, such as proximity of compatible wild relatives, fitness of hybrids, mating system and mode of pollination influence the extent of gene flow (Messeguer 2003). The possibility of pollen-based gene flow from the transgenic crops to other varieties is one of the major environmental concerns. Zhao *et al.* (2013) reported that risk probability of the pollen drift decreased with an increasing distance from the central planting area of transgenic *phyA* maize; pollen drift risk was 82.5% at 1 m and <8.7% at >5 m distance from the transgenic *phyA* maize.

In common with conventional agricultural pest control products – insecticides, biocontrol agents and conventionally bred resistant crops – one of the risks associated with the growing of Bt maize events is their potential effects on non-target arthropods, including biocontrol agents, pollinators and decomposers (Romeis *et al.* 2008; Sanvido *et al.* 2012). The Japanese ladybird beetle *Propylea japonica* is a common and abundant natural predator of eggs and larvae of many insect species. It also feeds on maize pollens and is likely to be exposed to Bt transgenic-produced insecticidal proteins by feeding on the targeted herbivores that ingest such insecticidal proteins. Transgenic maize pollens containing Cry1Ab/2Aj- or Cry1Ac toxin fed to larvae of the ladybird beetle did not adversely affect pupation, eclosion and fresh adult weights, but significantly shortened larval developmental time. This was attributed to unknown differences in the nutritional composition of pollens between Bt transgenic and non-Bt maize. The overall results of consumption of Bt transgenic maize pollens did not show any harm to larvae of the Japanese ladybird beetle (Liu *et al.* 2016b). The braconid parasitoid *Macrocentrus cingulum* is an abundantly occurring parasitoid of corn borers, suppressing their populations effectively in maize ecosystem. No difference in parasitism rate, cocoon weight and number of *M. cingulum* progeny produced per host was detected when *M. cingulum* adults were directly fed a 20% honey solution with or without Cry1Ac. However, life-table parameters of the parasitoid were significantly reduced when an *M. cingulum*-parasitised Asian corn borer Cry1Ac-susceptible strain was used in comparison to the Asian corn borer Cry1Ac-resistant strain. These adverse effects on the biology of a parasitoid might be explained by reduced host quality rather than by direct toxicity of Bt Cry proteins (Wang *et al.* 2017b). Such so-called prey quality-mediated effects have been observed in numerous tritrophic systems with Bt-transgenic plants (Han *et al.* 2015).

Chinese honey bee *Apis cerana cerana*, an important species for the balance of natural ecosystems and pollination of crops, depends on maize pollens as an essential food resource. Studies conducted under controlled laboratory conditions demonstrated that several Bt toxins (such as Cry1Ie and Cry1Ah) did not have any adverse effects on the pollen consumption and survival of the Chinese honey bee (Dai *et al.* 2012, 2016). A relatively minor difference in natural enemy communities between Bt and non-Bt maize was detected. Transgenic *cry1Ie* gene maize had little effect on arthropod natural enemy biodiversity (Guo *et al.* 2016).

Studies were conducted to assess the potential effects of transgenic Bt maize (Shuangkang12–5) expressing *cry1Ab* and *cry2Aj* genes on the soil-dwelling springtail *Folsomia candida*. Consumption of Bt maize pollen did not show any lethal or sublethal effects on springtails (Zhang *et al.* 2017a). A novel Bt maize event expressing *cry3* gene is toxic to corn rootworms *Diabrotica* spp., as Cry3 is specific to Coleoptera. Discussing risk assessment needs, it is suggested that the risk assessment should focus on non-target coleopteran species rather than species belonging to other taxonomic groups (Raybould *et al.* 2007). It is also suggested that if the proteins are not expressed in the pollen, there is no scientific justification for conducting pollen-feeding studies with bees or other arthropods (Romeis *et al.* 2008). Compared with non-transgenic maize, the transgenic HT maize CC-2 event with or without herbicide had no significant effects on composition and structure of the arthropod community and seasonal dynamics of dominant groups, leaving no risk to diversity of arthropods (Wang *et al.* 2014a). The arthropod community structures in the fields of dual stacked HT (for glyphosate) and IR (producing Cry1Ab) maize, DBN9936 and the conventional non-transgenic maize, were similar (Ren *et al.* 2017).

Resistance management strategy for transgenic maize

Laboratory selection experiments carried out by exposing larvae of Asian corn borer to an artificial diet incorporating the Bt toxins Cry1Ab, Cry1Ac, Cry1Ah and Cry1F have shown great potential for resistance evolution in the pest species (Wang *et al.* 2016d; Zhang *et al.* 2017b; Shabbir *et al.* 2018). The Cry1Ab-resistant strain, selected with trypsin-activated Cry1Ab for 142 generations, developed ~200-fold resistance to Cry1Ab and cross-resistance to Cry1Ac and Cry1F but not to Cry1Ie toxin (Zhang *et al.* 2017b). Cry1Ie toxin, which provided high levels of control against the Asian corn borer, did not show cross-resistance to commercialised Cry1A toxins, such as Cry1Ab and Cry1Ac (Wang *et al.* 2017a).

It is essential to determine the baseline susceptibility among populations in different geographic regions and to understand the genetic basis of resistance to Bt toxins in order to evaluate the risk of resistance and to implement strategies to establish an effective insect resistance management program. He *et al.* (2005) reported that all the tested Asian corn borer populations collected from major maize-producing areas of China showed differential susceptibility to Cry1Ab within natural range, with LC_{50} values of 0.10–0.81 μg toxin/g of diet. The bioassays of progenies of the reciprocal F_1 crosses to different Cry1 toxins indicated that inheritance of resistance trait in the Asian corn borer to any of Cry1Ab, Cry1F, Cry1Ah and Cry1Ie was autosomal without maternal effect or sex linkage. The effective dominance level (*h*) of F_1 offspring depended on the concentration of toxins, with resistance more dominant as concentration decreased. Finally, the analysis of actual and expected mortality of the progeny from a backcross (F_1 × resistant strain) indicated that the inheritance of resistance to Cry1Ab, Cry1F, Cry1Ah and Cry1Ie was due to more than one locus (Zhang *et al.* 2014; Wang *et al.* 2016d, 2017a; Shabbir *et al.* 2018). In some cases, the resistance was primarily monogenic, but was polygenic as resistance increased, such as in case of the Cry1Ac-resistant strain (Zhang *et al.* 2014).

The V-type proton ATPase catalytic subunit A and heat shock 70kDa proteins were identified as Cry toxins binding proteins in the brush border membrane vesicles of the Asian corn borer (Xu 2010). Significant down-regulation transcription level of aminopeptidase N (*apn1* and *apn3*), as well as a putative ATP binding cassette transporter group G (*abcg*) gene, may also be responsible for both Cry1Ab and Cry1Ac resistance in the Asian corn borer (Zhang *et al.* 2017b). Binding analysis showed that Cry1F receptor was either absent or mutated so that it no longer bound to Cry1F in the resistant larvae (Wang 2018).

Among the strategies for resistance management, application of the high-dose/refuge approach and pyramiding of two or more toxins with different modes of action have been most recommended. With the high-dose/refuge approach, a high dose must be combined with a non-Bt refuge in order to maintain a pool of susceptible homozygotes that would mate with rare homozygous-resistant individuals, thus diluting the resistance gene in the offspring population (Liu and Tabashnik 1997). Three conditions are especially important to promote the durability of Bt crop pyramids: each toxin must be high enough to kill all or nearly all susceptible insects, no cross-resistance occurs between toxins in the pyramid, and pyramids are not grown concurrently with single-toxin plants that produce one of the toxins in the pyramid (Zhao *et al.* 2005; Tabashnik *et al.* 2013; Carrière *et al.* 2015). Under an ideal scenario, combinations of two or more distinct Bt toxins can substantially slow evolution of pest resistance. However, unlike ideal conditions with theoretical models, each toxin in a pyramid does not kill nearly all susceptible insects and some cross-resistance occurs between the toxins in these pyramids. Given these uncertainties, Wang *et al.* (2019) set out selection experiments to mimic IR management strategies under laboratory conditions. Asian corn borer larvae were reared on an artificial diet with Bt toxins incorporated either singly (Cry1Ab, Cry1F, or Cry1Ie) or in pairs (Cry1Ab + Cry1F or Cry1Ab + Cry1Ie). The results suggest that evolution of resistance by Asian corn borer was not consistently delayed by pairs of Bt toxins relative to the same toxins used singly. This highlights the need to better understand biological factors leading to insecticidal protein resistance and to develop IR management strategies based on individual target insect species.

Experience with the commercial use of Bt cotton shows that the natural refuges derived from the mixed-planting system of cotton, corn, soybean and peanut on small-scale, single-family-owned farms play an important function in delaying Bt resistance in cotton bollworms (Jin *et al.* 2015). Additionally, studies are undertaken to identify significant alternative wild or cultivated host plants that serve as natural refuges for Asian corn borer in different maize-growing regions of China. If the alternate host crops (e.g. maize, sorghum, wheat, vegetables, soybean, peanut) and natural vegetation produce abundant primary maize pests, then refuge plantings of non-Bt maize may not be needed in some areas.

As the Chinese government has not yet approved the commercial planting of genetically modified maize with HT or IR/HT, it is impossible for weed to develop resistance to glyphosate/herbicide in maize in the field. But several whole-plant bioassays indicate that goosegrass *Eleusine indica* and red-root amaranth *Amaranthus retroflexus* have developed resistance to glyphosate, a herbicide widely used across the world (Yang and Bai 2014; Wang *et al.* 2016c).

Potential commercialisation of transgenic maize in China

Although multiple Bt maize events have been developed, and extensive tests have demonstrated their efficacy and benefits to human health and the environment (Pellegrino *et al.* 2018), no Bt maize event has yet been commercially planted in China. Lack of commercialisation of transgenic food crops is to some extent attributed to the fact that the public does not perceive the economic and environmental advantages, and has unjustified concern about safety despite rigorous tests (Li *et al.* 2016, 2018; Liu *et al.* 2016a). Bt transgenic cotton has been a shining example of enormous benefits in terms of economic returns as well as environmental safety in China. Fortunately, China's central government has put forward a series of policies to accelerate the development and application of transgenic crops in recent years and has shown commitment by approving transgenic maize BVLA430101 expressing *phytase* gene to enhance nutritional value. We believe that the cultivation of genetically modified maize may be approved in the near future as China has to depend upon maize imports to meet its domestic needs.[3] Its commercialisation will be of great strategic importance in increasing agricultural efficiency and farmers' incomes, and promoting competitiveness in an international market.

Acknowledgements

This research was funded as a National Science and Technology Major Project of China, No. 2016ZX080011003.

Endnotes

1. http://www.stats.gov.cn/tjsj/zxfb/201902/t20190228_1651265.html
2. http://www.moa.gov.cn/ztzl/zjyqwgz/zcfg/201007/t20100717_1601306.htm
3. http://www.globaltimes.cn/content/1146828.shtml

References

Bernardi O, Bernardi D, Horikoshi RJ, Okuma DM, Miraldo LL, Fatoretto J, Medeiros FC, Burd T, Omoto C (2016) Selection and characterization of resistance to the Vip3Aa20 protein from *Bacillus thuringiensis* in *Spodoptera frugiperda. Pest Management Science* **72**, 1794–1802. doi:10.1002/ps.4223

Carrière Y, Crickmore N, Tabashnik BE (2015) Optimizing pyramided transgenic Bt crops for sustainable pest management. *Nature Biotechnology* **33**, 161–168. doi:10.1038/nbt.3099

Chang XY, He KL, Wang ZY, Bai SX (2006) Evaluation of transgenic Bt maize for resistance to cotton bollworm. *Acta Phytophylacica Sinica* **33**, 374–378.

Chang X, Chang XY, He KL, Wang ZY, Bai SX (2007) Resistance evaluation of transgenic Bt maize to oriental armyworm. *Acta Phytophylacica Sinica* **34**, 225–228.

Chang X, Wang W, Shen ZC, Ye GY (2013) Evaluation of transgenic *cry1Ab/cry2Aj* maize for its resistance to *Ostrinia furnacalis. Acta Phytophylacica Sinica* **40**, 339–344.

Christou P (2002) No credible scientific evidence is presented to support claims that transgenic DNA was introgressed into traditional maize landraces in Oaxaca, Mexico. *Transgenic Research* **11**, 3–5. doi:10.1023/A:1013903300469

Conner AJ, Glare TR, Nap JP (2003) The release of genetically modified crops into the environment. Part II. Overview of ecological risk assessment. *Plant Journal* **33**, 19–46. doi:10.1046/j.0960-7412.2002.001607.x

Dai J, Li XY, Zhu L, Wang H, Zhang J, He KL, Lang ZH, Huang DF (2014) Molecular detection and agronomic traits analysis of insect-resistant transgenic maize harboring Bt *cry1Ah* gene. *Biotechnology Bulletin* **5**, 63–68.

Dai PL, Zhou W, Zhang J, Jiang WY, Wang Q, Cui HJ, Sun JH, Wu YY, Zhou T (2012) The effects of Bt Cry1Ah toxin on worker honeybees (*Apis mellifera ligustica* and *Apis cerana cerana*). *Apidologie* **43**, 384–391. doi:10.1007/s13592-011-0103-z

Dai PL, Jia HR, Jack CJ, Geng LL, Liu F, Hou CS, Diao QY, Ellis JD (2016) Bt Cry1Ie toxin does not impact the survival and pollen consumption of Chinese honey bees, *Apis cerana cerana* (Hymenoptera, Apidae). *Journal of Economic Entomology* **109**, 2259–2263. doi:10.1093/jee/tow204

Guo JF, He KL, Hellmich RL, Bai SX, Zhang TT, Liu YJ, Ahmed T, Wang ZY (2016) Field trials to evaluate the effects of transgenic *cry1Ie* maize on the community characteristics of arthropod natural enemies. *Scientific Reports* **6**, 22102. doi:10.1038/srep22102

Guo YH, Yang ZY, Sun HW, Li F, Yang SK, Lu XB (2014) Effect of phytase transgenic maize on the growing development and biochemical reaction of *Bombyx. Journal of Maize Sciences* **22**, 54–59.

Han Y, Wang H, Chen J, Cai WL, Hua HX (2015) No impact of transgenic *cry2Aa* rice on *Anagrus nilaparvatae,* an egg parasitoid of *Nilaparvata lugens,* in laboratory tests. *Biological Control* **82**, 46–51. doi:10.1016/j.biocontrol.2014.12.005

He FX, Lang ZH, Zhang J, Lu W, He KL, Huang DF (2012) Insect-resistant and glyphosate-tolerant transgenic maize transformed by Bt *cry1Ah* and *2mG2-epsps* genes. *Crops* **6**, 29–33.

He HP, Ren ZT, Shen WJ, Liu B, Xue K (2018) Effects of transgenic herbicide-tolerant maize on biodiversity of arthropod communities in the fields. *Journal of Ecology and Rural Environment* **34**, 333–341.

He KL, Wang ZY, Zhou DR, Wen LP, Song YY (2000) Methodologies and criterions for evaluating maize resistance to Asian maize borer. *Journal of Shenyang Agricultural University* **31**, 439–443.

He KL, Wang ZY, Zhou DR, Wen LP, Song YY, Yao ZY (2003) Evaluation of transgenic Bt corn for resistance to the Asian corn borer (Lepidoptera: Pyralidae). *Journal of Economic Entomology* **96**, 935–940. doi:10.1093/jee/96.3.935

He KL, Wang ZY, Wen LP, Bai SX, Zhou DR (2004) Transgenic maize evaluated for resistance to the Asian corn borer (Lepidoptera: Pyralidae). *Chinese Agricultural Science Bulletin* **20**, 240–246.

He K, Wang Z, Wen L, Bai S, Ma X, Yao Z (2005) Determination of baseline susceptibility to Cry1Ab protein for Asian corn borer (Lep. Crambidae). *Journal of Applied Entomology* **129**, 407–412. doi:10.1111/j.1439-0418.2005.00989.x

Jiang F, Liu KQ, Zhang TT, Bai SX, Wang ZY, He KL (2016) Tolerance of transgenic and herbicide- and insect-resistant corn to glyphosate. *Plant Protection* **42**, 74–78.

Jin L, Zhang HN, Lu YH, Yang YH, Wu KM, Tabashnik BE, Wu YY (2015) Large-scale test of the natural refuge strategy for delaying insect resistance to transgenic Bt crops. *Nature Biotechnology* **33**, 169–174. doi:10.1038/nbt.3100

Jing DP, Guo JF, Jiang YY, Zhao JZ, Sethi A, He KL, Wang ZY (2019) Initial detections and spread of invasive *Spodoptera frugiperda* in China and comparisons with other noctuid larvae in cornfield using molecular techniques. *Insect Science* doi:10.1111/1744-7917.12700

Li L, Wang J, Zhao Y, Liu HL (2015) Digestive stability of recombinant G10evo, Cry1Ab/Cry2Ab proteins of transgenic corn GAB-3 in simulated gastric and intestinal fluid. *Journal of Environmental Health* **32**, 112–115.

Li SK, Wang CT (2010) *Potential and Ways to High Yield in Maize.* Science Press, Beijing.

Li YH, Hallerman EM, Liu QS, Wu KM, Peng YF (2016) The development and status of Bt rice in China. *Plant Biotechnology Journal* **14**, 839–848. doi:10.1111/pbi.12464

Li YH, Hallerman EM, Peng YF (2018) How can China prepare for the domestic cultivation of Bt maize? *Trends in Food Science and Technology* **73**, 87–88. doi:10.1016/j.tifs.2018.01.012

Liu YB, Tabashnik BE (1997) Experimental evidence that refuges delay insect adaptation to *Bacillus thuringiensis. Proceedings. Biological Sciences* **264**, 605–610. doi:10.1098/rspb.1997.0086

Liu QS, Hallerman E, Peng YF, Li YH (2016a) Development of Bt rice and Bt maize in China and their efficacy in target pest control. *International Journal of Molecular Sciences* **17**, 1561–1575. doi:10.3390/ijms17101561

Liu YM, Liu QS, Wang YN, Chen XP, Song XY, Romeis J, Li YH, Peng YF (2016b) Ingestion of Bt corn pollen containing Cry1Ab/2Aj or Cry1Ac does not harm *Propylea japonica* larvae. *Scientific Reports* **6**, 23507. doi:10.1038/srep23507

Lu BR (2016) Challenges of transgenic crop commercialization in China. *Nature Plants* **2**, 16077. doi:10.1038/nplants.2016.77

Lu X, Song XY, Wu FC, Yang XY, Weng JF, Wang ZH, Li XH (2018) Effects of transgenic Bt maize CM8101 with *cry1Ab-Ma* gene on *Folsomia candida* and *Eisenia fetida. Journal of Environmental Entomology* **40**, 390–397.

Messeguer J (2003) Gene flow assessment in transgenic plants. *Plant Cell, Tissue and Organ Culture* **73**, 201–212. doi:10.1023/A:1023007606621

Montezano DG, Specht A, Sosa-Gómez DR, Roque-Specht VF, Sousa-Silva JC, Paula-Moraes SV, Peterson JA, Hunt TE (2018) Host plants of *Spodoptera frugiperda* (Lepidoptera: Noctuidae) in the Americas. *African Entomology* **26**, 286–300. doi:10.4001/003.026.0286

Nap JP, Metz PLJ, Escaler M, Conner AJ (2003) The release of genetically modified crops into the environment. Part I. Overview of current status and regulations. *Plant Journal* **33**, 1–18. doi:10.1046/j.0960-7412.2003.01602.x

Okuma DM, Bernardi D, Horikoshi RJ, Bernardi O, Silva AP, Omoto C (2018) Inheritance and fitness costs of *Spodoptera frugiperda* (Lepidoptera: Noctuidae) resistance to spinosad in Brazil. *Pest Management Science* **74**, 1441–1448. doi:10.1002/ps.4829

Pellegrino E, Bedini S, Nuti M, Ercoli L (2018) Impact of genetically engineered maize on agronomic, environmental and toxicological traits: a meta-analysis of 21 years of field data. *Scientific Reports* **8**, 3113. doi:10.1038/s41598-018-21284-2

Raybould A, Stacey D, Vlachos D, Graser G, Li X, Joseph R (2007) Non-target organism risk assessment of MIR604 maize expressing mCry3A for control of corn rootworm. *Journal of Applied Entomology* **131**, 391–399. doi:10.1111/j.1439-0418.2007.01200.x

Ren ZT, Shen WJ, Liu B, Xue K (2017) Effects of transgenic maize on biodiversity of arthropod communities in the fields. *Scientia Agricultura Sinica* **50**, 2315–2325.

Romeis J, Bartsch D, Bigler F, Candolfi MP, Gielkens MM, Hartley SE, Hellmich RL, Huesing JE, Jepson PC, Layton R, Quemada H, Raybould A, Rose RI, Schiemann J, Sears MK, Shelton AM, Sweet J, Vaituzis Z, Wolt JD (2008) Assessment of risk of insect-resistant transgenic crops to nontarget arthropods. *Nature Biotechnology* **26**, 203–208. doi:10.1038/nbt1381

Sanvido O, Romeis J, Gathmann A, Gielkens M, Raybould A, Bigler F (2012) Evaluating environmental risks of genetically modified crops: ecological harm criteria for regulatory decision-making. *Environmental Science and Policy* **15**, 82–91. doi:10.1016/j.envsci.2011.08.006

Shabbir MZ, Quan YD, Wang ZY, Bravo A, Soberón M, He KL (2018) Characterization of the Cry1Ah resistance in Asian corn borer and its cross-resistance to other *Bacillus thuringiensis* toxins. *Scientific Reports* **8**, 234–242. doi:10.1038/s41598-017-18586-2

Shen P, Zhang QY, Lin YH, Li WL, Li A, Song GW (2016) Thinking to promote the industrialization of genetically modified corn of our country. *China Biotechnology* **36**, 24–29.

Song M, Wang H, Zhang J, He KL, Liang GM, Zhu L, Huang DF, Lang ZH (2016) Resistance evaluation of Bt *cry1Ah*-transgenic maize to Asian corn borer, cotton bollworm and oriental armyworm. *Biotechnology Bulletin* **32**, 69–75.

Tabashnik BE, Brévault T, Carrière Y (2013) Insect resistance to Bt crops: lessons from the first billion acres. *Nature Biotechnology* **31**, 510–521. doi:10.1038/nbt.2597

Wang YQ (2018) Studies on evolution of resistance to Bt toxins in *Ostrinia furnacalis*. PhD thesis. Chinese Agricultural University, China.

Wang ZY, Lu X, He KL, Zhou DR (2000) Review of history, present situation and prospect of the Asian maize borer research in China. *Journal of Shenyang Agricultural University* **31**, 402–412.

Wang ZY, He KL, Wen LP, Zhang GY, Zheng L (2002) Spatial and temporal distributions of the fourth-generation cotton bollworm eggs on summer corn seeded at different times in North China. *Agricultural Sciences in China* **1**, 96–100.

Wang DY, Wang ZY, He KL, Cong B, Bai SX, Wen LP (2004) Temporal and spatial expression of Cry1Ab toxin in transgenic Bt corn and its effects on Asian corn borer, *Ostrinia furnacalis* (Guenée). *Scientia Agricultura Sinica* **37**, 1155–1159.

Wang ZY, Wang DY, He KL, Bai SX, Cong B (2005) Evaluation of the control effects of the transgenic *Bacillus thuringiensis* corn expressing Cry1Ab protein on the larvae of *Mythimna separata* (Walker) in laboratory. *Acta Phytophylacica Sinica* **32**, 153–157.

Wang ZY, He KL, Shi J, Ma SY (2006) Analysis of the heavy occurrence trend of the yellow peach borer in corn and management strategy for the insect pest. *Plant Protection* **32**, 67–69.

Wang P, He KL, Wang ZY, Wang YL (2012) Evaluating transgenic *cry1Ac* maize for resistance to *Ostrinia furnacalis* (Guenée). *Acta Phytophylacica Sinica* **39**, 395–400.

Wang S, Wang BF, Yan DS, Wang J, Wu FC, Xi JH, Song XY (2014a) Impacts of transgenic herbicide-resistant maize with *epsps* gene on arthropod biodiversity in the fields. *Journal of Biosafety* **23**, 271–277.

Wang YQ, He KL, Jiang F, Wang YD, Zhang TT, Wang ZY, Bai SX (2014b) Resistance of transgenic Bt corn variety Bt799 to the Asian corn borer. *Chinese Journal of Applied Entomology* **51**, 636–642.

Wang DM, Zhou L, Song XY (2016a) Evaluation on glyphosate tolerance and weediness risk of genetically modified maize CL38-1 with glyphosate-resistant character. *Journal of Weed Science* **34**, 23–28.

Wang J, Wu FC, Liu XY, Feng SD, Song XY (2016b) Evaluation of transgenic maize 'Shuangkang12-5' with complex traits of insect-resistance and glyphosate-resistance for the resistance to *Ostrinia furnacalis* and tolerance to glyphosate. *Plant Protection* **42**, 45–50.

Wang XL, Ma XY, Jiang WL, Ren XL, Ma YJ, Wang YY, Ma Y (2016c) Resistance of goosegrass [*Eleusine indica* (L.) Gaertn] to glyphosate in cotton fields. *China Cotton* **43**, 27–31.

Wang YQ, Wang YD, Wang ZY, Bravo A, Soberón M, He KL (2016d) Genetic basis of Cry1F-resistance in a laboratory selected Asian corn borer strain and its cross-resistance to other *Bacillus thuringiensis* toxins. *PLoS One* **11**, e0161189. doi:10.1371/journal.pone.0161189

Wang YQ, Yang J, Quan YD, Wang ZY, Cai WZ, He KL (2017a) Characterization of Asian corn borer resistance to Bt toxin Cry1Ie. *Toxins* **9**, 186–196. doi:10.3390/toxins9060186

Wang ZX, Li YH, He KL, Bai SX, Zhang TT, Cai WZ, Wang ZY (2017b) Does Bt maize expressing Cry1Ac protein have adverse effects on the parasitoid *Macrocentrus cingulum* (Hymenoptera: Braconidae)? *Insect Science* **24**, 599–612. doi:10.1111/1744-7917.12352

Wang YQ, Quan YD, Yang J, Shu CL, Wang ZY, Zhang J, Gatehouse AMR, Tabashnik BE, He KL (2019) Evolution of Asian corn borer resistance to Bt toxins used singly or in pairs. *Toxins* **11**, 461. doi:10.3390/toxins11080461

Xu LN (2010) Cross-resistance and molecular mechanism of Cry1Ab-selected *Ostrinia furnacalis* to other Bt toxins. PhD thesis. Chinese Academy of Agricultural Sciences, China.

Yang HN, Bai LY (2014) Resistance of *Amaranthus retroflexus* and *Portulaca oleracea* to glyphosate in cotton fields. *Cotton Science* **26**, 492–498.

Yang ZJ, Lang ZH, Zhang J, Song FP, He KL, Huang DF (2012) Studies on insect-resistant transgenic maize (*Zea mays* L.) harboring Bt *cry1Ah* and *cry1Ie* genes. *Journal of Agricultural Science and Technology* **14**, 39–45.

Yu SJ (1992) Detection and biochemical characterization of insecticide resistance in fall armyworm (Lepidoptera: Noctuidae). *Journal of Economic Entomology* **85**, 675–682. doi:10.1093/jee/85.3.675

Yue LL (2018) Impacts of transgenic insect-resistant maize (G1F-19) containing *mcry1F* gene on over-ground arthropod community structure. MSc thesis. Jinlin University, China.

Zhang DD, Wu KM (2019) The bioassay of Chinese domestic Bt-Cry1Ab and Bt-(Cry1Ab+Vip3Aa) maize against the fall armyworm, *Spodoptera frugiperda*. *Plant Protection* **45**, 54–60. doi:10.16688/j.zwbh.2019251

Zhang B, Yang Y, Zhou X, Shen P, Peng YY, Li YH (2017a) A laboratory assessment of the potential effect of Cry1Ab/Cry2Aj-containing Bt maize pollen on *Folsomia candida* by toxicological and biochemical analyses. *Environmental Pollution* **222**, 94–100. doi:10.1016/j.envpol.2016.12.079

Zhang TT, Coates BS, Wang YQ, Bai SX, Wang ZY, He KL (2017b) Down-regulation of aminopeptidase N and ABC transporter subfamily G transcripts in Cry1Ab and Cry1Ac resistant Asian corn borer, *Ostrinia furnacalis* (Lepidoptera: Crambidae). *International Journal of Biological Sciences* **13**, 835–851. doi:10.7150/ijbs.18868

Zhang TT, He MX, Gatehouse AMR, Wang ZY, Edwards MG, Li Q, He KL (2014) Inheritance patterns, dominance and cross-resistance of Cry1Ab- and Cry1Ac-selected *Ostrinia furnacalis* (Guenée). *Toxins* **6**, 2694–2707. doi:10.3390/toxins6092694

Zhang YW, Liu YJ, Ren Y, Liu Y, Liang GM, Song FP, Bai SX, Wang JH, Wang GY (2013) Overexpression of a novel *cry1Ie* gene confers resistance to Cry1Ac-resistant cotton bollworm in transgenic lines of maize. *Plant Cell, Tissue and Organ Culture* **115**, 151–158. doi:10.1007/s11240-013-0348-5

Zhao JZ, Cao J, Collins HL, Bates SL, Roush RT, Earle ED, Shelton AM (2005) Concurrent use of transgenic plants expressing a single and two *Bacillus thuringiensis* genes speeds insect adaptation to pyramided plants. *Proceedings of the National Academy of Sciences of the United States of America* **102**, 8426–8430. doi:10.1073/pnas.0409324102

Zhao SN (2016) Study on the glyphosate-resistance and living competitiveness of transgenic maize 'Shuangkang 12-6' and its effect on weeds variety. MSc thesis. Zhejiang University, China.
Zhao ZC, Guo WW, Su Y, Chen FJ (2013) Drift frequency and distance of transgenic *phyA* maize pollen. *Shengtaixue Zazhi* **32**, 3004–3009.
Zhou DR, He KL, Wang ZY, Ye ZH, Wen LP, Gao YX, Song YY (1995) *Asian Corn Borer and its Integrated Management*. Golden Shield Press, China.

9

Bringing Bt eggplant to resource-poor farmers in Bangladesh and the Philippines

Anthony M Shelton, Vijay Paranjape, Md J Hossain, Des Hautea, Md ZH Prodhan, Md A Hossain, Rituparna Majumder, Sayed H Sarwer, Dilafroza Khanam, Mohammad Kamrul Hasan, Abul K Azad, Bharat Char and Vijay Vijayaraghavan

Bt eggplant improves the lives of farmers in Bangladesh

The impact of biotechnology on the lives of resource-poor farmers is best told in a narrative. Below is the story of one famer in Bangladesh who chose to grow Bt eggplant, as described by Conrow (2019).

Mohammad Abul Hossain (Plate 1) is living proof that farmers can reap tremendous benefits from agricultural science. Like most of the ~150 000 smallholder farmers in Bangladesh, Hossain grows eggplant known as brinjal in Bangladesh. And like those other farmers, he used to experience tremendous losses due to the eggplant fruit and shoot borer (EFSB), a destructive pest that reduces eggplant yields and cuts into farmer profits.

Hossain had tried to fight the pest with insecticides. In Bangladesh, eggplant farmers typically spray insecticides two or three times per week — a rate that can increase to twice per day as harvest nears. Insecticides can account for 40% of a farmer's total production cost, while also inflicting severe impacts on human and environmental health. Worse, they're often ineffective, leaving farmers with yields reduced by 30–60%.

When Hossain heard about Bt eggplant, a genetically engineered (GE) eggplant variety that is resistant to the EFSB, he wanted to learn more. Developed by the Bangladesh Agricultural Research Institute (BARI) in cooperation with Cornell University, USAID and other partners, Bt eggplant is the first GE crop to be commercially released in Bangladesh. It contains a protein from a common bacterium that has been used to fight pests in organic crops around the world for more than 70 years.

Though his curiosity was piqued by the 2014 release of Bt eggplant, Hossain decided to hold back for a year and observe its performance in a relative's field. 'Realising its potential, I contacted BARI, met the officials and collected seed from the research station free of cost,' he recalled. After his first abundant harvest, he was so impressed that he began sharing what he knew with other farmers in his area.

'I have sold about 8056 kg (17 760 lbs) of Bt eggplant from my land,' Hossain told a group of more than 60 local farmers. 'There was no [borer] infestation in my crop, and I am expecting to harvest another 400 kg (881 lbs) of eggplant from the field with the same result.' Rather than spraying insecticides on a weekly basis, Hossain applied insecticides just two or three times during the entire season to protect against other insect pests.

In six months, Hossain earned about US$1600 growing Bt eggplant on a small piece of land. By comparison, he earned just $275 growing corn in the same field. Given the increased profits and reduced pesticide use, he thinks Bt eggplant will create hope among neighbouring farmers who struggle to earn a livelihood from their land.

Others are following his example. The number of Bangladeshi farmers growing Bt eggplant jumped from just 20 in 2014 to more than 27 000 in 2018, with more adopting the improved crop each day (Figure 9.1). Many of them have reported a dramatic reduction in insecticide use and a large increase in profit (Rashid *et al.* 2018; Ahmed *et al.* 2019).

This is not surprising since an analysis of the potential effect of Bt eggplant, if it were grown in Bangladesh, estimated that adoption of Bt eggplant would reduce insecticide use by 80% and increase the gross profit margin by nearly 45% (Islam and Norton 2007). In that same study, done more than a decade ago, it was estimated that widespread cultivation of Bt eggplant in Bangladesh would result in a benefit of more than US$200 million for the nation's smallholder eggplant farmers, a figure that would be much larger in 2019.

How this success came about in Bangladesh is a remarkable story that has been chronicled in several recent articles (Shelton *et al.* 2017, 2018, 2019; Mondal and Akter 2018; Prodhan *et al.* 2018). This chapter will highlight information from these publications but also include information on the situation in the Philippines where considerable work with Bt eggplant has been conducted and published, but where it has not yet been commercialised. References will also be made to India where Bt eggplant was evaluated for several years and passed the regulatory requirements but then was subjected to a moratorium for political reasons. Based on the success seen in Bangladesh, Indian farmers are now protesting this moratorium.

Based on the trials and safety of assessment of Bt eggplant, a strong argument can be made that Bt eggplant should be grown in more countries to improve the lives of farmers, the agricultural environment in which they farm, and the produce consumers purchase.

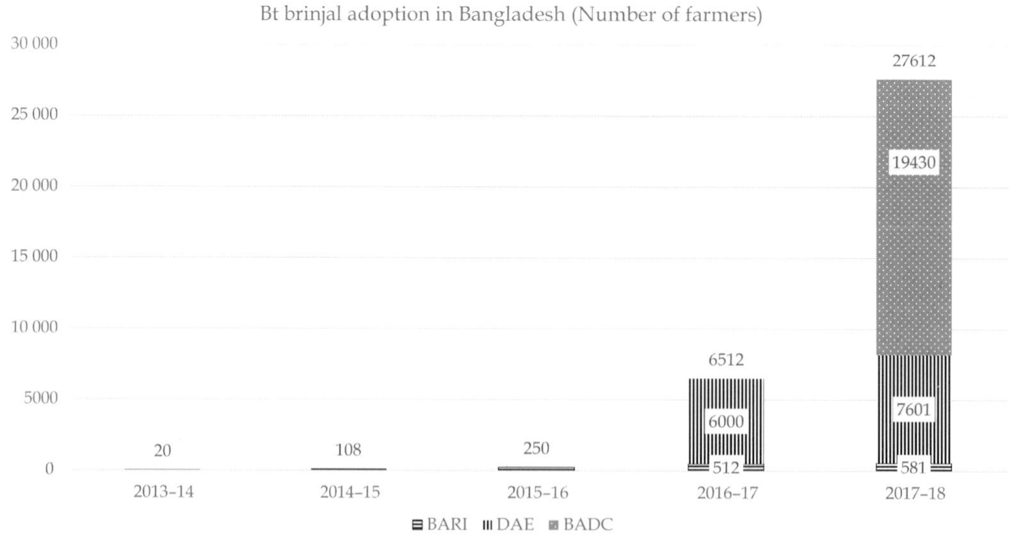

Figure 9.1 Adoption of Bt brinjal in Bangladesh. Farmers have three sources for obtaining the seeds or transplants: the Bangladesh Agricultural Research Institute (BARI), the Department of Agricultural Extension (DAE) and the Bangladesh Agricultural Development Corporation (BADC).

introgress EE-1 into them. In India, the ABSPII project had a vision to use two market channels to satisfy a diverse farming community – a 'pro-poor' channel for the distribution of open pollinated (OP) lines, and a commercial channel through which the higher-priced hybrid varieties would be sold. After extensive field testing in multi-location trials and safety evaluations conducted by the Indian biosafety body, the Genetic Engineering Appraisal Committee (GEAC) recommended commercialisation of Bt eggplant in India. However, responding to challenges from activist groups, on 9 February 2010 the Indian Minister of Environment and Forests imposed a moratorium on release, until political consensus could be reached (Shelton 2010). That moratorium was still in place as of April 2020. However, recent events in India may change this situation.

Commercialisation of Bt eggplant in Bangladesh

Bt eggplant was made available to growers for demonstration trials in 2014–15, when BARI provided seeds or transplants to its On-Farm Research Division (OFRD) to conduct research/demonstration trials on 108 farmer fields in 19 districts. In 2015–16 and 2016–17, demonstration trials were conducted in 250 farmer fields in 25 districts and 512 farmer fields in 36 districts, respectively. In 2017–18, BARI provided seeds to 581 farmers in 40 districts. In addition to distribution by BARI, the Department of Agricultural Extension distributed seeds to 6000 and 7601 farmers in 2016–17 and 2017–18, respectively. The Bangladesh Agricultural Development Corporation (BADC) sold seeds to an additional 19 430 farmers in 2018, making a total of 27 612 farmers in 2018 (the 2018 figures are slightly increased from previous reports by Mondal and Akter 2018 and Shelton *et al.* 2019) (Figure 9.1). Fruits from these fields were sold in the market and readily purchased by consumers. Because the four lines are OP varieties, farmers can keep their seed (or give or sell it to others) so the actual number of farmers growing Bt eggplant is likely higher than the statistics reported above.

Performance of Bt eggplant in Bangladesh

In a study conducted by BARI scientists in 35 districts during the 2016–17 cropping season using 505 Bt eggplant farmers and 350 non-Bt eggplant farmers, net returns per hectare were US$2151/ha for Bt eggplant compared to US$357/ha for non-Bt eggplant, a six-fold difference (Rashid *et al.* 2018). The study also indicated that farmers saved 61% of the pesticide cost compared to non-Bt eggplant farmers and experienced no losses due to EFSB. It is likely that pesticide use could be reduced further once farmers have more experience controlling EFSB with Bt eggplant, and thus economic returns would be even higher.

In a two-year study conducted by Prodhan *et al.* (2018), economic analysis revealed that all Bt lines had higher gross returns than non-Bt isolines whether they were sprayed or not. The non-sprayed, non-Bt isolines resulted in negative returns in most cases. Maximum fruit yield was obtained from sprayed plots compared to non-sprayed plots, indicating that other insects such as whiteflies, thrips and mites (which are not affected by Cry1Ac) can reduce plant vigour and subsequent fruit weight. Efforts by BARI scientists are underway to develop treatment guidelines for the complex of sucking insects that can reduce plant vigour.

A report by the International Food Policy and Research Institute (Ahmed *et al.* 2019) evaluated the impacts of the Bt eggplant technology on production systems, marketability and health. It compared results from 600 Bt eggplant farmers and 600 non-Bt eggplant farmers living in 200 villages. The study was limited to one of the four commercialised Bt varieties and to four districts in the north-west part of Bangladesh during the winter season of 2017–18. The findings were similar to those of other studies conducted in Bangladesh, with Bt plants showing nearly complete control of EFSB and farmers experiencing dramatically lower pesticide use, reduction in overall production costs, increased yields and higher profits. Their reduced exposure to highly toxic pesticides also resulted in decreased occurrences of pesticide-related symptoms and the subsequent need to seek medical care.

In a study conducted in 2019 in five districts in Bangladesh, the impacts of using Bt eggplant in regard to value and marketing were examined (Shelton *et al.* 2020). Based on a survey of Bt and non-Bt farmers, results indicated that Bt eggplant provided an average of 19.6% higher yield and 21.7% higher revenue compared to non-Bt varieties. On a per tonne basis, the revenue benefit of using Bt eggplant was 1.7%, reflecting different levels of acceptability among trade buyers and consumers. Some were prepared to pay higher prices for Bt eggplant compared to non-Bt eggplant because the fruit was less damaged, while others paid a price discount because the Bt eggplant was not available in preferred local varieties. Labour use, expressed in eight-hour days, for harvesting, grading and packaging of Bt eggplant was 14% higher for Bt eggplant, reflecting the increased yields of Bt eggplant. In terms of perceptions and experience of growing Bt eggplant, 83.1% of Bt eggplant growers were satisfied with the yields obtained, and 80.6% were satisfied with the quality of their fruit. This contrasts with non-Bt eggplant growers where 58.7% were satisfied with their yields and 28% indicated that a large portion of their fruit was infested (not a concern for Bt eggplant). Three-quarters of Bt eggplant growers planned to plant Bt eggplant next season because of the benefits of higher yields, revenue and fruit quality. Many also highlighted the benefits of reduced insecticide usage. Of the non-Bt growers, 39.6% had not heard of Bt eggplant. However, after hearing more about the impact of the technology, 71.4% of them indicated they planned to grow Bt eggplant next season. These findings confirm the significant benefits of Bt eggplant and highlight the importance of making the technology available in more varieties that are suitable to local conditions and consumer preferences. They also highlight the need for a more focused educational effort to increase farmer adoption.

Performance of Bt eggplant in the Philippines

Confined field trials conducted in the Philippines in 2010–12 (Plate 3) demonstrated that Bt lines provided essentially complete control of EFSB (Hautea *et al.* 2016) and no negative effect on non-target arthropods (Navasero *et al.* 2016). Data from the confined field trials were also used for *ex ante* analysis of costs and benefits from Bt eggplant adoption in the Philippines (Gerpacio and Aquino 2014). Results indicated that Bt eggplant adoption will have significant economic, health and environmental benefits due to lower production costs, increased marketable yield, higher net income and reduced insecticide use in eggplant production. Consumers may gain because more eggplants will be available at lower prices. Such results led to the expectation that UPLB's Bt eggplants would move through the regulatory process in a timely manner and become available to Filipino farmers. This has not yet occurred, because of the reasons outlined below.

Performance of Bt eggplant in India

In India, Bt eggplant with EE-1 was field tested from 2004 to 2008 in more than 50 locations representing the major eggplant-growing regions in India. Multi-location research trials, Indian Council of Agricultural Research trials and large-scale trials were conducted in addition to the greenhouse experiments and confined field trials. Trials were conducted for eight hybrids and 16 varieties developed by Mahyco; Tamil Nadu Agricultural University, Coimbatore; University of Agricultural Science, Dharwad; and Indian Institute of Vegetable Research, Varanasi. As summarised by Choudhary and Gaur (2009), the Bt eggplant hybrids provided nearly complete control of EFSB and required 77% less insecticides than non-Bt counterparts.

Rigorous food and feed safety studies, including toxicity and allergenicity evaluation and nutritional studies, confirmed that Bt eggplant is as safe as its non-Bt counterparts. Pollen flow, effects on soil microflora and non-target organisms, agronomy, germination and weediness studies indicate the safety of Bt eggplant with no unintended effects. The studies clearly indicated positive implications of Bt eggplant for human and environmental health and significant reduction in production costs.

Bt eggplant in the Philippines: the evolving regulatory framework

As mentioned above, biotech crops are grown under a country's regulatory system which may change over time. The evolution of the Philippine biosafety regulation is described in detail on the Philippine Biosafety Clearing House (BCH Pilipinas) website and is only briefly summarised here. Biosafety regulation in the Philippines has been promulgated through executive orders issued by the Office of the President of the Philippines. In 1990, Executive Order 430 (EO 430) created the National Committee on Biosafety of the Philippines in response to the urgent need to regulate the conduct of research on genetic engineering in the country. When the Philippines ratified the Cartagena Protocol on Biosafety, EO 514, which superseded EO 430, was issued in 2006. EO 514 established the National Biosafety Framework, which prescribes guidelines for its implementation, including provisions for more public participation, reorganising the National Committee on Biosafety of the Philippines, and other purposes. In 2002, the Department of Agriculture issued Administrative Order No. 8, implementing guidelines to regulate biotech crops intended for field trials, cultivation and importation for direct use as food and feed. Commercial planting of Bt corn in the Philippines started in 2003. Under EO 514, the Department of Science and Technology Biosafety Committee continued to supervise all regulated biotech activities for contained use, while the Bureau of Plant Industry, through DAO 08–2002, regulated the importation, field trials and cultivation of crops in the country. However, on 8 December 2015, the Philippine Supreme Court declared DAO 08–2002 null and void, temporarily stopping any application for contained use, field testing, propagation and commercialisation, and importation of GMOs. Although the Supreme Court reversed that decision on 26 July 2016, Joint Department Circular No. 1, Series of 2016 had been approved and signed on 7 March 2016 and taken effect on 15 April 2016. Due to recent changes in the regulatory system, there have been substantial delays in renewing permits and approving new applications for field trials, importation and cultivation of biotech crops. Applications for cultivation of Bt eggplants will be evaluated under the new regulatory system.

Challenges to Bt eggplant in the Philippines

Despite the fact that Bt maize has been cultivated and publicly accepted for more than 10 years in the Philippines, the development and commercialisation of Bt eggplant encountered serious challenges from active anti-GMO groups which have waged a relentless campaign against Bt eggplant since its early development. Like the Bt corn trials in the early 2000s, the Bt eggplant field trials were vandalised. In February 2011, Greenpeace South-east Asia and some Indian nationals forcibly entered the Bt eggplant confined field trial and illegally uprooted the plants. Because of the regulatory breach, UPLB was forced to request early termination of the field trial. The confined field trial in UP Mindanao suffered a similar fate.

When their negative media campaign and vandalism failed to stop the Bt eggplant field trials, the anti-GMO groups filed a lawsuit against the technology developers and regulators. In April 2012, they filed a petition in the Supreme Court of the Philippines for Writ of Continuing Mandamus and Writ of Kalikasan (Writ of Nature), urgently requesting immediate termination of the Bt eggplant field trials, which they alleged violated the petitioners' constitutional right to health and a balanced ecology. The case was heard by the Court of Appeals for more than a year; the Bt eggplant field trials were completed before all the evidence was presented. In September 2013, the Court of Appeals issued an order to permanently stop Bt eggplant field trials. The Supreme Court upheld the Court of Appeals decision on 8 December 2015 and ordered a stop to field testing Bt eggplant, declared DAO 08–2002 null and void, and temporarily stopped all GMO activities until a new administrative order was promulgated. The technology developers and other intervening parties appealed the decision and on 26 July 2016 the Supreme Court granted all motions for reconsideration and set aside its decision of 8 December 2015. Although the final decision effectively allowed for the continuance of the Bt eggplant project, the four years of court proceedings (April 2012–July 2016) resulted in substantial delays in the development and commercialisation of Bt eggplant in the Philippines.

The situation in India

Beginning in 2004, Bt hybrids were tested in several states of India and performed well, leading Krishna and Qaim (2007) to comment that 'several Bt hybrids have been tested in the field and are likely to be commercialised in the near future.' However, this has not happened. The technology was expected to make a noteworthy contribution towards alleviating poverty by increasing the income of resource-poor eggplant-growing farmers and providing consumers with a more affordable source of vegetables (Choudhary and Gaur 2009). The GEAC, which is the apex regulatory authority of India, had recommended the commercial release of Bt eggplant. However, anti-GMO organisations, primarily Greenpeace, had a different plan and for years fought against Bt eggplant in courts of law and in the court of public opinion. Largely as a result of anti-GMO activists' efforts, the Indian Minister of the Environment and Forests, the last gatekeeper before Bt eggplant would be commercialised, imposed a moratorium on Bt eggplant on 9 February 2010; this remains in force today (Shelton 2010). India's neighbour, Bangladesh, which also produces eggplant, decided to move forward, much to the consternation of Indian farmers.

From a regulatory standpoint, not much has changed in India since the moratorium went into effect. However, farmers started protesting and resorted to illegal cultivation of GE crops, including Bt eggplant and herbicide-tolerant (HT) cotton (Plate 4) (Deshpande 2019).). The current events – from farmers growing illegal GM eggplant in Haryana to those openly growing unapproved HT cotton in Maharashtra – is an indication that they are eager to evaluate new crops and technologies, even at the risk of facing legal action. The illegal cultivation of Bt eggplant defies the indefinite moratorium on the commercial release of Bt eggplant in India.

Growing unapproved varieties of crops is the outcome of a decade-long government paralysis (TOI Editorial 2019). The current situation may be perceived as a collective failure of the regulatory system and associated government institutions. The regulatory failure has not only created a deep mistrust about GM technology among the doubters, but has also potentially exposed farmers and consumers to unapproved technology. The case of GM eggplant is particularly disturbing. Bt eggplant has been grown in Bangladesh since 2013–14 and there was speculation that the seeds could have crossed to India from Bangladesh. However, the test report by the National Bureau of Plant Genetic Resources confirmed that the samples from Haryana are GM but do not have Mahyco's *cry1Ac* transgene, suggesting that the GM seeds used by farmers might have come straight from an Indian laboratory to the field, without entering the regulatory approval process (TOI Editorial 2019).

The regulatory delay for Bt eggplant is not unique in India, as shown by the example of GM mustard developed by an Indian public sector organisation. After GEAC approved GM mustard for commercial cultivation, it asked the developer to conduct a trial on the impact on honeybees, to allay any fears about damage to the environment. Such continuing requests in relation to GM mustard, after its approval, makes its ultimate fate uncertain (Bera 2019). Such developments may force the government and regulators to take a fresh look at the approval of new GM seeds by reviving the regulatory process and making approved and tested seeds available to Indian farmers at the earliest opportunity. The ball is now in the court of the new government that came to power in 2019.

Communications

Both ABSPII and Feed the Future South Asia Eggplant Improvement Partnership have devoted a major effort to communicate about the science involved in creating Bt eggplant, the need it fulfils and the success it has achieved. Communication activities focus on sensitising key stakeholder groups to the project; building alliances and communication capacity through social mobilisation; capacity-building with stakeholders and support networks; monitoring and analysing public knowledge, attitudes and opinions regarding agricultural biotechnology in general and Bt eggplant in particular; addressing the vigorous and well-resourced current and anticipated anti-biotech campaigns by undertaking communication activities to understand, monitor and influence public knowledge and opinion regarding modern biotechnology in general and Bt eggplant in particular; utilising media advocacy by developing proactive messaging campaigns and correcting misinformation; and promoting biotechnology through scientific publications, appropriate online tools, knowledge banks, data-sharing, reporting, newsletters, blogs and strategic use of social media.

The project website (http://bteggplant.cornell.edu/) serves as the home for information about the project and is updated regularly. It features a Q&A series about Bt eggplant, project updates and compelling stories told by Bangladeshi farmers who have benefited from it. A Facebook page and Twitter feed further help to inform anyone who is interested in the project. The site also hosts scientific papers for those wishing to dig deeper into the project.

The project partners, including the communication staff at USAID, have engaged in other outreach efforts through newsletters and various multimedia tools, including videos that BARI has produced to highlight the technology. The Bt eggplant project in the Philippines has worked continuously with the International Service for the Acquisition of Agri-biotech Applications (ISAAA) (http://www.isaaa.org) and its network of Biotechnology Information Centers and local partner organisations in implementing its communication and outreach strategy. ISAAA and its key partner, SEARCA-BIC, provide information and training in a multitude of formats. Its most important activities in support of Bt eggplant have focused on mobilising strong public support for Bt eggplant and coordinating a multi-stakeholder response during the court case, media advocacy, continuous outreach activities to key stakeholder groups, organising public consultations, producing 30 biotech-on-air episodes and videos, social media campaigns, and e-newsletters published through the SEARCA-BIC website (www.bic.searca.org).

Other partners include the Cornell Alliance for Science (https://allianceforscience.cornell.edu), which is funded by the Bill & Melinda Gates Foundation. The Alliance for Science 'seeks to promote access to scientific innovation as a means of enhancing food security, improving environmental sustainability, and raising the quality of life globally.' It devotes significant efforts to training people around the world about biotechnology and 'how to communicate effectively about science and promote evidence-based decision-making.' One highlight of Alliance programming is a 12-week intensive training course held at Cornell University. It focuses on strategic planning, grassroots organising and innovative communications with a strong emphasis on stories of science and effective communications around agricultural biotechnology. The project continues to work with ISAAA and other regional partners, providing information and training in a multitude of formats. Most recently, the Alliance has begun to work with Farming Future Bangladesh, an advocacy and communications initiative funded by the Bill & Melinda Gates Foundation to support improved understanding of science-based agricultural technology in Bangladesh.[3]

Protecting the technology

The financial investment and human capital involved in developing Bt eggplant and commercialising it in Bangladesh have been enormous and taken nearly two decades. Prior to its deployment, the project recognised the need to implement stewardship to ensure the product remains effective for an extended period. The evolution of resistance by the EFSB to Cry1Ac expressed by the plants is the main threat to the continued effectiveness of Bt eggplant in the field. There is a growing number of instances of insects evolving resistance to particular Bt proteins – the threat of control failure is real (Tabashnik *et al.* 2013). Implementing an insect resistance management (IRM) strategy is paramount for protecting the technology and a requirement for registration in the Philippines.

When the Bangladesh government approved the release of four Bt eggplant varieties in 2013, it did so as a limited release, and this will reduce the likelihood of resistance, for several reasons. First, Bt transplants or seeds were only available on a limited basis: 20 in 2014, 108 in 2015, 250 in 2016. Not until 2018 were the seeds widely used, when the Bangladesh Agricultural Development Corporation started distributing GM seed and the number of farmers using it increased to 27 612 in 2018 out of ~150 000 total eggplant farmers. Second, there are strong preferences for different types of eggplant in different regions, and the four released varieties are not preferred in all the regions. Third, most of the varieties that were released are not suitable for summer production, which is the time when EFSB populations are highest. These factors reduce the selection pressure for resistance. Three other strategies to reduce the likelihood for resistance are using a refuge of non-Bt plants to conserve susceptible alleles in the pest population, monitoring for any changes in susceptibility to Cry1Ac by periodic sampling of EFSB populations in the field, and developing plants that express multiple Bt proteins with different modes of action (Bates *et al.* 2005).

If resistance is a recessive characteristic (which it is in most cases), this can be an effective strategy because resistance alleles initially occur most frequently in the heterozygous form, and are rare. When these individuals, or the very rare individuals possessing the *rr* genotype, mate with individuals of the *ss* genotype and lay their eggs on the crop, the emerging larvae will be killed. Greenhouse studies (Tang *et al.* 2001) demonstrated the effectiveness of the refuge strategy for delaying resistance. This strategy has been proposed to be a major reason for the relatively few cases of field-evolved resistance. In Bangladesh, the recommendation is that 5% of the field be non-Bt eggplant and the refuge consist of a single row of non-Bt eggplant on all sides of the field.

The second strategy for monitoring resistance is commonly included in the registration requirements for Bt corn and Bt cotton in the USA and other countries. Monitoring methods can include bioassays of several doses or use of a discriminating dose (Bates *et al.* 2005). Monitoring studies have been conducted in Bangladesh on 18 regional populations of EFSB (Prodhan *et al.* 2019) and more are underway.

The use of pyramiding of Bt genes into a single variety has been promoted as an effective IRM strategy (Gould 1986; Roush 1998) and demonstrated by Zhao *et al.* (2003) to enhance durability. Mahyco has developed a dual Bt gene event that should be used for the second generation of Bt eggplant.

Excellence through stewardship and farmer training

Because BARI is currently the sole producer of Bt eggplant seed in Bangladesh (although the breeder seed is increased by BADC), it is imperative that it produce, package and distribute quality seed. Because Bt eggplant is the first biotech crop grown in Bangladesh, a major effort has been made to build BARI's capacity to produce quality seed with high genetic purity, viability and expression of Cry1Ac, in adequate amounts to meet grower demand, and to monitor performance of the crop. The capacity-building efforts follow the model of 'Excellence Through Stewardship,' a life-cycle approach to GE product management (www.excellencethroughstewardship.org). Training efforts have greatly increased the capacity of BARI scientists to produce, store and distribute quality seed. Additional efforts are needed to monitor growers' compliance with planting refuges to delay the evolution of resistance, especially since they can save seed from the four GM varieties for next year's planting. Seed-saving is formally discouraged by BARI because of the potential for outcrossing.

As with any new technology, farmer training is essential. Over the last few years in Bangladesh, BARI, OFRD and DAE officials have directly or indirectly trained over 22 000 farmers to grow Bt eggplant following proper agronomic and stewardship practices. In addition, over 650 sub-assistant agriculture officers and extension officials were trained as part of a 'train the trainer' program. They in turn each trained several farmers to grow Bt eggplant. Further, BARI organises farmer field days every year that bring in several farmers to experience the technology.

Public sector challenges

Globally, most successful commercial plant biotech products (e.g. GM cotton, corn and soybean) have come from the private sector. These technologies have been sustained over years due to a robust product pipeline and effective implementation of stewardship and communication strategies. Bt eggplant is the first product developed by a public institution (BARI, Bangladesh) with a technology licensed from a private company (Mahyco, India) and funded by USAID. The product was successfully released in 2014, and sustaining its benefits requires a well-developed strategy that encompasses good stewardship, a pipeline of products for the future, farmer training and public education. In Bangladesh this will require coordination between different departments involved in seed production, seed processing, packing, distribution, communication, extension, seed quality testing, farmer training and so on. This will be a challenge for the public sector. Furthermore, a public sector program often lacks the research needed for new product development and a pipeline to replace the current products. The private sector has more experience in bringing GM products to market and overseeing their performance in all phases of production, and developing a pipeline to meet the changing market demands. Public sector organisations will be largely dependent on the private sector for future products; these will come with conditions that protect the developers' intellectual property in the licensed technology.

Conclusion: looking forward

The development and commercialisation of Bt eggplant in Bangladesh was successful largely because of the severity of the problem caused by EFSB, the lack of effective alternatives, the scientific community's favourable experience with Bt crops, the support of USAID and the political will of influential people in government, especially Prime Minister Sheikh Hasina and Agriculture Minister Begum Matia Chowdhury. What will sustain it in Bangladesh will be adherence to proper stewardship practices in the laboratory and field, development of new lines with enhanced traits (e.g. multiple Bt genes, disease resistance, etc.), increased farmer training, enhanced involvement of the private sector and a well-functioning regulatory system. Most importantly, Bangladesh will have to take full responsibility and leadership for the project as USAID shifts its resources to other needs.

The fate of Bt eggplant in the Philippines will depend on its regulatory system and political pressure. At the time of writing, the regulatory dossier for food and feed and the regulatory dossier for cultivation were being finalised for submission. Both will be needed if Bt eggplant is to be grown in the Philippines. Pressure from parties that promote or oppose the commercialisation of GM crops will likely try to sway the legal channels and the attitude of the general public.

In India, the moratorium on growing Bt eggplant that was put into place in 2010 still exists at the time of writing. The frustration of Indian farmers who can look across the border and see their Bangladeshi neighbours increase their eggplant harvests and dramatically reduce their insecticide use, has led to civil unrest and pressure on the government to overturn the moratorium. A similar situation occurred in India with Bt cotton in 2001, described by Herring (2007). Bt cotton was illegally planted and was controlling bollworms while the neighbouring non-Bt fields were being devastated. The government ordered the Bt cotton fields to be burned but the orders were not carried out. Farmers from cotton-growing areas threatened to cultivate Bt cotton whether or not it was approved. The government quickly approved three Bt cotton varieties, and plantings of these have grown tremendously. In 2017, 93% of the cotton area in India was grown to Bt cotton, making it the world leader in Bt cotton production (ISAAA 2017). In that same year, India harvested nearly one-quarter of the world's production of eggplant. It is reasonable to predict that Bt eggplant, if approved in India, would have a similar adoption trend as Bt cotton.

Trials in Bangladesh, the Philippines and India have demonstrated the effectiveness of Bt eggplant to control EFSB. However, it is only in Bangladesh that Bt eggplant has been commercialised, resulting in increased income for farmers, dramatically reduced use of pesticides, and consumer benefits. Bangladeshi farmers have been pleased with the results (Plate 5). Whether other countries will choose to follow the lead of Bangladesh and reap these benefits is up to them. Given the documented benefits, one can ask – what other evidence is needed for more countries to adopt this technology?

Endnotes

1. http://www.ikisan.com/tg-brinjal-history.html
2. http://www.isaaa.org/gmapprovaldatabase/
3. https://farmingfuturebd.com/

References

Ahmed AU, Hoddinott JF, Islam KMS, Khan ASMMR, Abedin N, Hossain NZ (2019) *Impact of Bt Brinjal (Eggplant) Technology in Bangladesh.* Project Report prepared for the US Agency for International Development. International Food Policy Research Institute, Dhaka. <http://www.ifpri.org/>

Anderson JA, Ellsworth PC, Faria JC, Head GP, Owen MD, Pilcher CD, Shelton AM, Meissle M (2019) Genetically engineered crops need to be part of a diversified integrated pest management plan for improved durability and sustainability of agricultural systems. *Frontiers in Bioengineering and Biotechnology* 7, 24. doi:10.3389/fbioe.2019.00024

Bates SL, Zhao JZ, Roush RT, Shelton AM (2005) Insect resistance management in GM crops: past, present and future. *Nature Biotechnology* **23**, 57–62. doi:10.1038/nbt1056

Bera S (2019) Inside India's genetic crop battlefield. *Livemint*, 20 June. <https://www.livemint.com/industry/agriculture/inside-india-s-genetic-crop-battlefield-1561054298998.html>

Choudhary B, Gaur K (2009) *The Development and Regulation of Bt Brinjal in India (Eggplant/ Aubergine).* ISAAA Brief No. 38. ISAAA, Ithaca, NY. <https://www.isaaa.org/resources/publications/briefs/38/download/isaaa-brief-38-2009.pdf>

Conrow J (2019) *Crop technology Improves Farmers' Lives in Bangladesh.* <http://bteggplant.cornell.edu/content/news/blog/crop-technology-improves-farmers%E2%80%99-lives-bangladesh>

Del Prado-Lu JL (2015) Insecticide residues in soil, water, and eggplant fruits and farmers' health effects due to exposure to pesticides. *Environmental Health and Preventive Medicine* **20**, 53–62. doi:10.1007/s12199-014-0425-3

Deshpande A (2019) To protest ban, farmers to plant GM cotton, brinjal seeds today. *The Hindu*, 10 June. <https://www.thehindu.com/news/cities/mumbai/to-protest-ban-farmers-to-plant-gm-cotton-brinjal-seeds-today/article27707414.ece>

FAO (2013) *FAOSTAT Production Databases.* <http://www.fao.org/faostat>

FAO (2017) *FAOSTAT Production Databases.* <http://www.fao.org/faostat>

Francisco SR (2014) Socioeconomic impacts of Bt eggplant: evidence from multi-location field trials. In *Socioeconomic Impacts of Bt Eggplant: Ex-ante Case Studies in the Philippines.* (Eds RV Gerpacio and AP Aquino) pp. 205–232. International Services for the Acquisition of Agric.-Biotech Applications and the Southeast Asian Ministers of Education Organization–Southeast Asia Regional Center for Graduate Study and Research in Agriculture. Ithaca, NY and Los Banos, Laguna.

Fuchs R, Ream J, Hammond B, Naylor M, Leingruber RM, Berberich S (1993) Safety assessment of the neomycin phosphotransferase II (NPTII) protein. *Nature Biotechnology* **11**, 1543–1547. doi:10.1038/nbt1293-1543

Gerpacio RV, Aquino AP (Eds) (2014) *Socioeconomic Impacts of Bt Eggplant: Ex-ante Case Studies in the Philippines.* International Services for the Acquisition of Agric.-Biotech Applications and the Southeast Asian Ministers of Education Organization–Southeast Asia Regional Center for Graduate Study and Research in Agriculture. Ithaca, NY and Los Baños, Laguna.

Gould F (1986) Simulation models for predicting durability of insect resistant germplasm: a deterministic model. *Environmental Entomology* **15**, 1–10. doi:10.1093/ee/15.1.1

Grafius EJ, Douches DS (2008) The present and future role of insect-resistant genetically modified potato cultivars in IPM. In *Integration of Insect-resistant Genetically Modified Crops with IPM Systems.* (Eds J Romeis, AM Shelton and GG Kennedy) pp. 195–221. Springer, Berlin.

Gregory P, Potter RH, Shotkoski FA, Hautea DM, Raman KV, Vijayaraghavan V, Lesser WH, Norton G, Coffman WR (2008) Bioengineered crops as tools for international development: opportunities and strategic considerations. *Experimental Agriculture* **44**, 277–299. doi:10.1017/S0014479708006352

Hautea DM, Cruz VM, Hautea RA, Vijayaraghavan V (2011) Transgenic horticultural crops in Asia. In *Transgenic Horticultural Crops: Challenges and Opportunities.* (Eds B Mou and R Scorza) pp. 155–174. CRC Press, Boca Raton, FLA.

Hautea DM, Taylor LD, Masanga APL, Sison MLJ, Narciso JO, Quilloy RB, Hautea RA, Shotkoski FA, Shelton AM (2016) Field performance of Bt eggplants (*Solanum melongena* L.) in the Philippines: Cry1Ac expression and control of the eggplant fruit and shoot borer (*Leucinodes orbonalis* Guenée). *PLoS One* **11**, e0157498. doi:10.1371/journal.pone.0157498

Herring RJ (2007) Stealth seeds: bioproperty, biosafety, biopolitics. *Journal of Development Studies* **43**, 130–157. doi:10.1080/00220380601055601

IITA (2019) *Major Breakthrough for Farmers and Scientists as Nigerian Biotech Body Approves Commercial Release of Genetically Modified Cowpea.* Press release, 9 February. International Institute for Tropical Agriculture.<http://www.iita.org/news-item/major-breakthrough-for-farmers-and-scientists-as-nigerian-biotech-body-approves-commercial-release-of-genetically-modified-cowpea/>

ILSI (2010) A review of the environmental safety of the Cry1Ac protein. *Environmental Biosafety Research* **10**, 27–49.

ILSI (2016) *A Review of the Food and Feed Safety of the Cry1Ac Protein.* ILSI Research Foundation. <https://ilsirf.org/wp-content/uploads/sites/5/2017/02/cry1ac_en_ffs.pdf>

ISAAA (2017) *Global Status of Commercialized Biotech/GM Crops in 2017: Biotech Crop Adoption Surges as Economic Benefits Accumulate in 22 Years.* ISAAA Brief No. 53. ISAAA, Ithaca, NY.

Islam MN, Karim MA (1991) Management of brinjal shoot and fruit borer, *Leucinodes orbonalis* Guenee, (Lepidoptera: Pyralidae) in the field. In *Annual Research Report 1990–91*, pp. 44–46. Entomology Division, BARI, Gazipur, Bangladesh.

Islam SMF, Norton GW (2007) Bt eggplant for fruit and shoot borer resistance in Bangladesh. In *Economic and Environmental Benefits and Costs of Transgenic Crops: Ex-Ante Assessment.* (Eds C Ramasamy,

KN Selvaraj, GW Norton and K Vijayaraghavan) pp. 91–106. Tamil Nadu Agricultural University, Coimbatore, India.

Krishna VV, Qaim M (2007) Potential socioeconomic impacts of Bt eggplant in India. In *Economic and Environmental Benefits and Costs of Transgenic Crops: Ex-Ante Assessment.* (Eds C Ramasamy, KN Selvaraj, GW Norton and K Vijayaraghavan) pp. 57–71. Tamil Nadu Agricultural University, Coimbatore, India.

Lu JL (2012) Pesticide residues in eggplant during dry and wet seasons in Sta. Maria, Pangasinan. *Philippine Journal of Crop Science* **37**, 93–98.

Lu JL (2014) Insecticide residues in soil, water, and eggplant fruits: the case of Sta. Maria, Pangasinan. In *Socioeconomic Impacts of Bt Eggplant: Ex-ante Case Studies in the Philippines.* (Eds RV Gerpacio and AP Aquino) pp. 61–96. International Services for the Acquisition of Agric.-Biotech Applications and the Southeast Asian Ministers of Education Organization–Southeast Asia Regional Center for Graduate Study and Research in Agriculture. Ithaca, NY and Los Baños, Laguna.

Mondal MRI, Akter N (2018) *Success on Bt Brinjal in Bangladesh.* Asia-Pacific Consortium on Agricultural Biotechnology and Bioresources Asia-Pacific Association of Agricultural Research Institutions, Bangkok, Thailand.

Naranjo SE, Hellmich RL, Romeis J, Shelton AM, Velez AM (2019) The role and use of genetically engineered insect-resistant crops in IPM systems. In *Integrated Management of Insect, Mite and Nematode Pests in Agriculture. Vol 2: Current and Future Developments in IPM.* (Eds M Kogan and L Higley) pp. 283–340. Burleigh Dodds Science Publishing, Cambridge, UK.

NASEM (2016) *Genetically Engineered Crops: Experiences and Prospects.* National Academies of Sciences, Engineering, and Medicine. National Academies Press, Washington, DC. doi:10.17226/23395

Navasero MV, Candano RN, Hautea D, Hautea RA, Shotkoski FA, Shelton AM (2016) Assessing potential impact of Bt eggplants on non-target arthropods in the Philippines. *PLoS One* **11**, e0165190. doi:10.1371/journal.pone.0165190

Parker BL, Talekar NS, Skinner M (1995) *Field Guide: Insect Pests of Selected Vegetables in Tropical and Subtropical Asia.* Publication 94–427. AVRDC, Tainan, Taiwan.

Philippine Statistic Authority (2016) *Highlights of the Philippine Population 2015 Census of Population.* <https://psa.gov.ph/content/highlights-philippine-population-2015-census-population>

Philippine Statistic Authority (2019) *Employment Rate in April 2019 is estimated at 94.9 Percent.* <https://psa.gov.ph/content/employment-rate-april-2019-estimated-949-percent>

Prodhan MZH, Hasan MT, Chowdhury MMI, Alam MS, Rahman ML, Azad AK, Hossain MJ, Naranjo SE, Shelton AM (2018) Bt eggplant (*Solanum melongena* L.) in Bangladesh: fruit production and control of eggplant fruit and shoot borer (*Leucinodes orbonalis* Guenée), effects on non-target arthropods and economic returns. *PLoS One* **13**, e0205713. doi:10.1371/journal.pone.0205713

Prodhan ZH, Shirale DK, Islam Z, Hossain J, Paranjape V, Shelton AM (2019) Susceptibility of field populations of eggplant fruit and shoot borer (*Leucinodes orbonalis* Guenée) to Cry1Ac, the protein expressed in Bt eggplant (*Solanum melongena* L.) in Bangladesh. *Insects* **2019**(10), 198. doi:10.3390/insects10070198

Quicoy CB (2014) The eggplant subsector in Davao region, North Cotabato, Iloilo and Southern Leyte. In *Socioeconomic Impacts of Bt Eggplant: Ex-ante Case Studies in the Philippines.* (Eds RV Gerpacio and AP Aquino) pp. 97–125. International Services for the Acquisition of Agric.-Biotech Applications and the Southeast Asian Ministers of Education Organization–Southeast Asia Regional Center for Graduate Study and Research in Agriculture. Ithaca, NY and Los Baños, Laguna.

Ramasamy S (2009) *Insect and Mite Pests on Eggplant: A Field Guide for Identification and Management.* AVRDC Publication No. 09–729. World Vegetable Center, Shanhua, Taiwan.

Rashid MA, Hasan MK, Matin MA (2018) Socio-economic performance of Bt eggplant cultivation in Bangladesh. *Bangladesh Journal of Agricultural Research* **43**, 187–203. doi:10.3329/bjar.v43i2.37313

Rotino GL, Perri E, Acciarri N, Sunseri F, Arpaia S (1997) Development of eggplant varietal resistance to insects and diseases via plant breeding. *Advances in Horticulture Science* **11**, 193–201.

Roush RT (1998) Two-toxin strategies for management of insecticidal transgenic crops: can pyramiding succeed where pesticide mixtures have not? *Philosophical Transactions of the Royal Society of London. Series B, Biological Sciences* **353**, 1777–1786. doi:10.1098/rstb.1998.0330

Russell D, Uijtewaal B, Shelton AM, Chen M, Srinivasan R, Gujar GT, Rauf A, Grzywacz D, Gregory P (2008) Bt brassicas for DBM control: the CIMBAA public/private partnership. In *Management of Diamondback Moth and Other Crucifer Pests*. (Eds AM Shelton, HL Collins, YJ Zhang and QJ Wu) pp. 272–279. China Agricultural Science and Technology Press, Beijing.

Russell D, Huang D, Bhalla P, Singh M, Robin C, Golz J, Williams A, Good R, Edwards O, Carroll B, Bhatnagar RK, Bhattacharya RC, Kalia P, Gujar GT, Kalia V (2017) Progress in the development of transgenic cabbage, cauliflower and canola expressing stacked Bts for caterpillar control and RNAi for aphid suppression. *Mysore Journal of Agricultural Science* **51**(A), 159–167.

Shelton AM (2010) The long road to commercialization of Bt brinjal (eggplant) in India. *Crop Protection* **29**, 412–414. doi:10.1016/j.cropro.2010.02.016

Shelton AM (2012) Genetically engineered vegetables expressing proteins from *Bacillus thuringiensis* for insect resistance: successes, disappointments, challenges and ways to move forward. *GM Crops and Food: Biotechnology in Agriculture and the Food Chain* **3**, 175–183. doi:10.4161/gmcr.19762

Shelton AM, Hokanson KE, Hautea DM, Hossain MJ, Hossain MA, Paranjape V, Hautea RA, McCandless L, Sarwer SH (2017) Bt eggplant: a genetically engineered 'minor' crop comes of age in Bangladesh and the Philippines. *ISB News Report*, August. <https://vtechworks.lib.vt.edu/bitstream/handle/10919/78874/SHHHHPHMCS.pdf?sequence=1&isAllowed=y>

Shelton AM, Hossain MJ, Paranjape V, Azad AK, Raham ML, Khan ASMMR, Prodhan MZH, Rashid MA, Majumder R, Hossain MA, Hussain SS, Huesing JE, McCandless L (2018) Bt eggplant project in Bangladesh: history, present status, and future direction. *Frontiers in Bioengineering and Biotechnology* **6**, 106. doi:10.3389/fbioe.2018.00106

Shelton AM, Hossain MJ, Paranjape V, Prodhan MZH, Azad AK, Majumder R, Sarwer SH, Hossain MA (2019) Bt brinjal in Bangladesh: the first genetically engineered food crop in a developing country. In *Plant Genetics and Agriculture*. (Ed. P Ronald). Cold Springs Harbor Press.

Shelton AM, Sarwer SH, Hossain MJ, Brookes G, Paranjape V (2020) Impact of Bt brinjal cultivation in the market value chains in five districts of Bangladesh. *Frontiers in Bioengineering and Biotechnology* **8**, 498. doi:10.3389/fbioe.2020.00498

Shirale D, Patil M, Parimi S (2017) Insecticide resistance in field populations of *Leucinodes orbonalis* (Lepidoptera: Crambidae) in India. *Canadian Entomologist* **149**, 399–407. doi:10.4039/tce.2017.3

Tabashnik BE, Brévault T, Carrière Y (2013) Insect resistance to Bt crops: lessons from the first billion acres. *Nature Biotechnology* **31**, 510–521. doi:10.1038/nbt.2597

Taher D, Solberg SØ, Prohens J, Chou Y, Rakha M, Wu T (2017) World Vegetable Center eggplant collection: origin, composition, seed dissemination and utilization in breeding. *Frontiers in Plant Science* **8**, 1484. doi:10.3389/fpls.2017.01484

Tang JD, Collins HL, Metz TD, Earle ED, Zhao JZ, Roush RT, Shelton AM (2001) Greenhouse tests on resistance management of Bt transgenic plants using refuge strategies. *Journal of Economic Entomology* **94**, 240–247. doi:10.1603/0022-0493-94.1.240

TOI Editorial (2019) They want GM crops: farmers' revolt is the outcome of a decade-long political paralysis, which must end. *Times of India*, 26 June. <https://timesofindia.indiatimes.com/blogs/toi-editorials/they-want-gm-crops-farmers-revolt-is-the-outcome-of-a-decade-long-political-paralysis-which-must-end/>

USEPA (2001) *Biopesticides Registration Action Document – Bacillus thuringiensis Plant Incorporated Protectants*. <https://archive.epa.gov/scipoly/sap/meetings/web/pdf/brad1_execsum_overview.pdf>

World Bank (2018) *World Bank Open Data*. <https://data.worldbank.org>

Zhao JZ, Cao J, Li YX, Collins HL, Roush RT, Earle ED, Shelton AM (2003) Transgenic plants expressing two *Bacillus thuringiensis* toxins delay insect resistance evolution. *Nature Biotechnology* **21**, 1493–1497. doi:10.1038/nbt907

10

Adoption and advances in genetically modified oilseed crops in Australia

Neeta Lohani, Divya Jain, Mohan B Singh and Prem L Bhalla

Introduction

The Australian production of oilseed crops is estimated at 5 million tonnes each year (GAIN 2018), with canola and cotton accounting for over 90% of the local production (AOF 2018). Canola alone accounts for half of the oilseed crop production and cotton represents one-third of the total. Oilseed crops have a wide range of applications from processed foods to industrial applications for manufacturing soaps, lubricants and paints. Following extraction of oil, the seed meal is separately utilised as feed for livestock, providing another avenue of trade. Figure 10.1 provides a comparative analysis of Australian oilseed crops for 2017–18 in terms of production, export and area under cultivation.

Canola is the third most important crop in Australia after wheat and barley. Western Australia leads canola production, followed by Victoria, New South Wales and South Australia which also have large-scale productions (Table 10.1). The second major oilseed crop in Australia is cotton, which is classified as oilseed although it is mainly cultivated for fibre. Cottonseed oil has a high smoking point and low trans fat content, leading to its increasing use in restaurants and the fast food industry (GAIN 2018). Even though Australia is not the leading producer of cotton, it is one of the leading exporters, with ~90% of cotton produced being exported. New South Wales and Queensland account for almost all the cotton produced in Australia (ABARES 2018).

Apart from canola and cottonseed, relatively smaller quantities of soybean, peanut, sunflower, safflower and linseed are cultivated in Australia. Soybean cultivation extends from northern Queensland, through southern New South Wales and into Victoria. Soybean has become an integral part of the domestic market, along with international export (ABARES 2018). Peanut is mostly cultivated in Queensland, and its production was reported at 22 400 tonnes in 2018 (ABARES 2018). Safflower is grown largely for the food industry, with an increased demand for cultivars having high oleic acid content and lower levels of saturated fatty acids (GRDC 2017). Other minor oilseed crops like linseed, sesame and sunflower are also gaining popularity with the ever-increasing health-conscious and nutrition-oriented market. The geographical distribution of oilseed crops grown in Australia is shown in Figure 10.2.

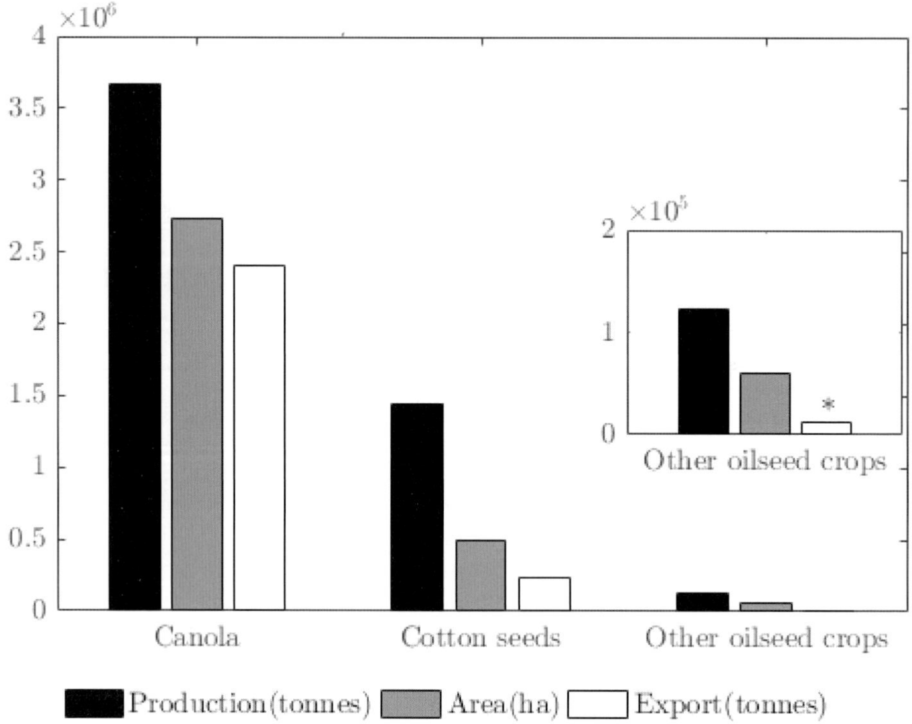

Figure 10.1 Australian oilseeds: production, area and export for 2017–18. Other oilseed crops = soybean, peanut, linseed, safflower and sunflower (* = no safflower and linseed export reported). Source: ABARES (2018).

Table 10.1. Oilseed cultivation in Australian states, 2018

	New South Wales	Victoria	Queens-land	South Australia	Western Australia	Tasmania	Total
Canola							
Area (ha)	650 000	450 000	2000	250 000	1 376 000	1000	2 729 000
Production (tonnes)	618 000	750 000	1200	320 000	1 978 000	1400	3 668 600
Cottonseed							
Area (ha)	310 000	N/a	190 000	N/a	0	N/a	500 000
Production (tonnes)	N/a	N/a	N/a	N/a	N/a	N/a	
Peanut							
Area (ha)	200	0	8000	0	0	N/a	8200
Production (tonnes)	300	0	22 000	0	0	N/a	22 300
Other oilseeds*							
Area (ha)	35 200	3000	20 000	4000	N/a	N/a	62 200
Production (tonnes)	62 000	1000	32 000	5000	N/a	N/a	100 000

* Other oilseeds = linseed, safflower seed, sesame seed, soybeans and sunflower seed.
Source: ABARES (2018).

Figure 10.2 Geographical distribution of oilseed crops in Australia. Source: ABARES (2018); GAIN (2018); GRDC (2018, 2019); AgriFutures Australia (2017a, b, c); PCA (2016).

The need for developing oilseed crops with desired agronomic traits of higher yields, enhanced nutritional quality, enhanced stress and pest tolerance is imperative in view of the economic impact, future food security and sustainability. Achieving these aims through conventional plant breeding system is laborious and time-consuming. The introduction of genetically modified (GM) oilseed crops has revolutionised the Australian oilseed industry. Over the past decades, the oilseed industry has grown into a major national agribusiness in Australia, making it one of the leading producers and exporters of oilseeds. It is the fifth-highest producer and second-largest exporter of canola and the eighth-highest producer of cotton (ABARES 2018). In 2016–17, canola and cotton exports were AUS$2128 million and AUS$86.1 million, respectively (ABARES 2017).

Australia has commercialised GM cotton since 1996 and GM canola since 2008; commercial production of GM safflower started in 2018. However, other oilseed GM crops, such as soybean, sunflower and peanut are yet to be commercialised in Australia. In this chapter, the GM oilseed crops grown in Australia are discussed, along with GM methods and their contribution to the Australian oilseed industry.

GM oilseed crops grown in Australia

GM crops and their products were introduced into global food markets in 1996, when the USA approved and began commercial cultivation of GM varieties of maize, soybean, cotton and canola. Australia was one of six countries that started commercial cultivation of GM crops in 1996. Since 1996, 24 countries have commercialised GM crops (ISAAA 2017). In 2017, Australia ranked 12th among the countries growing biotech crops, with 924 000 ha of GM cotton, canola and safflower crop cultivation (ISAAA 2017). Canola and cotton are the main GM crops cultivated in Australia, with 492 000 ha (53%) and 432 000 ha (47%), respectively. The commercial cultivation of GM crops in Australia is thus dominated by oilseed crops, which are modified for input traits – herbicide tolerance (HT) or insect resistance (IR), or both. One variety each of GM canola and GM safflower with enhanced oil quality were also commercialised in 2018. Enhanced quality traits, such as oil composition and increased nutritional value, may make GM crops more appealing to consumers in the future for consumption and industrial purposes.

GM cotton

The Australian cotton industry is one of the major contributors to the Australian agribusiness export sector. However, its success came at a cost. In the initial years of cultivation, the heavy dependence of cotton cultivation on insecticides and herbicides, along with the emergence of herbicide-resistant weeds and insecticide resistance in the insect pests, became a challenge for the cotton industry. The emergence of GM technologies in the late 1980s offered a potential opportunity for producing IR and HT cotton by altering the genetic make-up of the crop. GM IR cotton was the first GM crop introduced, in 1996. Since then, it has delivered considerable economic, social and environmental benefits to Australian cotton farmers. Continuous development of improved GM cotton varieties, availability of better germplasm and integration of new GM traits have made the Australian cotton industry one of the most efficient in the world. Over 99.5% of cotton currently grown in Australia is GM with either HT or resistance to major insect pests, or stacked with both traits (Table 10.2).

Table 10.2. Commercialised GM oilseed crops in Australia

S.no.	GM variety trade name	Modified trait/s
Cotton		
1	LibertyLink® and LibertyLink® x Roundup Ready®	Glufosinate and glyphosate HT
2	Bollgard II®, Roundup Ready®, Roundup Ready Flex®, Bollgard II® x Roundup Ready® and Bollgard II® x Roundup Ready Flex®	Glyphosate HT; IR
3	Widestrike™	IR; glufosinate HT
4	Pima cotton (1) Roundup Ready Flex®	Glyphosate HT e
5	Bollgard 3® and Bollgard 3® x Roundup Ready Flex®	Glyphosate HT; IR
6	GlyTol® and GlyTol® x TwinLink Plus®	IR; glufosinate and glyphosate HT
7	Xtendflex® and Bollgard 3® x Xtendflex®	IR; glufosinate, dicamba and glyphosate HT
8	COT102 (VIPCOT™ Cotton)	IR
Canola		
9	Roundup Ready® canola	HT; hybrid breeding system
10	InVigor® canola	HT; hybrid breeding system
11	InVigor® x Roundup Ready® canola	HT; hybrid breeding system
12	TruFlex™ Roundup Ready® canola (MON 88302)	HT
13	InVigor® x TruFlex™ Roundup Ready® canola	HT; hybrid breeding system
14	Optimum™ GLY canola	HT
15	DHA canola	Omega-3 oil content

Source: OGTR (2018e, f).

IR

The first IR GM trait approved for commercial cultivation in Australia was Ingard® cotton (known as Bollgard® globally). This GM cotton had a single gene, *cry1Ac*, sourced from a soil bacterium *Bacillus thuringiensis* (Bt) which produces an array of insecticidal proteins known as Bt toxins. These proteins include crystal (Cry) proteins and vegetative insecticidal proteins (Vip) (Estruch *et al.* 1996). Both protein classes are structurally distinct, having differential binding towards the receptors in the insect midgut (Sena *et al.* 2009). Within their classes and subclasses, the proteins show high specificity due to presence of specific binding sites in the insect midgut (Pardo-López *et al.* 2013). The Cry1Ac protein is toxic to a range of lepidopteran pests including cotton bollworms. Introduction of the *Bt cry1Ac* gene in cotton conferred resistance against the key pests. However, Ingard® cotton cultivation was capped to 30% of total cotton area in view of threat of bollworms evolving resistance to a single toxin. Bollgard II® cotton with two *Bt* genes – *cry1Ac* and *cry2Ab* – can produce toxins throughout the growing season, thereby providing better control and reducing the possibility of bollworms evolving significant further resistance. The introduction of dual-toxin GM cotton resulted in removal of the 30% cap on cultivated cotton area and drastic reduction in insecticide usage (Rossiter *et al.* 2008).

Alternative approaches for increasing the toxicity of Cry proteins involve interchanging the domains of different members of the Cry protein family, production of chimeric proteins by joining two or more proteins, site-directed mutagenesis, introduction of cleavage site or deletion of small fragments. An example is Widestrike™ GM cotton which has two synthetic *cry* genes: *cry1Ac* (synpro) synthesised by combining the active core from *cry1Ac1* and the carboxyl-terminal portion derived from *cry1Ca3* and *cry1Ab1*; and the other synthetic gene, *cry1F* (synpro), having a portion derived from *cry1Fa2* and a carboxyl-terminal portion derived from *cry1Ca3* and *cry1Ab1* (OGTR 2009). These synthetic genes are more effective and stable in controlling insect pests (APVMA 2011).

Another Bt toxin recently included in GM IR technology is the Vip3A protein. Vip3 proteins are toxic to many lepidopteran insect species including heliothine pests of cotton in Australia (Mahon *et al.* 2012). Bollgard 3® is a GM cotton variety which has the *vip3A* gene along with *cry1Ac* and *cry2Ab* (OGTR 2014a). The additional gene *vip3A* provides resistance against a key wet season pest, *Spodoptera exigua*, allowing farmers to grow cotton in the wet seasons (OGTR 2014a). There are no agronomic differences between Bollgard II® and Bollgard 3® cotton other than Bollgard 3® exhibiting a higher resistance to the cotton pests. COT102 is a recently approved IR GM cotton, which has only the *vip3Aa19* gene (a synthetic copy of the naturally occurring *vip3Aa1* gene) that provides resistance against many lepidopteran insect species (OGTR 2018b). COT102 has been approved for commercial release throughout Australia as a stack with other GM cotton lines, namely Bollgard 3® and Bollgard 3® x Roundup Ready Flex®, GlyTol® and GlyTol® x TwinLink Plus® and Xtendflex® and Bollgard 3® x Xtendflex®.

The significant threat to Bt-based insect control is the risk of evolution of IR. This is considerably reduced with stacking of insecticidal proteins and ensuring that resistance management tactics are prerequisites for cultivation of GM crops (Fitt 2000).

HT

Weeds are less of a threat to cotton production in Australia than cotton bollworms. However, some areas, such as formerly grazed lands or flood-prone lands, do have a high incidence of weeds. Weed management practices and control systems have evolved with the use of GM technology, providing greater flexibility for weed control. The second GM cotton crop Roundup Ready®, introduced in 2000, was tolerant to glyphosate. It has the enolpyruvylshikimate-3-phosphate synthase gene (*cp4 epsps*) which encodes the EPSPS enzyme. This gene is sourced from the CP4 strain of *Agrobacterium tumefaciens* and provides tolerance to glyphosate, a broad-spectrum herbicide (Constable *et al.* 2011). However, this gene had lower expression in reproductive tissues, which restricted application of the herbicide to the early growth stages of crop. Hence, Roundup Ready® cotton was superseded in 2006 by Roundup Ready Flex® cotton. This GM cotton variety had two *cp4 epsps* genes, one of which was highly expressed in plant reproductive tissues, imparting season-long HT to cotton (May *et al.* 2004).

In 2007, Liberty Link® cotton with a *bar* gene sourced from *Streptomyces hygroscopicus* was approved for cultivation. The *bar* gene encodes phosphinothricin acetyltransferase, which converts glufosinate ammonium into an inactive form and thus renders the plant tolerant to this herbicide. XtendFlex™ is another HT variety, which has the *dmo* gene in addition to the *cp4 epsps* gene. The *dmo* gene was derived from the bacterium *Stenotrophomonas maltophilia* (formerly *Pseudomonas maltophilia*) strain DI-6 (Herman *et al.* 2005). The *dmo* gene encodes for dicamba mono-oxygenase (DMO) that rapidly demethylates dicamba to non-herbicidal metabolites, namely, 3, 6-dichlorosalicylic acid and formaldehyde (OGTR 2006).

Glufosinate is a broad-spectrum herbicide like glyphosate, but it has lower efficacy towards grass weed species. For this reason, more than 98% of the GM cotton crop cultivated across Australia contains the Roundup Ready Flex® trait. This situation has resulted in over-reliance on glyphosate as the primary weed management tool in the farming system, with major implications like species shift in the weed spectrum and the development of herbicide-resistant weeds. Over 30 weed species around the world are reported to have resistance to glyphosate; 12 of those species are found in Australia. No glufosinate-resistant or dicamba-resistant weed species have been reported in Australia (Heap 2016). This highlights the need for increasing efforts for adopting an integrated weed management system.

Combined IR + HT trait

Conventional breeding practices along with GM technology have rendered it possible for developing GM varieties to harbour multiple traits. HT and IR varieties are developed by conventional cross-breeding methods. The GM traits can also be introduced by initial hybridisation and then backcrossing to transfer the traits from a GM cotton variety to a non-GM cotton variety. The majority of the commercially cultivated GM cotton varieties in Australia carry the combined/stacked traits (HT + IR).

GM canola

Canola is the most important oilseed crop in Australia. On average, the area under canola production has increased at a rate of 1 million ha producing up to 2.4 million tonnes of seed annually since the first commercial plantation in 1969. Weeds, insect pests and fungal diseases pose serious threats to canola production in Australia. Drought is also a challenge for canola production. While these factors contribute to canola being a risky rotation crop, its commercial cultivation has increased to the point where it is a crop of utmost economic importance. One of the major contributors in this positive change is the adoption of GM canola varieties. The Office of the Gene Technology Regulator (OGTR) gave approval for commercial release of GM canola in 2003; the majority of commercially cultivated GM canola is now modified for HT trait. Over the years there have been multiple licence applications for commercial release of GM canola with traits like herbicide resistance, disease resistance, high oil yield or oil with desired fatty acid composition. These efforts from the Australian and global scientific community have led to the commercial release of several GM canola varieties in Australia (Table 10.2).

Hybrid breeding system

The first licence for GM canola crop cultivation was issued in July 2003 in Australia. This led to the commercial release of seven GM InVigor® hybrid canola lines (Table 10.2). InVigor®

canola consists of a unique hybrid breeding system which uses a male sterility gene (*barnase*) and a fertility restorer gene (*barstar*), both derived from a common soil bacterium, *Bacillus amyloliquefaciens* (OGTR 2003). The InVigor® technology enables the production of hybrid seed, which allows increased plant vigour and seed production. The development of the pollen-producing parts of canola flowers (anthers) is suppressed in male sterile plants. Crossing a male sterile line with a restorer fertility line overrides the suppression and makes the progeny fertile. The progeny is expected to show enhanced agronomic performance. These GM lines also have the *bar* gene which provides the HT trait.

HT

Conventionally bred HT canola was introduced into commercial production in 1994 in Australia. These canola varieties were bred for tolerance to triazine herbicides using natural genetic variation for the trait. However, they had 10–30% lower yield and lower oil content than conventional varieties. In 2000, another canola variety (Clearfield®) with inbred tolerance to imidazolinone herbicides was introduced. With the advent of GM technology, in 2003Monsanto developed Roundup Ready® canola with tolerance against glyphosate. HT in Roundup Ready® was obtained by two genes: *cp4 epsps* gene used in HT cotton as it has low sensitivity to glyphosate, and *goxv247* from the soil bacterium *Ochrobactrum anthropi* encoding glyphosate oxidoreductase enzyme which breaks down glyphosate into non-herbicidal metabolites. The Roundup Ready® canola does not provide HT at the reproductive stage, and later application of glyphosate during flowering leads to yield loss due to phytotoxicity. To tackle this situation, another GM canola variety, TruFlex™ Roundup Ready® canola was introduced, with only one *cp4 epsps* gene under a modified promoter, *P-FMV/Tsf1*. This promoter contains enhancer sequences from the figwort mosaic virus (FMV) 35S promoter and DNA fragments from the promoter region of the *Arabidopsis thaliana Tsf1* gene (OGTR 2014b). The sequence derived from the *Tsf1* gene ensures strong and reliable constitutive expression of the *cp4 epsps* coding sequence in the GM canola and provides HT throughout the growing season (OGTR 2014b). Another GM canola variety, Optimum™ GLY, contains a different gene, *gat4621* which encodes a glyphosate acetyltransferase enzyme to confer tolerance to glyphosate. The gene was derived from three *gat* genes and optimised to create an enzyme with higher efficiency and increased specificity for glyphosate (OGTR 2016). The cultivation of HT canola has reduced the use of different types of herbicides and provided flexibility of timing for operations such as herbicide application and other cultural weed management practices, according to growing conditions. However, the drawbacks of over-reliance on one kind of herbicide (i.e. glyphosate) as the main solution for weed management remain a concern.

High omega-3 content

Omega-3 is a family of very long chain polyunsaturated fatty acids (VLC-PUFAs). All angiosperms possess the enzyme complex for synthesis of C18 PUFAs, linoleic and α-linolenic acid. However, these plants lack the ability to synthesise omega-3 VLC-PUFAs, such as eicosapentaenoic acid and docosahexaenoic acid (DHA), since they do not possess enzymes to desaturate and elongate C18 PUFAs to omega-3 VLC-PUFAs. DHA is required for healthy brain functioning and may confer other health benefits, such as enhanced cardiovascular health. The VLC-PUFAs pathway occurs in certain marine organisms. Therefore, some important VLC-PUFAs, such as DHA, are supplemented in the human diet through marine sources such as fish (Ruiz-López *et al.* 2012). The concerns related to overfishing and marine pollution led to research on the development of alternative sources, such as introduction of the VLC-PUFA biosynthetic pathways into oilseed crops.

DHA canola is a GM canola which was genetically engineered to produce DHA. It was recently approved for commercial cultivation in Australia. As discussed earlier, DHA is not naturally produced in non-GM canola. Seven genes sourced from yeast and microalgae, involved in the metabolism of VLC-PUFAs, were introduced in canola under seed-specific promoters (OGTR 2018a). These genes encode fatty acid desaturases and elongases. Two genes sourced from budding yeast (*Saccharomyces kluyveri*) catalyse the conversion of oleic acid to linoleic acid and then to α-linolenic acid. Five genes sourced from different microalgae lead to further desaturation and elongation of PUFAs, leading to the accumulation of DHA in canola (Ruiz-López *et al.* 2012).

GM safflower

GM safflower is a recent addition to the list of commercialised GM oilseed crops in Australia, licensed in June 2018. GM safflower expresses high oleic acid in its oil. In Australia, the commercial safflower cultivars are classified into two groups: linoleic varieties with 70–75% linoleic acid and oleic varieties with 70–80% oleic acid (Singh and Nimbkar 2006). An oleic acid class safflower cultivar, M1582 (originally from Mexico), was used as the parent cultivar for the released GM safflower. However, this particular cultivar is currently not cultivated in Australia (OGTR 2018c).

GM safflower was developed by modifying the fatty acid synthesis pathway to decrease the linoleic acid and increase the oleic acid in the safflower oil. This was done via modified expression of the two endogenous genes – *Palmitoyl-ACP thioesterase* (*CtFATB*) and *Δ12 desaturase* (*CtFAD2.2*) – involved in the fatty acid biosynthesis pathway. *CtFATB* encodes a carrier protein that mediates the export of saturated fatty acids from the plastid (Bonaventure *et al.* 2003) whereas *CtFAD2.2* encodes a desaturase protein that converts enzymatically oleic acid to linoleic acid (Harwood 1996). These genes were suppressed by using RNAi technology. Suppressing expression of *CtFATB* resulted in the formation of oleic acid by retaining the saturated fatty acids in the plastid until their desaturation. Suppression of expression of *CtFAD2.2* increased the oleic acid in the oil by preventing the conversion of oleic acid into linoleic acid (Wood *et al.* 2018).

GM safflower seeds produce oil which has ~92% oleic acid. This high-purity oleic oil has applications as an industrial raw material and a replacement for petroleum-based oils in the manufacture of plastics, lubricants and cosmetics (Vanhercke *et al.* 2013). It is approved for use in the industrial oil market, and not for human consumption. The meal produced after oil extraction can, however, be used as a stock-feed.

Genetic engineering techniques for improving oilseed crops

In the 1980s, the development of genetic engineering techniques presented an opportunity to revolutionise the process of integrating agronomically important traits in crops. *Agrobacterium*-mediated transformation is a leading technique employed in the production of GM crops commercially, including various IR and/or HT GM cotton and canola varieties. This method exploits the bacterium's natural ability to act as a vector by integrating its plasmid DNA into the crop genome. Modified transfer DNA (T-DNA) incorporated into the plant chromosomal DNA revolutionised genetic engineering for crop improvement. The ability to facilitate the transfer of large segments of DNA with a distinct insertion of the gene of interest into transcriptionally active regions of the chromosome is a major asset of this technology (Bakshi *et al.* 2011).

Biolistics is another method of genetic transformation which utilises a vector for delivering foreign genetic material into the plant genome. Here the vector system is a microprojectile, generally gold or tungsten particles 0.4–2.0 μm in diameter. The microprojectile crosses the physical barrier of cell wall, cell membrane and other tissue for integration with the host DNA. One great advantage of biolistics is that a variety of plant tissues like leaf discs and embryos can be used as explants for carrying out genetic transformations. Optimum^{TM} GLY canola is an HT commercial GM canola variety produced through the biolistic transformation of microspores (OGTR 2016). This technique is highly useful for recalcitrant plant species for which *Agrobacterium*-mediated transformation is not a viable option (Christou 1992). However, high cost, gene rearrangement, multiple insertions lead to high copy numbers through the biolistic approach, which often limits its use for plant genetic transformation.

Both methods offer practical application in production of transgenic plants. The constructs inserted varied from overexpression to RNA interference (RNAi) for silencing the gene of interest. Overexpression constructs rely either on constitutively driven promoters or tissue-specific promoters. Consecutive promoters such as *35s-CaMV* (cauliflower mosaic virus), *UBQ10* (*Ubiquitin-10*) and *35s-FMV* (figwort mosaic virus) are used to drive the gene expression throughout the plant (OGTR 2016). Anther-specific *PTA29* and seed-specific *NapA* promoters (*NapinA*) are utilised for the foreign gene overexpression in a particular tissue, making it highly efficient (OGTR 2003, 2018a). RNAi is based on the introduction of dsRNA with sequence complementarity to the gene of interest, inducing transcriptional or translational inhibition and thus silencing the targeted gene. *Agrobacterium*-mediated transformation of an RNAi construct was employed to genetically modify safflower that was accepted for commercial release in 2018 in Australia. With the alteration of the two genes, the oil content in safflower was modified to become highly rich in oleic acid and lower in linoleic acid (OGTR 2018d).

Over the years, techniques ranging from molecular markers, mutation breeding, *Agrobacterium*-mediated transformation and biolistics to genome editing have had varying impacts on crop improvement. The current shift towards more targeted modifications with the ability to focus at the nucleotide level is exciting. Genomics, transcriptomics and metabolomics have advanced the approach to crop improvement. Advancements in whole genome sequencing (WGS) techniques allowed the rapid identification of candidate genes and discovery of single nucleotide polymorphisms related to specific traits (Edwards and Batley 2010). Techniques such as high throughput RNA-seq and transcriptome assembly are contributing to our understanding of transcriptional regulations (Wang *et al.* 2016). The bioinformatic analysis of data generated through various forward and reverse genetic engineering techniques has made possible the application of site-specific nucleases for inducing DNA double-stranded breaks. These are then utilised to incorporate the desired sequence at the targeted site either being repaired through non-homologous end joining or homology-directed repair (Bonawitz *et al.* 2019). Zinc finger nucleases, transcription activator-like effector nucleases (TALENS) and clustered regularly interspaced short palindromic repeats/Cas9 (CRISPR associated 9-nucleases) (CRISPR/Cas9) are the nucleases most commonly used for crop modification (Osakabe and Osakabe 2015). ZFNs and TALENS are known as meganucleases and rely on DNA–protein interaction for sequence recognition whereas CRISPR/Cas9 employs a guide RNA for base pairing with the target sequence (Songstad *et al.* 2017). In comparison with TALENS and ZFN-mediated genetic engineering, CRISPR/Cas9 has an enhanced grade of precision. Also, the ability to carry out editing at more than one site of interest, using a combination of various guide RNAs, makes it easier to implement (Lozano-Juste and Cutler 2014). The versatility and adaptability of the system have ushered in a new era of developing GM crops. The CRISPR/Cas9 system has been employed in editing *CLATVA* genes for multi-locular silique development and increased seed production (Yang *et al.* 2018). The technique has a range of applicability from modifying *WRKY11 and WRK70* genes for enhanced resistance to *Sclerotinia*, to increasing oleic acid content through modification of an *FAD2* gene involved in fatty acid biosynthesis pathway in canola (Okuzaki *et al.* 2018; Sun *et al.* 2018). Similarly, in cotton, Wang *et al.* (2017) utilised the CRISPR/Cas9 system to target the arginase gene for enhanced lateral root formation. The improved lateral root system of these transgenic plants allowed for adaptation to cultivation in different soil types and for better nutrient absorption.

All these studies indicate that genetic engineering techniques are likely to influence future oilseed crop improvements. However, the level of acceptance by regulatory authorities, farmers and consumers still remains a major point of contention. The presence of selection markers such as antibiotics (kanamycin, hygromycin) or herbicide resistance (glufosinate) for screening the transformation events has also been perceived as issue of concern for the environment and human health.

Adoption of GM oilseed crops in Australia

The adoption and economic impact of GM crops in Australia have followed the overall global trends. Government regulations, consumer and farmer opinions have all played significant roles in the current status of GM crop adoption in the country. Since 1996, following the introduction of IR cotton in Australia, there has been a tremendous increase in the adoption rate of GM cotton and GM canola across the country. In 2017, the GM cotton adoption rate in Australia was close to 99% and GM canola comprised 24% of the total area under canola plantation (ISAAA 2017).

Various state governments have banned the commercial release of GM canola across Australia despite its approval by the central regulatory authority, OGTR. These bans were introduced due to concerns about negative market impacts and other social considerations. In 2003, licences for commercial release of two GM canola were issued by the OGTR. Except for Queensland and the Northern Territory, all other state and territory governments established GM-free areas and released new state-specific gene regulatory regulations. From 2008 to 2010, New South Wales, Western Australia and Victoria approved the cultivation of GM crops. However, the moratoria on commercial cultivation of GM crops are still in force in Tasmania, South Australia and the Australian Capital Territory. Table 10.3 provides an overview of the current Australian state and territory positions on GM crops (ABCA 2017).

Despite the increasing rates of GM canola adoption and canola meal production, Australia still has to rely on imported soybean meals for the expanding poultry and pig industries. Almost 730 000 tonnes of soybean meal was imported in 2014–15 to meet the industry demand for protein (ISAAA 2017). As mentioned earlier, soybean is planted from northern Queensland to the inland cropping regions of southern New South Wales and Victoria, but current soybean cultivation is not enough to meet Australian market demand. This issue can be resolved by growing GM soybean, with the desirable characteristics of higher yield of oil and protein, on a larger scale across Australia.

Conclusion

The impact of GM technology on the cultivation of oilseed crops in Australia has been undeniable. Over the years, there have been 167 licence applications for GM events in Australia, out of which 74 were for oilseed crops. The government and the regulatory authorities are continuously making efforts to improve the regulation of GM crops and their acceptance among farmers.

The majority of research for GM crop development is being carried out by universities, research councils, public institutions such as the CSIRO, and private industries (ABCA 2017). GM crop development is an expensive and lengthy process; hence, the most effective way to deliver the research conducted by public institutions is by public–private partnerships. Research for developing cotton with altered fatty acid composition, fungal resistance, drought tolerance, better IR and HT is continuing. Similarly, canola GM varieties producing essential fatty acids and better HT are being developed. Adoption of other important GM oilseed crops, such as GM soybean, would further reduce the overreliance on imported soybean meals to

Table 10.3. Overview of the status of Australian state and territory regulations on GM crops

State	Regulations on GM crops
Queensland	*Gene Technology Act 2001* • No GM crop moratorium • Large-scale commercial cultivation of GM cotton
Victoria	*Gene Technology Act 2001* *Control of GM Crops Act 2004* • No current orders in place • Commercial cultivation of GM canola since 2008
New South Wales	*Gene Technology (New South Wales) Act 2003* *Gene Technology (GM Crop Moratorium) Act 2003* • Moratorium on commercial cultivation of GM food crops • GM cotton exempt from moratorium and commercially cultivated • Exemption for commercial cultivation of GM canola granted in 2008
Northern Territory	*Gene Technology Act 2004* • No GM crop moratorium • No commercial cultivation of GM crops
Australian Capital Territory	*Gene Technology (GM Crop Moratorium) Act 2004* • Moratorium on commercial cultivation of all GM crops • Exemptions permitted for trials under specific conditions
Tasmania	*Gene Technology (Tasmania) Act 2012* *Genetically Modified Organisms Control Act 2004* • Moratorium on commercial cultivation of all GM crops • Whole of state designated GM-free area
South Australia	*Gene Technology Act 2001* *Genetically Modified Crops Management Act 2004* • Moratorium on commercial cultivation and transport of GM food crops and/or seeds • Whole state designated GM-free area • Exemptions granted for field trials under specific conditions
Western Australia	*Gene Technology Act 2005* • No GM crop moratorium

Source: ABCA (2017).

meet the protein requirements of the pig and poultry industries. These advances will not only enhance GM technology benefits but will also greatly influence oilseed crop cultivation in Australia, which will be vital in meeting future food demands under challenging climatic conditions.

Acknowledgements

We gratefully acknowledge financial support from the Australia–India Strategic Research Fund, GCF010009. NL and DJ were supported by a University of Melbourne Research Scholarship.

References

ABARES (2017) *Agricultural Commodity Statistics 2017*. Australian Bureau of Agricultural and Resource Economics and Sciences, Canberra. <http://data.daff.gov.au/data/warehouse/agcstd9abcc002/agcstd9abcc0022017_IugZg/ACS_2017_v1.1.0.pdf>

ABARES (2018) *Agricultural Commodity Statistics 2018*. Australian Bureau of Agricultural and Resource Economics and Sciences, Canberra. <http://www.agriculture.gov.au/abares/research-topics/agricultural-commodities/agricultural-commodities-trade-data" \l "2018">

ABCA (2017) *The Official Australian Reference Guide to Agricultural Biotechnology and GM Crops*. 3rd edn. Australia Biotechnology Council of Australia. <http://www.abca.com.au/materials/reports/>

AgriFutures Australia (2017a) *Cotton*. Rural Industries Research & Development Corporation, Canberra. <https://www.agrifutures.com.au/farm-diversity/cotton/>

AgriFutures Australia (2017b) *Soybeans*. Rural Industries Research & Development Corporation, Canberra. <https://www.agrifutures.com.au/farm-diversity/soybeans/>

AgriFutures Australia (2017c) *Canola*. Rural Industries Research & Development Corporation, Canberra. <https://www.agrifutures.com.au/farm-diversity/canola/>

AOF (2018) *Australian Oilseed Industry: Delivering High Quality Products to Local and Global Markets*. Australian Oilseed Federation. <http://www.australianoilseeds.com/oilseeds_industry>

APVMA (2011) *Public release summary on the evaluation of the new active Bacillus thuringiensis subsp. kurstaki Cry1Ac (synpro) gene and Bacillus thuringiensis subsp. aizawai Cry1F (synpro) gene and their controlling sequences in the product Widestrike Insect Protection Cotton Event 281–24–236/3006–210–23*. Australian Pesticides & Veterinary Medicines Authority. <https://apvma.gov.au/sites/default/files/publication/13601-prs-widestrike.rtf>

Bakshi S, Sadhukhan A, Mishra S, Sahoo L (2011) Improved *Agrobacterium*-mediated transformation of cowpea via sonication and vacuum infiltration. *Plant Cell Reports* **30**, 2281–2292. doi:10.1007/s00299-011-1133-8

Bonaventure G, Salas JJ, Pollard MR, Ohlrogge JB (2003) Disruption of the FATB gene in *Arabidopsis* demonstrates an essential role of saturated fatty acids in plant growth. *Plant Cell* **15**, 1020–1033. doi:10.1105/tpc.008946

Bonawitz ND, Ainley WM, Itaya A, Chennareddy SR, Cicak T, Effinger K, Jiang K, Mall TK, Marri PR, Samuel JP, Sardesai N, Simpson M, Folkerts O, Sarria R, Webb SR, Gonzalez DO, Simmonds DH, Pareddy DR (2019) Zinc finger nuclease-mediated targeting of multiple transgenes to an endogenous soybean genomic locus via non-homologous end joining. *Plant Biotechnology Journal* **17**, 750–761. doi:10.1111/pbi.13012

Christou P (1992) Genetic transformation of crop plants using microprojectile bombardment. *Plant Journal* **2**, 275–281. doi:10.1111/j.1365-313X.1992.00275.x

Constable G, Llewellyn D, Wilson L, Stiller W (2011) An industry transformed: the impact of GM technology on Australian cotton production. *Farm Policy Journal* **8**, 23–41.

Edwards D, Batley J (2010) Plant genome sequencing: applications for crop improvement. *Plant Biotechnology Journal* **8**, 2–9. doi:10.1111/j.1467-7652.2009.00459.x

Estruch JJ, Warren GW, Mullins MA, Nye GJ, Craig JA, Koziel MG (1996) Vip3A, a novel *Bacillus thuringiensis* vegetative insecticidal protein with a wide spectrum of activities against lepidopteran insects. *Proceedings of the National Academy of Sciences of the United States of America* **93**, 5389–5394. doi:10.1073/pnas.93.11.5389

Fitt GP (2000) An Australian approach to IPM in cotton: integrating new technologies to minimise insecticide dependence. *Crop Protection* **19**, 793–800. doi:10.1016/S0261-2194(00)00106-X

GAIN (2018) *Australia Oilseeds and Products*. Global Agricultural Information Network, USDA Foreign Agriculture Service. <https://gain.fas.usda.gov/Recent%20GAIN%20Publications/Oilseeds%20and%20Products%20Annual_Canberra_Australia_3-28-2018.pdf>

GRDC (2017) *GrowNotes: Safflower*. Grains Research and Development Corporation. <https://grdc.com.au/__data/assets/pdf_file/0031/238990/GRDC-GrowNotes-Safflower-Northern.pdf>

GRDC (2018) *GrowNotes: Peanuts Northern Region*. Grains Research and Development Corporation. <https://grdc.com.au/resources-and-publications/grownotes/crop-agronomy/peanutgrownotes?utm_medium=short_url&utm_term=North&utm_content=Peanuts%20Northern%20Region%20-%20GrowNotes%E2%84%A2&utm_source=website&utm_campaign=SBM00013-G>

GRDC (2019) *Grains Update Paper: Super High Oleic Safflower – A Game Changer for Grain Growers*. Grains Research and Development Corporation. <https://grdc.com.au/resources-and-publications/grdc-update-papers/tab-content/grdc-update-papers/2019/08/super-high-oleic-safflower-a-game-changer-for-grain-growers>

Harwood JL (1996) Recent advances in the biosynthesis of plant fatty acids. *Biochimica et Biophysica Acta -Lipids and Lipid Metabolism* **1301**, 7–56. doi:10.1016/0005-2760(95)00242-1

Heap I (2016) *International Survey of Herbicide Resistant Weeds: Weeds Resistant to EPSP Synthase Inhibitors*. <http://weedscience.org/Summary/MOA.aspx?MOAID=12>

Herman PL, Behrens M, Chakraborty S, Chrastil BM, Barycki J, Weeks DP (2005) A three-component dicamba O-demethylase from *Pseudomonas maltophilia*, strain DI-6: gene isolation, characterisation and heterologous expression. *Journal of Biological Chemistry* **280**, 24759–24767. doi:10.1074/jbc.M500597200

ISAAA (2017) *Global Status of Commercialized Biotech/GM Crops in 2017: Biotech Crop Adoption Surges as Economic Benefits Accumulate in 22 Years*. ISAAA Brief No. 53. ISAAA, Ithaca, NY. <http://www.isaaa.org/resources/publications/briefs/53/download/isaaa-brief-53-2017.pdf>

Lozano-Juste J, Cutler SR (2014) Plant genome engineering in full bloom. *Trends in Plant Science* **19**, 284–287. doi:10.1016/j.tplants.2014.02.014

Mahon RJ, Downes SJ, James B (2012) Vip3A resistance alleles exist at high levels in Australian targets before release of cotton expressing this toxin. *PLoS One* **7**(6), e39192. doi:10.1371/journal.pone.0039192

May OL, Culpepper AS, Cerny RE, Coots CB, Corkern CB, Cothren JT, Huber SA (2004) Transgenic cotton with improved resistance to glyphosate herbicide. *Crop Science* **44**, 234–240. doi:10.2135/cropsci2004.2340

OGTR (2003) *DIR 021/2002 Documents*. Office of the Gene Technology Regulator. <http://www.ogtr.gov.au/internet/ogtr/publishing.nsf/Content/DIR021-2002>

OGTR (2006) *DIR 062/2005 Documents*. Office of the Gene Technology Regulator. <http://www.ogtr.gov.au/internet/ogtr/publishing.nsf/Content/DIR062-2005>

OGTR (2009) *DIR 019 Documents*. Office of the Gene Technology Regulator. <http://www.ogtr.gov.au/internet/ogtr/publishing.nsf/Content/DIR091>

OGTR (2014a) *DIR 124 Documents*. Office of the Gene Technology Regulator. <http://www.ogtr.gov.au/internet/ogtr/publishing.nsf/Content/DIR124>

OGTR (2014b) *DIR 127 Documents*. Office of the Gene Technology Regulator. <http://www.ogtr.gov.au/internet/ogtr/publishing.nsf/Content/DIR127>

OGTR (2016) *DIR 139 Documents*. Office of the Gene Technology Regulator. <http://www.ogtr.gov.au/internet/ogtr/publishing.nsf/Content/DIR139>

OGTR (2018a) *DIR 155 Documents*. Office of the Gene Technology Regulator. <http://www.ogtr.gov.au/internet/ogtr/publishing.nsf/Content/DIR155>

OGTR (2018b) *DIR 157 Documents*. Office of the Gene Technology Regulator. <http://www.ogtr.gov.au/internet/ogtr/publishing.nsf/Content/DIR157>

OGTR (2018c) *DIR 158 Documents*. Office of the Gene Technology Regulator. <http://www.ogtr.gov.au/internet/ogtr/publishing.nsf/Content/DIR158>

OGTR (2018d) *Fact Sheet: Genetically Modified (GM) Safflower in Australia*. Office of the Gene Technology Regulator. <http://www.ogtr.gov.au/internet/ogtr/publishing.nsf/content/9AA09BB4515EBAA2CA257D6B00155C53/$File/24.%20%20Genetically%20modified%20(GM)%20safflower%20in%20Australia.pdf>

OGTR (2018e) *Fact Sheet: Genetically Modified (GM) Cotton in Australia*. Office of the Gene Technology Regulator. <http://www.ogtr.gov.au/internet/ogtr/publishing.nsf/content/9AA09BB4515EBAA2CA257D6B00155C53/$File/23.%20-%20Genetically%20modified%20(GM)%20cotton%20in%20Australia.pdf>

OGTR (2018f) *Fact Sheet: Genetically Modified (GM) Canola in Australia*. Office of the Gene Technology Regulator. <http://www.ogtr.gov.au/internet/ogtr/publishing.nsf/content/9AA09BB4515EBAA2CA257D6B00155C53/$File/12%20-%20Genetically%20modified%20(GM)%20canola%20in%20Australia.pdf>

Okuzaki A, Ogawa T, Koizuka C, Kaneko K, Inaba M, Imamura J, Koizuka N (2018) CRISPR/Cas9-mediated genome editing of the fatty acid desaturase 2 gene in *Brassica napus*. *Plant Physiology and Biochemistry* **131**, 63–69. doi:10.1016/j.plaphy.2018.04.025

Osakabe Y, Osakabe K (2015) Genome editing with engineered nucleases in plants. *Plant & Cell Physiology* **56**, 389–400. doi:10.1093/pcp/pcu170

Pardo-López L, Sóberon M, Bravo A (2013) *Bacillus thuringiensis* insecticidal three-domain Cry toxins: mode of action, insect resistance and consequences for crop protection. *FEMS Microbiology Reviews* **37**, 3–22. doi:10.1111/j.1574-6976.2012.00341.x

PCA (2016) *History of Australian Peanut Industry*. Peanut Company of Australia. <http://pca.com.au/pca-profile/history-of-the-peanut-industry/>

Rossiter L, Gunning R, McKenzie F (2008) Silver Anniversary of resistance management in the Australian cotton industry: an overview and the current situation for *Helicoverpa armigera*. In *Proceedings of the 2008 Australian Cotton Conference*, Broadbeach, Qld.

Ruiz-López N, Sayanova O, Napier JA, Haslam RP (2012) Metabolic engineering of the omega-3 long chain polyunsaturated fatty acid biosynthetic pathway into transgenic plants. *Journal of Experimental Botany* **63**, 2397–2410. doi:10.1093/jxb/err454

Sena JA, Hernández-Rodríguez CS, Ferré J (2009) Interaction of *Bacillus thuringiensis* Cry1 and Vip3A proteins with *Spodoptera frugiperda* midgut binding sites. *Applied and Environmental Microbiology* **75**, 2236–2237. doi:10.1128/AEM.02342-08

Singh V, Nimbkar N (2006) Safflower (*Carthamus tinctorius* L.). In *Genetic Resources, Chromosome Engineering and Crop Improvement: Oilseed Crops*. (Ed. RJ Singh) pp. 167–194. CRC Press, Boca Raton.

Songstad DD, Petolino JF, Voytas DF, Reichert NA (2017) Genome editing of plants. *Critical Reviews in Plant Sciences* **36**, 1–23. doi:10.1080/07352689.2017.1281663

Sun Q, Lin L, Liu D, Wu D, Fang Y, Wu J, Wang Y (2018) CRISPR/Cas9-mediated multiplex genome editing of the BnWRKY11 and BnWRKY70 genes in *Brassica napus* L. *International Journal of Molecular Sciences* **19**, 2716–2734. doi:10.3390/ijms19092716

Vanhercke T, Wood CC, Stymne S, Singh SP, Green AG (2013) Metabolic engineering of plant oils and waxes for use as industrial feedstocks. *Plant Biotechnology Journal* **11**, 197–210. doi:10.1111/pbi.12023

Wang J, Singh SK, Du C, Li C, Fan J, Pattanaik S, Yuan L (2016) Comparative transcriptomic analysis of two *Brassica napus* near-isogenic lines reveals a network of genes that influences seed oil accumulation. *Frontiers of Plant Science* 7(Sep), 1–14. doi:10.3389/fpls/2016.01498

Wang Y, Meng Z, Liang C, Meng Z, Wang Y, Sun G, Zhu T, Cai Y, Guo S, Zhang R, Lin Y (2017) Increased lateral root formation by CRISPR/Cas9-mediated editing of arginase genes in cotton. *Science China. Life Sciences* **60**, 524–527. doi:10.1007/s11427-017-9031-y

Wood CC, Okada S, Taylor MC, Menon A, Mathew A, Cullerne D, Stephen SJ, Allen RS, Zhou XR, Liu Q, Oakeshott JG (2018) Seed specific RNAi in safflower generates a superhigh oleic oil with extended oxidative stability. *Plant Biotechnology Journal* **16**, 1788–1796. doi:10.1111/pbi.12915

Yang Y, Zhu K, Li H, Han S, Meng Q, Khan SU, Fan C, Xie K, Zhou Y (2018) Precise editing of CLAVATA genes in *Brassica napus* L. regulates multilocular silique development. *Plant Biotechnology Journal* **16**, 1322–1335. doi:10.1111/pbi.12872

11

Bioefficacy testing of Bt corn against Asian corn borer in the Philippines

Merdelyn Caasi-Lit

Introduction

Genetically modified (GM) insect-resistant (IR) corn expressing a toxin gene from *Bacillus thuringiensis* (Bt) was first approved in December 2002 in the Philippines. Since then Bt corn has successfully controlled the menace of the most destructive corn pest, the Asian corn borer (ACB) *Ostrinia furnacalis* (Guenée). Bt corn has contributed significantly to the increase in total corn production in the country. The area for Bt corn has increased steadily since cultivation in 2003 and peaked in 2014 at 0.831 million ha, but declined to 0.702 million ha in 2015 because of drought. The area increased to 0.812 million ha in 2016 and again declined to 0.742 million ha in 2017 due to the use of counterfeit or unauthorised seeds known as *ukay-ukay* (ISAAA 2017). The significant decline in Bt corn area was probably also due to the unprecedented nullification of the Department of Agriculture Administrative Order 8 (AO8) by the Supreme Court of the Philippines in December 2015. AO8 is the national policy on biotechnology and GM crops in the Philippines, and for more than 20 years it had focused on regulating the importation and release of GM products in the environment (Dept of Agriculture 2002). As a result of the AO8 nullification, Joint Department Circular No. 1 was created in 2016 by the Department of Science and Technology, Department of Agriculture, Department of Environment and Natural Resources, Department of Health and Department of Interior and Local Government to be the new guideline for GM crops (Dept of Agriculture 2014). These events, however, did not stop farmers from planting and consumers from utilising Bt corn products.

The development of GM crops through importation and cultivation followed the established regulatory systems in the Philippines. The creation of the National Committee on Biosafety in the Philippines (NCBP) by virtue of Administrative Order No. 430 in 1990 was therefore strategic in dealing with the first Bt corn product to undergo strict and stringent regulations per AO8 in 2002. This regulation was further reinforced by the approval of Executive Order No. 514 in 2006, strengthening the National Biosafety Framework.

The policy in Circular No. 02, *Enhancing the Insect Resistance Management (IRM) Strategy for Bt Corn targeting Asian Corn Borer*, stated that new Bt corn products should undergo bioefficacy testing two years after the introduction of a commercially released Bt corn product (Dept of Agriculture 2014). Bioefficacy testing is therefore a continuing regulatory procedure for IRM for Bt corn testing and commercialisation for single and pyramided traits against the ACB.

As the leading national centre for research on host-plant resistance with state-of-the-art insect-rearing facilities, the Cereals Entomology Laboratory of the Institute of Plant Breeding (IPB) is mandated to work on bioefficacy studies. As a part of the regulatory procedures for the commercialisation of Bt corn for single and pyramided traits, field and laboratory bioefficacy studies are conducted to generate baseline response of Bt corn against the ACB and other lepidopteran pests of corn.

This chapter will focus on the protocols, both laboratory and field-based, associated with bioefficacy studies including mass rearing of ACB in the laboratory.

Asian corn borer *Ostrinia furnacalis*

The corn borer species in the Philippines was first named as *Pyrausta vastatrix* Schultze in 1906 by Charles S. Banks in 'Problems of economic entomology in the Philippines' (Banks 1906; Schultze 1908). The insect pest was recorded as corn moth borer. As damage became serious in the stalks, Jones (1913) called it 'maize stalk-borer' but continued to use the scientific name *P. vastatrix*. This was also the first record that this pest occurred only from Luzon. The name *P. vastatrix* was used from 1906 until Buligan changed it in 1929 to *P. nubilalis*, as advised by Carl Heinrich and DW Jones of the US Bureau of Entomology. The biology of the corn borer was studied by Buligan (1929) in view of its damage potential in corn-growing regions of the Ilocano and Visayan provinces, where corn was the principal staple food of the working classes. For years, *P. nubilalis* was an important pest of field corn (Capco 1957). From 1957, three names were used in several papers to refer to borers in corn: *Ostrinia damoalis*, *Pyrausta salentialis* and *Ostrinia salentialis* (Calora *et al.* 1965). Finally, in 1970, the taxonomic position of the corn borer as *O. furnacalis* (Guenée) was confirmed by Mutuura and Munroe (1970). Since then, *Ostrinia furnacalis* (Guenée), with the corrected common name Asian corn borer (ACB) has been used in all reports, notes and publications.

Life history

Based on previous records, Navarro (1911) was the first to describe the life-cycle of the corn borer moth (then known as *P. vastatrix*) and its characteristic damage to the corn plant. Buligan (1929) carried out excellent studies on biology of the corn borer, including detailed descriptions and illustrations of the different growth stages of the pest. Later, other studies on ACB biology in the Philippines were done by Calora *et al.* (1965), Camarao (1976, 1977) and Barrion *et al.* (1980). The total developmental period of the ACB from egg to adult emergence was noted as 26–41 days, with Camarao (1976) showing the shortest life-cycle and Buligan (1929) the longest. The number of larval instars was a bone of contention in some studies. Differences in the number of larval instars and duration of larval period were attributed to several factors such as maternal age, quality of the egg and the kind of host or diet, as reported by Esperk *et al.* (2007). Caasi-Lit and Sapin (2012) reported varying number of larval instars among five populations of the ACB.

Host-plant resistance of corn to ACB

Studies on host-plant resistance to ACB were formally started in 1975 with screening of corn germplasm from several local and foreign agencies at the University of Philippines College of Agriculture. In the project entitled 'Breeding for corn borer resistance', 14 parental lines, which included Antigua Grp 1, were found to be highly resistant to leaf-feeding damage by the ACB (IPB 1984; Caasi-Lit *et al.* 1987). This selected broad-based population was known as the corn borer resistant population (CBR 1). It was further improved and renamed as XV3. Unfortunately, corn XV3 was susceptible to downy mildew caused by *Perenosclerospora philippinensis*. Efforts were then made to screen cultivars for resistance to both ACB and downy mildew, which helped in the development of corn hybrid IPB Var 911. With the recent introduction of Bt corn into the country, the traditional breeding work was replaced with a Department of Agriculture research program, 'Corn germplasm utilization through advanced research and development'. This program aims to conserve and utilise traditional maize for the improvement and breeding of abiotic and biotic stress-tolerant varieties. The institute is now spearheading the work on traditional maize for resistance to ACB (Caasi-Lit *et al.* 2018a).

Genetic variation in ACB

Like other lepidopterans, corn borer populations from different locations in the Philippines showed diversity in terms of chromosomal polymorphism (Barrion *et al.* 1980), abdominal chaetotaxy (Barrion *et al.* 1981), genital morphometrics (Sapin *et al.* 2013), esterase isozymes (Mendoza *et al.* 1994) and larval instars and life-cycle duration (Caasi-Lit and Sapin 2012). Genetic variation has great relevance with respect to evolution of resistance to Bt corn. Benigno and Caasi-Lit (2018) compared two ACB populations in a simulation model and showed higher likelihood of the ACB population in Isabela (the region with the largest Bt corn planting) to develop resistance due to initial high resistant allele population (5%), high temperature, rainfall and continuous cropping over a long duration. Given the selection pressure that the ACB is facing with the introduction of Bt corn hybrids, it is necessary to proactively pursue IRM practices including resistance monitoring.

Basic behavioural studies on ACB

For IRM to be successful, among other prerequisites like high dose and refuges, the behaviour of ACB plays an important role in the context of corn agroecology. Studies on the pest behaviour, especially emergence pattern, mating behaviour, ovipositional preference and adult dispersal in the local ecosystem, were conducted under the project 'Basic studies on Asian corn borer' with funding from Biotechnology Coalition of the Philippines (Caasi-Lit 2017a), as required by Circular No. 2 on the implementation of new refuge, to generate local data on the ACB.

The emergence pattern and mating behaviour of the ACB were similar for both sexes of ACB in both laboratory and field situations. The results of this study strongly support the random mating of ACB in the field and reduction of the risk of asynchronous emergence of ACB adults from Bt corn *v.* non-Bt corn refuge for non-random mating.

Ovipositional preference of the ACB did not show any differentiation between Bt and non-Bt corn under the semi-field and smaller steel-cage conditions. Phenotypic characteristics and plant growth stages of host crop were shown to play an important role in ACB ovipositional preference. More egg masses were observed during the wet season as compared to dry season. These results provided important insights into refuge selection and planting.

The adult dispersal behaviour of field-released ACB was studied in South Cotabato on Mindanao and Isabela on Luzon using the mark–release–recapture method. Results showed that weather conditions and local habitat strongly influence the distance travelled and the direction of dispersal in adult ACB. The ACB exhibited non-random dispersal and propensity towards weedy areas, and few remained near the release point. The presence of weedy areas in field borders as aggregation sites favours assortative mating in adults emerging from Bt plants and non-Bt refuges. A shorter distance should be considered, to increase the chances of mating between potentially resistant ACB adults from Bt plants and susceptible ACB adults from the non-Bt refuges. Seed-blend refuges may also facilitate mating of ACB adults that emerge from Bt and non-Bt plants, especially short-range flying ACB adults.

Bioefficacy trials of approved Bt corn in the Philippines

Approved Bt corn hybrids

Since December 2002, several Bt corn products have been approved for commercialisation (see Chapter 13). The Bureau of Plant Industry (BPI), under the Dept of Agriculture-Biotech, records three single-event products with resistance to ACB and at least seven stacked Bt corn products with resistance to ACB and other lepidopteran pests, in combination with herbicide tolerance (HT).

IR evaluation of a transgenic crop was conducted by Fernandez *et al.* (1997). YieldGard™ was the first Bt corn product in the country, with the transformation event MON 810, expressing the *cry1Ab* gene. The Bt corn hybrid MON 810 is significantly resistant to the insect pest compared with its isogenic non-Bt line and the susceptible variety.

Bt11 (Agrisure LB/LL) was the second Bt corn that was introduced to the Philippines, in 2005, by Syngenta Philippines. The Bt11 corn also expresses *cry1ab* gene and *pat* gene, the former offering IR and the latter offering tolerance to glufosinate ammonium (Rasco *et al.* 2008). This Bt corn, an IR–HT hybrid, has been in the commercial market for more than 14 years. It is included in the Lists of Approval Registry prepared by the Dept of Agriculture-BPI. Another Bt corn, Bt11 × TC1507 × GA21 (Agrisure EZ Refuge), was released in October 2015. Bioefficacy testing showed that this stacked Bt corn hybrid is resistant to ACB and to other lepidopteran pests of corn.

Pioneer Hi-Bred Philippines spearheaded the hybrid corn industry in the Philippines and has developed Bt corn products for cultivation. These include Bt corn event TC1507 (Herculex) expressing the Cry1F protein, which is effective against ACB and other lepidopterans (Dalmacio 2003; Thompson *et al.* 2010). Another new corn hybrid, TC1507 × MON 810 × NK603 (Intrasect®), was approved for commercialisation in 2014 and was found effective against ACB, true armyworm and corn earworm (Caasi-Lit, 2016). Combined with transformation event NK603, it was also HT.

The transgenic corn hybrid MON 89034 and IR–HT stacked hybrid MON 89034 × NK603, each expressing Bt Cry1A.105 and Cry2Ab2 proteins, have consistently provided excellent protection against ACB (Samson *et al.* 2009). The effects on ACB larval survival or mortality were initially confirmed in the greenhouse by Solsoloy *et al.* (2007). Field efficacy data also demonstrated resistance of MON 89034 against ACB during the wet (2008) and dry (2009) cropping seasons in different corn production areas in the country. This Bt corn hybrid continues to show high efficacy in all corn-growing regions. Bioefficacy tests using both leaf-disc and whole-plant bioassays confirmed the excellent performance of the Bt corn hybrid MON 98034 in effectively controlling ACB, based on the Laguna and Isabela populations (Caasi-Lit *et al.* 2018b).

Methods of bioefficacy testing

Several methods for bioefficacy testing have been developed since the introduction of Bt corn products in the country (Table 11.1). A proposed bioefficacy protocol is submitted by the technology developer to the BPI for approval.

The first Bt corn hybrid, YieldGard™, was evaluated using whole-plant assay in the greenhouse (Fernandez *et al.* 1997). Test plants at 25, 41 and 56 days after planting (DAP) were infested with 40 neonates per plant. Leaf-feeding damage was assessed 14 days after infestation (DAI) using the rating scale of Guthrie *et al.* (1960) and stalk and ear shank damage were evaluated upon harvesting. The same method was used by Rasco *et al.* (2008) in the resistance screening of Bt corn Bt11 against ACB. On the other hand, Bt corn TC1507 was evaluated using the above method with modification by Dalmacio (2003). The first set was evaluated using 24-day old plants infested with 20 neonates, then leaf and stalk-feeding damage assessed at 17 DAI. Ear shank damage was assessed by infesting 20 neonates in the earshot on another set of plants at 55 DAP.

Table 11.1. Summary of bioefficacy testing protocols for single and stacked Bt corn products with resistance to ACB in the Philippines

Bt corn event(s)	Non-Bt control	Bioassay method	Test insects and protocols	References
Corn MON 810 (Bt-transformed HI-2 hybrid Bt corn)	Isogenic line Supersweet corn	Whole-plant assay Greenhouse Leaf feeding damage at 34, 48 and 90 DAP Number of tunnels, stalk damage and shank damage at 90 DAP	40, 50 and 50 neonates/plant, respectively at 25, 42 and 56 DAP 10 replicates per treatment with 1 plant per replicate	Fernandez *et al.* (1997)
Corn Bt11 (NK5445Bt)	Isogenic hybrid of NK5445 Sweet corn hybrid Sugar 73 NK5468	Whole-plant assay Greenhouse Leaf feeding damage at 41, 52 and 55 DAP Number of entrance holes and length of feeding tunnels at harvest	Three egg masses (100 eggs) per plant at 27 and 42 DAP while 15 neonates at 58 and 62 DAP 10 replicates per treatment with 1 plant per replicate	Rasco *et al.* (2008)
Corn Bt11 (NK5445Bt)	DK818YG Isogenic hybrid DK9051 NK8840	Field assay Wet and dry season Leaf feeding damage at vegetative stage Tassel damage at reproductive stage Number of holes at harvest	Natural infestation	Rasco *et al.* (2010)
Corn TC1507 ^{32}P75 Bt	^{32}P75 Non-Bt 30D55	Whole-plant assay Greenhouse Leaf feeding damage at 37 and 54 DAP Number of borer holes at 37 and 54 DAP	Two egg masses/plant at 24 DAP, 20 neonates at 37 DAP and 20 neonates at 56 DAP 10 replicates per treatment with 1 plant per replicate	Dalmacio (2003)
Corn TC1507 (QZ…826; QZ…782)	Isogenic hybrids QZ…588 QZ…544 Hybrid DK818	Field assay Wet and dry season Foliar injury before tassel emergence Number and length of tunnels at harvest	Natural infestation 10 plants each from the inner 2 rows of the 4-row plots	Thompson *et al.* (2010)
Corn MON 89034		Whole-plant assay Greenhouse Leaf feeding damage at 10 and 90 DAI Number of borer tunnels, stalk damage and shank damage at 80 DAP	5–10 egg masses/plant, respectively at V5 and V10 6 replicates per treatment with 1 plant per replicate	Solsoloy *et al.* (2007)

Continued

Table 11.1. continued

Bt corn event(s)	Non-Bt control	Bioassay method	Test insects and protocols	References
Corn MON 89034		Field assay Wet and dry season Leaf feeding damage at 10–15 DAI Borer holes, number and length of tunnels at harvest	1st ACB infestation 10 insects per plant at 27–29 DAP 20 plants/treatment 2nd infestation 10–20 insects per plant at 55–58 DAP 20 plants/treatment	Samson *et al.* (2009)
Corn MON 89034	NK603	Leaf-disc assay Larval survival every day for 7 days Whole-plant assay Leaf feeding damage at 1, 3 and 5 DAI	1 2nd instar per leaf disc (30 DAP) 50 replicates 2 ACB populations (Laguna and Isabela) 30 2nd instar per plant (30 DAP) 30 replicates	Caasi-Lit *et al.* (2018b)
Corn TC1507 × MON 810 × NK603 (Intrasect) TC1507 MON 810	Isogenic hybrid 30B80	Leaf disc assay Larval mortality at 3, 5 and 7 DAI Whole-plant assay Screenhouse Leaf damage at 3, 5 and 7 DAI Larval survival and larval weight at 7 DAI	1 2nd instar per well 160 replicates, 10 bioassay trays (16 wells) per treatment 2 populations (Laguna and Isabela 1 2nd instar per plant 200 replicates per ACB population; 2 populations (Laguna and Isabela)	Caasi-Lit (2017b)
Corn Bt11 × GA21 (NK8840BtGT)	Corn hybrid NK8840GT, GA21	Field assay Wet and dry season Leaf feeding damage at 5 and 10 DAI Larval and pupal survival at 15 DAI Stalk damage, number of borer holes and tunnel length at harvest	Artificial/natural infestation 50 4-day old 2nd instar at 30 and 40 DAP Caged plots RCBD, 3 replications 5 m row in 7 rows 24 plants per row 100 plants per replicate 3 replications 2 ACB populations (Laguna and Isabela)	Caasi-Lit *et al.* (2019)
Corn Bt11 × TC1507 × GA21 (Agrisure EZ Refuge) Bt11 × Ga21		Field assay Wet and dry season Leaf feeding damage at 5 and 10 (DAI Larval and pupal survival at 15 DAI Stalk damage, number of borer holes and tunnel length at harvest	50 4-day old 2nd instar at 45 DAP Caged plots RCBD, 3 replications 5 m row in 7 rows 24 plants per row 100 plants per replicate 3 replications 2 ACB populations (Laguna and Isabela)	Caasi-Lit (2017c)

Table 11.1 enumerates the approved single and stacked Bt corn products in the Philippines with ACB resistance and bioassay methods. Field assaying was also employed to confirm the reaction of the single-trait Bt corn hybrids to natural infestation of ACB. For Bt11, field trials were conducted during dry and wet seasons in one location (Rasco *et al.* 2010), whereas field trials for TC1507 were done in several locations in Isabela and South Cotabato for both wet and dry seasons (Thompson *et al.* 2010).

The stacked-trait Bt corn hybrids were evaluated using several methods. For MON 89034, greenhouse experiments were done by Solsoloy *et al.* (2007) following the same procedures as for single-trait Bt corn, described above. Field screening was also performed by Samson *et al.* (2009) in several locations under natural infestation of ACB. Recent screening was done by Caasi-Lit *et al.* (2018b) using leaf-disc assay of one-month-old corn test plants of Bt and non-Bt corn hybrids.

For the other stacked trait Bt corn hybrid, TC1507 × MON 810 × NK603, resistance screening was done using whole-plant assay with 50 second instar larvae infested onto the whorl of test plants (Caasi-Lit 2017b). Leaf-disc assay was used in the bioefficacy testing for the pyramided Bt corn hybrid Bt11 × TC1507 × GA21 (Caasi-Lit 2017c).

Field bioefficacy

The efficacy of Bt corn products is best evaluated under field conditions, as it is the most reliable method to generate valuable data. It also shows the field constraints under which the trial is done. This chapter discusses the field protocol of Syngenta Philippines (Caasi-Lit *et al.* 2019). The field trial generated robust data and the results are very significant. The infestation technique developed during the trial is very effective in producing uniform damage among the test plants, especially on the susceptible control, thereby giving outright significant difference among treatments.

Design, treatments and field layout

The location of the bioefficacy study is crucial in conducting field screening. For this reason, a pre-visit to the experimental site is done to ensure soil fertility, plain topography and availability of irrigation facility.

The experimental design is laid out in a randomised complete block design with minimum error degree of freedom being 12 (the higher the better), as replicates are varied depending upon treatments. Depending on the kind of data to be generated, plots may receive artificial and/or natural infestation. Each treatment plot has at least seven rows 4.6 m long (Plate 6). The two outer rows (Rows 1 and 7) are considered the border rows and the data are collected from the inner rows. The plant spacing should follow recommended agronomic practices such as 0.70 m between rows and 0.20 m between hills. Each row comprises 24 plants and only plant numbers 3–22 serve as the test plants. For an example plot of seven rows, 4.6 m long and 0.2 m between hills, each plot would contain a total of 168 plants; however, only 100 plants would be treated as test plants, as represented by the yellow block in Plate 6A. The actual field layout of the plot is presented in Plate 6B.

At first, screening is done during the vegetative and reproductive stages. The idea is to determine the efficacy of the Bt corn product, specifically in these two phenological stages. However, this requires a higher number of plots and the study consequently becomes more expensive. As the protocol is evaluated, the choice of one phenological stage is proved to be better and the screening is done at mid-whorl stage (40 DAP), where insect infestation is done at 30–40 DAP. Both leaf-feeding and stalk-feeding damage rating, as well as destructive sampling of the test plants for the data on tunnel length, are done as part of the screening method.

Use of insect net cages in the experimental plots

Insect cages measuring 2.5 × 2.5 × 3 m are installed before artificial infestation of ACB neonates, using bamboo poles and other wooden (sapling) materials (Plate 7). Prior to installation, ACB egg masses, lepidopteran pests and other insect pests or predators of ACB are manually removed. ACB egg mass counting, scraping and removal of other lepidopteran pest larvae are done every three days starting at 21 DAP. For experiments that include plots for natural ACB infestation, the number of egg masses on these plots is recorded every three days without scraping. Possible natural enemies were also noted.

Insect source and mass rearing

Initial populations of insects are collected from a specific location depending on the requirements from the regulatory agency. For bioefficacy testing, at least two populations of insects are used, mainly from the major corn-growing areas of the country: Isabela on Luzon and South Cotabato on Mindanao, and lately, Pangasinan, also on Luzon. Laguna population is included as the control population. Egg masses, larvae and pupae are collected from corn plants, documenting the part of the plant collected, at generally 35–45 DAP. The excess portion of the leaves is cut, lined in petri dishes with moistened filter paper, and transported to the laboratory for mass rearing. The test insects are mass reared in the Entomology Laboratory at IPB, Los Baños College. The parent populations are reared with neonates seeded in IPB modified artificial diet (components in Table 11.2 and diet preparation in Caasi-Lit *et al.* 2015). The newly hatched (F1 generation) neonates are used as test insects for the bioassay experiments.

Artificial infestation

The infestation technique using biodegradable skewers developed by Caasi-Lit *et al.* (2017) is used for ACB infestation for field bioefficacy. Fresh whorl leaves and stalks are harvested from the greenhouse and prepared in the laboratory. Fresh stalks of isoline maize, aged 30–40 DAP, are sliced into pieces each ~1 cm thick, and skewered onto pre-thinned barbecue sticks. The skewered corn stalks are placed into the infestation cup and the newly hatched larvae are transferred using a camelhair brush. Fifty ACB neonates are transferred to each infestation cup containing pre-cut leaf tissues with corn stalk barbecues for feeding. This also serves as the transfer medium of larvae during the actual plant infestation. Pre-infested leaf tissues and cornstalk barbecues are placed inside the whorl of the test plants. This infestation technique improves handling of test insects, which therefore allows greater larval survival and decreases variation in the data collected.

When screening for both vegetative and reproductive trials, in each treatment, all the plants in each plot are artificially infested with 50 four-day-old second-instar larvae per plant at

Table 11.2. Components of the IPB modified artificial diet for ACB

Ingredient	Amount	Ingredient	Amount
Ground corn	96 g	Wheat germ	2 g
Ground soybean	50 g	Methyl paraben	2 g
Brewer's yeast	40 g	Choline chloride	2 g
Vitamin complex	5 g	Gelatin	2 bars (9 g each)
Casein	2 g	Distilled water	700 mL
Ascorbic acid	4 g	Vitamin E (Myra E)	1/2 capsule (30 mg)
Sorbic acid	125 g		

vegetative V4–V6 stage (30–35 DAP) and 30 four-day old second-instar larvae per plant at early reproductive V10–V11 (45–50 DAP) stage. However, if the trial is only for one stage, 50 four-day old second-instar larvae per plant are used to infest maize plants at V5–V6 (35–40 DAP).

Damage rating

All infested plants in each treatment are examined for leaf feeding at 5 DAI and 10 DAI at vegetative and reproductive stages. Data are gathered only in the inner potion of each plot, and not from the rows and plants designated as border. Leaf-feeding damage assessment of ACB is done following the harmonised rating scale by Guthrie *et al.* (1960) and Wiseman *et al.* (1966).

Destructive sampling is done at 15 DAI at vegetative- and reproductive-stage set-ups to record the larval and pupal survival. All plants are dissected and examined for the presence of borer holes and stalk damage. The number of borer holes per plant in each treatment is counted and recorded. For the reproductive stage infestation, all plants that have tassel and ear injuries and stalk tunnelling are carefully examined and counted. Plants are dissected carefully to record and count the number of individual feeding galleries. After dissection, the length of the tunnel is measured. Larval and pupal survival are recorded, together with ear and stalk damage assessment. The larvae and pupae that survived in Bt plants are collected and further reared in artificial diet for susceptibility confirmation through leaf-disc assay.

Sample results from field assay

Leaf-feeding damage of ACB at vegetative stage is significantly different between the Bt and non-Bt corn hybrids. The mean damage rating of Bt11 and non-Bt (GA21) were 1.34 and 3.90, respectively. Leaf damage progressed significantly as shown by the rating scale at 5 DAI and 10 DAI.

Sample destructive sampling

At 15 DAI in the vegetative- and reproductive-stage set-ups, destructive sampling is done to record the larval survival. All plants are examined for the presence of borer holes and stalk damage. The number of borer holes per plant in each treatment is also counted and recorded.

Developing cobs are also observed to be damaged after infestation. ACB larvae showed tunnelling of the ear of non-Bt corn, whereas no damage was observed in Bt corn.

Laboratory bioassay

Leaf-disc assay is the easiest and fastest method to conduct bioefficacy testing. Compared with whole-plant assay in the greenhouse, leaf-disc assay uses only the fresh leaves of the treatments, then exposes them to the test insects. The results are generated after several days. Although the whole-plant assay provides more realistic conditions (environmental variability, intact plants) leaf-disc assays enable greater precision in estimating efficacy as the conditions are controlled and uniform. Whole-plant assay needs enough space and should be insect-proof to avoid infestation of other pests and natural enemies that will complicate the design and increase the cost.

For laboratory bioefficacy testing, the methods in screening the resistance of MON 89034 against ACB are presented (Caasi-Lit *et al.* 2018b).

Test insects

The initial populations of ACB for laboratory screening are derived from the Laguna or Isabela populations. The collected insects are reared in the laboratory under controlled conditions using the IPB modified artificial diets. Laboratory mass rearing follows the procedures outlined in Caasi-Lit *et al.* (2015). For the bioassay, newly hatched second-generation ACBs are grown initially in the artificial diet until they moult to second instar, then the day-old larvae are used as the test insects.

Plant materials and leaf-disc preparation

Corn plants at V6 stage (30 DAP) grown in the greenhouse are used as test plants. Young, pale green-coloured leaves of the Bt and non-Bt plants are cut from the middle portion and placed in separate containers before transferring to the laboratory for leaf-disc preparation. Fifty leaf discs each of the Bt and non-Bt plants are cut using a small, cylindrical metal leaf disc cutter with a diameter of 2.3 cm. The leaf discs are then laid on petri dishes with moist filter paper to keep the leaves fresh.

Leaf-disc assay

Leaf-disc assay procedures are adapted from the improved leaf disc-assay technique developed from several trials. Using a fine camelhair brush, one second-instar ACB larva is transferred onto each leaf disc. The petri dish is carefully covered and properly labelled, including the replicate numbers. The petri dishes (8.3 cm in diameter, 1.2 cm in thickness) are wrapped with clear plastic film before covering. The dishes are placed on a rearing rack under ambient room conditions (28.9°C, 68% RH) with 12 h daylight. The set-up is observed daily and photographed for the next seven days.

Data-gathering and statistical analysis

The number of surviving larvae is counted for the seven-day observation period. Larval survival is computed as percentage of larvae that survived over the total number of infested larvae multiplied by 100. Using MS Excel 2007, data on per cent larval survival are analysed using one-way ANOVA. Treatment differences are determined using *t*-test at $\alpha = 0.05$ level.

Sample results from leaf-disc bioassay

The overall mean survival of ACB larvae fed with leaf whorl of Bt (MON 89034) and non-Bt corn (NK603) for the seven-day period was analysed. From the second day onward, survivals differed significantly. The per cent survival of larvae on Bt declined from 98% on the first day to 10% on the second day and thereafter to 0%, while that on non-Bt corn continued to be 100%. These studies show that observations until second day are good enough.

Whole-plant bioassay

Test insects

Newly hatched second-generation ACB larvae fed with IPB modified artificial diets reared in the laboratory under controlled conditions are used as the test insects for the efficacy of Bt corn using whole-plant bioassay. The initial populations are collected in sentinel sites in Laguna and Isabela provinces. For the bioassay, the egg masses of ACB are placed on petri dishes and the newly hatched larvae are collected.

Plant materials and larval infestation

Seeds of the Bt and non-Bt corn hybrids are planted on black plastic pots with sterilised garden soils and placed on metal benches inside the greenhouse. Corn plants are maintained in the greenhouse until V6 stage (30 DAP) before the conduct of the assay. Thirty second-instar ACB larvae are infested onto the Bt and non-Bt corn test plants. The larvae are infested directly onto the whorl of each plant.

12

Insect resistance management of Bt corn in the Philippines

Edwin P Alcantara, Arian J Jacildo, Jomar F Rabajante, Lilia A Portales, Joel L Adorada and Barbara L Caoili

Introduction

Corn is the second-most important crop, following rice, in the Philippines. Yellow corn in particular is consumed by the poultry and livestock industry (Gerpacio *et al.* 2004). However, insect pests pose a major problem in corn production. The Asian corn borer (ACB) *Ostrinia furnacalis* (Guenée, 1854), is a major insect pest of corn in the Philippines (Caasi-Lit and Sapin 2012). Prior to the advent of *Bacillus thuringiensis* (Bt) corn there was no single effective control of ACB, often resulting in 20–80% farm yield losses (Morallo-Rejesus 1985; Schreiner and Nafus 1988; Nafus and Schereiner 1991; Velasco 1992). Consequently, corn production remained low at 2 t/ha (Mutuc *et al.* 2012). For this reason, the use of Bt corn was considered by the Philippine government as a viable option for effective management of ACB infestation.

A joint committee on biosafety from the University of Philippines Los Baños and the International Rice Research Institute in Los Baños was formed in 1987 to formulate guidelines on biosafety. The initiative, spearheaded by scientists from both institutions, eventually led to the establishment of the National Committee on Biosafety of the Philippines. In 1998, the National Committee formulated and released the biosafety 'guidelines on planned release of genetically modified organisms and potentially harmful exotic species' (Ramirez 2009). Another milestone in the regulation of genetically modified organisms (GMO) in the Philippines is the issuance of Administrative Order No. 8 (AO 8) in 2002 by the Department of Agriculture 'which effectively transferred the assessment and monitoring of genetically modified (GM) crops in field tests as well as commercial production to the BPI' (Ramirez 2009). Thus, the legal framework required for the introduction of transgenic crops in the country is established.

Bt corn MON 810 is the first GM corn introduced in the Philippines. Prior to commercialisation, MON 810 was subjected to rigorous risk assessment in order to establish its safety. The risk assessment consisted of greenhouse evaluation, limited field tests and multi-location tests. Results of greenhouse evaluation demonstrated the efficacy of MON 810 to its target insect pest, the ACB. The limited field tests revealed the absence of any adverse effect of MON 810 to the adjacent environment, including non-target insects, at the test site. The multi-location tests were conducted to determine the efficacy of MON 810 under varying degrees of pest infestation in different environments. As with the limited field tests, the performance of

MON 810 Bt corn was encouraging. The environmental risk assessment before approval of MON 810 for commercial planting in the Philippines has been described in detail by Ramirez (2009) and Cariño (2009). Additional evidence supporting claims of the environmental safety of MON 810 was later generated using both tiered (Alcantara *et al.* 2010; Alcantara 2012) and ecological approaches (Lit *et al.* 2011a, b, 2012, 2016).

Backed up by the legal framework for adoption of transgenic crop technology and bioefficacy data (Fernandez *et al.* 1997), the Department of Agriculture approved the commercialisation of MON 810 expressing a single insecticidal protein, Cry1Ab, in 2002 (Ebora *et al.* 2005). In 2003, the Philippines grew an estimated 11 000 ha of Bt corn. This modest area increased dramatically to ~600 000 ha as of 2018, as a result of significant increase in yield and other tangible economic benefits to farmers. For example, yields obtained by Filipino farmers growing Bt corn hybrids averaged 41–60% higher than those for non-Bt corn farmers, and net income rose by ~34% (Huesing and English 2004). In terms of environmental benefits, Bt corn does not pose a significant harm to non-target organisms, specifically natural enemies and pollinators (Reyes 2005; Alcantara *et al.* 2010; Alcantara 2012; Lit *et al.* 2011a, b, 2012, 2016). In addition, significantly lower insecticide use was required in Bt corn than the conventional corn (Yorobe and Quicoy 2006).

The presence of Bt corn throughout the year in major corn farms in northern and southern Philippines represents high selection pressure to ACB. The possibility of ACB resistance development in Bt corn is a real threat (Afidchao *et al.* 2013; Caoili *et al.* 2017; Wang *et al.* 2017) that, if left unchecked, could result in failure of the transgenic crop in the field. Thus, there is an urgent need to establish and implement clear and effective insect resistance management (IRM) for ACB.

In this chapter, we present what has been accomplished on Philippine IRM, evaluate the current status of ACB resistance management initiatives and suggest possible mitigating measures. Specifically, we outline the process of developing an IRM plan for Bt corn using existing information from relevant sources. After presentation of the outline, we discuss the outcome and impact of IRM initiatives and challenges in IRM implementation. We also describe how computer simulation tools could be successfully used in identifying the relative importance of various factors in mitigating the ACB resistance to Bt corn in the Philippines.

IRM initiatives

IRM is an essential tool for effective stewardship of Bt corn. The ultimate goal of IRM is to slow down the evolution of insect resistance to Cry toxins and sustain insect-resistant transgenic technology (Head and Greenplate 2012). Following the government approval for cultivation of MON 810 in December 2002, farmers began to adopt the transgenic crop technology in the following year. The Bureau of Plant Industry (BPI) in collaboration with various stakeholders crafted an IRM strategy suitable to Philippine Bt corn landscape conditions. The components of the IRM strategy are embodied in the policy guidelines contained in several Memorandum Circulars issued from 2003 to 2014 by the BPI under the Department of Agriculture. These IRM components includes high-dose plus refuge (HDR), use of pyramided events, resistance monitoring, farmer education and introduction of mitigating measures.

HDR constitutes the core component of the Philippine IRM strategy. High-dose is defined as 25 times the amount of insecticidal protein needed to kill 99.99% of a susceptible population (US EPA 2001). The high-dose expression in the Bt crop should eliminate the target heterozygous individuals produced by the mating of rare resistant individuals from the Bt crop with the susceptible adults from the nearby refuge. Since the first commercial planting of Bt corn MON 810 in 2003, there has been no report of field failure in the Philippines. The approval of pyramided Bt corn further added to the arsenal of IRM tools to manage ACB resistance. In the Philippines, the first pyramided event approved for commercial propagation was MON 89034, which produces Cry1A.105 and Cry2Ab2 insecticidal proteins. The other commercially available pyramided Bt corn varieties in the Philippines are Corteva's TC1507 × MON 810 and Syngenta's Bt11 × TC1507; both express unrelated Cry1Ab and Cry1F toxins. Modelling studies showed that pyramiding insecticidal proteins in a single product is the most effective strategy in delaying insect resistance as compared to alternating or sequentially deploying the same proteins (Tabashnik 1989; Roush 1998; Caprio 1998). These pyramided events are required to provide blended (refuge-in-bag) 5% refuge. The chronology of approvals for propagation of pyramided events in the Philippines is shown in Table 12.1. Recently, Onstad *et al.* (2018) demonstrated that the use of blended refuge delayed resistance evolution in ACB and ensured refuge compliance by farmers.

In consideration of farmer concerns about refuge size and the availability of alternate hosts (Lit and Fernandez 2004), the refuge size for single event Bt crop was reduced to 10% from the previous 20%. The modified refuge requirement is indicated in Department of Agriculture Memorandum Circular No. 2, series of 2014 (www.biotech.da.gov.ph). Needless to say, the importance of refuge and refuge compliance cannot be ignored, as the recent experiences in India have demonstrated (see Chapter 4 in this book).

Monitoring is another critical element of an effective IRM strategy. It includes monitoring for the adoption rate of Bt corn, refuge compliance and resistance/susceptibility status. This can be emphasised from the experience in South Africa, where the African stem borer *Busseola fusca* developed field resistance to Bt maize (MON 810) expressing *cry1Ab* within seven years of its introduction in 1998 (van Rensburg 2007). Larvae of the African stem borer, being relatively less susceptible to Cry1Ab during their second generation, survived up to second instar by feeding on soft silk of tassels of maize crop that contained a lower concentration of toxin. This helped them to tolerate higher concentrations inside the stem of Bt maize and to complete larval and subsequent stages, and thus evolve field resistance. The goal of resistance monitoring is to initiate timely mitigating measures before field control failure occurs (Huang 2006; Farinós *et al.* 2018). For ACB resistance monitoring, a diagnostic concentration has been developed for MON 810 Cry1Ab (Alcantara *et al.* 2011).

Farmer understanding of the concepts behind IRM is crucial to the long-term success of Bt corn technology in the Philippines. These IRM options have been officially adopted and are being implemented by BPI in collaboration with the industry, third-party partners from the academe and Bt corn farmers.

Monitoring of ACB resistance

For the past 10 years, resistance of ACB to Bt corn was monitored pursuant to the IRM policies. The rapid expansion of Bt corn area to support and sustain the growing livestock industry, along with the shift from the traditional corn–rice cropping system into a rice–corn–corn cropping system, Bt corn plantings in new areas, the shift in susceptibility level of ACB from the baseline in single event Bt corn, and farmers' practices of planting Bt corn seeds in localities other than the place of purchase, made it imperative to implement the refuge-in-bag IRM strategy nationwide. In this connection, Department of Agriculture issued Memorandum Circular No. 3, Series of 2012, which BPI implemented on 1 September 2012. All technology developers and seed producers were directed to package refuge seeds in a properly labelled white bag containing 20% non-Bt corn seeds inside a larger bag containing 80% Bt corn seeds. This strategy is intended to ensure that non-Bt corn seeds are made available to farmers and a structured refuge is implemented in all farms planted to Bt corn. However, under Circular No. 2, Series of 2014 titled 'Enhancing the Insect Resistance Management (IRM) Strategy for Bt Corn Targeting Asian Corn Borer (ACB)', the structured bag-in-bag refuge requirement was reduced from 20% to 10%. The reduction of non-Bt corn refuge to 10% was based on the recommendation of the

Table 12.1. Approved GM corn for environmental release in the Philippines

Event	Gene	Trait	Company	Year approved	Year(s) renewed
MON 810	*cry1Ab*	Resistance to ACB	Monsanto	2002	2007, 2012, 2018
Bt11	*cry1Ab* & pat	Resistance to ACB and phosphinothricin herbicide	Syngenta	2005	2010, 2015
MON 89034	*cry1A.105* & *cry2Ab*	Resistance to lepidopteran pests*	Monsanto	2010	2015
NK603	CP4EPSPS	Tolerance to glyphosate	Monsanto	2005	2010, 2015
GA21	EPSPS	Tolerance to glyphosate	Syngenta	2009	2014
TC1507	*cry1F*	Resistance to ACB	Pioneer Hi-bred	2013	2019
MON 810 × NK603	*cry1Ab* & CP4EPSPS	Resistance to ACB and tolerance to glyphosate	Monsanto	2005	2010, 2015
Bt11 × GA21	*cry1Ab*, pat & CP4EPSPS	Resistance to ACB and phosphinothricin herbicide, tolerance to glyphosate	Syngenta	2010	2015
MON 89034 × NK603	*cry1A.105* & *cry2Ab*, CP4EPSPS	Resistance to lepidopteran pests*; tolerance to glyphosate	Monsanto	2011	2016
TC1507 × MON 810 × NK603	*cry1F*, pat, *cry1Ab* & CP4EPSPS	Resistance to ACB and tolerance to glyphosate	Pioneer Hi-bred	2014	2019
TC1507 × MON 810	*cry1F*, pat & *cry1Ab*	Resistance to ACB and tolerance to glyphosate	Pioneer Hi-bred	2014	
TC1507 × NK603	*cry1F*, pat & CP4EPSPS	Resistance to ACB, tolerance to glufosinate and glyphosate	Pioneer Hi-bred	2014	
Bt11 × TC1507 × GA21	*cry1Ab* & pat, *cry1F*, pat, CP4EPSPS	Resistance to ACB and tolerance to glufosinate and glyphosate	Syngenta	2015	
MIR 162	*vip3Aa20*	Insect resistance	Syngenta	2018	
Bt11 × MIR162 × MON 89034 × GA21	*cry1Ab*, pat, pmi, *vip3Aa20*, mepsps, *cry2Ab2*, *cry1A.105*	Insect resistance and tolerance to glufosinate and glyphosate	Syngenta	2020	

* ACB, European corn borer, fall armyworm and corn earworm.
Source: BPI, Manila.

IRM Advisory Team, which reviewed recent developments (stacked/pyramided transgenics) in crop biotechnology and related fields, the presence of natural host crops of ACB including non-Bt open-pollinated corn, agronomic practices and concerns of small and marginal farmers suffering from significant yield loss, with a 20% refuge need. Consequently, the Advisory Team considered refuge reduction the most appropriate step under local settings. Lit and Fernandez (2004) had earlier reported the availability of natural refuges compensating reduction in refuge size in the corn and surrounding environment. Fallow periods and crop rotations also provide temporal refuges for susceptible insects.

Structured refuge bag-in-bag survey results from 2017, which covered 10 municipalities in regions CAR, 1, 2, 10 and 12 with 947 farmer respondents, showed that 99.8% knew about refuge plants that support Cry protein-susceptible ACB individuals to delay ACB resistance development. These farmers realised the importance of planting structured refuge, as reflected by 97.1% positive answers in the survey (Manalo 2018). Meanwhile, the market availability of pyramided Bt corn with provision of seed blended refuge-in-bag (RIB) (5% non-Bt refuge seed mixed with 95% Bt corn seed) ensured farmers' compliance by default for those products.

The specific monitoring schedules in the identified sentinel areas (Figure 12.1) in region 2 (Cauayan and Tumauini, Isabela), region 3 (Magalang, Pampanga), region 10 (Baungon, Bukidnon) and region 12 (Banga and Koronadal City, South Cotabato) for both Bt corn and herbicide-tolerant (HT) corn are coordinated with the BPI-Biotech Core Team Post Approval

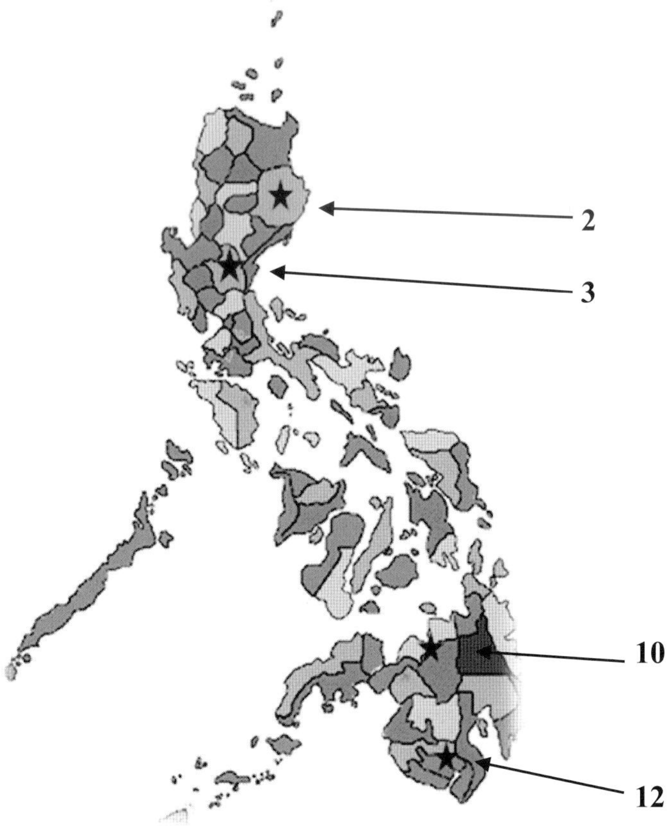

Figure 12.1 Bt corn sentinel areas currently in use for Asian corn borer resistance monitoring.

Monitoring Group and the Regional Field Office Monitors. The schedules are planned with the technology developers, based on the growth stages of corn (seedling, vegetative, reproductive stage and post-reproductive stages), noting the presence of weeds, target pest occurrence and their damage to the crop, and the presence of non-target beneficial and useful organisms. Monitoring of sentinel sites for HT corn stacked with insecticidal Bt protein after harvesting or before land preparation is necessary to check weed dominance and diversity. The specific protocols and guidelines in data monitoring are strictly followed as indicated in the data-gathering procedures.

Random field monitoring in GM corn areas at different growth stages is necessary to check and validate the efficacy of Bt corn in the field against target lepidopteran insect pests. Similarly, random field monitoring of HT corn-growing areas with the continuous application of glyphosate or glufosinate is important to verify performance claims and to monitor possible build-up of resistant weeds.

As mentioned earlier, baseline susceptibility of ACB to Cry1Ab was studied in 2002–04 for many populations, and median lethal concentration (LC_{50}) of Cry1Ab was found to range from 0.42–2.32 ng/cm^2 of diet. The studies showed natural variation among insect populations in terms of susceptibility towards Cry1Ab. Using upper limit LC_{99} value of 104 ng/cm^2 diet as a diagnostic concentration of MON 810 Cry1Ab protein, resistance monitoring of 11 ACB populations collected from the regions where Bt corn was grown for three years was carried out in 2009. Studies showed some larval survivals at the diagnostic concentration; however, none of progenies survived in the next generation (Alcantara *et al.* 2011). Further confirmation showed no survivals in the Bt corn MON 810 leaf tissue assay.

Monitoring adoption rate and IRM survey in sentinel sites

Biosafety permits for both Bt and HT corn propagation are valid for five years, and renewable every five years subject to the terms and conditions stated in AO 8. In the case of Bt corn, the permit-holder must comply with the following requirements:

- approved stewardship program consisting of enhanced IRM strategy;
- approved monitoring activities;
- regular reports of seed sales;
- conduct of required technical studies.

Pursuant to the implementation of post-commercial monitoring of Bt and HT corn, the Department of Agriculture issued several policy guidelines embodied in several Circulars and Special Orders. This includes Circulars No. 17, Series of 2003; No. 8, Series of 2005; No. 1, Series of 2006; No. 4, Series of 2007; and No. 3, Series of 2012.

Circular No. 17 pertains to the additional requirements for IRM strategy for Bt corn. It stipulates the structured refugia to be set up when the Bt corn adoption rate in a contiguous 200 ha or more exceeds 80%. It also states that monitoring of resistance development should be done in coordination with the Department of Agriculture. It requires the strengthening of stakeholder education on IRM as well as support of technical studies on IRM to fill in information gaps.

Circular No. 8 of 2005 amends section IIIa(ii) of the 2003 Circular No. 17, which strengthened the decision-making of the Department of Agriculture through science-based IRM for Bt corn. Circular No. 8 reiterates the scientific basis of the high-dose refuge IRM strategy for Bt corn, including whether it is to be structured or unstructured depending on considerations such as:

- the level of Bt corn adoption in an area;
- the size of Bt corn farms;

- the presence of alternate host crops and weeds around the corn fields which can support the population of susceptible ACB;
- the farmer's practice of crop rotation or diversified farming; for example, intercropping, which can serve as a temporal or spatial refuge.

The planned structured refuge, which is planting 80% Bt corn and 20% non-Bt corn plants in the same field as per Circular No. 17 section IIIa(i), is implemented in the following conditions:

- a minimum contiguous production area/cluster of 200 ha planted to corn;
- at least 80% adoption rate of Bt corn in the contiguous 200 ha area;
- non-practice of crop rotation.

On the other hand, temporal refuge through crop rotation involves planting of a crop other than Bt corn in one planting season within the year to break the cycle of selection for Bt corn-resistant ACB individuals. Nonetheless, when no ACB resistance to Bt corn and presence of unstructured refuge were observed from locally generated data, the implementation of a structured refuge after the two-year period as stipulated in Circular No. 17 section IIIa(ii) should be lifted.

Circular No. 1 from 2006 focuses on the procedural guidelines and formats for Bt corn IRM monitoring for industry technology developers. It discusses compliance monitoring for the Bt corn planted in selected areas and ACB resistance monitoring, such as IRM awareness briefing, Bt corn adoption or planting, and monitoring of ACB, other corn pests, beneficial organisms and other soil organisms. ACB resistance monitoring sites were identified in the municipalities of Sta Maria and Malasiqui in Pangasinan, Jones and Cauayan in Isabela, Solana in Cagayan, Mexico in Pampanga, Sablayan in Mindoro Occidental, General Santos City in South Cotabato and Bagumbayan in Sultan Kudarat. The monitoring activities are done by the technology developer monitoring team together with Department of Agriculture Regional Crop Protection Center (DA-RCPC) technical officers.

In 2006, the Department of Agriculture issued SO No. 7, Series of 2006 to create the IRM Advisory Team, which is composed of scientists and technical experts. The IRM Advisory Team, as a scientific and advisory body, provides technical expertise related to IRM issues and concerns to the Department of Agriculture and BPI. Its composition was indicated in SO No. 589, Series of 2012 then changed again by SO No. 24, Series of 2017 and SO No. 1051, Series of 2018.

Meanwhile, in 2007 Circular No. 4 revised the guidelines for insect resistance monitoring as well as the templates for Bt corn IRM monitoring and reports by the technology developers and DA-RCPC. It improved the guidelines for the monitoring compliance for Bt corn planted in each of the identified areas, monitoring of ACB resistance, data-gathering and verification of ACB resistance.

To date, Bt corn is planted in 14 regions of the Philippines (Figure 12.2). Bt corn adoption is highest in region 2, with ~294 686 ha based on industry sales. This is followed by regions 1 and 12 with 65 411 ha and 34 346 ha, respectively. The details on the relative area planted to different Bt corn events could not be disclosed, since these are considered by seed companies to be confidential business information.

Deployment of transgenic corn plants

The BPI maintains the official registry of approved GM plants and plant products, which have undergone thorough safety assessment by experts and regulatory agencies (Table 12.1). Currently, there are 13 approved GM crops for commercial propagation in the country. In December 2002, the first GM corn, MON 810, was approved for propagation. This GM corn produces Cry1Ab toxin, which confers resistance to ACB. Seven approved GM corns have stacked traits, five have single traits (either with IR or HT trait) and one pyramided product.

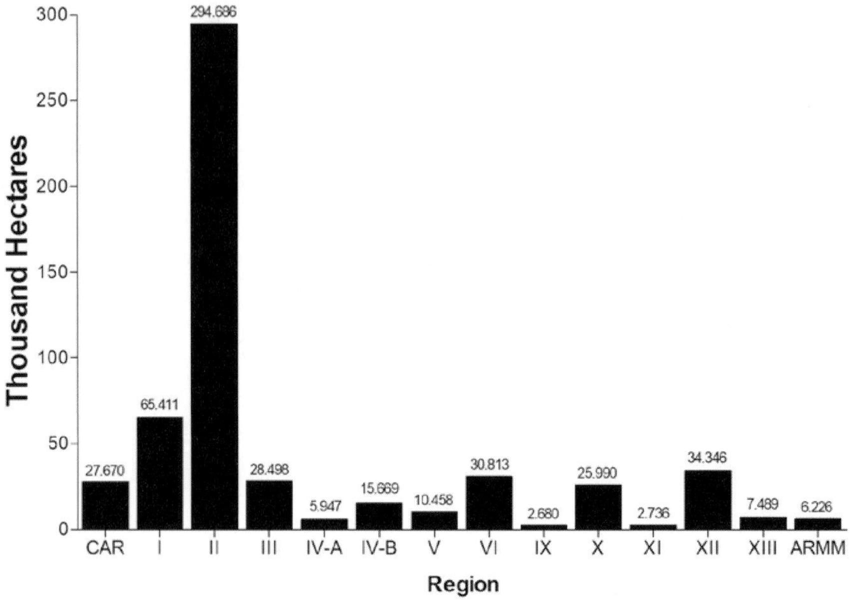

Figure 12.2 GM corn regional adoption in the Philippines (CY April 2017–March 2018).

Stacked products contain traits attacking different pests while the pyramided products contain traits attacking the same pest (Hurley and Sun 2019).

Industry compliance for IRM

In 2012, the Asian Farmers Regional Network co-organised the National Agricultural Biotechnology Farmers Conference with the Department of Agriculture, International Service for the Acquisition of Agri-biotech Applications and South-east Asian Regional Center for Graduate Study and Research in Agriculture. The conference produced a resolution to address farmer needs for science-based information on biotechnology to aid them to make informed choices, and a well-defined IRM program specifically for corn (Navarro *et al.* 2013). The main objective of farmer education in the context of GM crops is to guide them in implementing measures to prolong the durability of Bt technology. From 2015 to 2016, an IRM awareness program comprising meetings, field and demonstration tours, a harvest festival, corn fiesta, crop tour and information campaigns on structured bag-in-bag refugia was conducted by the technology developers. A total of 117 918 participants were recorded and submitted to BPI as part of their stewardship compliance requirements (L. Portales, pers. comm.).

Circular No. 2 in 2014 issued guidelines on compliance to formulate more effective strategies to avoid insect resistance development. It requires studies for each technology developer to guide them in refuge deployment, prediction modelling to estimate any shift in Bt crop susceptibility to ACB, and the bioefficacy of Cry1Ab after more than 16 years of commercialisation. The required studies are as follows.

1. General ACB data.
   a. Adult dispersal.
   b. Mating behaviour.
   c. Oviposition behaviour.
2. Product-specific data to demonstrate dose/efficacy.
   a. Bioassay of second-instar ACB larvae on Bt corn plant at vegetative stage.

b. Bioassay of second-instar ACB larvae on Bt corn plant using leaf disc.
c. Field efficacy based on reduction in ACB tunnels (length and number of tunnels) versus the non-Bt control.
d. Evidence that each protein provides at least 90% mortality of the target insect and pyramided product provides at least 99% mortality.
e. Evidence that the proteins expressed in the pyramid product show low potential for cross-resistance, as indicated by certain parameters such as differences in protein structure and binding sites.

3. Required data to support seed blending for pyramids.
4. Basic studies on larval movement (instar, rate, distance).
   a. Movement and survival.
   b. ACB production on refuge plant (block versus blend).
   c. ACB survival in Bt plants after movement from refuge plants.

As part of the industry compliance with Circular No. 2, basic studies on ACB adult dispersal, mating and oviposition behaviour were undertaken by the technology developers from July 2015 to June 2017. These studies aimed to support the IRM strategy implementation for Bt corn in the Philippines. The results suggest that random mating between rare resistant ACB and predominant susceptible ACB is highly probable, given the observed wide age window for mating. The studies also confirmed the usefulness of synchronous planting of non-Bt corn as the priority refuge for ACB oviposition and of planting non-Bt corn in very close proximity to Bt corn to serve as an effective refuge.

IRM implementation challenges

The Philippines is the first Asian country to establish a functional biosafety regulation mechanism that approved a GM corn event for cultivation in 2002. It is guided by a responsive and sound national policy on the safe use of GM products implemented by the concerned regulatory agencies. The commercial propagation of IR and HT corn in the Philippines is governed by AO No.8, issued in 2002. After corn events have been granted a permit for propagation or commercialisation, an IRM plan has to be submitted and approved by the regulatory authorities. The plan includes monitoring of ACB resistance in identified sentinel sites with high Bt corn adoption.

Since Bt crops are capable of season-long expression of the Bt protein, precautionary steps have to be taken in order to avoid the development of insect resistance (James 2015). Since the start of commercial propagation, Bt corn plantings increased from 10 769 ha in 2003 to 728 078 ha in 2013. Due to this drastic increase, the use of structured refuge with an 80:20 structured bag-in-bag deployment scheme was implemented in 2012, first in Luzon then nationwide in 2013. In 2014, a new refuge scheme was set in place using 90:10 structured bag-in-bag refuge for a single-toxin Bt corn, and 95:5 structured bag-in-bag and 95:5 seed blended refuge-in-bag for dual-toxin pyramided Bt corn. The updated policy was crafted and implemented after consultation between the regulators, technology developers, policy-makers and other stakeholders.

The regulators face challenges during the implementation of IRM, such as lack of full-time technical staff for post-approval monitoring. Further, some corn stakeholders need regular IRM awareness programs that entail additional costs. Another challenge is the implementation of refuge schemes, particularly the use of structured refuges in the field. Based on the 2017 structured bag-in-bag survey, ~25% of the farmer respondents did not plant the refuge seeds as shown by the resulting ACB damage that entailed cost for insecticide use.

In 2016, IRM concerns were raised due to the proliferation of unregistered seeds in the market. These seeds are conventionally bred and contain the Cry1Ab protein, but have not undergone the process of GM crop registration. They are sold as conventional seeds or hybrids and thus their producers do not implement any IRM strategy like the refuge system. The method of seed production and non-compliance with IRM requirements pose a serious threat to product durability and sustainable use of Bt corn technology in the Philippines. Furthermore, the probability of selecting for Bt-resistant ACB is more likely in the areas where these seeds are planted since the dynamics of ACB in the Bt corn field will be distracted. This is a major concern for the technology developers and the government.

Currently, there is no IRM plan for secondary insect pests such as earworm and cutworm nor for other Bt crop events exhibiting low bioefficacy levels. Devising these will be a challenge for the future. Joint Department Circular No. 1, Series of 2016 states that deregulation of a registered Bt event can be implemented once guidelines are crafted – this will affect IRM implementation since once a Bt product is deregulated the technology developers may not implement IRM strategies. Regulators may not be able to enforce authority over deregulated GM events. IRM for Bt corn should be a concerted effort of all stakeholders and, for the Bt technology to remain sustainable, seed and trait quality must be strictly monitored. In the long run, this will benefit farmers in the Philippines.

Application of computer modelling to Philippine IRM initiatives

IRM is a very challenging scientific area, given the complex nature of the Philippine corn farming landscape. The myriad of biotic and abiotic factors including the important role of farmer practices, notwithstanding the long span of time required to observe any significant change in susceptibility of ACB in the field, clearly present difficult challenges for experimentation. Given this reality, it is vital to consider the role of computer modelling in building a conceptual framework for insect resistance. The first computer modelling study was reported by Benigno in 2013 (unpubl.). It showed that the current high-dose/refuge strategy for a single Bt corn event could be maintained for as long as 20 years given 10% refuge, 0.01% alternate host and 0.001 initial resistance allele frequency. The study led to a government decision to reduce the required structured refuge from 20% to 10% bag-in-bag mode of deployment. Further local modelling efforts should be carried out to address other important IRM issues such as the impact of alternate host in synchronous ACB mating and the development of a sensitive monitoring tool to estimate resistance allele frequencies and ACB family genotypes to predict the status of proliferation of heterozygotes, which are the common carriers of resistance in the field (Li *et al.* 2016). Follow-up modelling studies were conducted to determine the favourable effect of refuge size in decreasing the population of resistant ACB as long as there exists a relatively huge population of susceptible ACB (Rabajante *et al.* unpublished). The model implied that refuge is useless in inhibiting the evolution of resistance in ACB if a sufficient initial population of Bt-susceptible corn borers is absent. An agent-based modelling approach was tested by Jacildo *et al.* (unpublished) to simulate the development of resistance in ACB to Bt corn and to look further into the spatial dynamics of refuge areas and pyramid Bt events.

Conclusion: recommendations

Signs of shifts in susceptibility and farmer compliance with refuge requirements have driven IRM to favour the deployment of pyramided Bt corn varieties in the Philippines Bt corn landscape. Results of our modelling studies with a single Bt event showed the importance of refuge size in containing the rapid development of resistance. However, the practicality of implementing increased refuge size >20% is questionable in light of the difficulty of compliance by Bt corn farmers. A more feasible approach would be the conservation of nearby alternate host such as grasses to serve as sinks for susceptible ACB, assuming that their development is in synchrony with resistant ACB in Bt corn. The use of pyramided Cry proteins as an added mortality factor is favoured to suppress the proliferation of resistance allele in the population.

Monitoring for early signs of resistance development should be continued, especially in populations where a shift in susceptibility has been observed for several years. Monitoring should be expanded to adjacent farms to determine the extent of the shift in susceptibility. This is done through inter-agency cooperation involving the developers, farmers, local government units, regulatory bodies and academics.

In view of the mosaic of IR corn events (producing toxins like Cry1Ab, Cry1A.105, Cry1F and Cry2Ab) in the Philippines, resistance monitoring in terms of multiple and cross-resistance evolution in ACB should be carried out. Shabbir *et al.* (2018) reported cross-resistance in Cry1Ah-resistant ACB to other toxins like Cry1F, Cry1Ab and Cry1Ac but none to Cry1Ie.

Empirical estimates of resistance allele frequency of ACB for all Bt toxins present in the Bt corn events in the market should be made, if possible using a genomic approach that is less labour-intensive but is accurate in predictive capability. At present, there is no local information on the genetic properties of Cry protein-resistant ACB. Isolation of field-evolved resistant ACB is desirable to provide information for accurate modelling of IR development.

Computer modelling can provide useful insights on the evolution of insect resistance, when empirical studies are difficult or impractical to implement. Appropriate models that are relevant to the Bt corn landscape in the Philippines should be developed, to assist regulators in making science-based decisions for proactive IRM.

References

Afidchao MM, Musters CJM, de Snoo GR (2013) Asian corn borer (ACB) and non-ACB pests in GM corn (*Zea mays* L.) in the Philippines. *Pest Management Science* **69**, 792–801. doi:10.1002/ps.3471

Alcantara EP (2012) Post commercialization monitoring of the long-term impact of Bt corn on non-target arthropod communities in commercial farms and adjacent riparian areas in the Philippines. *Environmental Entomology* **41**, 1268–1276. doi:10.1603/EN10297

Alcantara EP, Caoili BL, Javier PA, Mostoles MDJ (2010) A multilayer, large scale comparison of arthropod communities in commercially managed Bt and non-Bt corn fields. *Philippine Entomologist* **24**, 150–164.

Alcantara E, Estrada A, Alpuerto V, Head G (2011) Monitoring Cry1Ab susceptibility in Asian corn borer (Lepidoptera: Crambidae) on Bt corn in the Philippines. *Crop Protection* **30**, 554–559. doi:10.1016/j.cropro.2010.12.019

Caasi-Lit MT, Sapin GD (2012) Intraspecific life and larval instar variations among local populations of Asian corn borer, *Ostrinia furnacalis* (Guenée) (Lepidoptera: Crambidae) in the Philipines. *Philippine Entomologist* **26**, 13–27.

Caoili BL, Sapin GD, Cayabyab BF, Alcantara EP (2017) *Screening for Bt Resistance Allele in the Philippine Populations of Ostrinia furnacalis (Guenée) (Lepidoptera: Crambidae) Collected from Bt Corn Sentinel Sites.* Terminal Report. DA-BIOTECH R1203. University of the Philippines Los Baños.

Caprio MA (1998) Evaluating resistance management strategies for multiple toxins in the presence of external refuge. *Journal of Economic Entomology* **91**, 1021–1031. doi:10.1093/jee/91.5.1021

Cariño FA (2009) Marked "Approved:" BT Corn was first. In *Modern Biotechnology and Agriculture: A History of the Commercialization of Biotech Corn in the Philippines.* pp. 21–43. STRIVE Foundation, Los Baños, Laguna.

Ebora RV, Ampil AC, Palacpac MC, Custodio CG (2005) *Commercialization of Bt corn in the Philippines.* Status Report. APCoAB. ICRISAT. NASC Complex. Dev Prakash Shastri Marg, Pusa Campus, New Delhi.

Farinós GP, Hernandez-Crespo P, Ostego F, Castañera P (2018) Monitoring of *Sesamia nonagrioides* resistance to MON 810 maize in the European Union: lessons from a long-term harmonized plan. *Pest Management Science* **74**, 557–568. doi:10.1002/ps.4735

Fernandez EC, Dalmacio SC, Hautea RA (1997) Bioefficacy of transgenic corn expressing the *Bacillus thuringiensis Cry1A(b)* gene against Asiatic corn borer, *Ostrinia furnacalis* (Guenée). *Philippine Entomologist* **11**, 119–130.

Gerpacio RV, Labios JD, Labios RV, Diangkinay EI (2004) *Maize in the Philippines: Production Systems, Constraints, and Research Priorities.* International Maize and Wheat Improvement Center, Mexico.

Head GP, Greenplate J (2012) The design and implementation of insect resistance management programs for Bt crops. *GM Crops and Food: Biotechnology in Agriculture and the Food Chain* **3**, 144–153. doi:10.4161/gmcr.20743

Huang FN (2006) Detection and monitoring of insect resistance to transgenic Bt crops. *Insect Science* **13**, 73–84. doi:10.1111/j.1744-7917.2006.00070.x

Huesing J, English L (2004) The impact of Bt crops on the developing world. *AgBioForum* **7**, 84–95.

Hurley TM, Sun H (2019) Softening shock and awe pest management in corn and soybean production with IPM principles. *Journal of Integrated Pest Management* **10**, 1–8. doi:10.1093/jipm/pmz001

James CI (2015) *50 Biotech Bites.* ISAAA, Ithaca, New York.

Li GP, Reisig D, Miao J, Gould F, Huang FN, Feng HQ (2016) Frequency of Cry1F non-recessive alleles in North Carolina field populations of *Spodoptera frugiperda* (Lepidoptera: Noctuidae). *PLoS One* **11**(4), e0154492. doi:10.1371/journal.pone.0154492

Lit MT, Fernandez EC (2004) *Alternate Hosts Study of the Asian Corn Borer, Ostrinia furnacalis (Guenée) in Major Corn Growing Areas of the Philippines.* Terminal Report. University of Philippines-Los Baños. College, Laguna.

Lit IL, Caasi-Lit M, Ramal AFB, Yap SA, Sapin GD, Eusebio OL, Larona AL (2011a) Non-target organisms on Bt corn MON 89034 and MON 89034/NK603: Part 1. Abundance and diversity of canopy- and ground surface-dwelling arthropods in regulated field trial sites during dry season in Luzon and Mindanao, Philippines. *Philippine Entomologist* **25**, 131–155.

Lit IL, Caasi-Lit M, Ramal AFB, Benigno EA, Yap SA, Sapin GD, Eusebio OL, Larona AL (2011b) Non-target organisms on Bt corn MON 89034 and MON 89034/NK603: Part 2. Functional guilds of arthropods in regulated field trial sites during dry season in Luzon and Mindanao, Philippines. *Philippine Entomologist* **25**, 131–155.

Lit IL, Caasi-Lit M, Benigno EA, Ramal AFB, Yap SA, Sapin GD, Eusebio OL, Larona AL (2012) Non-target organisms on Bt corn MON 89034 and MON 89034/NK603: Part 2. Abundance and diversity of canopy- and ground surface-dwelling arthropods in regulated field trial sites during dry season in Luzon and Mindanao, Philippines. *Philippine Entomologist* **26**, 28–53.

Lit IL, Caasi-Lit M, Abenis K, Manalo N, Larona AL, Eusebio OL, Benigno E (2016) Non-target organisms on Bt corn hybrids MON 89034 and MON 89034/NK603: Part 3. Functional guilds of arthropods in regulated field trial sites during the wet season in Luzon and Mindanao, Philippines. *Philippine Entomologist* **30**, 29–51.

Manalo A (2018) *Results of the National IRM Survey 2017.* Paper presented during the IRM Capacity Building and Workshop at TREES, College of Forestry and Natural Resources, Philippines, 13 June 2018.

Morallo-Rejesus B (1985) Research advances and the needs in the control of Asiatic corn borer, *Ostrinia furnacalis* (Guenée), in the Philippines. *Philippine Entomologist* **6**, 406–426.

Mutuc MEM, Rejesus RR, Pan S, Yorobe JM (2012) Impact assessment of Bt corn adoption in the Philippines. *Journal of Agricultural and Applied Economics* **44**, 117–135. doi:10.1017/S1074070800000201

Nafus DM, Schereiner IH (1991) Review of the biology and control of the Asian corn borer, *Ostrinia furnacalis* (Lep: Pyralidae). *Tropical Pest Management* **37**, 41–56. doi:10.1080/09670879109371535

Navarro M, Tome K, Ginutao K (2013) *From Monologue to Stakeholder Engagement: The Evolution of Biotech Communication.* ISAAA Brief No. 45. ISAAA, Ithaca, NY.

Onstad DW, Crespo AL, Pan Z, Crain PR, Thompson SD, Pilcher CD, Sethi A (2018) Blended refuge and insect resistance management of insecticidal corn. *Environmental Entomology* **47**, 210–219. doi:10.1093/ee/nvx172

Ramirez D (2009) The R&D to commercialization continuum: biosafety risk assessment of Bt corn (MON 810). In *Modern Biotechnology and Agriculture: A History of the Commercialization of Biotech Corn in the Philippines*, pp. 1–20, STRIVE Foundation, Los Baños, Laguna.
Reyes SG (2005) Wet season population abundance of *Micraspis discolor* (Fabr.) (Coleoptera: Coccinellidae) and *Trichomma cnaphalocrosis* Uchida (Hymenoptera: Ichneumonidae) on three transgenic corn hybrids in two sites in the Philippines. *Asia Life Sciences* **14**, 217–224.
Roush RT (1998) Two-toxin strategies for management of insecticidal transgenic crops. Can pyramiding succeed where pesticide mixtures have not? *Philosophical Transactions of the Royal Society of London. B* **353**, 1777–1786. doi:10.1098/rstb.1998.0330
Schreiner IH, Nafus DM (1988) No-tillage and detaselling: effect on the Asian corn borer *Ostrinia furnacalis* and ants. *Philippine Entomologist* **7**, 435–442.
Shabbir MZ, Quan Y, Wang Z, Bravo A, Soberon M, He K (2018) Characterization of the Cry1Ah resistance in Asian corn borer and its cross resistance to other *Bacillus thuringiensis* toxins. *Scientific Reports* **8**, 234. doi:10.1038/s41598-017-18586-2
Tabashnik BE (1989) Managing resistance with multiple pesticide tactics: theory, evidence, and recommendations. *Journal of Economic Entomology* **82**, 1263–1269. doi:10.1093/jee/82.5.1263
US EPA (2001) *Biopesticides Registration Action Document: Bacillus thuringiensis Plant-incorporated Protectants. E: Benefit Assessment.* <http://www.epa.gov/oppbppd1/biopesti- cides/pips/Bt_brad2/5-benefits.pdf>
Velasco LRI (1992) Questions about detasseling as a control technique against the Asian corn borer (*Ostrinia furnacalis* (Guenee), Pyralidae, Lepidoptera). *Philippine Entomologist* **8**, 1103–1116.
Wang Y, Yang J, Quan Y, Wang Z, Cai W, He K (2017) Characterization of Asian corn borer resistance to Bt toxin Cry1le. *Toxins* **9**, 186. doi:10.3390/toxins9060186
Yorobe JM, Quicoy CB (2006) Economic impact of Bt corn in the Philippines. *Philippine Agricultural Scientist* **89**, 258–267.

13

Socioeconomic and environmental impacts of GM corn in the ASEAN: the case of the Philippines

Leonardo A Gonzales and Joyce Louise Ignacio-Castillo

Introduction

The year 2015 earmarked the operationalisation of the Association of South-east Asian Nations (ASEAN) Economic Community (AEC). The AEC blueprint 2015 exudes the ASEAN's determination to achieve higher levels of sustainable economic prosperity, inclusive growth and integrated development. It recognises the increasing interdependence of the ASEAN member states (AMS) within the region and the rest of the world, stressing the importance of narrowing the development gaps within the AEC (ASEAN 2008). The ASEAN is a huge market of 10 member countries totalling around 617 million people. The ASEAN region has a total area of 440 million ha and is a global player in agro-food trade of over US$150 billion in annual exports. The region is also a global leader in the production of palm and coconut oil, rice, coffee, banana, tuna, pepper, cassava and aquaculture. According to the AEC blueprint, ASEAN shall have five core elements in the 'free flow' of goods, services, investment, capital and skilled labour. This includes two important components: the priority integration sectors and food, agriculture and forestry (ASEAN 2008). Under the food, agriculture and forestry component, the AEC will enhance intra- and extra-ASEAN trade and long-term competitiveness. This can be achieved through several actions like the establishment of good agricultural and aquacultural practices, harmonisation of maximum residue limits of commonly used pesticides, harmonisation of biosafety guidelines and research protocols on biosafety for genetically modified (GM) agricultural and fishery products, and harmonisation of the safety and quality standards for horticultural produce and agricultural products of economic importance in the ASEAN region (ASEAN 2008).

Agribiotechnology in the ASEAN region

The importance of agribiotechnology in expanding food production frontiers and achieving food security has increased over decades of rising global demand and declining arable land. The world population reached 7.6 billion in 2018, having more than doubled over the last 50 years. The rapid growth of population, coupled with increasing incomes in the developing countries, fuelled a rising demand for food which is expected to grow by 59–98% by 2050 (Elferink and Schierhorn 2016). The same trend could be expected for Asia, whose population is projected to grow to ~4.5 billion in 25 years (ADB 2014). In the ASEAN, the population has increased by 50% in a span of 20 years (ADB 2017) while incomes have grown at a yearly average of 4.1% for the last two decades (ASEAN 2017). Along with resource constraints and rising demand, the ASEAN is disadvantaged in sustaining food supplies due to the challenge of climate change, to which its member states are among the most vulnerable (Iha *et al.* 2015).

Global area planted to GM crops, since the first commercialisation in 1996, increased to ~191.7 million ha in 2018 (ISAAA 2018). However, only <1 million ha of GM crops is grown in the ASEAN member countries, with the Philippines alone contributing ~0.6 million ha.

A cursory look at agribiotechnology profiles among AMS indicate that a major constraint in the adoption of GM technology is the uneven biotechnology development among the 10 AMS. In terms of these countries' involvement in biosafety management and agribiotechnology development, AMS can be divided into three groups. Group 1 includes countries (Indonesia, Malaysia, the Philippines, Thailand and Vietnam) that have an existing regulatory framework and assessment protocols. Two countries (the Philippines and Vietnam) in this group have already commercialised production of GM crops. Group 2 consists of member countries (Cambodia, Lao PDR and Myanmar) with draft or in progress regulatory framework and assessment protocols, while Group 3 is composed of member countries (Brunei Darussalam and Singapore) with no existing biosafety framework or assessment protocols and not signatory to the Cartagena Protocol on Biosafety (Gonzales *et al.* 2018). Further, the clustered growth in the region in terms of the development of laws and regulations related to biosafety and biotechnology development is due to diversity among AMS in population, food demand, technology readiness and capacity for innovation, government prioritisation of biotechnology development and biosafety management, and overall public acceptance of GM products. Under these circumstances – unless the AMS fast-track the regional harmonisation of biosafety guidelines and research protocols – attainment of the major goals of eradication of absolute poverty and inequality, ensuring food security and global competitiveness as envisioned in *ASEAN Economic Community into 2025 and Beyond* (Sta Maria *et al.* 2018) will be difficult.

Global benefits of GM crops

The global benefits of GM crop adoption are well documented. The adoption of GM crops has generated significant and multiple benefits to farmers. Global crop productivity has improved by 675.6 million tons and resulted in savings of 183 million ha of land due to reduced cultivation, and an economic gain of US$186.1 billion among smallholder farmers (Brookes and Barfoot, cited in James 2017). Likewise, the Philippines gained a total of US$724 million since the introduction of GM crops in 2003 (Brookes and Barfoot, cited in James 2017).

In terms of GM specific traits, the socioeconomic impacts of insect resistant (IR) corn on a global scale were reported by Brookes and Barfoot (2014). These included lowering of pest damage and insecticide cost in comparison to the actual cost of IR seed technology. Globally from 1996 to 2016, the cumulative net benefit was estimated at US$36.9 billion. In 2016 alone, net benefits from the planting of IR corn were valued at US$3.7 billion (Brookes and Barfoot, cited in James 2017). Brookes and Barfoot (2014) also estimated the farm-level economic impact of herbicide-tolerant (HT) corn on global and country scales. According to their study, the main benefits of using HT corn were reduced expenditure on labour and herbicide, which translated into reduced costs and improvement in corn profitability as against the seed cost premium due to HT technology. On a global perspective, the cumulative economic benefit of HT corn was estimated at US$5.4 billion from 1997 to 2012, roughly equivalent to US$337 million per year during the period. Of this cumulative value, 74% was attributed to cost savings, and 26% to yield gains. The accumulated net economic benefits from HT corn for the period 1997–2015 was valued at US$11.1 billion. In 2016 alone, economic benefits gained from the adoption of HT corn were estimated at US$2.1 billion (Brookes and Barfoot, cited in James 2017).

Philippines GM corn impact assessment studies

In the Philippines, several studies reported socioeconomic assessments of GM corn (Gonzales 2005, 2007, 2019; Yorobe and Quicoy 2006; Yorobe *et al.* 2010; Yorobe and Smale 2012; Gonzales *et al.* 2013; MASIPAG 2013; Afidchao *et al.* 2014). Gonzales *et al.* (2013) estimated that GM corn planted on ~643 800 ha in 2011 brought in additional gains of US$400 million. These macroeconomic impacts were disaggregated into incremental net farm incomes of GM production as compared to non-GM corn; pre-harvest labour savings of land preparation, weed and insect control, income from seed and fertiliser sales; and post-harvest labour income multiplier effect due to higher yields of GM corn as compared to its non-GM counterpart. The stack corn (IR/HT) had the highest share of these macro incremental benefits at 89%, followed by HT at 10% and IR at 1% (Gonzales *et al.* 2013).

At the farm level, almost all of these socioeconomic impact assessments indicated relatively better performance of GM corn over non-GM corn in terms of yield, cost efficiency and profitability. One study (MASIPAG 2013) indicated a negative impact on net farm income when adopting GM corn technology in the Philippines. However, the methodology used in the socioeconomic assessment is still being debated among socioeconomic researchers in the country.

Philippines agriculture and the Philippines Development Plan

Philippines agriculture is faced with at least four development challenges:

- unbridled population growth that demands higher levels of food and agricultural services;
- the need to produce more with less, limited and deteriorating natural resource frontiers;
- enhancing the efficiency of the supply value chains of agricultural products to ensure sustainable household incomes of participants;
- development of a set of comprehensive enabling and appropriate policies to promote competitive advantage in the global market (Gonzales 2019; Gonzales *et al.* in press).

Historically, this set of challenges has haunted Philippines agriculture and the economy in search of inclusive growth and sustainable development. The Philippines Development Plan 2017–2022 outlines these development concerns (NEDA 2017). The Plan also envisions a long-term version entitled *Ambisyon Natin* 2040, with four areas for strategic policies, programs and projects over a period of 25 years. These development areas include:

- building a prosperous, predominantly middle-class society where no one is poor;
- promoting long and healthy life;
- becoming smarter and more innovative;
- building a high-trust society (NEDA 2017).

By 2022, the Plan also aims to lay the foundation for inclusive growth, a high-trust resilient society and a globally competitive economy. One of the three pillars of this development strategy is increasing growth potential focused on the adoption of modern technology, especially for production. The adoption of agribiotechnology in Philippines agriculture and natural resource sectors is clearly a central strategy in the attainment of inclusive growth, competitiveness and sustainable development in the country. The challenge is to subject this new technology to evidence-based socioeconomic impact assessment to determine whether, in the medium to long term, there are positive socioeconomic impacts for society.

Research framework

Given the development goals of the Philippines government, this socioeconomic impact analysis began with the establishment of a research evaluation framework. Since agribiotechnology is a seed technology innovation, its adoption within the existing agricultural production systems should be assessed with selected protocols or verifiable indicators that can lead to socioeconomic and environmental effects and outcomes on food security, poverty eradication and environmental sustainability and eventually to impacts on sustainable development (Figure 13.1).

The protocols or standard indicators being measured are technical efficiencies in terms of yields or productivity, input/output patterns and level of adoption. Socioeconomic protocols exhibited by allocative efficiency, in the efficient use of inputs or resources and environmental protocols in the use of physical resources like land, water and energy, were also estimated. The framework for analysis also included socioeconomic and environmental effects on productivity, profitability and resource use efficiency leading to food security, poverty alleviation and environmental sustainability. These broad development parameters were all determinants of overall impacts on sustainable development (Figure 13.1).

Data sources

There were two basic data sources used in the socioeconomic impact assessment of IR corn and HT corn from 2003 to 2011 and stack corn (IR/HT) from 2007 to 2011. The secondary data came from the datasets published by the Philippines Statistics Authority.[1] They included data on production, area and yields of total corn and yellow corn from 1995 to 2017; and farm-level input utilisation levels (fertiliser, labour and farm chemicals) for total corn and yellow corn from 2002 to 2011. The primary data were collected by SIKAP/STRIVE through its socioeconomic household monitoring surveys of GM corn for 2003–04, 2004–05, 2007–08 and 2011–12 across the nine major corn-producing provinces (Isabela, Pangasinan, Tarlac, Camarines Sur, Misamis Oriental, Bukidnon, Lanao del Sur, North Cotabato and South Cotabato) of the Philippines. The surveys were conducted for yellow GM and non-GM corn in eight cropping seasons during a four-year intermittent period. The aggregate number of household respondents across the cropping season, crop year and seed variety totalled 3505. SIKAP/STRIVE's household surveys included basic information on the sample farmer and their household demographic profile, basic characteristics of the corn farm, input–output coefficients of corn production, prices of inputs and corn grain, and other pertinent information at the household level. The household surveys of SIKAP/STRIVE had information on household poverty levels, income sources, food consumption and nutritional status. This survey involved 441 farmers for wet season and 568 farmers for dry season.

Socioeconomic impact indicators

The socioeconomic impacts of GM corn technologies were assessed by comparing the 10-year performance of GM corn and ordinary hybrid (OH) corn. The comparison was applied using five socioeconomic impact indicators: farm yield, farm production cost, net farm income, subsistence-level and poverty-level carrying capacities (CCs), and return on investment (ROI). The impacts were represented by the percentage difference of annual trends of GM corn and OH corn.

Farm yield

Corn yield is defined as the total grain output (kg dry weight) per hectare of farmland. The farm outputs that were sold on a fresh weight or corn-on-cob basis were converted to the corresponding grain dry weight basis. This procedure was done to maintain product consistency and for ease of data aggregation.

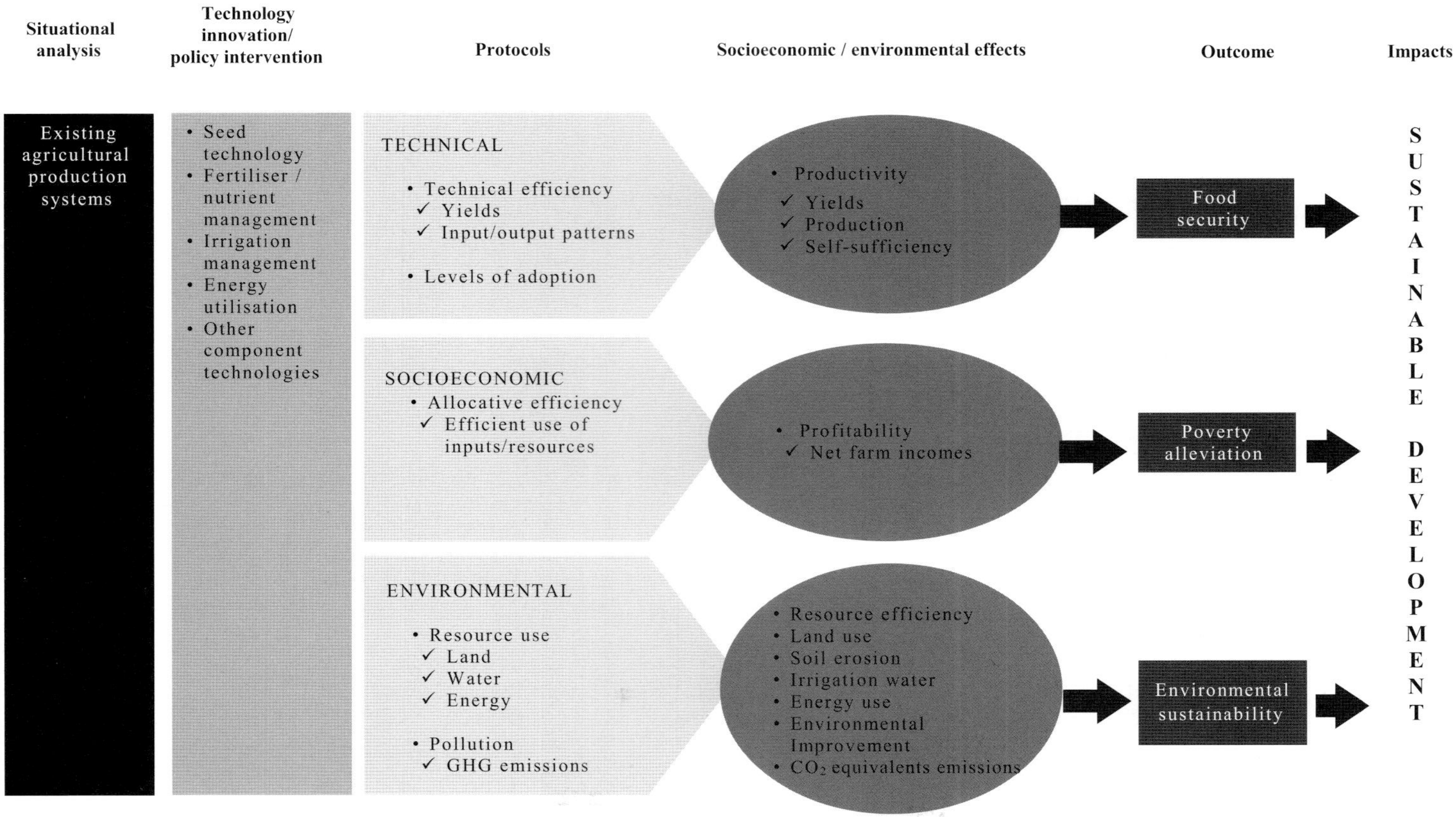

Figure 13.1 System framework for assessing technology innovation/policy interventions to enhance productivity, cost efficiency, profitability, food security and environmental sustainability.

Farm production cost

Farm-level production cost was the summation of costs incurred during land preparation, crop establishment and management, and harvest and postharvest operations. Typical farm budgets had the following major cost components: seed, fertiliser, land preparation, pre-harvest, harvest and post-harvest labour costs.

Net farm income

The net farm incomes of GM corn and OH corn were obtained by constructing farm-level cost and return partial budgets. Gross return on a per hectare basis was the gross receipts from the sale of agricultural produce on-farm. It was defined as the product of farm yield and farm gate price. Net farm income was computed as the difference between gross revenue and farm production cost. The farm activity generated profit for the farmer if the net farm income was positive.

Subsistence and poverty level CCs

The concept of economic CC was first used by the Philippines Rice Research Institute (PhilRice) as a policy impact indicator in 2000 (Gonzales *et al.* 2002). The major objective of the PhilRice exercise was to determine whether net farm incomes of rice farming households in the Philippines were enough to cover their expenses on food (subsistence) and food plus basic necessities (poverty). Given the size of farmers' landholdings, cropping intensities and the type of rice technologies adopted by rice farmers, multiplied by their net farm incomes derived from these activities as the numerator, a CC ratio can be generated if the numerator is divided by the food threshold or poverty threshold times 15% level of savings.

The CC methodology developed in rice production was used in the current assessment to measure the impact of GM corn technology adoption on subsistence (food) and poverty (food and basic needs) expenditures (Table 13.1).

Using the same formula and substituting subsistence expenditures (SE) with poverty expenditures (PE), we can arrive at the same decision in determining whether the corn farm households are above the poverty threshold (CC >1), neutral (CC = 1), or below the poverty line (CC <1).

Therefore, in this analysis, there were two CC indicators, classified by type of household expenditures. One is food subsistence CC, to show the ability of net income from corn farming to cover basic food needs. The other is poverty CC, which includes coverage of both food and non-food subsistence needs.

ROI

The ROI on-farm, defined as the ratio of net farm income over farm production cost expressed in per cent, was an indicator of business profitability. A 10-year comparison of the ROIs of GM corn and OH corn seed users was included in the analysis.

Environmental impact indicators

The environmental impact indicators in the study were represented by the resource use efficiency of corn production. Resource use efficiency was defined as the amount of a specific

Table 13.1. CC and household food security

If the CC is >1	Food secure, net farm income covers food thresholds
If the CC = 1	Neutral, net farm income is just enough to cover food expenditures
If the CC is <1	Food insecure, net farm income is not enough to cover the food expenses of the household

resource or input needed to produce 1 metric ton of corn grain. So this estimate was really a ratio of the physical unit of input over the physical unit of output. Field Market: The Keystone Alliance for Sustainable Agriculture has nominated four indicators of resource use efficiency (FMKASA 2009): land, fertiliser, labour and pesticide use efficiencies.[2]

Land use efficiency

Land resource use efficiency was the land area needed to produce 1 metric ton of corn grain. The area harvested was expressed in hectares and corn grain output was expressed in dry weight (containing 14% or less moisture) of corn grain. Harvested area and corn grain yield data were gathered during the eight seasons surveyed by SIKAP/STRIVE.

Fertiliser use efficiency

The household socioeconomic surveys of SIKAP/STRIVE (2012) also provided data on fertiliser application rates. Corn farmers applied six types of fertiliser during the survey period: urea, complete, ammonium sulphate, ammonium phosphate, muriate of potash and sulphate of potash. All fertiliser types were converted to their corresponding nitrogen (N), potassium (P) and phosphorus (K) mineral equivalents in order to account for the NPK fertiliser utilisation rates of corn. The ratio of NPK fertiliser application to grain yield was expressed as the fertiliser use efficiency of corn through time.

Labour use efficiency

Labour use included all forms of human labour depending on the mode of payment: hired, family, exchange, contractual and paid in kind. In hired labour, work was done by hired and paid labourers on a per hour and/or per day basis. Family labour and exchange labour entailed all farm activities done by unpaid family household members and unpaid non-household members, respectively. Contract labour was contracted labourers who were paid a fixed amount to do one or several farm activities, regardless of the number of persons hired or the number of days worked. Finally, in-kind labour was paid using a portion of the agricultural produce. In this study, labour use efficiency was represented by the proportion of total human labour to corn grain yield.

Pesticide use efficiency

The environmental impact of GM corn adoption in terms of pesticide use efficiency was represented by the ratio of pesticide application rates with corn grain yield. There were different types of pesticides depending on the prevalent pest in the area, farmer preference for particular brands, and availability of pesticides in the local market. All pesticides used by farmers during the study period were aggregated on a mL/ha basis.

Philippines corn production system

In the Philippines context, corn grain has many versatile uses. It is used as food (white corn) and animal feed (yellow corn). The contribution of total corn production to agriculture gross value added (GVA) was estimated at 7% in 2017.[3] However, the strategic importance of corn, particularly yellow corn, is its structural economic linkage with the poultry and livestock industries. Together, three subsectors contributed an aggregate 38% of agriculture GVA in 2017.[4]

Evolution of corn seed technologies

From 1975 to 2019, the Philippines corn production system was driven by at least three types of seed technologies: from 1975–79, open-pollinated varieties (OPVs); 1980–2002, an era of corn hybridisation; and 2002–present, hybridised GM corn technologies in seed, concentrated mainly in yellow corn production (Gonzales 2019; Gonzales *et al.* in press). Bt corn MON 810 event was the first to be approved for control of the Asian corn borer (ACB), in 2002, followed by Bt11 and NK603, the latter exclusively HT corn in 2005. At the time of writing, 13 GM corn events are approved in the Philippines.

Trends in corn harvested area, production and yields

A cursory look at the total area harvested to white and yellow corn, segmented by seed technology regimes, indicated that total area harvested for corn started at 2.4 million ha in the period of OPVs in 1970, peaked at 3.8 million ha in 1992 during the corn hybridisation period and tapered back to 2.5 million ha in 2018 during the GM corn period. Total corn production reached 7.9 million tons in 2018, from a low of <5 million tons in 2002 (GM corn period). The share of yellow corn production in 2018 was 5.8 million tons (73%), in contrast to the white corn production share of 2.1 million tons (27%).

The two major drivers of corn yield increase in the Philippines were the adoption of hybrid corn seeds exhibiting hybrid vigour, and genetic engineering via the genetic transformation of IR and HT traits among the existing yellow corn hybrids. The GM corn era from 2002 to the present in terms of yield increases, was basically a GM yellow corn production phenomenon.

Yellow corn yields grew on average from <1 metric ton/ha in 1980 to around 4.14 t/ha in 2018. This impressive annual growth rate of 21% also increased the average yield of 3.09 t/ha of total corn at an average growth rate of 3% during the period. White corn, on the other hand, although growing at a positive annual rate of 2%, reached an average of 1.86 t/ha in 2018 (Figure 13.2).

GM yellow corn

GM corn technology in the Philippines is concentrated on yellow corn hybrids, which are used primarily for animal commercial feeds. In 2002, Monsanto Philippines was the first company

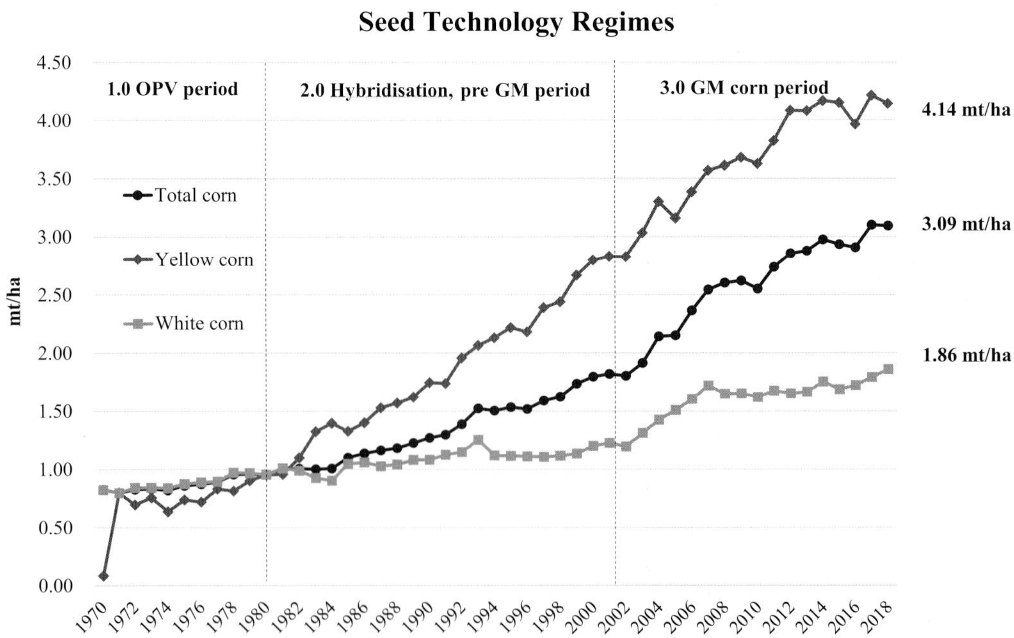

Figure 13.2 Trends in corn yield by technology regimes, 1975–2018. Source of basic data: SIKAP/STRIVE (2012); Bureau of Plant Industry and ISAAA 2003–17.

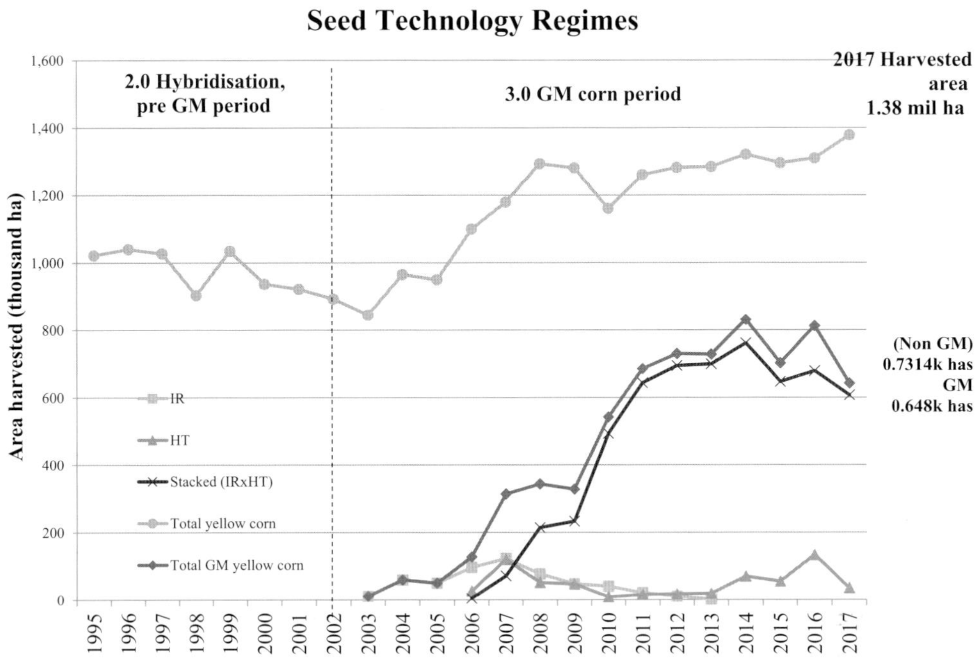

Figure 13.3 Trends in area harvested to yellow corn by seed technology, 1995–2017. Source of basic data: PSA (2019).

to be given a biosafety permit by the Bureau of Plant Industry (BPI)[5] to commercially propagate MON 810 corn event expressing *Bacillus thuringiensis* (Bt) Cry1Ab toxin that is effective against ACB. After 2002, other corn traits such as HT and several combinations of traits called stack (IR/HT) corn events were likewise commercialised. The trend in area harvested for first generation (F1) GM hybrid yellow corn is shown in Figure 13.3. From a low of 10 000 ha in 2003, total GM corn reached its peak harvested area of 800 000 ha in 2014. The HT and stack (IR/HT) corn were the major drivers of total F1 GM corn harvested area during the period.

There was decline in GM corn harvested area from 2014 to 2015, then from 2016 to 2017. These declines were due in part to the moratorium from the Supreme Court ruling in 2014 on all agribiotechnological activities (RP SC 2015). The other reason was adjustment in GM corn plantings due to the passage of a new Joint Department Circular by the Department of Science and Technology (DOST), Department of Agriculture (DA), Department of Environment and Natural Resources (DENR), Department of Health (DOH) and Department of Interior and Local Government (DILG) on biosafety regulations (DOST-DA-DENR-DOH-DILG 2016).

Socioeconomic and environmental impacts of GM corn

The socioeconomic and environmental impact assessment of GM corn in the Philippines was done within the context of the Cartagena Protocol on Biosafety (SCBD 2000). The Philippines Department of Science and Technology that chairs the National Committee on Biosafety of the Philippines ensures strict compliance with the provisions of the Cartagena Protocol. Unlike other countries that may require compulsory socioeconomic assessment as part of the mandatory risk assessment of biosafety prior to approval of a GM event (Horna *et al.* 2013), the Philippines recognises the voluntary nature of Article 26.1 of the Cartagena Protocol that deals with socioeconomic considerations. The National Committee on Biosafety allows the inclusion of socioeconomic, cultural and ethical assessments of living modified organisms in the decision-making process in regard to possible impacts on conservation and the sustainable use of biological diversity. However, the socioeconomic assessment is separate from and independent of the compulsory biosafety risk assessment of a GM event before its regulatory approval.

The analysis of socioeconomic and environmental impacts was primarily focused at the farm level operation of corn farmers, then aggregated at the national level. Trend analysis from 2004 to 2011 on the impact indicators was made possible through the extrapolation of primary panel survey data generated by SIKAP/STRIVE from 2004, 2005, 2008 and 2012, and the time series secondary data by the Philippines Statistics Authority during the same years.

Socioeconomic impacts

Farm yield

The results showed that IR corn had significantly higher farm yields than OH corn from 2003 to 2011. The average yield advantage of IR corn over OH corn was 19% over eight years. The ACB resistance agronomic trait of IR corn largely contributed to the yield difference. Crop damage due to ACB was in the range of 6–12% based on the results of the field surveys from 2003 to 2011. The expected higher contribution of IR corn to national corn yield was dampened by the unfavourable effects of the El Niño phenomena in 2004–05 and 2007 (Figure 13.4).

Farm production cost and farm income

The yearly comparison of per unit cost in peso (at constant 2003 prices) between GM corn and OH corn seed users showed favourable results for GM corn seed users. On average from 2003 to 2011, IR corn and stack (IR/HT) corn had a cost advantage of 10% and 12% respectively, relative to OH corn. The cost advantage can be attributed to a higher yield with a lower production cost. The lower production cost was due to the negligible cost of spraying insecticide against ACB and the associated labour requirement (Figure 13.5).

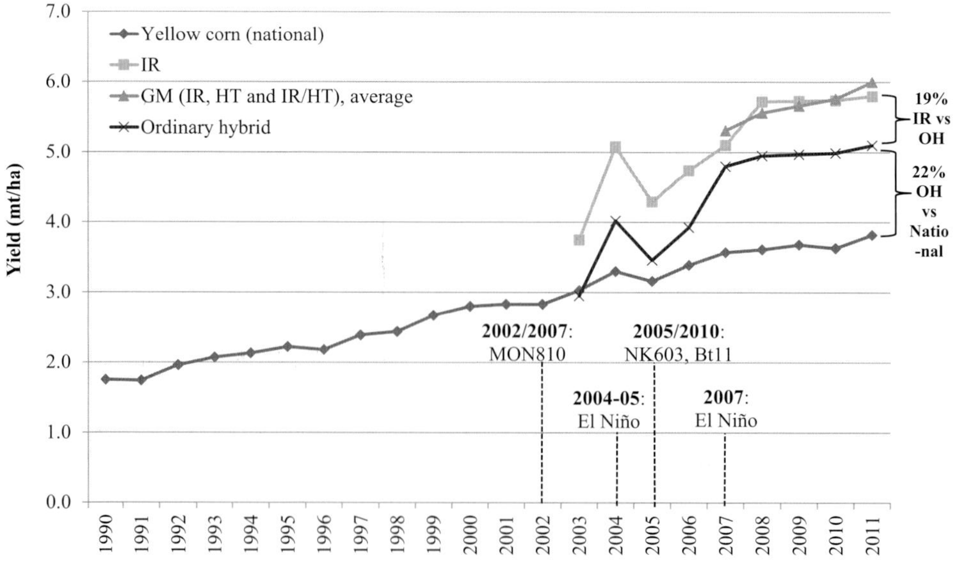

Figure 13.4 Trends in yellow corn yield by seed technology with corn events MON 810, NK603 and Bt11. Source of basic data: SIKAP/STRIVE (2012); Gonzales *et al.* (2013).

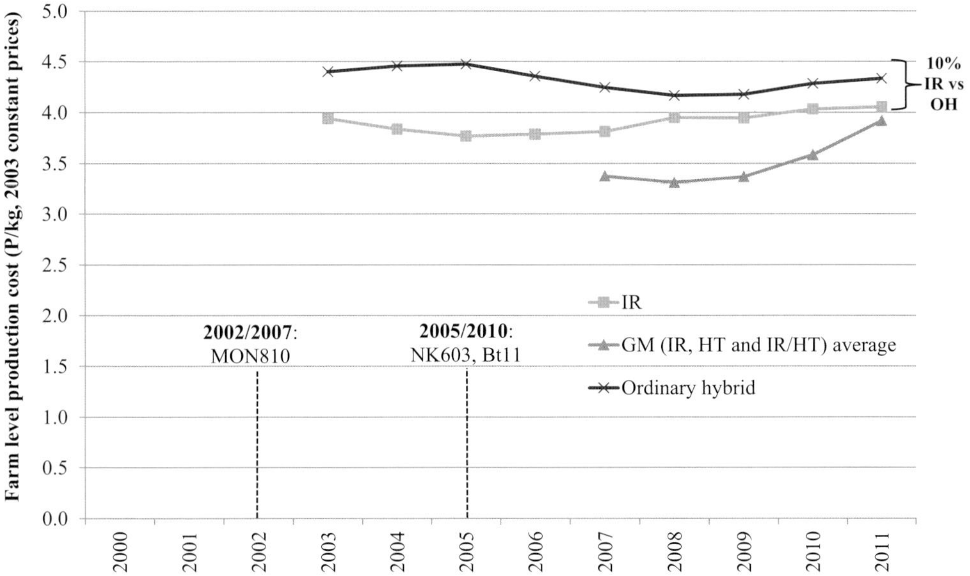

Figure 13.5 Trends in yellow corn farm production cost by seed technology. Source of basic data: SIKAP/STRIVE (2012).

Net farm income

The analysis showed that IR corn had a higher per kilo income (at constant 2003 prices) than OH corn from 2003 to 2011. The average real Peso/kg income advantage of IR corn and of stack IR/HT corn over OH corn was 8% per annum. The major probable reason for the income advantage was the combined effects of high yield due to added agronomic trait and cost superiority due to cost savings from insecticide non-use.

Subsistence and poverty level CCs

A detailed examination of the ability of corn farming households in the Philippines to cover basic subsistence needs in food and the poverty threshold revealed that a family of five needs at least 4 t/ha and 5 t/ha of corn yield in order to live above the food and poverty thresholds, respectively. Any corn yield less than the respective 4 t/ha and 5 t/ha benchmarks would signify household malnourishment and poverty. In terms of CC indicators, GM corn consistently outperformed OH corn by at least 29% in meeting the food and poverty thresholds, from 2003 to 2011 (Figure 13.6).

ROI

The study showed that IR corn users had a 42% higher ROI than OH corn users from 2003 to 2011. In order to recover the investments at the farm level, corn yield must be at least 4 t/ha. The higher ROI was positively correlated with higher yield levels. This implied that, in order to get more value for money, future investment streams must be allocated to GM corn technologies (IR, HT and IR/HT corn) (Figure 13.7).

Environmental impacts

Land use efficiency

The land use efficiency of yellow corn, defined as the land resource required to produce 1 t of corn grain, significantly improved from 2000 to 2011. The improvement in the national land

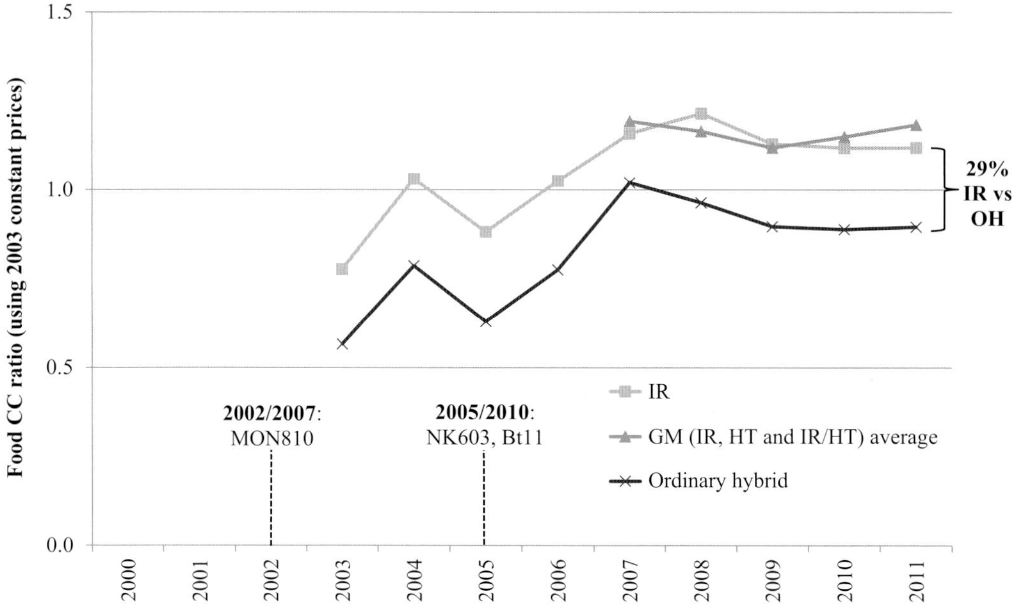

Figure 13.6 Trends in yellow corn food CC by seed technology. Source of basic data: SIKAP/STRIVE (2012); Gonzales *et al.* (2013).

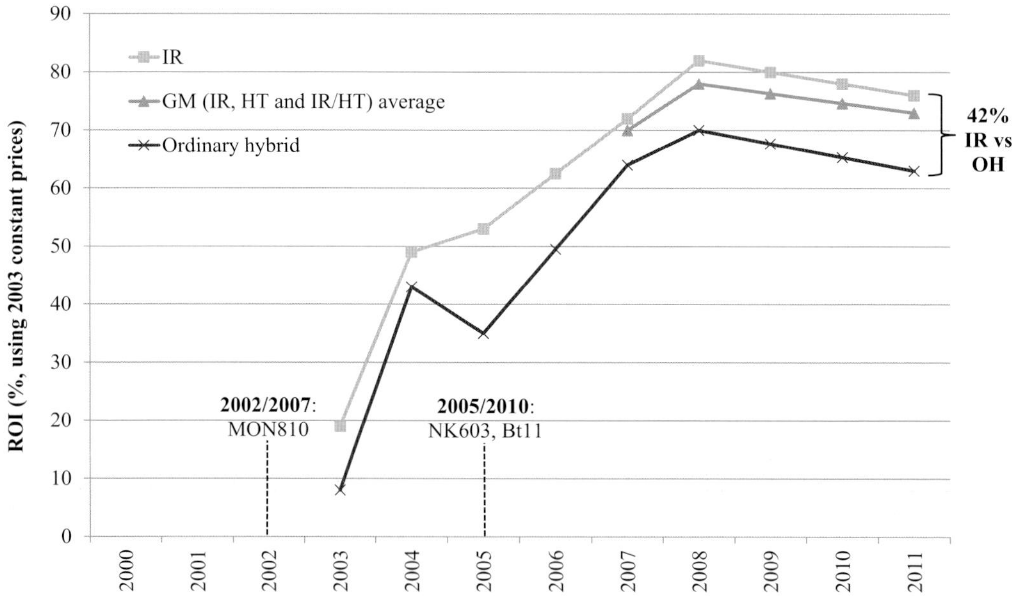

Figure 13.7 Trends in yellow corn ROI by seed technology. Source of basic data: SIKAP/STRIVE (2012).

use efficiency of yellow corn was 27% during the 11-year period, primarily due to marked improvements of yellow corn yields. Land use efficiencies of various seed technologies including OH, IR and GM (average of IR, HT and IR/HT) also increased significantly. IR corn seed users required 15% less land than OH corn seed users to generate 1 t of corn grain from 2003 to 2011 (Figure 13.8).

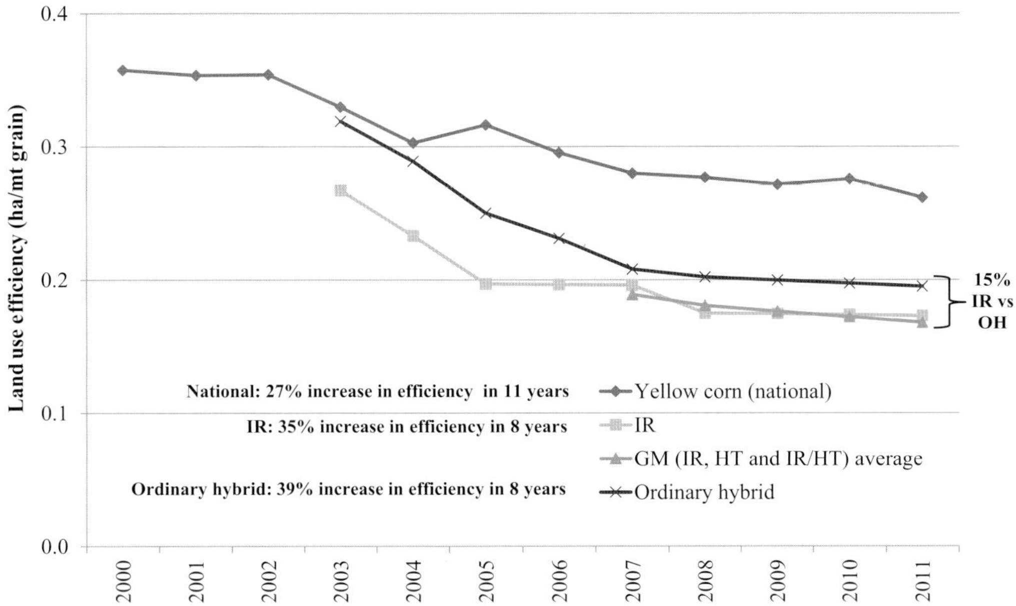

Figure 13.8 Land use efficiency of yellow corn by seed technology. Source of basic data: SIKAP/STRIVE (2012); Gonzales *et al.* (2013).

Fertiliser use efficiency

Macronutrients in the form of NPK are vital determinants of corn crop growth. Due to the relative volatility of fertiliser prices, it is crucial to optimise fertiliser use efficiency. Across different seed technologies, fertiliser use efficiency generally improved from 2003 to 2011. The El Niño years of 2004–05 and 2007 temporarily slowed the yearly efficiency improvements. On average, IR corn seed users applied less fertiliser than OH corn seed users (91% compared to 100% usage) to produce 1 t of corn grain from 2003 to 2011 (Figure 13.9).

Labour use efficiency

Human labour was represented by the amount of hired, family, exchange, contractual and paid in kind labour dedicated to yellow corn production. Establishing the ratio of labour use in man-days per hectare and corn grain yield in metric ton per hectare produces the human labour use efficiency indicator. There was a general labour use efficiency improvement across seed technologies (OH, IR, HT and IR/HT) from 2003 to 2011 in the range of 30–50%. IR corn was estimated to involve 26% less labour usage (basically pre-harvest labour) than OH corn, due to labour savings in crop care (spraying insecticide) and associated activities (Figure 13.10).

Pesticide use efficiency

Pesticide use efficiency of yellow corn, as represented by the ratio of pesticide application rates to grain yield, significantly improved from 2003 to 2011. For IR corn, OH corn and GM corn, increasing yellow corn grain yield through time was associated with decreasing pesticide application rates, leading to consistent improvement in pesticide use efficiencies. The results showed that IR corn required 54% less pesticides than OH corn to produce the same amount of corn grain from 2003 to 2011 (Figure 13.11).

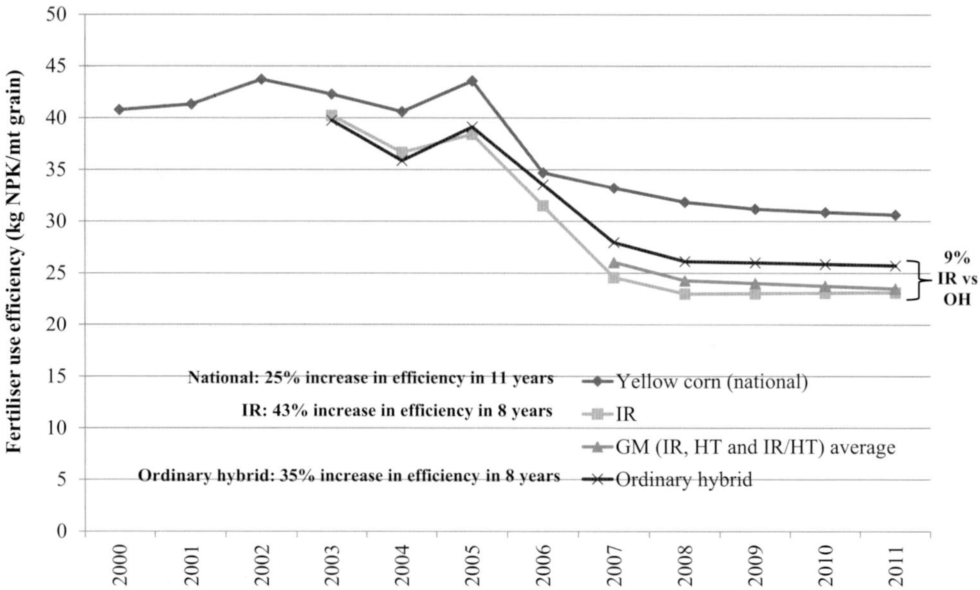

Figure 13.9 Fertiliser use efficiency of yellow corn by seed technology. Source of basic data: SIKAP/STRIVE (2012); Gonzales *et al.* (2013).

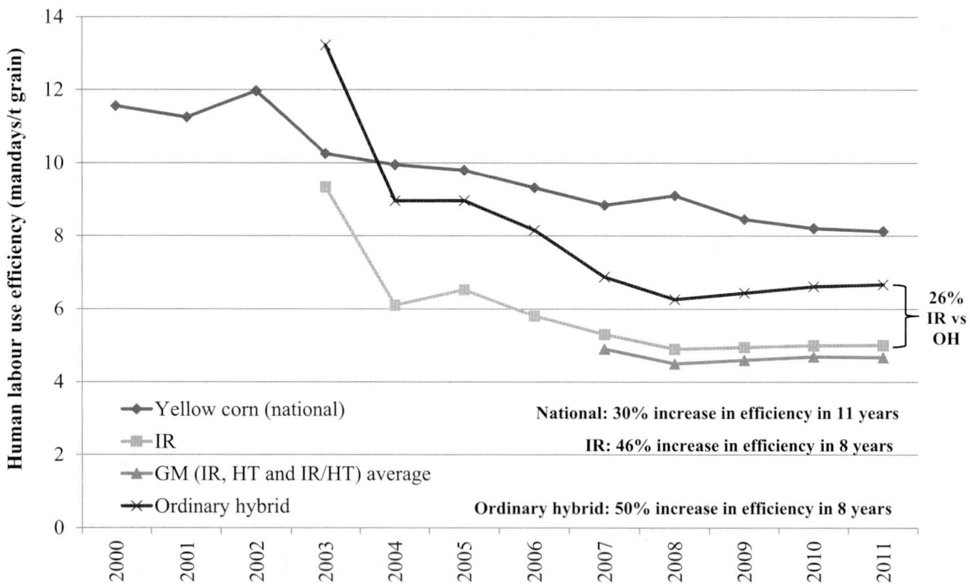

Figure 13.10 Labour use efficiency of yellow corn by seed technology. Source of basic data: SIKAP/STRIVE (2012); Gonzales *et al.* (2013).

Synthesis

Analysis of more than a decade of GM corn commercialisation in Philippine showed that GM corn performed better than OH corn at the farm level in terms of productivity, cost efficiency, net profitability, food security cum alleviation of poverty and ROI. In regard to environmental impact, analysis showed that GM corn had better performance than OH corn in terms of more efficient use of land, labour, fertiliser and pesticide as inputs to production. The findings

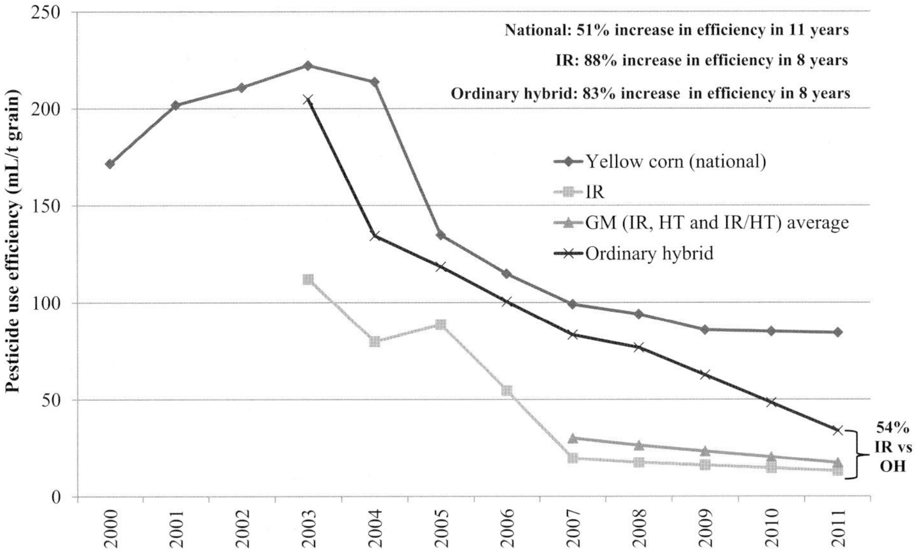

Figure 13.11 Pesticide use efficiency of yellow corn by seed technology. Source of basic data: SIKAP/STRIVE (2012); Gonzales *et al.* (2013).

demonstrated the relatively better economic performance of GM corn over OH corn across a decade. This trend of better socioeconomic and environmental impact performance is expected to continue in the future with the development of new combinations of agronomic traits in GM stack corn.

Potential benefits of GM adoption under the AEC 2015 framework

The Philippines GM experience demonstrated that GM corn technology had robust socioeconomic and environmental impacts at the farm level. However, the policy decision to follow the GM technology development pathway was never uncontroversial. As the Philippines continues to expand its GM coverage to other crops, such as Bt eggplant, golden rice, Bt cotton and delayed ripening papaya, the struggle for dominance continues unabated between those who oppose and those who favour GM technologies.

Vietnam and Indonesia are closest in following the policy path that the Philippines has taken in adopting GM technology. Hopefully, they can learn from the Philippines experience. But will other ASEAN member states follow suit?

Under the AEC Blueprint of 2015, there is growing interest in fast-tracking GM crop technology through regional harmonisation of biosafety guidelines and research protocols. If this happens among ASEAN member states, two potential scenarios will emerge that will bring benefits to the AEC. These include the expansion of ASEAN internal and rest-of-the-world trade, and improvement of productivity in the feedgrain–animal industries in the ASEAN region (Gonzales *et al.* 2018).

Expanding ASEAN and rest-of-the-world trade in agricultural products

The ASEAN region is a net exporter of food and agricultural feed products both within the ASEAN region and to the rest of the world. However, its trade balance with the rest of the

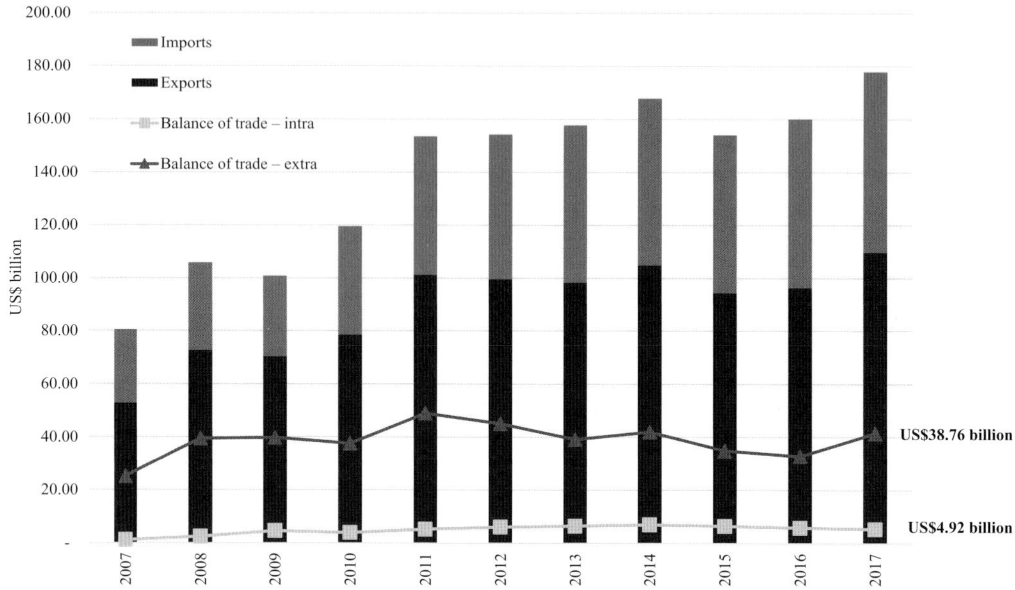

Figure 13.12 ASEAN trade in agricultural goods within the region and with the rest of the world, 2000–17. Source of basic data: ASEAN (2018).

world, averaging US$36.81 billion from 2011 to 2016, has declined, while the intra-ASEAN trade balance, with an average value of US$4 billion, remained constant (Figure 13.12). Effective and harmonised biosafety guidelines and protocols can trigger commercial adoption of GM food and feed crops and thereby competitively expand intra-ASEAN trade and its trade with the rest of the world.

Enhancing the productivity of the feedgrain–livestock sectors

The ASEAN region is a net importer of feedstuff, primarily corn and soybeans. The ratio of corn imports to corn production averaged 53% (USDA 2017) across 2013–17. Soybean meal imports (~17 million t), on the other hand, were much higher than the domestic production of ~3 million t. The import production ratio averaged 398% from 2013 to 2017 (USDA 2017). Corn import and production was estimated at 34.5 million t and 16.1 million t, respectively – a significant gap in the ASEAN region.

ASEAN corn and soybean meal imports in 2016 were estimated at US$5.7 billion and 2.3 billion, respectively. If the region developed effective harmonised biosafety guidelines and protocols, it could save foreign exchange through import substitution activities.

The feedgrain–livestock subsectors are structurally integrated in ASEAN. The intra trade tariff rates of feed products averaged to zero, in contrast to the most-favoured-nation tariff of 1.27% on the World Trade Organization database. Likewise, the most-favoured-nation tariff rate for livestock products is 10% while under ASEAN it is only 3% (WTO 2017). The integration of feedgrain would be an efficient economic activity in the region, given the high productivity and cost efficiency of GM corn production.

Endnotes

1. https://psa.gov.ph/
2. https://fieldtomarket.org/our-program/sustainability-metrics/

3. https://psa.gov.ph/
4. https://psa.gov.ph/
5. http://bpi.da.gov.ph/bpi/index.php

References

ADB (2014) *ASEAN 2030: Toward a Borderless Economic Community.* Asian Development Bank, Tokyo. <https://www.adb.org/sites/default/files/publication/159312/adbi-asean-2030>

ADB (2017) *Key Indicators for Asia and the Pacific.* Metro Manila, Philippines. Asian Development Bank. <https://www.adb.org/sites/default/files/publication/357006/ki2017.pdf>

Afidchao MM, Musters CJM, Wossin A, Balderama OF, de Snoo GR (2014) Analysing the farm level economic impact of GM corn in the Philippines. *NJAS Wageningen Journal of Life Sciences* **70–71**, 113–121. doi:10.1016/j.njas.2014.05.008

ASEAN (2008) *ASEAN Economic Community Blueprint 2015.* Association of Southeast Asian Nations Secretariat, Jakarta.

ASEAN (2017) *ASEAN Plan of Action on Science, Technology and Innovation (APASTI) 2016–2025.* Association of Southeast Asian Nations Secretariat, Jakarta.<http://asean.org/storage/2017/10/01-APASTI-2016-2025-FINAL.pdf>

ASEAN (2018) *ASEAN Trade in Goods (in US$).* Association of Southeast Asian Nations Secretariat, Jakarta.<https://data.aseanstats.org/trade.php>

Brookes G, Barfoot P (2014) GM crops: Global socio-economic and environmental impacts 1996–2012. *GM Crops Food* **5**(2), 149–160. doi:10.4161/gmcr.28449

DOST-DA-DENR-DOH-DILG (2016) *Rules and Regulations for the Research and Development, Handling and Use, Transboundary Movement, Release into the Environment, and Management of Genetically-modified Plant and Plant Products Derived from the Use of Modern Biotechnology, Philippines, Manila.* Joint Department Circular No. 1, Series of 2016. Department of Agriculture, Department of Environment and Natural Resources, Department of Health and Department of Interior and Local Government. <http://biotech.da.gov.ph/upload/Final_DOST-DA-DENR-DOH-DILG_JDCs2016.pdf> <https://dilg.gov.ph/issuances/mc/Guidelines-on-the-Implementation-of-DOST-DA-DENR-DOH-DILG-Joint-DC-No-1-s-2016/2772>

Elferink M, Schierhorn F (2016) *Global Demand for Food is Rising. Can We Meet It?* Harvard Business Publishing, New York. <https://hbr.org/2016/04/global-demand-for-food-is-rising-can-we-meet-it>

FMKASA (2009) *Environmental Resource Indicators for Measuring Outcomes of On-farm Agricultural Production in the United States.* First Report. Field to Market: The Keystone Alliance for Sustainable Agriculture. <https://fieldtomarket.org/our-program/sustainability-metrics/>

Gonzales LA (2005) *Harnessing the Benefits of Biotechnology: The Case of Bt Corn in the Philippines.* SIKAP/STRIVE, Laguna.

Gonzales LA (2007) *Four Seasons of Commercialization: Monitoring and Evaluating the Socio-Economic Impact of Bt Corn in the Philippines.* SIKAP/STRIVE, Laguna.

Gonzales LA (2019) *Socio-economic Impact of Technology Innovation and Policy Reform in the Philippines Corn Industry.* Paper presented during the Regional Corn Conference 2019, Penang, Malaysia, 26–28 February.

Gonzales LA, Sebastian L, Francisco S, Gonzales VA, Casiwan C, Cataquiz G, Elca C, Lapina G, Banta S (2002) *Food Carrying Capacity and Income-Generating Capacities of the Rice Sector.* SIKAP/STRIVE and Philippines Rice Research Institute, Los Baños.

Gonzales LA, Elca C, Paningbatan B, Umali R, Gonzales A, Ignacio JL (2013) *Micro-Macro Impacts of Technological Change: The GM (Bt) Corn Experience in the Philippines.* Final Report. SIKAP/STRIVE, Los Baños.

Gonzales LA, Ramirez D, Cariño F, Gonzales AV, Gonzales A, Ignacio JL, Evangelista G, Pablo N (2018) *Development of Framework for Harmonizing Biosafety Guidelines and Research Protocols on Biosafety in the ASEAN Region.* Terminal Report. SIKAP/STRIVE, Los Baños.

Gonzales LA, Gonzales VA, Ignacio JL, Gonzales A, Evangelista G, Cajipe J, Elca C, Umali R, Encarnacion J, Pablo N (in press) *The GMO Effect: The Socioeconomic, Environmental, and Intersectoral Impacts of GM Corn in the Philippines*. SIKAP/STRIVE, Los Baños.

Horna D, Zambrano P, Falck-Zepeda J (2013) *Socioeconomic Considerations in Biosafety Decision-making: Methods and Implementation*. International Food Policy Research Institute, Washington, DC. doi:10.2499/9780896292079

Iha K, Poblete P, Panda D, Sebastian W (2015) A footprint analysis of ASEAN: ensuring sustainable development in an increasingly resource constrained world. *Asian Biotechnology and Development Review* **17**(2), 57–67. <https://www.footprintnetwork.org/content/documents/ABDR_Footprint_of_ASEAN_07_2015.pdf>

ISAAA (2018) *Global Status of Commercialized Biotech/GM Crops in 2017: Biotech Crops Continue to Help Meet the Challenges of Increased Population and Climate Change*. ISAAA Brief No. 54. ISAAA, Ithaca, NY.

James C (2017) *Global Status of Commercialized Biotech/GM Crops in 2017: Biotech Crop Adoption Surges as Economic Benefits Accumulate in 22 Years*. ISAAA Brief No. 53. ISAAA, Ithaca, NY.

MASIPAG (2013) *Socio-economic Impacts of Genetically Modified Corn in the Philippines*. Magsasaka at Siyentipiko para sa Pag-unlad ng Agrikultura, Laguna. <http://masipag.org/wp-content/uploads/2013/05/MASIPAG-Book-low-res.pdf>

NEDA (2017) *Philippines Development Plan 2017–2022*. Abridged Version. National Economic Development Authority, Pasig. <http://www.neda.gov.ph/wp-content/uploads/2018/01/Abridged-PDP-2017-2022_Updated-as-of-01052018.pdf>.

PSA (2019) *OpenStat: Agriculture, Forestry, Fisheries – Crops*. Philippines Statistics Authority, Quezon City.<http://openstat.psa.gov.ph/PXWeb/pxweb/en/DB/DB__2E__CS/?tablelist=true&rxid=bdf9d8da-96f1-4100-ae09-18cb3eaeb313>

RP SC (2015) *General Register No. 209271, 8 December*. Republic of Philippines, Supreme Court, Manila.

SCBD (2000) *Cartagena Protocol on Biosafety to the Convention on Biological Diversity: Text and Annexes*. Secretariat of Convention on Biological Diversity, Montreal.

SIKAP/STRIVE (2012) *Household Corn Survey for Crop Years 2004, 2005, 2008 and 2012*. SIKAP/STRIVE, Los Baños.

Sta Maria R, Urata S, Intal PS Jr (2018) *The ASEAN Economic Community into 2025 and Beyond*. ASEAN@50, vol. 5.

USDA (2017) *Feed Grains Database*. US Department of Agriculture. <https://www.ers.usda.gov/data-products/feed-grains-database/>

WTO (2017) *Tariff Download Facility*. World Trade Organization. <http://tariffdata.wto.org/ReportersAndProducts.aspx>

Yorobe J, Quicoy C (2006) Economic impact of Bt corn in the Philippines. *Philippines Agricultural Scientist* **89**, 258–267.

Yorobe J, Smale M (2012) Impacts of Bt maize on smallholder income in the Philippines. *AgBioForum* **15**(2), 152–162.

Yorobe J, Birol E, Smale M (2010) Farmer preferences for Bt maize, seed information and credit in the Philippines. In *Choice Experiments in Developing Countries: Implementation, Challenges and Policy Implications*. (Eds JW Bennett and E Birol) pp. 225–243. Edward Elgar, Cheltenham.

14

Biosafety management for genetically modified crops in Vietnam: from policy to practice

Ngo Xuan Quy, Le Huy Ham and Pham Hanh Nguyen

Introduction

Agriculture is one of the most important economic sectors in Vietnam. Food and fibre annual crops occupy 9.9 million ha (~20%) of the cultivated land, while perennial industrial crops like coffee, rubber and coconut constitute ~80%. Over the years, Vietnam has become an important exporter of agricultural and horticultural products. With intensification and diversification of agriculture due to adoption of high-yielding short-duration varieties, along with advanced farm technology, net income from this sector plus fisheries and forestry is remarkable: it now constitutes ~23% of the national GDP. In absolute monetary terms, this sector is growing robustly with the help of biotechnology and related technologies.

Biotechnology policy in Vietnam

Biotechnology is considered one of the four priority technologies in Vietnam's industrialisation and modernisation. In 2006, the Prime Minister approved the biotechnology program in various fields with an emphasis on its development and application in agriculture and rural development up to 2020 (Decision No. 11/2006/QD-TTg of 12 January 2006). This biotechnology program is targeted at cultivation of genetically modified (GM) crops and promotion of GM products, both as feed and food. One of its objectives envisaged cultivation of GM crops over 30–50% of total arable land by 2020, with agricultural biotechnology comprising over 50% of the total contribution of science and technology in terms of value in agriculture. The program is an initiative of the Ministry of Agriculture and Rural Development (MARD), one of the National Competent Authorities on biosafety management in Vietnam.

To date, the biotechnology program has achieved many encouraging results such as application of molecular assisted selection technology and creation of new rice varieties carrying genes for aroma, insect and disease resistance, and drought tolerance. Many of the plant varieties resulting from this program are of great potential and are being further bred for the purpose of new varietal registration. The standard processes for micropropagation on a large scale have been established and transferred to on-farm sites for production of seedlings for marketing by different companies and enterprises. The program implemented several production trials for micropropagation of cut flowers (e.g. *Gerbera jamesoii*, *Diathus caryophyllus* and Dendrobium) and of eucalyptus and acacia hybrids in Quang Ninh, Yen Bai province.

Biosafety management in Vietnam

Cartagena Protocol on Biosafety

The Cartagena Protocol on Biosafety to the Convention on Biological Diversity was finalised and adopted in Montreal on 29 January 2000. To date, 172 countries have ratified the Cartagena Protocol. Its objective is to ensure an adequate level of protection for the safe transfer, handling and use of living modified organisms resulting from modern biotechnology that may have adverse effects on the conservation and sustainable use of biological diversity, taking into account risks to human health. It specifically focuses on trans-boundary movements, as stated in Article 4. The Protocol does not apply to living modified organisms that are used for pharmacy or medicine.

Vietnam ratified the Cartagena Protocol on 21 January 2004. Currently, the Ministry of Natural Resources and Environment (MONRE) is assigned as the National Focal Point for implementation of the Protocol.

Biosafety policy for GMOs

In 2008, the National Assembly approved the Biodiversity law, which included provisions for the risk management of GM organisms (GMOs), genetic specimens and GMO products. After the Biodiversity law came into force, the government and relevant ministries developed and issued decrees and circulars to guide and manage the biosafety of GMOs. These documents are as follows:

The Law on Biodiversity (No. 20/2008/QH12 of 13 November 2008) regulates risk management of GMOs, and genetic specimens of GMOs in biodiversity.

- Defining concepts: a GMO is an organism whose genetic structure is modified by gene transfer technology. Risk assessment of a GMO involves identifying the potential risk and damage level to biodiversity that may occur due to activities related to the use and release of the GMO and its genetic specimens. Release of a GMO is the intended introduction into the environment. Risk management is the implementation of safety measures to avoid, handle and remedy risks to biodiversity to a GMO and its genetic specimens.
- The main requirements for a risk assessment report of a GMO to biodiversity include description of risk assessment measures, the level of risk to biodiversity and the risk management plan.
- Risk management responsibility: organisations and individuals who research and develop a GMO and its genetic specimens shall register with the Ministry of Science and Technology (MOST) and satisfy the conditions of infrastructure, science and technology expertise, material and technical foundations, technologies and professional staffs. Organisations and individuals who develop, import, purchase, trade or release a GMO or its genetic specimens granted approval by a National Competent Authority shall make reports on risk assessment of the GMO to biodiversity and shall publicly inform the level of risk and the risk management measures.
- Management of GMO databases and its genetic specimens: MONRE is assigned to manage the GMO databases and develop the biosafety clearing-house website on the GMO and its specimens.

Governmental Decree No. 69/2010/ND-CP on 21 June 2010: The Decree stipulates the following main elements: (i) risk assessment and risk management of a GMO; (ii) research and development of technique of a GMO and its product; (iii) trial of a GMO;(iv) biosafety certificate; (v) conditions that must be satisfied for a GMO to be used as food or feed; (vi) production, trading, import, export, transport and storage of a GMO and its genetic specimens; (vii) information on a GMO and its products.

- Regulations of risk assessment and risk management of a GMO: the Decree stipulates the principle of risk assessment for the environment, biodiversity, human and animal health. The risk assessment shall be scientific and transparent. It shall be conducted on case-by-case basis subject to the GMO, the purpose of use and the receiving environment.
- Regulations on scientific research and technology development of a GMO and its product: these activities must be conducted under the framework of the projects which are approved in accordance with the regulations. Confined research activities, development, analysis and trials must be done in the accredited laboratory which is recognised by MOST.
- Regulations of conducting GMO trials: trials of a GMO must be conducted step by step, from confined trials to large-scale trials under isolated conditions. GMO trial content shall include objectives such as their potential to become a weed and disease, potential risk to non-target organisms, possibility of adverse effect to surrounding ecosystem and other adverse effects.
- Regulations on biosafety certificate: the biosafety certificate is an official document which certifies that the specified GMOs are safe to the environment and biodiversity, and are allowed to be released into the environment under specific conditions. For a GMO to be granted a biosafety certificate it must satisfy the following conditions: (i) the GMO must have undergone the trials in Vietnam conditions and the trial results have been recognised by MARD; (ii) the National Biosafety Committee concluded that the said GMO is safe for the environment and biodiversity.
- Regulations of the GMO that fulfilled conditions to be used as food and feed: the GMO that is granted a certificate to be used as food and feed shall satisfy one of the following requirements: (i) the GMO Food and Feed Safety Committee has appraised the application dossier for issuing the GMO food and feed safety certificate and concluded that the said GMO does not pose any uncontrollable risk to human and animal health; (ii) the GMO has been approved to be used as food and feed and has not posed any risk in five developed countries.
- The Decree stipulates production and business conditions for release or production of GMO, its products to be used for food and feed, export, import, storage, packaging and transportation of GMO and products thereof. It stipulates that goods which contain a GMO or its products at the threshold higher than 5% of each composition shall provide information related to GMO on the label, and comply with other current labelling regulations. The Decree also regulates the confidentiality and disclosure of information, and management of databases on GMO.
- Because of Law on Food Safety (No. 55/2010/QH12 of 17 June 2010) this Decree is amended in respect of some articles issued by the government (Decree No. 108/2011/ND-CP of 30 November 2011): changing the authorisation for order procedures of granting and revocation of GMO food certificate from the Ministry of Health.

Decree No. 155/2016/ND-CP of 18 November 2016.

- The Decree stipulated in Article 15 the punishment for the violation of regulations on research and technology development of GMO, its genetic specimens and GMO trial: in which case, the maximum fine of up to VND50 million (US$2174) shall apply to one of the activities, as follows: (i) failure to strictly comply with regulations on isolation distance causing the unintended release of a GMO in the environment during research and trial;(ii) failure to apply the urgent remedy to treat and completely destroy GMO material, once it was found to have risk to environment, biodiversity and human and animal health; and was not able to be controlled; (iii) unintended release of a GMO into the environment during the research and conducting trial of that GMO.

- The Decree stipulated in Article 46 violation of the sanction regulating production, trading, import, storage and transportation of a GMO and products of a GMO. In such cases, the highest punishment up to VND100 million (US$4348) shall apply to the following violations: (i) raising, planting or intentionally releasing a GMO and its specimens (which have not yet been granted the biosafety certificate) into the environment and adversely affecting the environment and biodiversity; (ii) illegally importing a GMO and its genetic specimens with adverse effect on the environment and biodiversity.

Following the government Decree 69 and Decree 108, relevant ministries developed and issued circulars for guidance and biosafety management of GMOs according to their functions and responsibilities (Table 14.1).

Table 14.1. Responsibilities of Vietnam authorities in biosafety management

Authorities	Role	Responsibilities
MONRE	Lead government agency in biosafety management	1. To issue a biosafety certificate 2. To withdraw a biosafety certificate 3. To develop a list of GM products granted biosafety certificates 4. To develop regulations on the storage, package and transportation of GM products 5. To develop and manage a database on GM products
MARD	Regulate field trials of GM crops and approve food and feed safety of GM products	1. To issue permits for field trials of GM crops 2. To accredit MARD's agencies for conducting field trials of GM crops 3. To conduct field trials of GM crops 4. To approve GM products for food and feed safety use
MOST	Key government agency managing the research and development of GM products	1. To accredit GM research laboratories 2. To manage GM research projects 3. To develop labelling regulations for GM products and goods
MOH, MOIT	Supporting agency in biosafety and food/feed safety assessment	1. To participate in the Biosafety Committee 2. To participate in the Food and Feed Safety Committee
MOF	Regulates fee payment for risk assessment	1. To develop regulations on fees and fee management in the field of biosafety management 2. To develop regulations on fees and fee management in the field of food/feed management

Source: GAIN Report (2018) <https://apps.fas.usda.gov/newgainapi/api/report/downloadreportbyfilename?filename = Agricultural%20Biotechnology%20Annual_Hanoi_Vietnam_10-22-2018.pdf>

Ministry of Natural Resources and Environment:

- Circular No. 09/2012/TT-BTNMT of 22 August 2012 stipulates the GMO information and data supplement and exchanging.
- Circular No. 08/2013/TT-BTNMT of 16 May 2013 stipulates the order and procedures on granting and revocation of biosafety certificates for GM crops.
- Circular No. 13/2013/TT-BTNMT of 21 June 2013 stipulates technical procedures and norms of technical and finance on GMO detection by means of qualitative and quantitative analysis of deoxyribonucleic acid.

Ministry of Science and Technology:

- Circular No. 20/2012/TT-BKHCN of 20 November 2012 regarding the clear guidance on conditions, procedures and the process to grant accreditation to a laboratory for conducting GMO research.
- Circular No. 21/2012/TT-BKHCN of 20 November 2012 regulates general biosafety in research and technology development related to GMOs.

Ministry of Agriculture and Rural Development:

- Circular No. 69/2009/TT-BNNPTNT of 27 October 2009 regulates the process of conducting trials for risk assessment of GM crops to biodiversity and environment.
- Circular No. 72/2009/TT-BNNPTNT of 17 November 2009 relates to the list of permitted GM species for conducting risk assessment of GM crops to the environment and biodiversity for the purpose of plant varieties in Vietnam.
- Circular No. 02/2014/TT-BNNPTNT of 24 January 2014 stipulates the order, procedures on granting and revocation of GM food and feed certificates for GM plants that fulfilled the requirements to be used as food and feed.
- Circular No. 29/2014/TT-BNNPTNT of 5 September 2014 amends and supplements Article 7 of Circular No. 23/2010/TT-BNNPTNT of 7 April 2010 on the recognition of technical advances of biotechnology in the field of agricultural and rural development.
- Joint Circular No. 45/2015/TTLT-BNNPTNT-BKHCN of 23 November 2015 between MARD and MOST for guidance on labelling of packaged GM food. GM food is defined as one containing 5% and above GM content.

Ministry of Finance:

- Circular No. 36/2014/TT-BTC of 24 March 2014 stipulates the rates and the regime of collection, remittance, management and use of the verification fee of application for a biosafety certificate for GM crops. Amended and supplemented in Circular No. 186/2016/TT-BTC of 8 November 2016.
- Circular No. 106/2014/TT-BTC of 8 August 2014 stipulates the rates, the regime of collection, remittance, management and use of the verification fee of application for certification for GM plants for use as food and feed. Amended in Circular No. 225/2016/TT-BTC of 11 November 2016.

Stakeholder involvement in biosafety management of GMOs

An applicant intending to develop a GMO for commercial purposes must first seek accreditation of their laboratories/research facilities from MOST (Figure 14.1). MOST is responsible for the management and regulation of laboratories conducting research on GMOs, and coordination with other relevant ministries in stipulating the labelling regulation of GMOs.

As the research on a GMO makes satisfactory progress and proof of concept is established and found useful, the GMO products are allowed to be further developed. These products

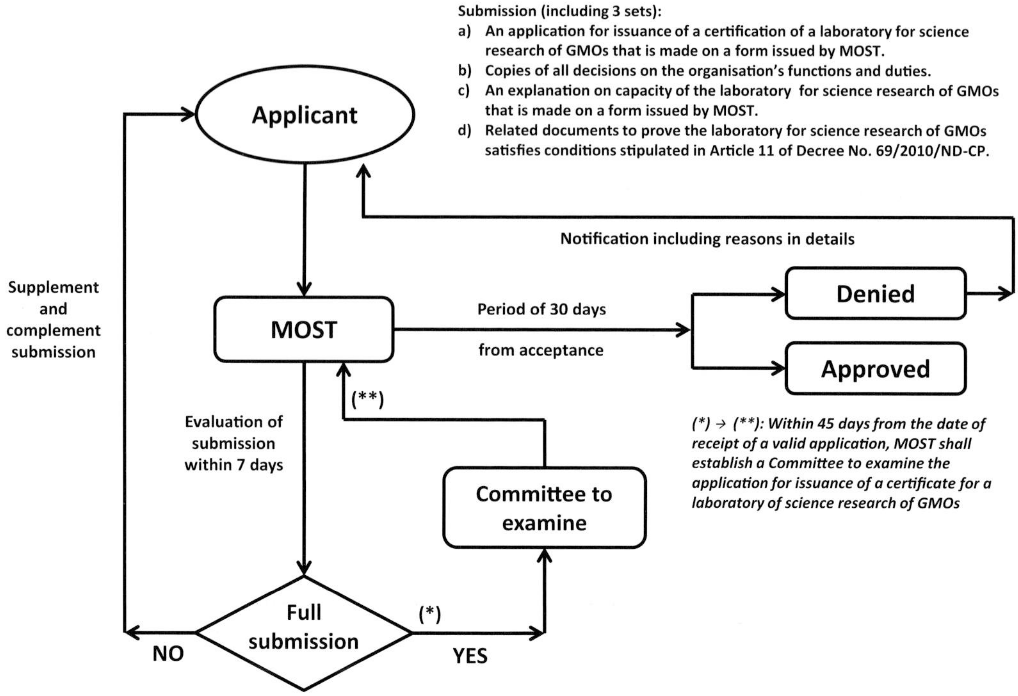

Figure 14.1 Procedure of provision for laboratory GMO. Source: Vietnam Environment Administration, MONRE <antoansinhhoc.vn>.

undergo extensive testing for their trait characterisation, efficacy and safety at various levels from confined controlled conditions to the open field.

MARD is responsible for managing and regulating the GMO efficacy trials and for granting certificates that allow the GMO to be used as food and feeds. Figure 14.2 shows activities under the control of MARD.

MONRE is responsible for issuance of the biosafety certificate, storage, packaging and transportation of GMOs, and for management of information and database of GMOs (Figure 14.3).

Integration of GMO research and development as per regulations is coordinated by MONRE, MARD and MOST. Figure 14.4 shows the different departments and committees involved.

Other ministries (Ministry of Industry and Trade, Ministry of Health, Ministry of Finance, Ministry of Public Security, and the local People's Committees) command and operate risk management and report to MONRE if a risk has occurred, scheduling an inspection for the implementation of risk management to promptly handle violations according to the current regulations (Table 14.1).

Biosafety management results

Research and development activities

After the promulgation of Circular No. 20/2012/TT-BKHCN of 20 November 2012, MOST accredited 10 research laboratories on GMO at biotechnology laboratory level 2. To date, MOST has not received any notifications of biosafety violations in research and technological

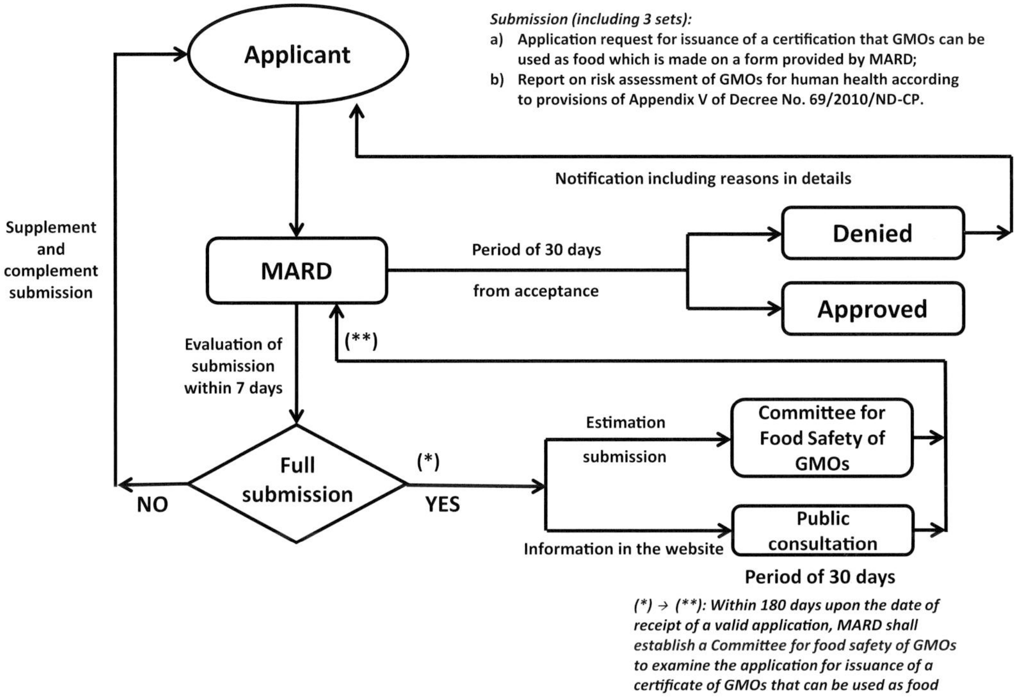

Figure 14.2 Procedure for assessment of a GMO for use as food and feed. Source: Vietnam Environment Administration, MONRE <antoansinhhoc.vn>.

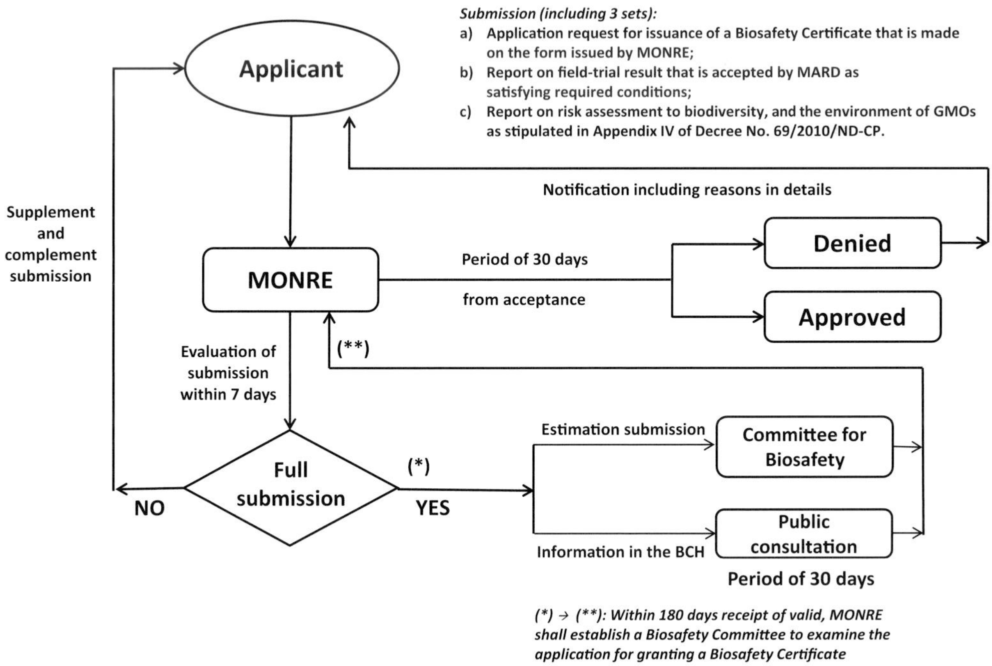

Figure 14.3 Procedure for biosafety certification. Source: Vietnam Environment Administration, MONRE <antoansinhhoc.vn>.

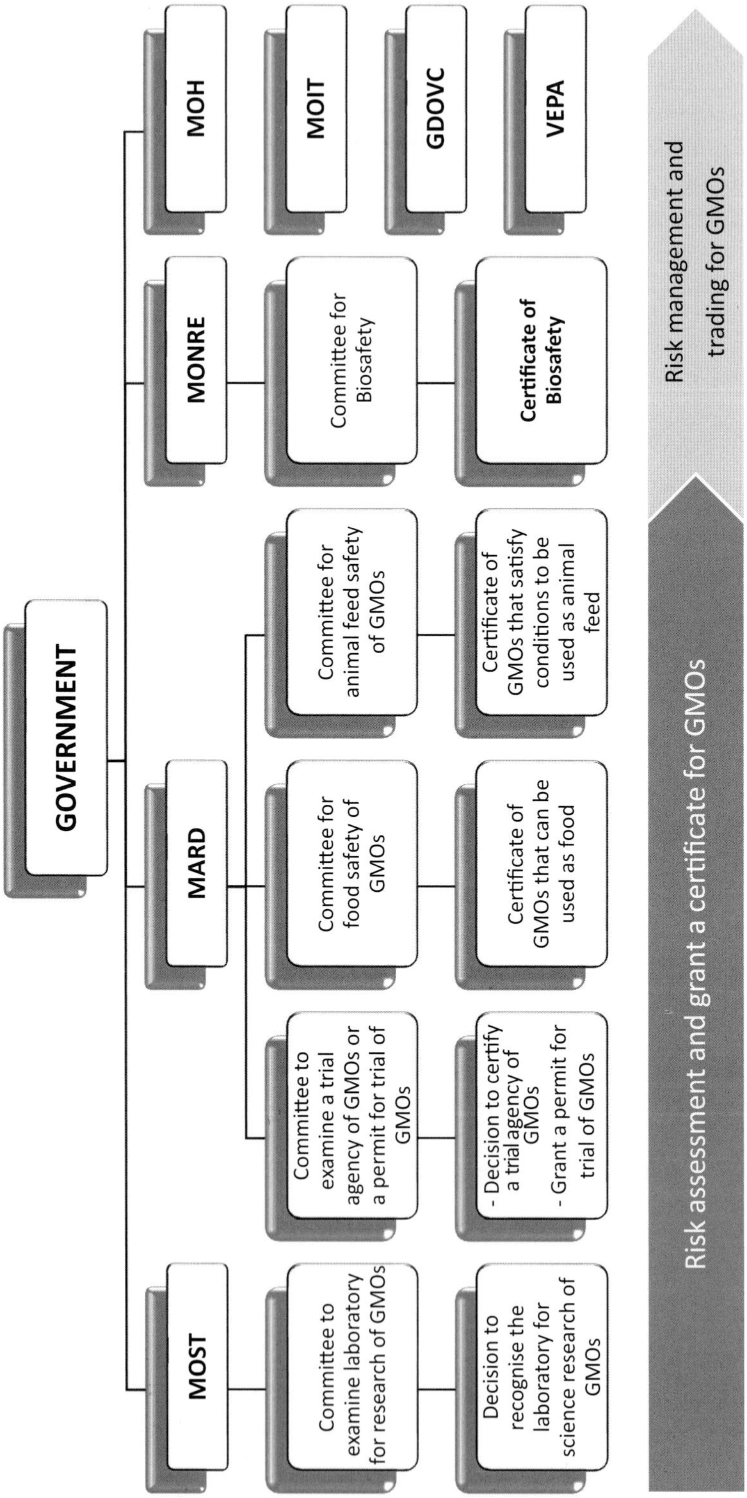

MONRE : Ministry of Natural Resources and Environment
MARD : Ministry of Agriculture and Rural Development
MOST : Ministry of Science and Technology

MOH : Ministry of Health
MOIT : Ministry of Industry and Trade
GDOVC : General Department of Vietnam Customs
VEPA : Vietnam Environment Police Agency

Figure 14.4 Vietnam Biosafety Management Chart. Source: Vietnam Environment Administration, MONRE <antoansinhhoc.vn>.

development of GMOs. However, the resources for managing biosafety of GMOs in this area are limited, and monitoring and supervision are not properly implemented. The number of accredited laboratories is limited compared to the number of biotech laboratories operating nationwide.

Confined and large-scale field trials

According to Circular No. 72/2009/TT-BNNPTNT of 17 November 2009 from MARD, GM crop plants allowed field trials for risk assessment to biodiversity and the environment are maize, soybean and cotton. At the same time, MARD accredited the organisations qualified to conduct risk assessments. These organisations include Agricultural Genetics Institute at Hanoi, Plant Protection Research Institute at Hanoi, Nha Ho Research and Development of Cotton Institute at NHon Son (Ninh Thuan province) and the Southern Center for Trial and Testing of Plant, Plant Products and Fertilisers at Ho Chi Minh City.

Since 2010, MARD has granted permission for conducting confined and large-scale trials for seven events of GM maize, which express insect resistance and herbicide tolerance. It has recognised the results of field trials of five maize events; namely, insect- resistant MON89034, herbicide-tolerant NK603 (Dekalb Vietnam Co.); dual-stacked insect-resistant and herbicide-tolerant Bt11, herbicide-tolerant GA21 (Syngenta Vietnam Co.); and insect-resistant TC1507 (Pioneer Hi-Bred Vietnam Co.).

Environmental release

After MARD has validated the trial results, applications are submitted to MONRE. To date, MONRE has received, appraised and issued biosafety certificates for five events of GM maize, including lepidopteran-resistant MON89034, glyphosate-tolerant GA21, glyphosate-tolerant NK603, lepidopteran-resistant Bt11 and lepidopteran-resistant and glyphosate-tolerant TC1507. In 2015 GM maize varieties were commercially grown in Vietnam, marking this country the 29th country in the world to allow the planting of GM maize for commercial purpose.

Based on the biosafety certificates granted by MONRE, MARD has approved 16 GM maize varieties including NK66Bt, NK66Bt/GT, NK66GT, K4300Bt/GT, NK67Bt/GT, NK7328Bt/GT (Syngenta Vietnam Co.); C919S, C919R, DK9955S, DK9955R, DK6818S, DK6818R, DK6919S, DK6919R, DK8868S (Dekalb Vietnam Co.); and CP501S (CP Vietnam Seed Co.) (Table 14.2). GM maize varieties were approved in recognition of technical advances in biotechnology in the agriculture and rural development sector, specifically in relation to higher efficiency in terms of socioeconomic benefits, environmental protection and food hygiene and food safety over the current techniques, technologies and products commonly used in production; meeting the needs of production, and with GM maize morphology equivalent to the conventional background variety.

By 2017, the total amount of GM maize seeds imported was ~1500 t, enough to plant an area of ~80 000 ha.

Food and feed use

To date, MARD has received and processed 48 application dossiers and has approved and certified 26 events of GM crops to be used for food and feed (Table 14.2): 11 events of GM soybean (Dekalb Vietnam Co. has six events, Bayer Vietnam Co. has two events, Pioneer Hi-Bred Vietnam Co. has one event, Syngenta Vietnam Co. has one event, BASF Vietnam Co. has one event); and 15 events of GM maize (Dekalb Vietnam Co. has six events, Bayer Vietnam Co. has one event, Pioneer Hi-Bred Vietnam Co. has two events and Syngenta Vietnam Co. has six events).

Table 14.2. Approved events of GM crops in Vietnam

Sr No.	Event	Organisation	Year	Trait	Gene(s)	Target insects/ herbicides	Approval for food, feed, environmental release
Maize							
1	MIR604	Syngenta Vietnam Co.	2016	Rootworm resistance	*mcry3A, pmi*	Rootworm belongs to lepidopteran resistance	Food/feed
2	5307	Syngenta Vietnam Co.	2016	Rootworm resistance	*ecry3.1Ab, pmi*	Rootworms	Food/feed
3	MON810	Dekalb Vietnam Co.	2015	Stem borer resistance	*cry1Ab*	Stem borers	Food/feed
4	MON88017	Dekalb Vietnam Co.	2015	Herbicide tolerance, root worm resistance	*cp4 epsps, cryBb1*	Herbicide tolerance, rootworms	Food/feed
5	MON87460	Dekalb Vietnam Co.	2015	Drought tolerance	*cspB*	Drought tolerance	Food/feed
6	MON87427	Dekalb Vietnam Co.	2015	Herbicide tolerance	*cp4 epsps*	Glyphosate	Food/feed
7	T25	Bayer Vietnam Co.	2015	Glufosinate ammonium tolerance	*pat*	Glufosinate ammonium	Food/feed
8	MIR162	Syngenta Vietnam Co.	2014	Lepidopteran resistance	*Vip3Aa20 and pmi*	Lepidopterans	Food/feed
9	TC1507	Pioneer Hi-Bred Vietnam Co.	2016	Lepidopteran resistance	*cry1F, pat*	Lepidopterans	Food/feed Environmental release
10	GA21	Syngenta Vietnam Co.	2014	Herbicide tolerance	*cp4epsps*	Glyphosate herbicide	Food/feed Environmental release
11	Bt11	Syngenta Vietnam Co.	2014 (food/feed) 2015 (environmental release)	Lepidopteran resistance Herbicide tolerance (containing glufosinate ammonium)	*Btk, pat*	Lepidopterans Glufosinate ammonium	Food/feed Environmental release
12	NK603	Dekalb Vietnam Co.	2014	Herbicide tolerance (Roundup from glyphosate)	*cp4epsps*	Glyphosate	Food/feed Environmental release
13	MON89034	Dekalb Vietnam Co.	2014	Lepidopteran resistance	*cry1A.105, cry2Ab2*	Lepidopterans	Food/feed Environmental release

14	DAS-59122–7	Pioneer Hi-Bred Vietnam Co.	2018	Lepidopteran resistance Glyphosate herbicide tolerance	*cry34Ab1, cry35Ab1*	Lepidopterans Glyphosate	Food/feed
15	3272	Syngenta Vietnam Co.	2019	Expression of heat-resistant α-amylase enzyme (AMY797E)	*amy797E*	AMY797E enzyme	Food/feed
Soybean							
16	MON87769	Dekalb Vietnam Co.	2015	Rich omega-3	*Pj.D6D, Nc.Fad3*	Rich omega-3	Food/feed
17	A2704–12	Bayer Vietnam Co.	2015	Glufosinate ammonium tolerance	*pat*	Glufosinate ammonium	Food/feed
18	A5547–127	Bayer Vietnam Co.	2015	Glufosinate ammonium tolerance	*pat*	Glufosinate ammonium	Food/feed
19	MON87708	Dekalb Vietnam Co.	2015	Herbicide (Dicamba) tolerance	*dmo*	Dicamba	Food/feed
20	MON87701	Dekalb Vietnam Co.	2015	Lepidopteran resistance	*cry1Ac*	Lepidopterans	Food/feed
21	MON87705	Dekalb Vietnam Co.	2015	Higher content of oleic acid and herbicide tolerance (Roundup from glyphosate)	*cp4 epsps*	Glyphosate Oleic acid	Food/feed
22	40–3-2	Pioneer Hi-Bred Vietnam Co.	2015	Herbicide tolerance (Roundup from glyphosate)	*cp4 epsps and uidA*	Glyphosate	Food/feed
23	MON89788	Dekalb Vietnam Co.	2014	Herbicide tolerance (Roundup from glyphosate)	*cp4 epsps*	Glyphosate	Food/feed
24	DP305423-I	Pioneer Hi-Bred Vietnam Co.	2018	Higher content of oleic acid	*gmfad2–1, gm-hra*	Oleic acid	Food/feed
25	SYHTOH2	Syngenta Vietnam Co.	2019	HPPD inhibiting and glufosinate-ammonium herbicide tolerance	*pat, AvHPPD-03*	Herbicide	Food/feed
26	CV127	BASF Vietnam Co.	2019	Herbicide tolerance (imidazolinone)	*csr1–2*	Imidazolinone herbicide	Food/feed

Current cultivation status of GM maize

Vietnam started growing GM maize from 2015 when it was planted on ~3500 ha. The area increased to 35 000 ha in 2016 and then to ~45 000 ha in 2017 and ~49 000 ha area in 2018. About 37 500 farmers are cultivating GM maize. A total of 22 GM events were approved by 2017, 14 for maize and eight for soybean for food, feed and cultivation. Presently 15 maize and 11 soybean events are approved for food, feed and cultivation. GM technology may boost productivity of crops as Vietnam imports ~50% of its maize and to some extent soybean meals for meeting its animal and aquaculture needs.

Report and monitoring

As Vietnam became a signatory to the Cartagena Protocol on Biodiversity in 2004, by 2006 MONRE had developed and put into operation a Biosafety Clearing House. This is connected to the Global Biosafety Clearing House operated by the Secretariat of the Convention on Biological Diversity, which shares national implementation activities on the biosafety of GMOs.

In 2010, Article 46 of Decree 69 stipulated the responsibility of MONRE to manage a GMO database, maintain a website on management of biosafety, and supervise the ministries which provide relevant information and databases to MONRE. MONRE issued Circular No. 09/2012/TT-BTNMT of 22 August 2012, which clearly stipulates required GMO data, information, development and management, public information, and provision and use of information.

MONRE has regularly updated the biosafety information of Vietnam to the Global Biosafety Clearing House.

In recent years, after the Law on Biodiversity came into force, ministries and related authorities have been coordinating with national and international organisations such as the Centre for Environmental Risk Assessment and the International Food Policy Research Institute in the USA, the European Food Safety Authority in Italy and CropLife International in Belgium to conduct workshops and training to enhance capacity and awareness in stakeholders, including management staff and researchers at the central and local levels. Information on biosafety is also published widely through mass media that includes documentary films shown on national television. Other publications like FAQs, pocket books, leaflets and brochures are also developed and distributed to update biosafety knowledge.

Following the environmental release of five events of GM maize for cultivation, Quy *et al.* (2016) determined that assessment of GM maize's insect-resistant trait for its effect on environment and biodiversity in Vietnam is a complicated issue. They proposed an elaborate criterion for the environmental risk assessment of insect-resistant GM maize in Vietnam (Quy *et al.* 2017). Table 14.3 details the process of environment risk assessment. Earlier, Andow and Hilbeck (2004) elaborated broad principles of risk assessment of GM crops for non-target organisms, with methodology by Hilbeck *et al.* (2006) and Romeis *et al.* (2008).

One of the ecological consequences of GM crops is the evolutionary pressure that the toxins exert on the target pests to develop resistance (Andow 2008). As Bt crop sustainability is very important in view of the importance of Bt toxins as a natural resource for pest management, resistance monitoring and its management has become a prerequisite for approval of Bt crops. In Vietnam, Bt maize resistance management in the target insects is a necessity for maize sustainability.

Challenges, lessons learnt and recommendations

Challenges

There is a lack of baseline database on biodiversity and the provision of information and exchange of data on biosafety have not yet been fully implemented. In Vietnam, biodiversity

Table 14.3. Criteria for environmental risk assessment of GM maize in Vietnam

Criteria group	Hazard	Assessment criteria	Criteria assessment requirement	Detailed criteria
Criteria group 1. Assess the hazards: gene flow, potential for weediness, persistence or becoming invasive species of insect-resistant GM maize	*Hazard 1.1*: The GM maize cannot be grown under Vietnam conditions. There are unintended differences in growth characteristics in comparison to the conventional counterpart	*Criterion 1.1*: Field trial of distinctness (including germination, phenotype, flowering, maturity, plant height etc.) of the GM maize	• Confined and large-scale field trials of the GM maize in comparison to the conventional counterpart • Comparison with national technical regulation on testing of distinctness, uniformity and stability of maize varieties	Comparison of germination, phenotypic, flowering, maturity and plant height characteristics of the GM maize in comparison to the conventional counterpart
	Hazard 1.2: The GM maize cannot itself or its progeny overwinter under Vietnam conditions. There are unintended differences in overwintering ability in comparison to the conventional counterpart	*Criterion 1.2*: Assessment of distinctness (including germination, phenotype, flowering, maturity, plant height etc.) of the GM maize in winter field trial	• Confined and large-scale field trials of the GM maize in comparison to the conventional counterpart at least 1 winter crop for each type of trial • Comparison with national technical regulation on testing of distinctness, uniformity and stability of maize varieties	Comparison of germination, phenotypic, flowering, maturity and plant height characteristics of the GM maize in comparison to the conventional counterpart at least 1 winter crop
	Hazard 1.3: The GM maize can reproduce and hybridise with compatible relatives under Vietnam conditions. There are unintended differences in hybridisation potential in comparison to the conventional counterpart	*Criterion 1.3*: • Monitoring dispersal of the GM maize's pollen to nearby plants • Monitoring the flowering time of the GM maize	• Confined and large-scale field trials of the GM maize in comparison to the conventional counterpart • Comparison with national technical regulations on testing of distinctness, uniformity and stability of maize varieties	• Comparison of dispersal time of GM maize's pollen with that of nearby plants. • Comparison of flowering time of GM maize with the conventional counterpart

Continued

Table 14.3. continued

Criteria group	Hazard	Assessment criteria	Criteria assessment requirement	Detailed criteria
	Hazard 1.4: The GM maize is more persistent than conventional counterparts under Vietnam agricultural conditions	***Criterion 1.4***: Comparison of persistence time in environment of the GM maize and the conventional counterpart after cropping	Confined and large-scale field trials of the GM maize in comparison to the conventional counterpart after cropping	Comparison of persistence time in the environment of the GM maize and the conventional counterpart after cropping
	Hazard 1.5: The GM maize's trait increases the fitness or compatible relatives in Vietnam agricultural system	***Criterion 1.5***: Monitoring the increase in the population of the GM maize	Confined and large-scale field trials of the GM maize	Assessment of the ratio of normally grown maize seeds and cultivated maize seeds
	Hazard 1.6: The GM maize can form feral populations under Vietnam conditions	No hazards		
	Hazard 1.7: Recombinant DNA coding insect-resistant trait of the GM maize can transfer into soil microorganisms	***Criterion 1.7***: • Comparison recombinant DNA coding insect-resistant trait of the GM maize with the total DNA of soil microorganisms • Testing presence of antibiotic resistance marker genes in soil microorganism populations	Lab testing of soil samples from confined field trials of the GM maize	• Comparison recombinant DNA coding insect-resistant trait of the GM maize with the total DNA of soil microorganisms • Testing presence of antibiotic resistance marker genes in soil microorganism populations
Criteria group 2. Assess the interactions of the GM plant with non-target organisms (NTO) in the Vietnam ecosystem	***Hazard 2.1***: The GM maize impacts the ability to attract NTOs compared with the conventional counterpart	***Criterion 2.1.*** Assessment of NTOs • **Herbivores**: *Rhopalosiphum padi, Phenice moesta, Phloeothrips* sp. (*Ph. oryzae* Matsumura), *Monolepta signata, Sitophilus zeamais, Euproctis scintillans, Helicoverpa armigera, Mythimna separata, Agrotis ipsilon.* • **Predators**: *Paederus fuscipes, Propylea japonica, Menochilus sexmaculatus, Araneus inustus, Argyrodes* sp., *Anisolabis annulipes, Ischiodon scutellaris* • **Parasitoids**: *Trichogramma* sp., *Lysiphlebus* sp., Apanteles sp. • **Pollinators**: Honeybee	Confined and large-scale field trials of the GM maize	Comparing the species' biodiversity index in confined and large-scale field trials of the GM maize **Herbivores**: *Rhopalosiphum padi, Phenice moesta, Phloeothrips* sp. (*Ph. oryzae*), *Monolepta signata, Sitophilus zeamais, Euproctis scintillans, Helicoverpa armigera, Mythimna separata, Agrotis ipsilon.*

		• **Decomposers**: Earthworm, termine, rolling, springtails, colembolans, roundworms, *Oppia nitens* (Acari)		• **Predators**: *Paederus fuscipes*, *Propylea japonica*, *Menochilus sexmaculatus*, *Araneus inustus*, *Argyrodes* sp., *Anisolabis annulipes*, *Ischiodon scutellaris* • **Parasitoids**: *Trichogramma* sp., *Lysiphlebus* sp., Apanteles sp. • **Pollinators**: Honeybee • **Decomposers**: Earthworm, termine, rolling, springtails, colembolans, roundworms, *Oppia nitens* (Acari)
Criteria group 3. Assess the interactions of the GM plant with target organisms (TO)	***Hazard 3.1***: Development of the TO's resistance	***Criterion 3.1***: Development of TO resistance management strategy	Confined and large-scale field trials of the GM maize	• Comparison of TO population size with that of the conventional counterpart • Development of TO resistance management strategy
Criteria group 4. Assess impacts of GM maize cultivation, management and harvesting techniques on the environment and biodiversity in Vietnam	***Hazard 4.1***: Cultivation of the GM maize will impact on cultivation, management and harvesting techniques of the traditional maize	***Criterion 4.1:*** • Comparison of GM maize cultivation, management and harvesting techniques with that of the conventional counterpart during confined and large-scale field trials of GM maize After releasing the GM maize into environment: • Changes of the GM maize's TO and NTO during the process of cultivation, including crop rotation • Changes of integrated pest management in the GM maize cultivation • Monitoring pesticides use in the GM maize cultivation	• Confined and large-scale field trials of GM maize • After releasing the GM maize into the environment	• Comparison of GM maize cultivation, management and harvesting techniques with that of the conventional counterpart during confined and large-scale field trials of the GM maize • Changes of the GM maize's TO and NTO during the process of cultivation, including crop rotation • Changes of integrated pest management in the GM maize's cultivation • Monitoring pesticides use in the GM maize cultivation

databases are kept by different agencies and organisations, and there is no aligned database on biodiversity. Therefore, it is difficult to assess the risk of GM crops to biodiversity as there is no baseline database for comparison before and after trials.

There is limited awareness and understanding of biosafety. Over the past few years, Vietnam has implemented capacity-building programs and increased biosafety awareness across ministries, research and academic institutions, and many other non-government and government organisations. However, the level of awareness and understanding of biosafety is still limited.

There is a lack of information on rising issues. At its recent technical meetings, the Convention on Biological Diversity raised the need for further guidance on risk assessment and risk management of GMOs, and the issue that synthetic biology should be taken into consideration and discussed in order to identify a proper management approach. For that reason, it will be useful to keep updated on the situation and integrate the issues into national legislations and policies.

There is a lack of national technical guidelines for GMOs (first focus on GM maize with each event) approved by competent authorities.

Lessons learnt

There needs to be closer collaboration of line ministries including MONRE, MARD, MOST, Ministry of Health, Ministry of Industry and Trade, Ministry of Finance and the Vietnam Academy of Science and Technology for the process of developing and implementing policies and regulations on biosafety. It is vital to continue raising awareness at all levels about the roles, benefits and contributions of GMOs.

Recommendations

There needs to be continued integration of biodiversity consideration and biosafety management issues into national and sector strategies and action plans. The Vietnamese government encourages the application of biotechnology to improve the livelihoods of its people but, at the same time, it must ensure the safety of human health, animal and biodiversity. Biosafety management has been integrated into the National Strategy for Biodiversity and biotechnology application programs in agriculture. It should also be included in the strategies and action plans for national socioeconomic development, biodiversity, agricultural and rural development. These are reviewed and updated every five to 10 years.

There should be development and promulgation of legal documents on the biosafety of GMOs. In Vietnam, GMOs have been legally regulated but as biotechnological applications evolve we must update the information and amendment of national legislation and policies. In Vietnam, the sole document on biosafety management is Decree 69, with an update on the role of MARD in Decree 108.

The capacity of National Competent Authorities and technical working groups must be enhanced, including those of MONRE, MARD and MOST. Effective coordination mechanisms have been established by sending representatives to the National Biosafety Committee, and Committee for Food and Feed Safety of GM Plants. A Technical Advisory Team with specialist members from research institutes and universities is responsible for examining technical details and supporting the National Biosafety Committee in evaluating application dossiers, replying to public comments and providing summaries for the benefit of other agencies. It is important to continue the collaboration with national and international organisations via training courses to strengthen the capacity of the National Biosafety Committee, and Committee for Food and Feed Safety of GM Plants.

There need to be planned regions of GM maize across the country, including monitoring of target and non-target organisms in GM maize-growing areas based on selection of focal species.

References

Andow DA (2008) The risk of resistance evolution in insects to transgenic insecticidal crops. *Collection of Biosafety Reviews* **4**, 142–199.

Andow DA, Hilbeck A (2004) Science-based risk assessment for nontarget effects of transgenic crops. *Bioscience* **54**, 637–649. doi:10.1641/0006-3568(2004)054[0637:SRAFNE]2.0.CO;2

Hilbeck A, Andow DA, Arpaia S, Birch ANE, Fontes EMG, Lovei GL, Sujii E, Wheatley R, Underwood E (2006) Methodology to support non-target and biodiversity risk assessment. In *Environmental Risk Assessment of Genetically Modified Organisms*, Vol 2. (Eds A Hilbeck, DA Andow, E Fontes) pp. 108–132. CABI Publishing, Wallingford, UK.

Quy NX, Thanh Thuy NT, Cuong PA (2016) Development of criteria for risk assessment of genetically modified maize to environment and biodiversity in Viet Nam. *Journal of Vietnam Agricultural Science and Technology* **3**(64), 42–47.

Quy NX, Thanh Thuy NT, Cuong PA (2017) Establishing the procedure to assess impacts of genetically modified maize (GM maize) to environment and biodiversity in Viet Nam. *Vietnam Environment Magazine* **1**, 64–70.

Romeis J, Bartsch D, Bigler F, Candolfi MP, Gielkens MM, Hartley SE, Hellmich RL, Huesing JE, Jepson PC, Layton R, Quemada H, Raybould A, Rose RI, Schiemann J, Sears MK, Shelton AM, Sweet J, Vaituzis Z, Wolt JD (2008) Assessment of risk of insect resistant transgenic crops to nontarget arthropods. *Nature Biotechnology* **26**, 203–208. doi:10.1038/nbt1381

15

Biotechnology and its regulatory system in Japan

Mieko Kasai and Ryo Ohsawa

Introduction: history of the Japanese regulatory system

The Ministry of Education and Science Technology Agency (currently, Ministry of Education, Culture, Sports, Science and Technology, MEXT) first developed the Guidelines for Recombinant DNA Experiments in 1979 in response to emerging progress in biotechnology inside and outside Japan.[1] The guidelines were intended to ensure safe use of the techniques by universities and national research organisations. Following announcement of the experimental guidelines, other ministries started discussion on the regulatory aspect of biotechnology and developed guidelines in each area of jurisdiction: the Guidelines for Industrialization of Recombinant DNA Techniques by the Ministry of Economy (MOE) in 1986, the Guidelines for Application of Genetically Modified Organisms in Agriculture, Forestry, Fisheries etc. by the Ministry of Agriculture, Forestry and Fisheries (MAFF) in 1989, the Guidelines for Safety Assessment of Foods and Food Additives Produced by Recombinant DNA Techniques by the Ministry of Health and Welfare (MHW) in 1991, and the Guidelines for Safety Assessment of Feed made from Genetically Modified Organisms by MAFF in 1996. Each ministry established an expert review committee/council under these guidelines, and safety assessment of genetically modified organisms (GMO) begun on a voluntary basis.

By 1999, 29 genetically modified foods (GM foods) had been reviewed and approved by the Food Safety Investigation Council under the voluntary guideline system.[2] All the technology developers followed the voluntary review system, and the system seemed to be well functioning. However, to cope with rapid development of commercial use of GM foods in the international market, the Ministry of Health, Labour and Welfare (MHLW), the successor to the MHW, decided to legally impose the safety assessment of GM foods to ensure that no GM foods were distributed in Japan without safety assessment. Hence, in May 2000, the safety assessment of GM foods became a mandatory requirement under the Food Sanitation Law.[3] After the mandatory food safety review started, the Food Safety Commission (FSC) was established under the Cabinet Office in July 2003 and tasked with the responsibility of reviewing food safety of GM foods; the responsibility of managing GM foods remained with MHLW. MAFF also revised the Feed Safety Law, and feed safety assessment also became mandatory in 2003.

In regard to environmental safety, a virus-resistant GM tomato was first approved under the voluntary guideline system in 1992. Since then, GM tomato, rice, soybean, canola, corn, carnation etc. were reviewed and approved under the voluntary system. Like the food safety review and approval system, all the technology developers followed the voluntary review system. In January 2000, however, the Cartagena Protocol on Biosafety to the Convention on Biological Diversity was adapted and entered into force in September 2003. The Cartagena Protocol establishes an advance informed agreement procedure to ensure that countries are provided with the information necessary to make informed decisions before agreeing to the import of GMOs into their territory. The Protocol is a legally binding international agreement and therefore countries which ratify it have an obligation to establish a domestic law to enforce its requirements. Consequently, before ratification of the Cartagena Protocol in November 2003, the Japanese government had considered domestic measures. As a result of the considerations, the Law concerning the Conservation and Sustainable Use of Biological Diversity through Regulations on the Use of Living Modified Organism (the Cartagena Law) was promulgated in June 2003 and entered into force in February 2004.[4]

When the Cartagena Law was enforced in 2004, the MAFF voluntary guidelines were abolished. Compatibility between approvals under the voluntary guidelines system and those under the new Cartagena Law was evaluated. It was determined that there was a need for re-evaluation of the already approved GM crops and plants from a viewpoint of impact to biodiversity. Hence, all GM crops and plants went through re-evaluations by generating additional data during the grace period. Some of the GM crops that did not go through the re-evaluation did not have their registrations renewed.

Today, a developer of GMOs is required to receive three different approvals on food, feed and environmental safety (i.e. impact on biodiversity) before commercial distribution of the products in Japan under the Food Sanitation Law, the Feed Safety Law and the Cartagena Law, respectively. Figure 15.1 summarises each law and competent ministries.

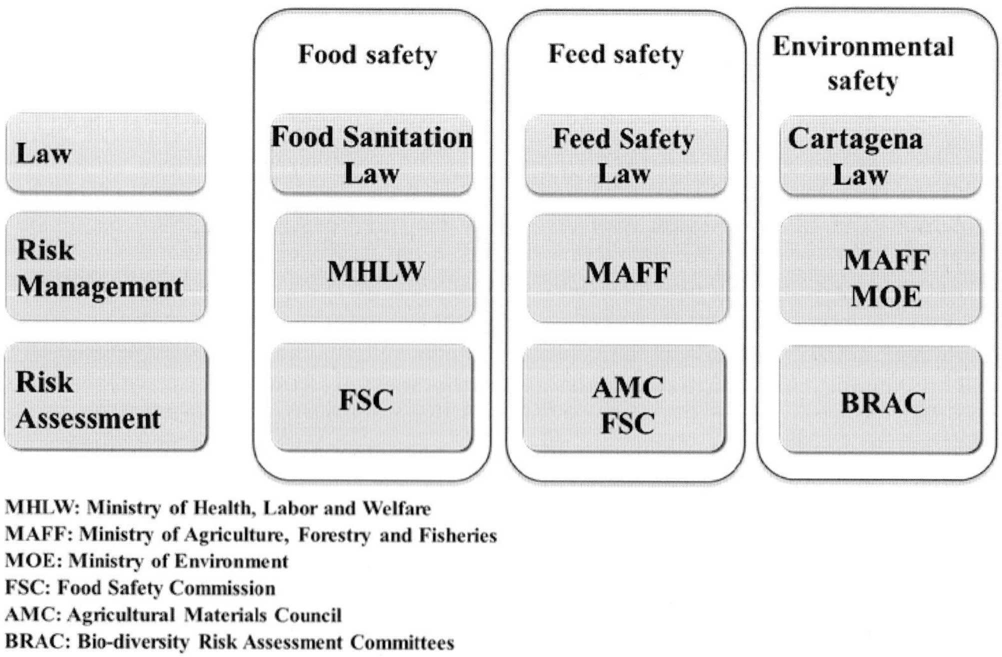

Figure 15.1 GMO-related law and competent authorities.

Current regulations on GMOs

Food safety

Parental GM event

GM foods are required to receive food safety approval from MHLW before distribution in the Japanese market by following the process described in Figure 15.2 under the Food Sanitation Law. An applicant submits a dossier for each parental GM event to MHLW. After MHLW has conducted a completeness check of the dossier, it requests FSC to conduct assessment for food safety. The actual food safety assessment is conducted by the Expert Committee on Genetically Modified Foods which consists of medical doctors, university professors and scientists from national research organisations (Figure 15.1). After the Expert Committee completes the food safety assessment, FSC opens a public comment period for 30 days. After reviewing the public comments, if no new safety concern is identified, FSC responds back to MHLW the result of the food safety assessment and MHLW notifies the completion of the safety assessment, in the Official Gazette.

The food safety assessment is based on the Standards for the Safety Assessment of Genetically Modified Foods (Seed Plants).[5] The standards are built on the concept of substantial equivalence in accordance with the Codex Guideline for the Conduct of Food Safety Assessment of Foods Derived from Recombinant-DNA Plants.[6] Japan served as the chair of the Codex *ad hoc* Intergovernmental Task Force when the guideline was developed. Safety assessment takes into account both intended and unintended changes. The following information is required to be included in the dossier:

- properties of the host used as a counterpart in safety assessment and its differences from the GM crop;
- purposes and usages of the GM crop;
- history of safe consumption of the host plant including information on production of harmful physiological substances;
- safety of the inserted DNA or gene(s) and the gene product(s);
- food safety of the GM crop;
- additional information when safety cannot be confirmed based on the information listed above.

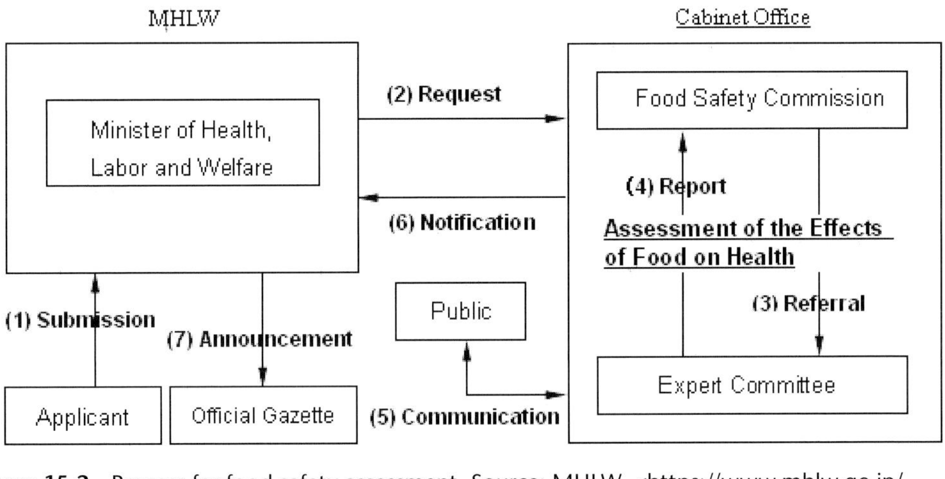

Figure 15.2 Process for food safety assessment. Source: MHLW, <https://www.mhlw.go.jp/english/topics/foodsafety/dna/01.html>

For assessment of safety of the inserted DNA or gene(s), the property of the donor organisms and the inserted DNA or gene(s) must be articulated as well as the method of transformation. When an antibiotic resistance gene is used, the administration route and the usage of the antibiotic should be described. Also, a concern that the insertion of the antibiotic resistance gene(s) might increase intestinal bacteria resistant to the antibiotic must be addressed. With regard to safety of the gene product(s), the function of the inserted DNA or gene(s) and the properties and functions of its product(s) (i.e. RNA or protein) should be clarified together with scientific rationale demonstrating that expression of the protein(s) has no adverse human health effects. In principle, the expressed protein(s) should not have significant homology to known toxins or allergens.

For safety assessment of the GM crop, a comprehensive molecular characterisation and compositional analysis as described below are required:

- the structure and copy number of the DNA inserted into the host genome;
- analysis for open reading frames (ORF);
- expression levels of the gene product(s) in the GM crop;
- daily intake of the gene product/protein(s) and its significance;
- allergenicity of the gene product/protein(s);
- stability of the gene(s) introduced into the GM crop;
- effect of the expressed gene product/protein(s) on the plant metabolic pathways (including its potential to react with substrates in the host);
- analysis of nutrients, anti-nutrients and toxins; i.e. comparative data between the GM crop and its non-GM counterpart.

For the molecular characterisation, the generation of the GM crop used for the characterisation is critical because the food safety approval is granted only to the progenies of the generation for which the molecular characterisation is done.[7] For example, when the molecular characterisation is conducted with T2 generation, the food approval is limited to the T2 generation and its progenies. In other words, the progenies developed directly from a T1 generation are outside the scope of the approval.

For analysis of ORFs, in principle, the DNA or gene(s) inserted into the host genome should not contain an ORF that expresses an unintended protein. If an ORF that may express an unintended protein has been identified, scientific rationale should be presented to demonstrate that the ORF and its expressed protein does not cause adverse human health effects.

A step-wide approach is applied for assessment of allergenicity of the expressed protein(s), and assessment is made based upon the first four bullet points listed below. An evaluation of IgE-binding activity of the expressed protein(s) with known allergens (point five) is required only if allergenicity cannot be judged based upon the information from the first four points:

- information on allergenicity of the donor(s) of the introduced gene(s);
- information on allergenicity of the expressed protein(s);
- sensitivity of the expressed protein(s) to physicochemical treatments; i.e. data on the alterations in the molecular weight, enzymatic activity and immunoreactivity of the expressed protein(s) after treatment of the simulated gastric fluid, the simulated intestinal fluid, or heat;
- amino acid sequence homology of the expressed protein(s) with known allergens (including the proteins involved in gluten-sensitive enteropathy);
- an evaluation of IgE-binding activity of the expressed protein(s) using the validated human serum, if the safety is not confirmed by the above information.

The assessment of effect of the expressed gene product/protein(s) on the plant metabolic pathways is to address a concern on whether or not the inserted DNA or gene(s) might have caused unintended effects by interacting with endogenous metabolic pathways. If the gene product is an enzyme, its substrate specificity should be determined. In principle, the substrate specificity of the inserted enzyme should be high so that it only catalyzes the intended substrate. If the inserted enzyme interacts with an unintended metabolic pathway in the host and causes alterations in the related components of the pathway, scientific rationale should be presented to demonstrate that the alterations do not cause any adverse human health effects.

For analysis of nutrients, anti-nutrients and toxins, comparative data between the GM crop and its non-GM counterpart are required and statistically analysed. In principle, there should be no significant difference between the GM crop and its counterpart. If any significant differences are revealed, scientific rationale should be presented to demonstrate that the values are within the reported variations of host crop varieties with history of safe use and thus the compositional differences do not cause or are not expected to cause any adverse human health effects.

It should be noted that information regarding residues of the herbicide and its metabolite(s) has to be included for a herbicide-tolerant crop if the crop, gene and herbicide combination has not been evaluated in the past. It is also noted that when a novel metabolite(s) is produced and/or herbicide residue level exceeds the current maximum residue limits (MRLs) in the crop as a result of the insertion of the herbicide-tolerant gene, an application has to be submitted in another dossier to the Food Safety Standards and Evaluation Division, Pharmaceutical Safety and Environmental Health Bureau of MHLW for the establishment or revision of MRLs for the herbicide.[8]

If food safety cannot be confirmed based on the information already listed, an animal toxicological study may be required.

Stacked event

A stacked event is a product produced by combining individual parental GM events through conventional breeding. On the other hand, a molecular stack, which is produced by co-transformation, is handled as a new GM parental event. Regulatory policy for stacked events was built on the Stance on the Safety Assessment of GM Plant Generated through Cross Breeding, which was developed by MHLW (Public Notice No. 233 in 2000).[9] In the public notice, MHLW defined that the progenies developed through traditional cross-breeding between or among approved GM crops are regarded as an approved event when they fulfil the following conditions:

- the properties newly acquired through recombinant DNA technique in the parental GM crops have not altered in the progeny;
- no cross has been made between subspecies;
- neither the dietary intake, the edible part nor the processing methods etc. has changed.

FSC announced its stance on safety assessment of stacked events in 2004, by taking account of the MHLW's stance.[10] An unique point of the FSC stance is categorisation of GM crops into the following three in accordance with the introduced traits and the necessity for safety assessment of a stacked event based upon the category into which the parental GM crops fall. The three categories are as follows:

Category I: A trait such as herbicide tolerance, insect resistance, or virus resistance in which a plant metabolic pathway is not altered by the inserted gene(s).

Category II: A trait such as increased amount of a certain nutrient or altered contents of cell walls as a result of alteration of the plant metabolic pathway by the inserted gene(s).

Category III: A trait in which a novel component is produced as a result of the inserted gene(s) using a part of the plant metabolic pathway.

Necessity for safety assessment of a stacked event was determined as summarised in Table 15.1. That is, a stacked event between or among GM crops classified into Category I is regarded as an approved product and thus does not have to go through the safety assessment process by FSC as long as it meets the following two criteria:

- no cross has been made between subspecies;
- neither the dietary intake, the edible part nor the processing methods etc. has changed.

On the other hand, if a Category I crop is stacked with a Category II or III crop, FSC determined that safety assessment was necessary for such stacked events until decided otherwise. The required information for safety assessment of a stacked event between Category I and II crops was streamlined in 2017 and a simplified review approach has been taken since 2017 for these stacked events (Table 15.1). With regard to a stacked event between Category II crops, between Category II and III crops, and between Category III crops, it was determined that such stacked events are subjects of the safety assessment without setting a time limit. That is, those stacked events are handled as new GM events.

For safety assessment of a stacked event, regardless of which category the stacked event falls into, the key question is whether or not interactions between or among proteins expressed from the inserted genes would occur. If the protein expressed by the inserted gene in the parental GM crop does not possess an enzymatic activity and thus does not alter a plant metabolic pathway, there would be no concern that stacked genes interact each other and result in an unexpected alteration of a plant metabolic pathway in the stacked event. This is the scientific rationale that a stacked event between or among Category I crops is regarded as an approved event and no safety assessment is conducted. It is also considered that it is unlikely that the expressed protein with no enzymatic activity will unexpectedly interact with another expressed protein with enzymatic activity of which the substrate specificity is clearly understood. Therefore, FSC sets a time limit for necessity of safety assessment for a stacked event between Category I and Category II or III crop.

As briefly mentioned above, although the basic stance on safety assessment of stacked events has remained unchanged since its first announcement in 2004, the required data and information as well as the safety assessment practice have been revised over time. Initially,

Table 15.1. Progression of the stack safety assessment over time

Categories	Stance in 2004	Current stance
I × I	• Regarded as an approved event • A confirmation dossier with bio-efficacy data	• Regarded as an approved event • Notification on a commercial plan to MHLW
I × II	• Handled as a new parental GM event • A full dossier	• Handled as a new parental GM event • A simplified dossier with bio-efficacy and compositional analysis data
I × III II × II II × III III × III	• Handled as a new parental event • A full dossier	No changes since 2004

even though FSC determined that the food safety assessment of a stacked event between or among Category I crops was not required, an applicant was required to submit a simple format dossier as shown in Table 15.1. The purpose of the submission is to confirm that the stacked event is actually a cross between or among Category I crops and meets the above two criteria; thus, it can be regarded as an approved GM event. Bio-efficacy data are required to demonstrate that the newly acquired traits from the parental GM crops are stably inherited in the stacked event without interacting, as expected.

After MHLW receives a simple format dossier, the FSC Expert Committee reviews it to confirm that no safety evaluation is necessary. In 2011, with seven years' experience reviewing simple format dossiers, the Expert Committee concluded that such confirmation review of a stacked event between or among Category I crops was no longer necessary.[11] As a result, the confirmation review came to be conducted only by FSC commissioners without asking the experts' opinions. Subsequently, in 2014 MHLW amended Public Notice No. 233, issued in 2000, and defined that stacked events between or among Category I crops are regarded as approved GM events.[12] MHLW posted a list of GM parental crops which were classified into Category I so that developers can self-determine whether or not a newly developed stacked event requires submission of the simple format dossier for confirmation. As of January 2019, 98 GM crops are listed as Category I crops.[13] Although these stacked events are no longer required to go through the confirmation review process, a developer is required to notify its intent to commercialise a stacked event between or among Category I crops so that MHLW can keep track of stacked events on the market (Table 15.1).

For a Category I crop stacked with a Category II or III crop, FSC determined initially that safety assessment was necessary until decided otherwise, as mentioned above. Therefore, although the key question is whether or not interactions between or among proteins expressed from the inserted genes would occur in the stacked event, an applicant is required to submit a dossier following the format used for the safety assessment of the parental GM crops. Some of the data, such as comprehensive molecular analysis and the allergenicity and toxicity of the newly expressed proteins generated for the safety assessment of the individual parental GM crops, were applied in safety evaluation of the stacked events. Since the safety assessment of several stacked events between Category I and II crops revealed that no interaction of the stacked genes was observed, FSC announced a new stance and streamlined the data requirement for a stacked event between Category I and II crops in 2017.[14] Under this new stance, information on only four aspects is required. The rest of the information which is required for parental GM events can be omitted. The following information is still required:

- basic information for the safety assessment of a stacked event; i.e. traits added in the parental GM crops, no cross made between subspecies, and no change in the dietary intake, edible part nor processing methods, etc.;
- stable inheritance of the traits in the stacked event, shown by analysis of DNA sequence, analysis of the gene products, or confirmation of the traits;
- scientific rationale demonstrating that no interactions among the stacked genes has occurred;
- comparison of the intended compositional changes between the parental GM crop and the stacked event.

The stance on a stacked event between Category I and III remained unchanged because there has been no such case. The changes on safety assessment of stacked events show that Japanese regulatory policy is science-based and progressive. It is expected that the data requirements and review process on a stacked event between Category I and III will be revisited after an experience and as knowledge accumulates.

Feed safety

Parental GM event

GM feeds must receive feed safety approval from MAFF before distribution in the Japanese market by following the process described in Figure 15.3. The feed safety assessment is based upon the following two points of view:

- safety of the GM crops to animal health;
- safety of the animal products such as meat, milk, and eggs produced by animals fed on the GM crop to human health.

An applicant submits a dossier for each parental GM event to MAFF. After MAFF has conducted a completeness check of the dossier, it requests the Feed Subcommittee of Agricultural Materials Council, which consists of experts, to conduct the safety assessment of the GM crops to animal health. MAFF also requests FSC to conduct the safety assessment of the products from animals fed on the GM crop. After both the Agricultural Materials Council and FSC complete the safety assessment, MAFF invites public comments for 30 days. After reviewing the public comments, if no new safety concern is identified, MAFF notifies the completion of the safety assessment in the Official Gazette.

As with the food safety assessment, the feed safety assessment is built on the concept of substantial equivalence and takes into account both intended and unintended changes. The required data and information for the feed safety assessment are similar to those required for the food safety assessment.[15] The key elements of the assessments are as follows:

- safety of the inserted genes;
- safety of the gene product/protein(s);
- sensitivity of the expressed protein(s) to physicochemical treatments;
- comparison of nutrients and anti-nutrients between the GM crop and its non-GM counterpart.

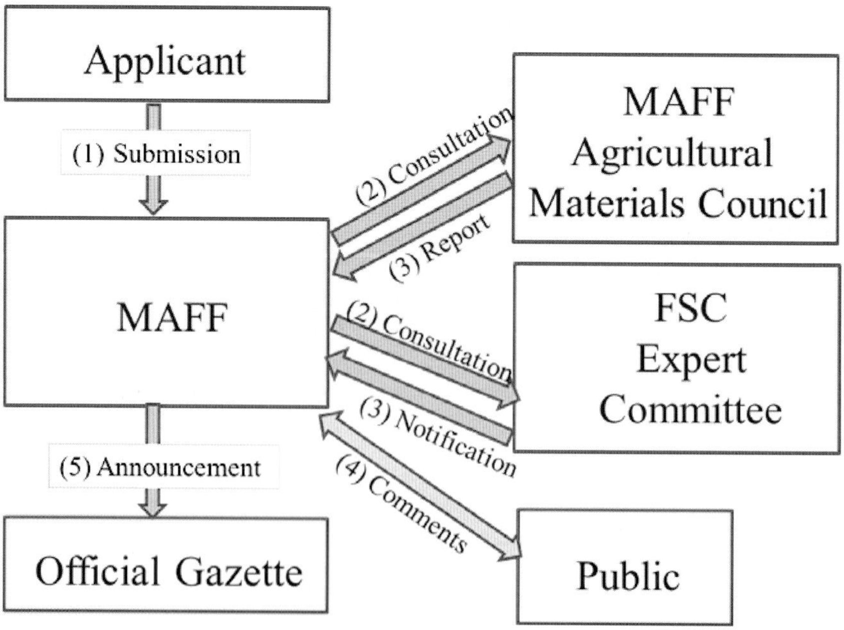

Figure 15.3 Process for the feed safety assessment.

The safety assessment of products from animal fed on the GM feed to human health is conducted by the FSC Expert Committee on Genetically Modified Foods. Since most of the GM crops are used for both food and feed and therefore are assessed for food safety, the following three potential risks are considered in the safety assessment of the animal products fed on GM feed:[16]

- possibility that a new harmful substance(s) is produced in the GM feed and transferred into animal products such as meat, milk, and eggs;
- possibility that the gene product(s) is converted into a harmful substance in the animal products and that harmful substance is accumulated in the animal products;
- possibility that the gene product(s) interacts with an animal metabolic pathway(s) and produces a new harmful substance(s).

An applicant must assess those three possibilities and submit a dossier demonstrating why such possibility is negligible, based on a scientific rationale. In general, it is not known that a gene product/protein expressed in a GM crop by an inserted gene is transferred into animals fed on the GM crop. It is very unlikely that a gene product/protein is converted into a harmful substance in animal products and that such a harmful substance can therefore accumulate in the animal products, nor that it interacts with an animal's metabolic pathway and produces a new harmful substance. In addition, safety of the expressed protein is evaluated during the food safety assessment of the GM crop and the expression of the protein is concluded to be of no concern. Therefore, it is considered that there would be no human health concern from consuming products from animals fed on GM crops unless a scientific rationale identifies otherwise. In the absence of such further possibilities, no further assessment is conducted.

Stacked event

The feed safety assessment of stacked events is not a mandatory requirement, but has been conducted under the voluntary system. The required data and information for the voluntary feed safety assessment are very similar to those for the food safety assessment. That is, a notification of a commercial plan is expected for a stacked event between or among Category I crops so that MAFF can keep track of stacked events on the market. If a Category I crop is stacked with a Category II or III crop, a simple format dossier confirming the absence of any interactions between or among proteins expressed from the stacked genes is submitted.

Environmental safety: impact on biodiversity

Parental GM event

Prior to environmental release, GM crops are required to receive Type I-use approval from MOE and MAFF by following the process described in Figure 15.4 under the Cartagena Law.

The Type I-use approval is for non-confined use of GM crops; i.e. domestic cultivation and/or import for food, feed and processing. The Type II-use approval is for contained use in greenhouses or laboratories with containment measures. For Type I-use approval, an applicant submits a dossier for each event to MAFF. After MOE/MAFF have done a completeness check of the dossier, they request the Biodiversity Impact Assessment Study Committee, which consists of experts from the universities and national research organisations, to assess the biodiversity impact of the GM crop. After the Committee has completed the risk assessment, the ministries jointly open a public comment period for 30 days. After reviewing the public comments, if no new safety concern is identified, the ministries notify the completion of the safety assessment in the Official Gazette. For food and feed crops, the environmental safety approval is put on hold until food and feed safety approvals have been granted. The standard

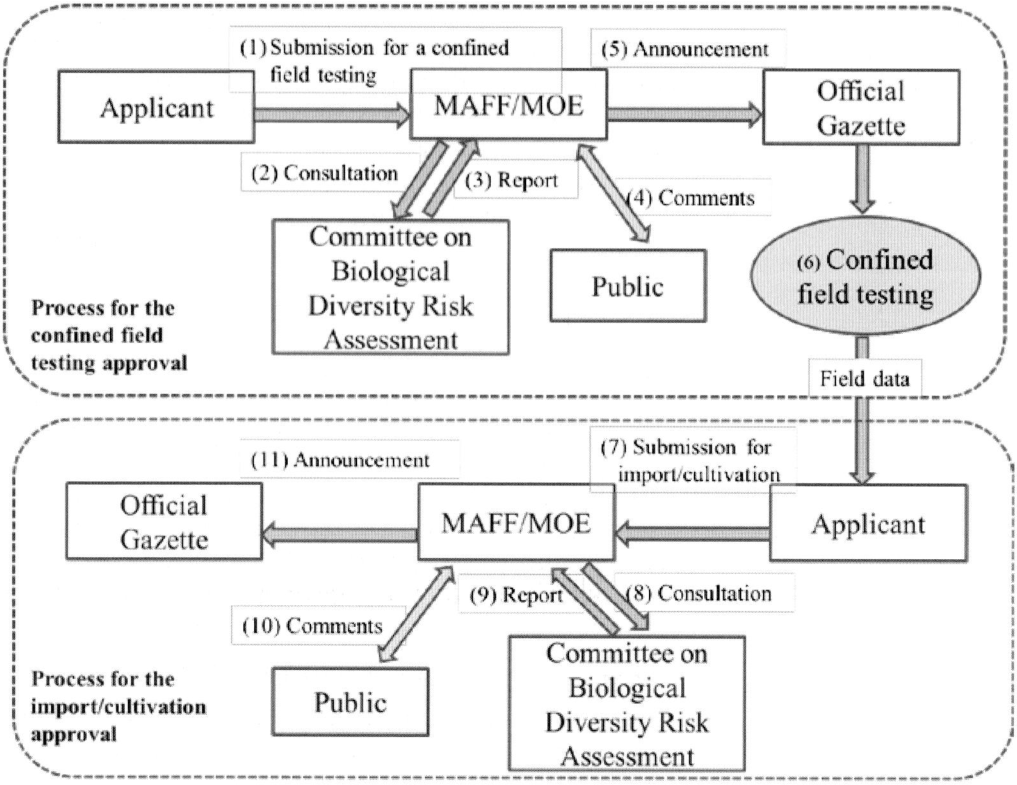

Figure 15.4 Process for the environmental safety assessment.

operating procedures regarding the biodiversity impact assessment and management of GM crops is posted on the MAFF website. For Type II-use approval, a developer is required to receive confirmation from the ministries that the proposed containment measures are appropriate for the contained use of GM crops.

One of the unique features of the Japanese biodiversity impact assessment process is that even if Type I-use approval is given for import of a GM crop for food, feed and processing (FFP), agronomic characteristic data must be attached in a dossier by conducting confined field testing at one location in one season in Japan (Figure 15.4). The need for local confined field testing is determined case-by-case and crop-by-crop. Confined field testing is required for imported GM crops for FFP to address the concern that viable crops may fall from a truck during a transportation and grow in the Japanese environment, and to address the concern that how the GM crop might grow in the Japanese environment is unknown. That is, even if there are data and knowledge on the growth of a GM crop in its native country, the crop might show a different growth pattern in Japanese environmental conditions. Confined field testing also evaluates whether unintended changes have occurred in the GM crop.

An applicant must first obtain approval to conduct confined field testing by following the full approval process described in Figure 15.4. To prevent GM crops/pollens distributing outside a confined field site, the field must be surrounded by a fence, shelterbelt and ditch. The detailed requirements for confined field testing site are available in the guidance document concerning application for Type I-use approval.[17] Before conducting confined field testing, an applicant must hold a neighbourhood meeting to explain the testing plan and ensure there are no objections to the testing. In the year following confined field testing, an applicant submits another dossier that includes the field testing results. Normally an applicant asks for an approval which covers both cultivation and import of the GM crop.

As described above, the fundamental framework of the environmental safety assessment established by MAFF was based on familiarity of the GM crop to its conventional counterpart. That is, if the environmental impact of the GM crop on the receiving environment does not exceed that of the conventional counterpart, it is regarded as safe as the conventional crop. Risk assessment under the Cartagena Law was built upon the same principle. However, since the purpose of the Cartagena Law is conservation and sustainable use of biodiversity, more emphasis is put on the impact of GM crops on Japanese biodiversity.

Upon enforcing the Cartagena Law, the Japanese government clearly defined 'biodiversity to be protected' as the state of biodiversity that existed before the Meiji Period, which began in 1868. Japan is an island country and closed its ports against foreign countries from 1639 to 1858. Consequently, plants naturalised after 1868 are not regarded as part of the Japanese biodiversity. The Japanese government defined an adverse biodiversity impact as reduction in the diversity of wild plants due to GM crops gaining competitive advantages, reduction or extinction of wild plants, insects, animals and microorganisms due to GM crops producing a harmful substance, and replacement of wild relatives due to outcrossing between GM crops and wild plants, as shown in Plate 9.

A dossier for biodiversity impact assessment consists of three sections.[18] The major requirements are as follows.

Section 1. Information gathered upon assessment of biological diversity impact

1. Information concerning a host organism such as taxonomy, state of distribution in natural environment, usage, physiological and ecological properties, etc.
2. Information concerning preparation of the GM crop such as donor organisms, vectors, a method of transformation, molecular analysis, a detection method, and differences between the host and the GM crop such as added traits and agronomic characteristics.
3. Information concerning the usage of the GM crop such as contents and methods of the usage, a method of collecting information, emergency measures, the results of testing in a laboratory or similar environment, approvals in other countries etc.

Section 2. Assessment for items relating to biological diversity impact

1. Competitiveness.
2. Production of harmful substances.
3. Crossability.
4. Other properties, where appropriate and necessary to be considered.

Section 3. Comprehensive evaluation on biological diversity impact of the GM crop

Upon preparing Section 1, sources of the information have to be included. When confined field testing is conducted, the data and results are described in Section 1. A statistical analysis should be conducted on the field data where possible. When statistically significant differences are observed, the issue of whether or not the difference is biologically significant must be discussed. For some items, the data collected in other countries can be used.

The biodiversity impact of the GM crop is assessed in Section 2 by applying the data and information collected in Section 1. The first step of assessment is identification of wild animals, plants etc. that might be impacted by the GM crop in terms of competitiveness, production of harmful substances or crossability. If such wildlife is identified, potential adverse effects of the GM crop on the identified wildlife, and the likelihood of that effect, are discussed. With regard to competitiveness, the key question is whether or not the GM crop competes with surrounding wild plants for fertiliser, sunlight and habitat, and interferes with the ecosystem and impacts on wildlife. The data on morphological and reproductive characteristics collected during the confined field testing are applied, along with whether or not the traits added to the GM crop, such as herbicide tolerance etc., would give a competitive advantage to the GM crop. Assessment is also made on whether or not the GM crop produces harmful substances and as a result interferes with growth of wild plants, insects, animals or microorganisms. The three types of data, with soil microbial test, plough-in test and post-production test, are collected during confined field testing for the evaluation. When the added trait to the GM crop is insect resistance, via expression of a Bt protein etc., assessment is also made of the impact of the expressed protein on non-target insects.

With regard to crossability, outcrossing with wild relatives is not regarded as an adverse biodiversity impact because conventional crops also outcross with wild relatives in nature. Hence, the key question is not whether the GM crop outcrosses with wild relatives, but whether those hybrid plants replace the wild relatives and as a result reduce the population of the wild relatives. Japan grows a wild relative of soybean, *Glycine soja* (OECD 2000). Therefore, a prudent approach has been taken to the biodiversity impact assessment of GM soybeans. It is known that both soybean and *G. soja* are self-pollinating plants and their flowering times do not overlap (Yoshimura *et al.* 2006). Consequently, outcrossing rates between GM soybean and *G. soja* was expected to be negligible. On the basis of several studies, the National Institute for Agro-Environmental Sciences reported that the crossability is <0.1% even if GM soybean and *G. soja* are grown side-by-side and have their flowering timing overlapped by adjustment of the seeding time (Mizuguti *et al.* 2010; Goto *et al.* 2017).

With accumulation of knowledge and experience, discussion on data transportability for Type 1-use approval for FFP has begun (Nakai *et al.* 2015), especially on corn since there are no wild relatives which outcross with corn in Japan. After careful discussion among the experts, MOE/MAFF amended the Notification regarding the application of GM crops for Type I-use in 2014. GM corn was exempted from the requirement of data collection in the Japanese environment when it fulfils the following two conditions, significantly reducing the time required for obtaining import approvals:[19]

- the mode of action of the inserted gene(s) is clearly defined in a peer-reviewed paper and/or through common understanding of multiple experts at a national review meeting etc.;
- it is considered that the biodiversity impact from the inserted gene(s) would not exceed that of GM corns which have been approved for Type I-use.

The amended Notification states that it is obvious that how such GM corn grows in Japanese environmental conditions from the scientific point of view when the above information is available and thus there is no need to collect additional data in the Japanese environment. This amendment applies not only to the Type I-use approval for import for FFP, but also to the cultivation approval. A newly developed GM corn is assessed on a case-by-case basis, using the MAFF lists of inserted genes and added traits for which confined field testing is not necessary so that developers can self-determine.[20] In 2019, cotton was added to the list of crops for which data collection in the Japanese environment is not required when the GM cotton fulfils the above conditions.[21]

Stacked event

Assessment of the biodiversity impact of stacked events started in 2004. Unlike the food and feeds safety approvals, the Cartagena Law does not distinguish stacked events based upon categories of the parental GM crops. The data requirement is the same for any combination of parental GM crops. The key question is the same as that of the food safety approval; i.e. whether or not the proteins expressed from the inserted genes in the parental GM crops interact in the stacked event. An applicant must submit a simple dossier to MOE/MAFF with bio-efficacy data which demonstrate that the traits newly acquired through recombinant DNA in the parental GM crops are stably inherited in the stacked event and that the traits added in the parental GM crops have not been altered in the stacked event.

In 2012, after over 100 stack events had been reviewed and approved with no observation of interaction among the stacked genes/proteins, the Biodiversity Impact Assessment Study Committee acknowledged that scientific knowledge and review experience showed that bio-efficacy data were no longer necessary to confirm no interactions when an adequate scientific rationale is available.[22] That is, when the mode of actions of the inserted genes are clearly defined and known to be independent, a conclusion can be drawn that they do not interact. In 2014, the review process was further simplified. Up to 2014, to obtain environmental safety approval for all commercial stacked events applicants had to submit separate dossiers for each event. For example, if an applicant planned to commercialise three different stacks developed with five parental GM crops (e.g. A × B × C × D × E, A × B × C, and C × D × E), they had to submit three dossiers for each stack. In 2014, it was decided that approval of the highest combination stacked event (e.g. A × B × C × D × E) would cover any lower combination of those parental GM crops, as long as it was demonstrated that no interactions among the proteins expressed from the stacked genes would occur with the highest combination stacked event.[23]

Simplification of the biodiversity assessment of stacked events continued through introduction of simplified dossier format, electric submissions etc., to allow Committee members and regulators to focus on new and more critical issues of concern.

Monitoring

As described above, all the approved GM crops imported into Japan have been confirmed to have no adverse impact on Japanese biodiversity. Moreover, all are highly domesticated and are known to require cultivation. However, MAFF has monitored any growth of GM crops at the ports and along roadsides, to allay public concern about unintended spread. The results of the monitoring, posted on the MAFF website, are briefly discussed below.

Corn

An investigation was conducted for three years from 2013 to 2015 at seven ports and along roads to three processing factories.[24] The area of the investigation at the ports was within 5 km radius of the unloading point and a total of 238 km of road edge was investigated. The results showed that although some spillage of seeds was found and several seeds germinated, spontaneously grown corn was not observed.

Cotton

An investigation was conducted for three years from 2014 to 2016 at seven locations (three warehouses, three feed processing plants and one oil processing plant) and along a road edge from one warehouse to a feed processing plant.[25] The result showed that although several seeds germinated, spontaneously grown cotton was not observed.

Soybean

Due to concern about outcrossing with wild relative *G. soja*, monitoring of GM soybean has been conducted for a longer period of time. The investigation began in 2009 at 10 ports, with investigation of an area within 5 km radius of the unloading point. The latest report from 2018 showed that there were eight GM soybeans grown from seed spillages at one port.[26] However, no expansion of the growth area of GM soybean nor outcrossing with the wild relatives were observed. Outcrossing among different GM soybean events was also not observed.

Canola

Although canola is not considered as a part of Japanese biodiversity to be protected, monitoring on canola has been conducted since 2006 due to a concern on outcrossing with related varieties. In 2018, an investigation was conducted at 17 ports[27] within 5 km radius of the unloading point. A total of 105 GM canola plants had grown from seed spillages at a total of 10 ports. However, no expansion of the growth area of GM canola was observed. The outcross rate with wild oilseed rapes did not exceed the reported values.

All these investigations supported the conclusions of the biodiversity risk assessment. MAFF plans to continue monitoring on soybean and canola.

Events approved in Japan

The GM crops approved for food safety are posted at the MHLW website (https://www.mhlw.go.jp/stf/seisakunitsuite/bunya/kenkou_iryou/shokuhin/idenshi/index_00002.html). As of August 2019, 320 GM crops including parental GM and stacked events were approved for food safety in eight different crops – alfalfa, canola, corn, cotton, papaya, potato, soybean and sugar beet. Corn has the highest numbers of approvals (206) followed by cotton (47), soybean (28) and canola (21). No Japanese company has received food safety approvals.

The events approved for feed safety are posted at the MAFF website (https://www.maff.go.jp/j/syouan/tikusui/siryo/index.html, https://www.maff.go.jp/j/syouan/tikusui/siryo/attach/pdf/index-64.pdf) and the English-language version is available at the Food and Agricultural Materials Inspection Center website (http://www.famic.go.jp/ffis/feed/r_safety/r_feeds_safety33.html). As of November 2019, 93 GM crops (parental GM events only) were approved for feed safety on seven crops – alfalfa, canola, corn, cotton, potato, soybean and sugar beet. Corn has the highest numbers of approvals (29), followed by cotton (21), soybean (18) and canola (18). No Japanese company has received feed safety approvals.

The events approved for environmental safety are posted at the Japan Biosafety Clearing House (https://www.biodic.go.jp/bch/english/e_index.html) and MAFF websites (http://www.maff.go.jp/j/syouan/nouan/carta/torikumi/index.html#1). As of May 2019, 175 GM crops including parental GM and stacked events were approved for environmental safety in seven different crops – alfalfa, canola, corn, cotton, papaya, soybean and sugar beet. GM carnation and roses also have environmental safety approval. 130 crops are approved for both domestic cultivation and import for food, feed and processing. Corn has the highest numbers of approvals (88), followed by cotton (34), soybean (30) and canola (16).

The large number of food safety approvals is due to approval being granted for each combination stacked event. In contrast, environmental safety approvals for the highest combination stacked event cover the lower combination stacked events, as mentioned above. With feed safety approvals, the list contains only the parental GM event since the stacked events are not regulated for feed safety.

Local GM research efforts and consumer acceptance

Although Japan is a major importer of GM crops, as of 2019 there had been no commercial production of domestically developed GM crops, except for GM rose. The major reason for the private sector not developing GM crops is consumer concern about GM foods. Since 2004, FSC has been conducting consumer surveys on food safety matters with 'food safety monitors' who have registered with FSC to offer their opinions on food safety matters. [28] In the first survey, 26% and 48% of respondents stated that they were very or somewhat concerned about the safety of GM foods, respectively. GM foods were ranked as the 6th highest item of concern. The major reason for the stated concern was doubt about scientific evidence relating to safety.

The concern decreased from 2010, by which time GM food safety ranked below the top seven concerns. In 2019, only 12% and 28% respondents were very or somewhat concerned about safety of GM foods, respectively, and the total number of respondents who were not concerned about safety of GM foods (59%) exceeded those who were concerned (40%).

Another consumer survey conducted in 2017 by the Council of Biotechnology Information Japan showed that 64% of consumers had heard about GM food but did not have good understanding of it.[29] Only 15% stated they felt they knew enough about GM foods, and 60% stated that they had eaten GM food. The survey also showed that 15% and 47% of consumers had negative or somewhat negative perceptions of GM foods, respectively. The top three reasons for the negative perceptions were aversion to artificial things, lack of knowledge, and concern about the food and environmental safety. Major sources of information on which consumers based their negative opinions were voluntary non-GMO food labelling and TV programs. Consumers who had positive opinions of GM food had taken into account the need for sustainable agriculture, expectation of a new innovative technology, trust in the regulatory review process etc. Over 30% of consumers intended to purchase GM foods if they offered benefits such as increased vitamins and minerals, were good for weight loss, had low use of pesticides and could mitigate pollen allergy.

Although there have been not many research and development efforts in the private sector, several research projects are in progress regarding rice and tomato. The National Agriculture and Food Research Organisation is the leading organisation on GM rice research. Traits that have been tested are rice blast- and rice leaf blast-tolerance, high yielding, high tryptophan, and mitigation of Japanese cedar pollinosis. High tryptophan rice is for feed use to increase Japanese self-sufficiency with feeds. The mitigation of Japanese cedar pollinosis, a major pollen allergen, could be the first Japanese GM rice which has a direct consumer benefit (Takagi *et al.* 2005, 2006; Hirose *et al.* 2006). The research aims to mitigate the allergenic reaction via consumption of the GM rice which expresses the peptide of the allergen, thus functioning as an edible vaccine. Since the GM rice is considered as a medicine, tests for medical use are required before commercialisation of the product.

There is ongoing research of a miraculin protein in tomatoes, developed by University of Tsukuaba, which will be used as an alternative form of low-calorie sweetener (Hirai *et al.* 2010; Yano *et al.* 2010). FSC began safety assessment of the tomato in December 2017.

Several research and development programs are also underway on flowers and silkworms.

Conclusion

Japan has taken a cautious approach to the safety assessment of GM crops, to cope with consumer concern. The three most notable features of the Japanese regulatory system are the transparency of the assessment processes, the science-based assessments by expert review committees, and the progressive amendments of data requirements and review practices.

In terms of transparency, anyone can participate as an observer in the upper-level committee meetings which endorse the assessment conclusions made by the experts. Consumers can provide comments on the results of the safety assessment during the public comment period. Also, applicants are invited to an expert meeting and given the opportunity to explain the contents of their dossier and respond directly to experts' questions.

As described, safety assessments are science-based and product-based in accordance with the concept of substantial equivalence and familiarity. With accumulation of scientific knowledge and review experience, discussion takes place and data requirements and review practices can be streamlined over time, where appropriate. Good examples are simplification of the data requirements for stacked events, exemption from the safety review for certain stacked events, and exemption of local field testing for corn and cotton when they fulfil specific conditions. Simplification of the review practices allows the review committee and regulators to allocate their time proportional to risk and to prioritise new and more critical events of concern.

The Japanese experience of science-based and product-based assessment also applies to its regulatory considerations on genome editing techniques. In 2019, MHLW and MOE/MAFF announced the stance on products developed through genome editing techniques (MHLW 2019; MOE 2019; MAFF 2019). An SDN-1 product (a double-stranded break in the genome of a plant without addition of foreign DNA) is not subject to the GMO regulations when it is confirmed that the final product does not contain residual foreign DNA. Handling of an SDN-2 product (a small nucleotide template is supplied which is complementary to the area of the break) and an SDN-3 product (a template containing a gene or other sequence of genetic material) is determined a case-by-case basis depending on the source of the template. Since this is a new technology and innovations are rapidly occurring, developers are requested to voluntarily provide a notification document of products prior to commercialisation. The experience with GM crops shows that proactive and courteous communication efforts by the government and academia are essential for the value of this technology to be recognised and accepted by Japanese consumers.

Acknowledgements

This chapter was prepared by referring to the websites of the Food Safety Commission, Japan Biosafety Clearing House, Ministry of Agriculture, Fisheries and Forestry, Ministry of Environment, and Ministry of Health, Labour and Welfare (accessed on 1 November 2019).

Endnotes

1. https://www.naro.affrc.go.jp/archive/nias/gmo/cartagena/pubcomme_cartagena_20090828_1.pdf
2. https://www.mhlw.go.jp/english/topics/food/3-8.html
3. https://www.mhlw.go.jp/english/topics/food/3-2.html
4. https://www.biodic.go.jp/bch/english/e_index.html
5. https://www.fsc.go.jp/english/standardsforriskassessment/gm_kijun_english.pdf
6. http://www.fao.org/fileadmin/user_upload/gmfp/docs/CAC.GL_45_2003.pdf
7. http://www.fsc.go.jp/senmon/idensi/index.data/For_HP_Stance_on_the_Safety_Assessment_of_GM_Plants_for_Food_Use.pdf
8. https://www.mhlw.go.jp/stf/seisakunitsuite/bunya/kenkou_iryou/shokuhin/zanryu/index_00016.html
9. https://www.mhlw.go.jp/file/06-Seisakujouhou-11130500-Shokuhinanzenbu/1_11.pdf
10. http://www.fsc.go.jp/hyouka/index.data/GM_plants_through_cross-breeding.pdf
11. https://www.mhlw.go.jp/topics/idenshi/anzen/pdf/sankou6.pdf
12. https://www.mhlw.go.jp/file/06-Seisakujouhou-11130500-Shokuhinanzenbu/0000049695.pdf
13. https://www.mhlw.go.jp/content/11130500/000470369.pdf
14. http://www.fsc.go.jp/hyouka/index.data/For_HP_Approach_on_the_Safety_Assessment_of_cross-breeding.pdf
15. http://www.maff.go.jp/j/kokuji_tuti/kokuji/k0000543.html
16. https://www.fsc.go.jp/senmon/idensi/gm_siryoukijyun_english.pdf
17. https://www.biodic.go.jp/bch/english/law.html
18. https://www.biodic.go.jp/bch/english/law.html
19. http://www.maff.go.jp/j/syouan/nouan/carta/c_data/notice/pdf/01_tree_20141205.pdf
20. http://www.maff.go.jp/j/syouan/nouan/carta/tetuduki/attach/pdf/index-19.pdf
21. http://www.maff.go.jp/j/syouan/nouan/carta/c_about/attach/pdf/reg_2-28.pdf
22. http://www.affrc.maff.go.jp/docs/commitee/diversity/121120/pdf/121120_gijiroku.pdf
23. http://www.affrc.maff.go.jp/docs/commitee/diversity/140203/pdf/shiryou4.pdf

24. https://www.maff.go.jp/j/syouan/nouan/carta/torikumi/attach/pdf/index-63.pdf
25. http://www.maff.go.jp/j/press/syouan/nouan/171129.html
26. http://www.maff.go.jp/j/press/syouan/nouan/181220.html
27. http://www.maff.go.jp/j/press/syouan/nouan/181220.html
28. https://www.fsc.go.jp/monitor/monitor_report.html
29. https://cbijapan.com/wp-content/uploads/2018/04/2017%E5%B9%B4CBIJ%E6%B6%88%E8%B2%BB%E8%80%85%E8%AA%BF%E6%9F%BB_%E5%85%A8%E8%B3%AA%E5%95%8F_180223_final.pdf

References

Goto H, McPherson M, Comstock B, Stojšin D, Ohsawa R (2017) Likelihood assessment for gene flow of transgenes from imported genetically modified soybean (*Glycine max* (L.) Merr.) to wild soybean (*Glycine soja* Seib. et Zucc.) in Japan as a component of environmental risk assessment. *Breeding Science* **67**, 348–356. doi:10.1270/jsbbs.16134

Hirai T, Fukukawa G, Kakuta H, Fukuda N, Ezura H (2010) Production of recombinant miraculin using transgenic tomatoes in a closed cultivation system. *Journal of Agricultural and Food Chemistry* **58**, 6096–6101. doi:10.1021/jf100414v

Hirose S, Takagi H, Kawakatsu T, Wakasa Y, Domon E, Endo Y, Muraoka K, Hirai K, Watanabe T, Hattori M, Tateishi K, Takaiwa F (2006) Trials for safety assessments of the transgenic rice expressing T cell epitopes derived from Japanese cedar pollen allergens: environmental assessments for biodiversity and large scale cultivation under the isolated field. *Ikushugaku Kenkyu* **10**, 23–30.

MAFF (2019) Regulatory stance on genome editing. Webpage on the handling policy (Japanese only) <http://www.maff.go.jp/j/syouan/nouan/carta/tetuduki/nbt.html>. Webpage on the notification process (Japanese only) <https://www.maff.go.jp/j/syouan/nouan/carta/tetuduki/attach/pdf/nbt_tetuzuki-2.pdf>. MAFF notification document on the policy and notification procedure (Japanese only) <https://www.maff.go.jp/j/syouan/nouan/carta/tetuduki/attach/pdf/nbt-4.pdf>

MHLW (2019) Regulatory stance on genome editing. Webpage on the handling policy (Japanese only. English documents are posted) <https://www.mhlw.go.jp/stf/seisakunitsuite/bunya/kenkou_iryou/shokuhin/bio/genomed/index_00012.html>. MHLW decision document on handling procedure (English) <https://www.mhlw.go.jp/content/000550824.pdf>

Mizuguti A, Ohigashi K, Yoshimura Y, Kaga A, Kuroda Y, Matsuo K (2010) Hybridization between GM soybean (*Glycine max* (L.) Merr.) and wild soybean (*Glycine soja* Sieb. et Zucc.) under field conditions in Japan. *Environmental Biosafety Research* **9**, 13–23. doi:10.1051/ebr/2010004

MOE (2019) Regulatory stance on genome editing. Webpage on the handling policy (Japanese only. English documents are posted) <https://www.env.go.jp/press/106439.html>. Handling policy and notification procedure (English) <https://www.env.go.jp/press/2_2_%20genome%20editing_En.pdf>

Nakai S, Hoshikawa K, Shimonot A, Ohsawa R (2015) Transportability of confined field trial data from cultivation to import countries for environmental risk assessment of genetically modified crops. *Transgenic Research* **24**, 929–944. doi:10.1007/s11248-015-9892-6

OECD (2000) *Consensus Document on the Biology of Glycine max (L.) merr. (Soybean)*. Series on Harmonization of Regulatory Oversight in Biotechnology No. 15. ENV/JM/MONO (2000) 9. Organisation for Economic Co-operation and Development, Paris.

Takagi H, Saito S, Yang L, Nagasaka S, Nishizawa N, Takaiwa F (2005) Oral immunotherapy against a pollen allergy using a seed-based peptide vaccine. *Plant Biotechnology Journal* **3**, 521–533.

Takagi H, Hirose S, Yasuda H, Takaiwa F (2006) Biochemical safety evaluation of transgenic rice seeds expressing T cell epitopes of Japanese cedar pollen allergens. *Journal of Agricultural and Food Chemistry* **54**, 9901–9905. doi:10.1021/jf061848v

Yano M, Hirai T, Kato K, Hiwasa-Tanase K, Fukuda N, Ezura H (2010) Tomato is a suitable material for producing recombinant miraculin protein in genetically stable manner. *Plant Science* **178**, 469–473. doi:10.1016/j.plantsci.2010.02.016

Yoshimura Y, Matsuo K, Yasuda K (2006) Gene flow from GM glyphosate-tolerant to conventional soybeans under field conditions in Japan. *Environmental Biosafety Research* **5**, 169–173. doi:10.1051/ebr:2007003

Plate 1. Mohammad Abul Hossain, a Bt brinjal farmer in Bangladesh reaping the benefits of biotechnology. Photo credit: Md A Hossain.

Plate 2. Brinjal damaged by the eggplant fruit and shoot borer in Bangladesh. Photo credit: A Shelton.

1st season confined field trial -Pangasinan

Plate 3. First-season confined field trial with event-EE1-derived UPLB Bt eggplant open-pollinated (OP) lines in Bgry Paitan, Sta Maria, Pangasinan, Philippines, 13 June 2010. Photo: Courtesy of DM Hautea.

Plate 4. Farmers in India protesting the government's policies by planting unapproved genetically modified cottonseed. Source: Vivian Fernandes, 2019, www.smartindianagriculture.com

Plate 5. Bangladeshi farmer and family with harvest of Bt eggplant. Photo credit: Md A Hossain.

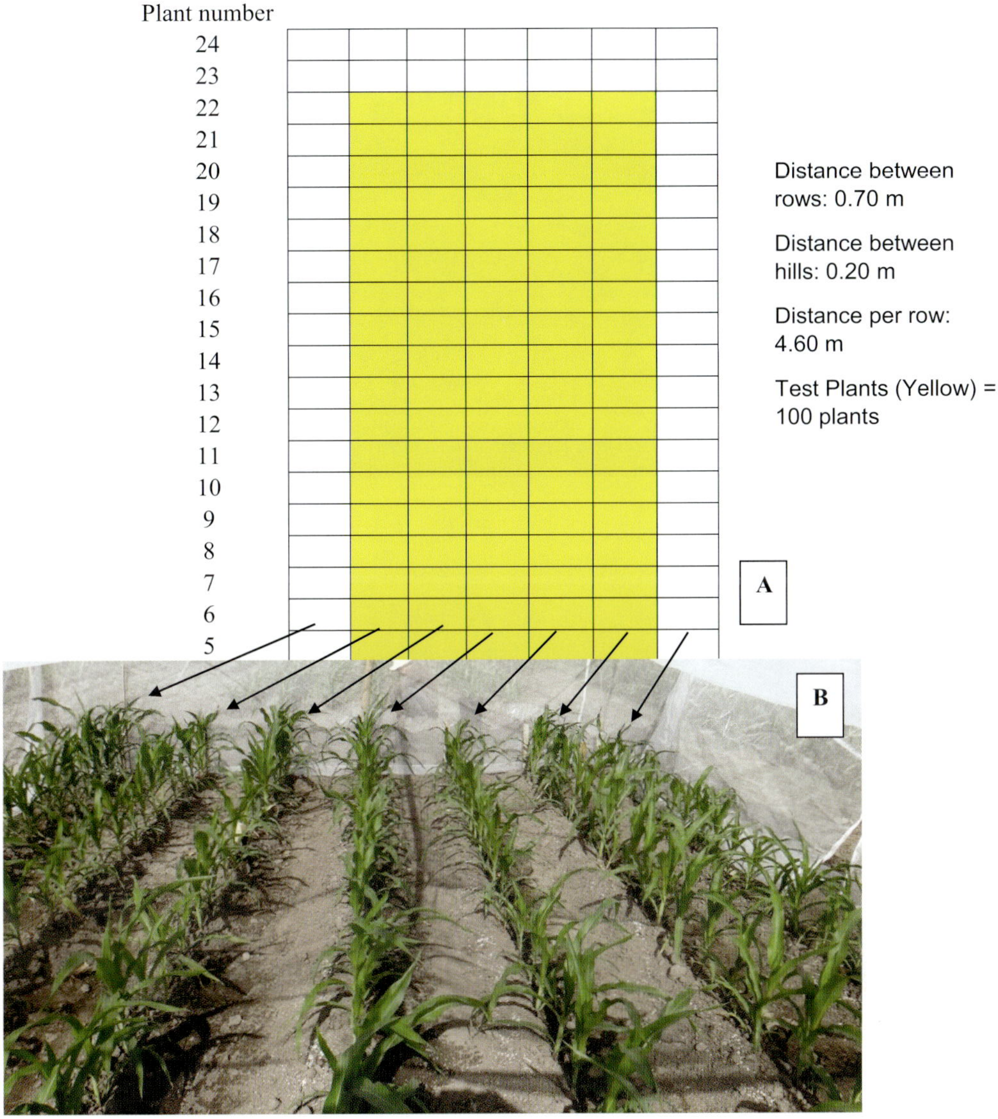

Plate 6. Field layout of a single plot. (A) Graphical illustration of the number of plants per row, number of rows per plot and distance between rows. (B) Actual layout in the field.

Plate 7. Installation of insect net. (A) Bamboo poles and strips used for installing the net cages. (B) Net cages installed in each plot.

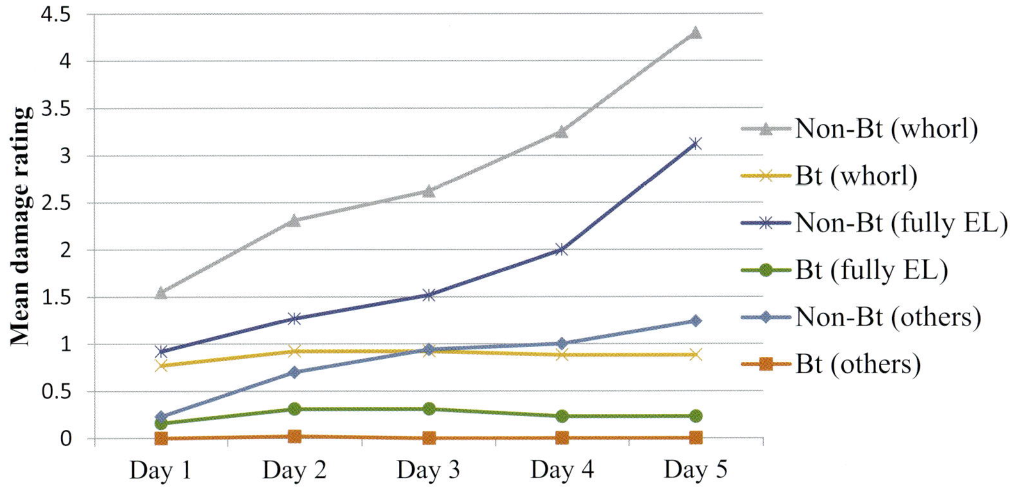

Plate 8. Mean damage rating due to leaf feeding by Asian corn borer (Laguna population) on several leaf parts of the whole vegetative plants of Bt corn (MON 89034) and non-Bt corn (NK603) for five days. EL = expanded leaf.

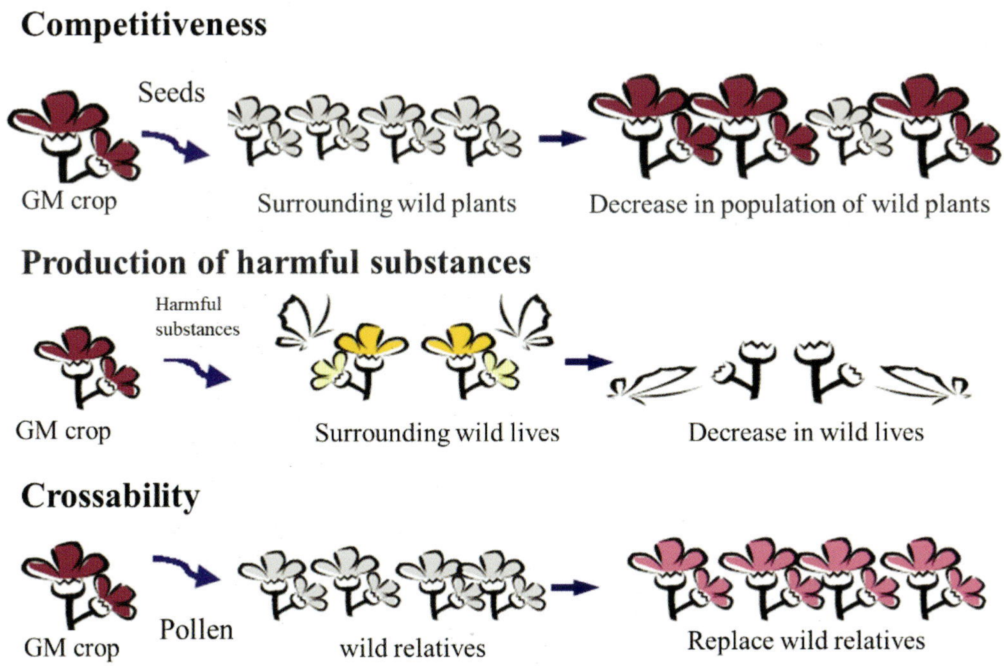

Plate 9. Assessment endpoints for an adverse biodiversity impact. Source: MOE, <www.biodic.go.jp/bch/download/lmo/hyoka_komoku.jpg> (translated by the authors).

Plate 10. *Phytophthora infestans* infection responses of potato varieties Atlantic (74), Granola (76) and Katahdin (75) at 61 days after planting and hybrid clones (66,62) Granola × Katahdin SP951 at 51 days after planting from unsprayed plot in CFT IVEGRI, Lembang (2011–12).

Plate 11. Phenotypic response of hybrid clones (above) of Granola GE Katahdin SP951 and non-GE Granola (below) to *P. infestans* in CFT Pangalengan (2013) at 49 days after planting in the absence of fungicide spraying.

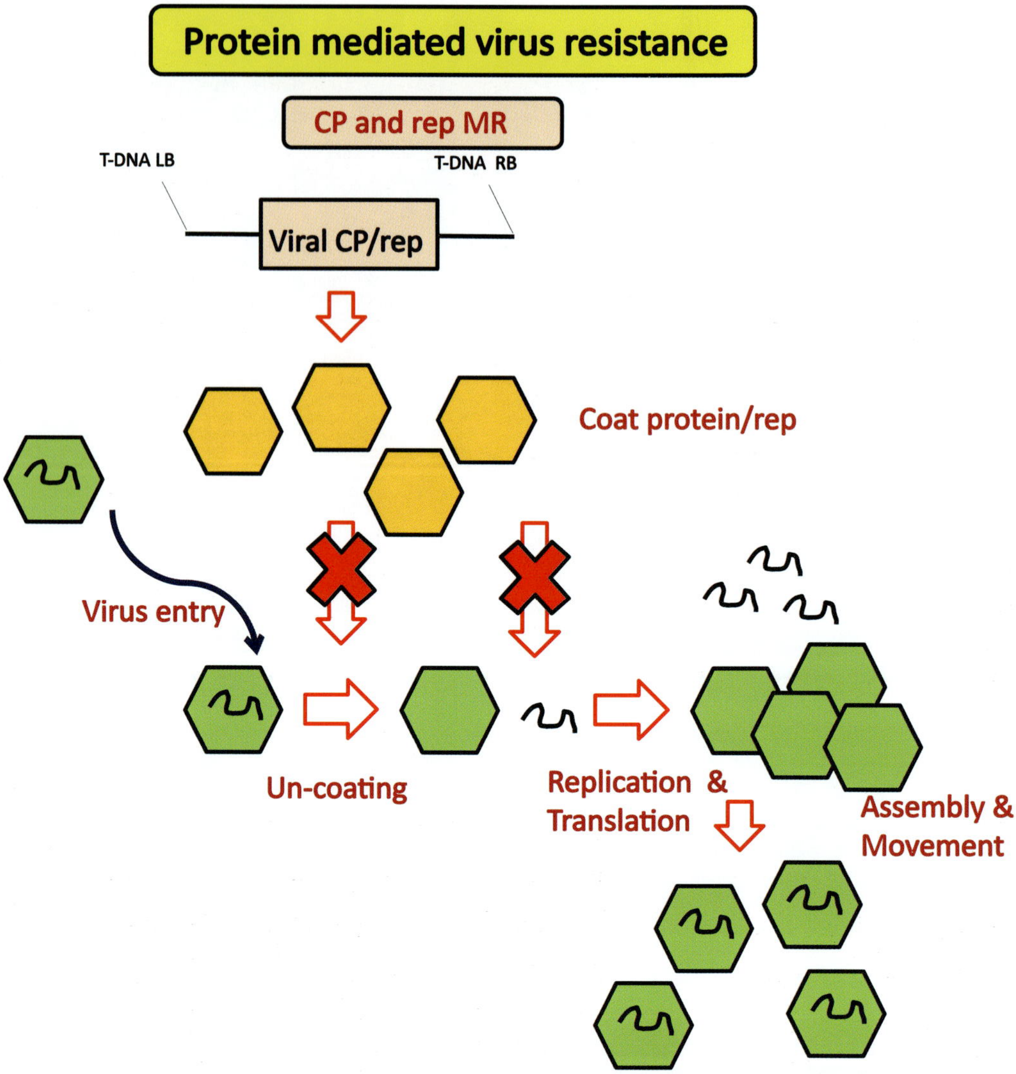

Plate 12. Protein-mediated virus resistance involves transgenic expression of virus-derived coat protein (CP) and/or replicase (rep) protein. Expressed CP and Rep interfere in the uncoating, replication and translation processes of virus particles during the natural infection process, thereby providing resistance against the invading cognate virus.

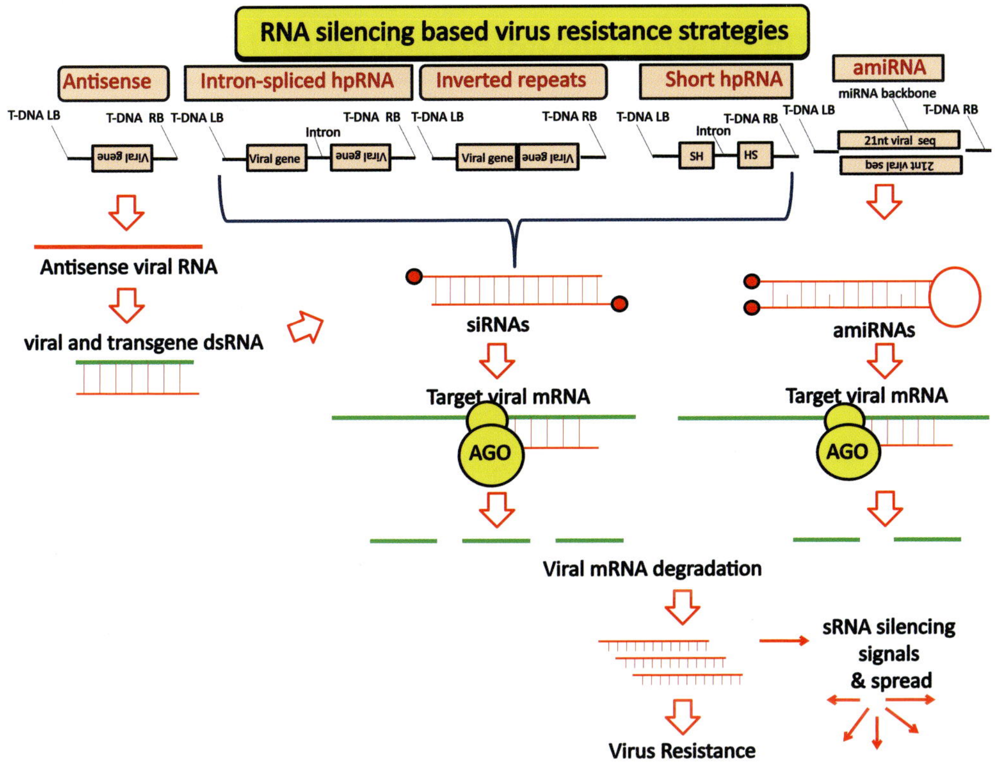

Plate 13. RNA silencing-based virus resistance strategies involve production of small non-coding RNAs (sncRNAs) like small interfering RNAs (siRNAs) or microRNAs (miRNAs) targeted against viral transcript. Various gene construct design strategies like antisense RNA, intron-spliced hairpin RNA (ihpRNA), inverted repeats (IR) and artificial miRNA (amiRNA) are depicted. The silencing of viral transcript occurs due to a sequence homology dependent silencing process called RNA interference (RNAi). The effector molecules of silencing (siRNA and amiRNA), their amplification and spread of silencing signals are also presented.

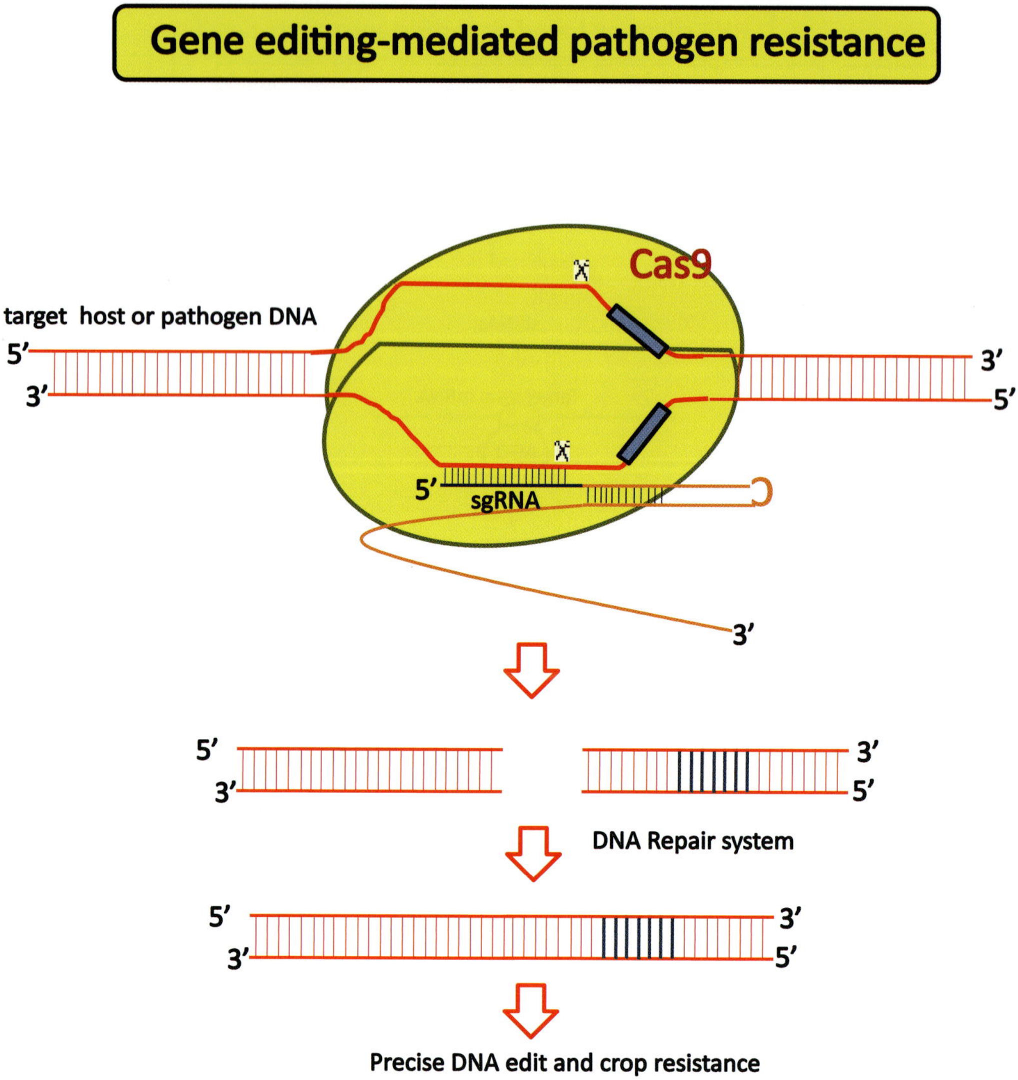

Plate 14. Gene editing technology utilises Cas9 protein and small guide RNA (sgRNA) to effect desired changes in the nucleotide sequences. The DNA repair system ensures that precise DNA edit is accomplished in either host (genes encoding susceptibility factors, transcription or translation factors) or pathogen DNA (avirulence genes) that is ultimately manifested as crop resistance phenotype.

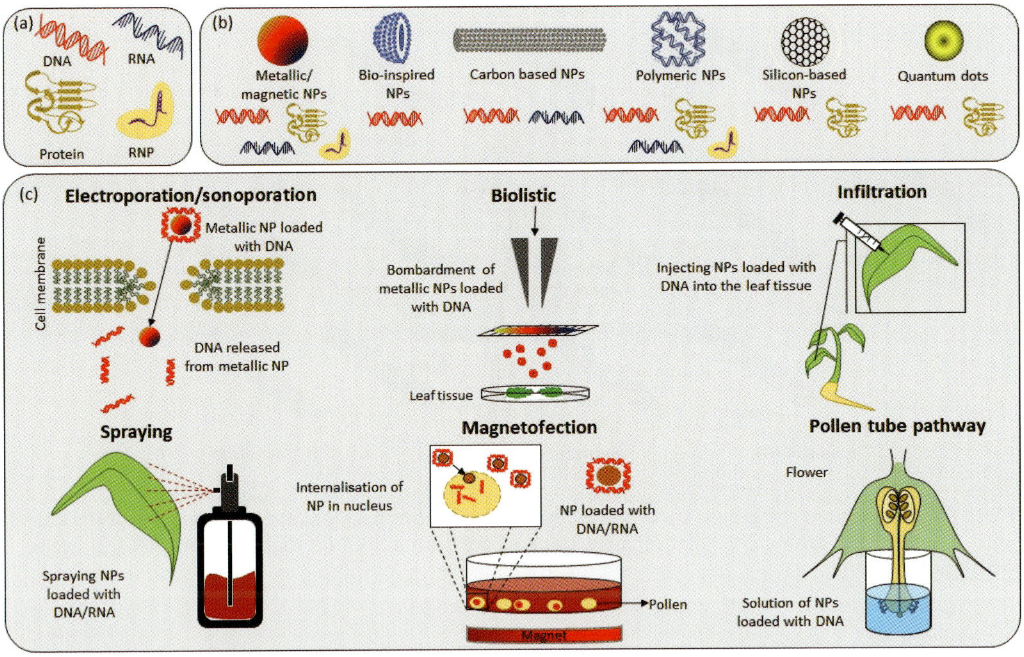

Plate 15. Nanoparticle-mediated biomolecular cargo delivery into plants for genetic engineering. (A) Different types of cargo used for genetic engineering in plants. (B) Nanoparticles used for the delivery of biomolecules (DNA, RNA proteins and RNPs-ribonucleoproteins) in plant systems cover five categories: metallic/magnetic, bio-inspired, carbon-based, polymeric, silicon-based and quantum dots. (C) Modes of cargo delivery for nanoparticle-mediated genetic transformation: electroporation/sonoporation, biolistics, infiltration, spraying, magnetofection and pollen tube pathway.

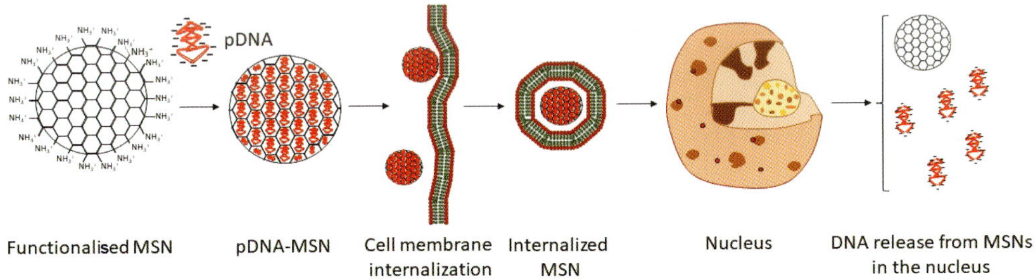

Plate 16. Mesoporous silica nanoparticles-based delivery of plasmid DNA (pDNA).

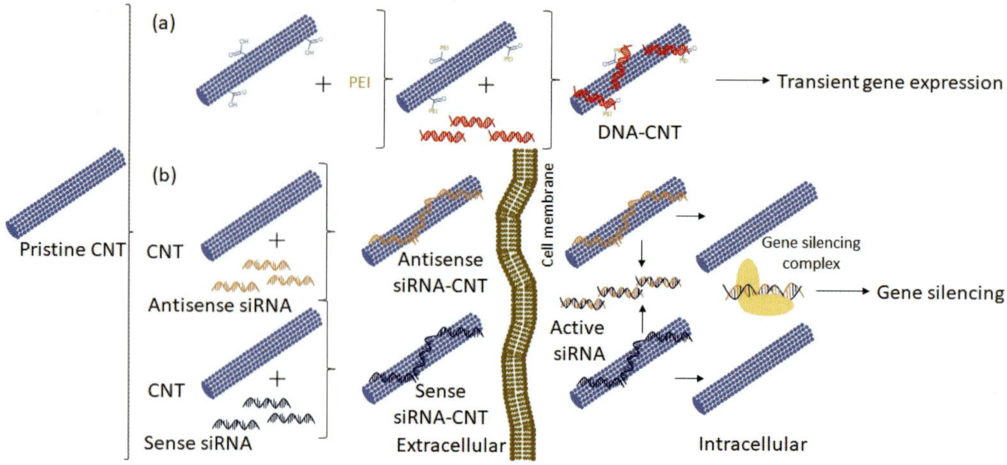

Plate 17. Genetic engineering by pristine carbon nanotubes (CNTs). (a) Carboxylated CNT coated with polyethylene amine (PEI) for transient gene expression. (b) CNTs loaded with siRNA for gene silencing.

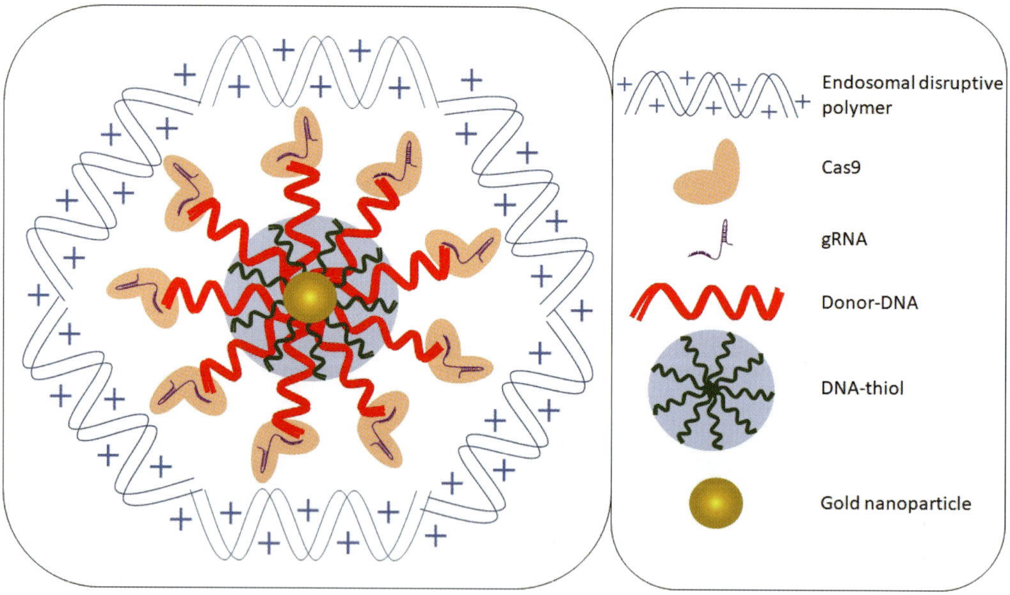

Plate 18. Endosomal disruptive complex as a potential strategy for delivery of CRISPR/Cas in plants.

16

Development of yellow stem borer resistant rice varieties in Indonesia

Satya Nugroho, Inez H Slamet-Loedin and Amy Estiati

Introduction

Rice is the primary staple food in Indonesia, which is the fourth most populous country in the world. In concurrence with the global trend, with an increase in per capita income the average annual consumption per capita of rice decreased in Indonesia, to 114.6 kg in 2017. However, with the population increasing at the rate of 1.36% per year[1] the need for more rice is obvious. In 2017 Indonesia produced 81.38 million tons of rice and had to import ~305 275 tons of rice.[2] In the future, further imports may be needed if the current production technologies cannot fulfil demand against the backdrop of climate change and other agricultural challenges.

There are many abiotic and biotic stresses that impact rice productivity in Indonesia. Among these, the most important are rice yellow stem borer (*Scirpophaga incertulas* Walker) and brown planthopper (*Nilaparvata lugens* Stål). On average, the rice stem borer reduces rice production by 5–10%. However, in a severe case, the loss can be as high as 95% (Wunn *et al.* 1996). In recent years, rice stem borer and brown planthopper have been two of the most damaging insect pests in Indonesia. The Indonesian Ministry of Agriculture reported that the rice stem borer affected 102 430 ha and 126 086 ha in 2015 and 2016, respectively, as compared to 28 653 ha and 44 415 ha affected by the brown planthopper in the same period (CADIS 2017). In 2017, rice stem borer-affected areas declined to 81 174 ha, less than that affected by the brown planthopper (85 989 ha). Although the percentage of rice growing area affected by rice stem borer was low, comprising ~0.5% of the total rice growing area (15.7 million ha) in 2017 (www.bps.go.id), the impact was significant since the area affected was mostly in Central and West Java (40%), which are the main rice producing areas.

In Indonesia, there are different species of rice stem borer belonging to two different taxonomic families: rice yellow stem borer, rice stripe stem borer (*Chilo suppressalis* Walker), rice white stem borer (*Scirpophaga innotata* Walker) and black-headed rice stem borer (*Chilo polychrysus* Meyrick) belonging to Pyralidae; and pink rice stem borer (*Sesamia inferens* Walker) in the Noctuidae family (Kalshoven 1981). The two most damaging species are *S. incertulas* and *S. innotata*. *S. incertulas* is the most damaging in paddy fields as it can infest rice plants in all plant growth stages (Baehaki 2013). Infestation during the vegetative stages causes drying of the central shoot called 'dead heart', while during the generative or reproductive stage it may cause 'white head' wherein the emerging panicles are whitish and empty of developing grains. When infested during the vegetative stage, however, rice can compensate loss by producing new tillers (up to 30% recovery). Therefore, the damage is less severe than infestation during the generative stages (Wunn *et al.* 1996)

Since there is no known naturally occurring resistant relative, development of a stem borer resistant rice cultivar through conventional breeding is not possible (Bennett *et al.* 1997). The most common approach to control stem borer is the application of pesticides, which is expensive, unsustainable and detrimental to animal health and the environment. The only sustainable approach appears to be development of transgenic rice through genetic engineering (GE), also called genetic modification (GM), to express lepidopteran-specific insecticidal *cry* genes isolated from *Bacillus thuringiensis* (Bt) to protect the crop during the whole season. This approach has been successful in developing insect-resistant Bt transgenic crops like Bt corn, Bt soybean, Bt cotton and Bt eggplant (www.isaaa.org; Shelton *et al.* 2018; Perlak *et al.* 1990; Cohen *et al.* 2008).

Many studies on rice transformations with Bt genes have been reported. In China alone, dozens of Bt rice events have been developed under the government-supported research project, the National GMO New Variety Breeding Program (Liu *et al.* 2016). The events include overexpression of single *cry* genes, fusion of *cry* genes and stacking of *cry* genes with other transgenes, such as trypsin inhibitor, herbicide-tolerant and disease-resistant genes. Examples were synthetic *cry2Aa* gene transformed indica rice effective for lepidopteran pests (Chen *et al.* 2005), synthetic *cry1Ca* for indica rice (Tang *et al.* 2006), synthetic *cry1Ab* and *cry1Ac* genes for striped stem borer and yellow stem borer control (Cheng *et al.* 1998) and synthetic *cry1Ab* for eight lepidopteran pest species (Shu *et al.* 2000) for the single gene overexpression. There was also overexpression of *cry1Ab-1Ac* fusion genes (Wang *et al.* 2016) and stacked events *cry1Ac* and *CpTI* (cowpea tripsin inhibitor) (Zhao *et al.* 2004). Liu *et al.* (2016) have comprehensively reported rice transgenics. Two Bt rice events – Huahui 1 and the hybrid ride Bt Shanyou 63 (Tu *et al.* 2000), both co-expressing *cry1Ab* and *cry1Ac* genes with toxin levels in stems and leaves in the middle ranges in comparison to those of the other Bt rice lines – were approved with biosafety certificates in China. The efficacy of these lines was high, in the range of 91.4–99.6% against yellow stem borer in the field. However, they were never commercially released (Liu *et al.* 2016). The International Rice Research Institute developed transgenic IR72 rice events overexpressing hybrid *cry1Ab* and *cry1Ac* (Tu *et al.* 1998). In Vietnam a fusion gene *cry1Ab-cry1B* was used for transformation of elite Vietnamese rice to control yellow stem borer (Ho *et al.* 2006). Bt rice (a local self-pollinating aromatic variety called Tarom Molaii) expressing *cry1Ab* gene was commercially grown in Iran over several thousand hectares in 2004 after ~10 years of thorough studies on various aspects including animal feed safety, composition analysis, field trial for yield and environmental impact, and other risk assessments.[3]

In Indonesia, no Bt rice event has been submitted for biosafety certification. However, several activities to develop Bt rice and to gather data for biosafety clearance have been performed.

Development of stem borer resistant rice (Bt rice) in Indonesia

Several attempts have been made to develop yellow stem borer resistant GM rice by transforming it with *cry* genes in Indonesia. The first attempt was initiated in 1995–96 at the Research Center for Biotechnology-Indonesian Institute of Sciences (LIPI) (Slamet-Loedin *et al.* 2003). The target rice cultivars were indica rice cultivars, IR64 and Cisadane; and javanica (tropical japonica) rice cultivar, Rojolele KA rice. Two resistant genes were used for transgenesis. The first was *cry1Ab* fused to the maize *ubiquitin* promoter to target the rice yellow stem borer, and the second was *GNA* (*Galanthus nivalis* agglutinin or snowdrop lectin) fused to the maize *ubiquitin* promoter (Rao *et al.* 1998) to target rice brown planthopper. The selectable marker gene was *hygromycin phophotransferase* (*hpt*) obtained from plasmid pCAMBIA1200 (Roberts *et al.* 1997). In this activity, the co-transformations were performed by particle bombardment to obtain resistant transgenic cv Rojolele rice plants (Slamet-Loedin *et al.* 2003; Satoto *et al.* 2008). Transgenic rice cv Rojolele harbouring *cry1Ab* only, *GNA* only and the co-integration of *cry1Ab* and *GNA* was obtained (Satoto 2003). Integration analyses until T3 generations of one *cry1Ab* and two *GNA* lines showed that the *cryIAb* and *GNA* were stably integrated. The *cry1Ab* and *GNA* of the two lines harbouring both genes showed segregation patterns indicating integration in different loci. Later, stacked events of Rojolele harbouring both *cry1A*b and *GNA* were developed by crossing the *cry1Ab* and *GNA* lines. Efficacy studies of transgenic events harbouring *cry1Ab* against rice yellow stem borer in the glasshouse showed ~30–60% increased resistance. GNA was found to be ineffective against Biotype 2 BPH (Baehaki *et al.* 1991), which was indicated by non-significant resistant improvement of the transgenic rice events harbouring *GNA* compared to the non-transgenic control lines (Satoto 2003). Those transgenic lines are pending for further development.

The method of *Agrobacterium*-mediated transformation in rice standardised by Hiei *et al.* (1994) was routinely used to develop Indonesian Bt rice. The introduction of selectable marker-free CaMV 35S promoter-driven *cry1Ab* gene into indica rice cultivar Ciherang by *Agrobacterium*-mediated transformation was intended to integrate clean *cry1Ab* gene free of *hpt* selectable marker (Sulistyowati *et al.* 2011). At the same time, the *cry1B-cry1Aa* fusion gene controlled by maize *ubiquitin* promoter into rice variety Rojolele (Rahmawati and Slamet-Loedin 2006) and *cry1B* driven by the wound-inducible promoter *mpi* (maize proteinase inhibitor) (Breitler *et al.* 2000, 2004) into rice varieties Rojolele and Ciherang, were carried out successfully by *Agrobacterium* mediated transformation (Estiati *et al.* 2007). The *cry1B-cry1Aa* fusion gene was an optimised version for rice (Bohorova *et al.* 2001; Anderson *et al.* 2005; Ho *et al.* 2006). The application of the *cry* fusion gene and the inducible promoter is an attempt to reduce the risks of plants' resistant breakdown. At LIPI, six transgenic Rojolele events overexpressing the fusion gene *cry1B-cry1Aa* and four events overexpressing *cry1B* under *mpi* inducible promoter are being selected to be pursued for release. These six transgenic Rojolele rice events overproducing the fusion toxin protein Cry1B-Cry1Aa have been registered at the Ministry of Agriculture under the names Biotama 1, 2, 3 to Biotama 6. Four transgenic rice cv Rojolele overproducing the toxin protein Cry1B under *mpi* inducible promoter are being registered under the name of Biotama 7, Biotama 8, Biotama 9 and Biotama 10. The development of Bt rice at LIPI is summarised in Table 16.1. In addition to activities at LIPI, preliminary work to introduce *cry1Ac* and *cry1Ab* was initiated by ICCABIOGRAD, the Indonesian Ministry of Agriculture (Somantri *et al.* 2008).

Contained and confined efficacy testing of Bt rice

Efficacy studies of Bt rice lines have been performed in both contained facilities and confined field trials (Table 16.1). *In planta* experiments on transgenic javanica rice lines cv Rojolele overexpressing different types of *Bt* genes using different constructs were performed following the previously published protocol (Heinrich *et al.* 1985). They showed improved resistance of 90–100% (most often 100%) to the stem borer artificial infestations in the contained facilities (Usyati *et al.* 2009; Estiati *et al.* 2013). Consistent results were obtained from repeated experiments across several generations. The results of *in vitro* efficacy studies on stem samples of the transgenic rice supported the *in planta* experiments. Subsequent experiments on the advance generation of T9 proved that the traits were stably inserted and expressed.

Table 16.1. Timelines of the development of Bt rice at LIPI, Indonesia

	Transformation events		
Year	Cry1Ab-GNA	Mpi::Cry1B	Cry1B-Cry1Aa
1996	Rice transformation		
2002			Rice transformation
2004	Environment safety test (1 location)		
2005–06	Environment safety test (1 location)	Rice transformation	
2012–14		Environment safety test (3 locations)	Environment safety test (4 locations)
2014–18		Food safety tests (acute and subchronic test, simulated gastric fluid and simulated intestinal fluid tests, allergenicity)	Food safety tests (acute and subchronic test, simulated gastric fluid and simulated intestinal fluid tests, allergenicity)
2018			Feed safety test (feeding studies to chicken and lamb)

In addition, field testing under confined field trials (CFT) have been conducted on two traits, the *cry1B-cry1Aa* fusion and *mpi:cry1B* lines, for efficacy against rice yellow stem borer, in collaboration with the Center for Rice Research in Sukamandi, West Java. The results indicated 0–1.5% infestations on Bt rice compared to 20.51% in non-Bt control, which was a significant improvement in the resistance level to yellow rice stem borer. The results were comparable to the WT control lines treated with pesticides (1%).

Biosafety regulation and policies on the release of GM products

As one of the mega-biodiverse countries, Indonesia ratified the Convention on Biological Diversity by issuing Legislation no. 5/1994. The Convention regulates the safety of the application of modern biotechnology by developing a protocol for the regulation of cross-boundary movement, handling and utilisation of GM products. The Convention was the driving force for the development of the Cartagena Protocol on Biosafety to the Convention of Biological Diversity in 2000, which was ratified by Indonesia through Legislation no. 21/2004.

In 1996, in anticipation of the importation of GM products into Indonesia, the government issued Legislation no. 7/1996 on food. Section 3 Article 13 specifically mentioned that anyone who produces food or uses raw materials, food additives and auxiliary materials in the food production process that result from GE, is required to assess their safety to human health prior to distribution. The government established requirements and principles/guidelines for research, development and application of GE methods in the production process, and established requirements for GE food examination by issuing regulations on food and feed safety.

In 1999, a Joint Decree by the Minister of Agriculture, Minister of Forestry and Plantations, Minister of Health and State Minister of Food and Horticulture (No. 998.1/Kpts/OT.210/9/99; 790.a/KptsIX/1999; 1145A/MENKES/SKB/IX/99; 015A/ NmenegPHOR/09/1999) on biosafety and food safety of GE agricultural product was enacted to replace the Ministry of Agriculture Decree no. 856/Kpts/HK.330/9/1997 on the biosafety of agricultural biotechnology products.

The Joint Decree established a Biosafety and Food Safety Committee. The Committee then established a Technical Team for Biosafety and Food Safety, which consisted of Food, Plant, Animal, Fish and Microbial groups (Estiati and Herman 2016).

In 2004, the government enacted Regulation no. 28/2004 on safety, quality and food nutrition. It states in Section 3 Article 14 that, similar to Legislation no. 7/1996, anyone who produces food or uses raw materials, food additives and auxiliary materials in the food production process that result from GE, is required to assess their safety to human health before distribution. In the same year, the government ratified the Cartagena Protocol by issuing Legislation no. 21/2004.

In 2005, the 1999 Joint Decree on the Biosafety and Food Safety of Agricultural Products Derived from GE was updated to Regulation no. 21/2005. The regulation defined the biosafety of GM products as comprising environmental, food and/or feed safety. Environmental safety is conditions and efforts required to prevent any risk to biodiversity from the application of GE product. Food safety is conditions and efforts required to prevent any risk that may harm or endanger human health due to production, preparation, storage, distribution and utilisation of GM food. Feed safety is condition and efforts required to prevent any impact that may harm or endanger animal and fish health due to production, preparation, storage, distribution and utilisation of GM feed. In response to Regulation no. 21/2005, in 2012 the National Agency of Drug and Food Control issued Regulation no. HK.03.1.23.03.12.1563/2012 to guide the assessment of food safety. In the same year, the Minister of Environment released Regulation no. 25/2012, Guidance on Environmental Risk Analysis and Safety Assessment of GM products. In 2016, the Minister of Agriculture issued Regulation no. 36/2016 on feed safety assessment of GM products. The Ministry of Agriculture is developing another regulation, on post-release monitoring. Biosafety recommendations can be obtained from the respective ministry or agency. The Biosafety Committee, assisted by three technical teams, provides recommendations to the ministry and agency responsible for a decision on biosafety.

Biosafety aspect of Bt rice for environmental release

The traits that are being sought for release are the lines overexpressing the fusion protein Cry1B-Cry1Aa and the Cry1B driven by the inducible promoter *mpi* gene. To fulfil the biosafety requirements of Bt rice resistant to yellow stem borer, data on food, feed and environment safety need to be provided for assessment by the regulatory authorities. However, not many reports on these aspects are available in the public domain. Many are unpublished official work reports, currently restricted to internal consumption.

The most common questions among three biosafety aspects relate to the genetic information. The description of genetic modification should involve details of the transformation method, detailed description of the gene of interest and other details like origin, expression, regulation, transferability of vector to different hosts and the presence of plasmids backbone in GM crops. Questions about the character of the genetic modification relate to expression and its regulation in the GM crops, including the new phenotypes, and the genetic stability. Other information required is specific to each biosafety aspect. Information specific to environmental safety includes impact to non-target organisms and biodiversity, weediness potential, possibility of gene flow, strategy for environment risk communication, and strategy for management and monitoring for environmental safety and agronomic performance. For food safety, the required information includes substantial equivalence, nutritional value, allergenicity, toxicity and other considerations that may impact on food safety. For feed safety, the specific information required is very similar to that for food safety. In the case of GM rice, feed safety is focused on the secondary products of rice plants that are commonly used as feed components, such as rice straw which is commonly used as a source of fibre and rice brands as a source of proteins. Data to support those three biosafety aspects have been obtained for the *cy1B-cry1Aa* rice lines.

Environmental safety

Environmental safety assessments are required prior to the release of GM plants. Two guidelines are in place to be followed by developers of GM plants: the Minister of Environment Regulation no. 25/2012, Guidelines for the Compilation of Environment Risk Assessment of GM plants; and the Minister of Environment and Forestry Regulation no. P.69/MENLHK/ SETJEN/KUM. 1/8/2016, Guidelines for Environmental Safety Testing of GM Products in Biosafety Containment. The guidelines state that, to obtain an environmental safety certificate, field tests have to be conducted in four confined field trials in four locations with different agroecology. The first reported environmental safety trials of Bt rice were the Rojolele overexpressing both *cry1Ab* and *GNA* at the Indonesian Center for Rice Research field trial in Karawang in West Java in 2004 (Mulyaningsih *et al.* 2009) and Sukamandi-Subang in West Java in 2005. Bt rice overexpressing the *cry1B-cry1Aa* fusion genes and the *mpi::cry1B* gene have undergone environmental safety tests in four and three locations, respectively. The field tests were located in rice paddy fields in Sukamandi-Subang, Muara-Bogor and Kuningan in West Java, Indonesian Center for Rice Research, and in Banten under the Assessment Institute for Agricultural Technology.

The tests were designed to observe the diversity and abundance of organisms that may be affected by the presence of the Bt toxin directly or indirectly, ranging from insects and microbes to macrovertebrates, which were usually found in rice paddy fields especially during the early stages of plant development when irrigation is available. The insect fauna being assessed in the field included rice yellow stem borer and other pests such as brown planthopper, whitebacked planthopper, rice black bug (*Scotinophara coarctata*), and non-pest insects and arthropods like spiders, Paederus, Coccinella, Ophionea, Cyrtorhinus and Collembola. Laboratory studies on the impact of Bt rice on predator *Verania lineata* were also reported (Usyati *et al.* 2015). Other non-target organisms include soil microbes, such as those of nitrogen-fixing bacteria (*Azotobacter, Beijerinckia, Chromatium, Rhodopseudomonas, Rhodospirillum, Rhodomicrobium, Chlorobium, Chloropsedomonas, Desulfovibrio, Desulfotomaculum, Klebsiela, Bacillus, Clostridium, Azospirillum, Pseudomonas, Vibrio, Thiobacillus* and *Methanobacillus*), phosphate-solubilising bacteria (*Pseudomonas, Bacillus, Mycobacterium, Micrococcus, Flavobacterium, Citrobacterium* and *Enterobacter*), cellulolytic bacteria (*Achromobacter, Angiococcus, Bacillus, Cellulomonas, Cytophaga, Clostridium, Cellivibrio, Flavobacterium, Pseudomonas, Poliangium, Sorangium, Sporocytophaga, Vibrio* and *Cellfalcicula*), mould and fungi. The results indicate that Bt rice had no significant effects on the biodiversity and abundance of the non-target organisms observed.

Food safety

Several food safety tests have been performed on both Bt rice overexpressing the *cry1B-cry1Aa* fusion genes and *mpi::cry1B* following the guidelines of GM food safety assessment by Badan POM. The test includes analysis of the nutritional composition of Bt rice compared to its wild type, acute and subchronic toxicity test for the transgenic rice expressing *cry1B-cry1Aa* fusion genes, and simulated gastric fluid and simulated intestinal fluid tests to assess the stability of expressed proteins (Cry1B and Cry1Aa).

Nutritional composition

The nutritional composition of the rice grains grown in CSC-LIPI in Cibinong-Bogor was analysed. The analyses, performed by a certified laboratory (Saraswanti Indo Genetech), showed that the crude protein, crude fat, crude ash, carbohydrates, starch, sucrose, amino acid composition (19 amino acids) and vitamin compositions of Bt rice expressing the *cry1B-cry1Aa* fusion genes did not differ from those of its non-transgenic WT.

Toxicity and allergenicity tests

In silico studies to assess the possibility of Cry proteins acting as allergens performed using AllergenOnline version 19 (allergenonline.org) belonging to the University of Nebraska-Lincoln (Goodman *et al.* 2016), showed that neither Cry1B nor Cry1Aa has any allergenic fragments. Toxicity studies were performed on mice via acute and subchronic toxicity tests in collaboration with the Faculty of Veterinary, Bogor Agricultural University. Observations during the acute toxicity test on male and female DDY mice aged ~2–2.5 months under five different doses of transgenic grains/bodyweight indicated that there were no changes in behaviour, appetite, skin and hair, breathing, heartbeat, vision, faeces, muscle and joint, reflex, tremor, convulsion, urine, salivation and paralysis. Observations on the pathological anatomy did not show any adverse changes in liver, spleen, heart, lung or kidney. Further observations on total bodyweight, mortality, haematological response and liver and kidney functions after subchronic toxicity tests on male and female Spraque Dawley mice aged six to seven weeks under metabolic cage with controlled environment, and fed with food composition containing transgenic rice grains as source of carbohydrate for 90 days, also resulted in no observed adverse effects of the fusion Bt protein on mice. Studies indicated that various organs like liver, spleen, heart, lung and kidney functioned normally in animals fed a Bt rice diet similar to those fed the WT non-Bt rice diet.

Simulated gastric and intestinal fluid tests

These experiments aimed to understand the stability of Cry1B and Cry1Aa proteins to digestion under simulated gastric fluid and intestinal fluid conditions. The experiments were done separately for Cry1B and Cry1Aa. The proteins were overexpressed in *Escherichia coli* and purified using the standard procedures. The tests were performed following the previously published protocols (Farias *et al.* 2015). The results showed that both proteins were degraded quickly, indicating instability under gastric or intestinal conditions.

Feed safety

Feeding studies were performed according to the guidelines of the Ministry of Agriculture. The aim of the studies was to understand the effects of the transgenic feed on cattle and poultry, which in Indonesia are fed the by-products of rice like rice bran and rice straw as a source of fibre and proteins. Studies were performed in collaboration with the Faculty of Veterinary, Bogor Agriculture University on two animal models: broiler chickens, representing monogastric animals that feed on rice bran as the source of amino acids; and lambs, representing ruminants that consume rice straw as the main source of fibre. The feed compositions for the feeding studies were designed to meet Indonesian National Standards (SNI 01–3931–2006) for both broiler chicken and lamb.

The broiler chicken feeding study was conducted in a pathogen-free environment. Day-old chicks (DOC) broiler chicken strain Cobb were fed with the feed as per specifications for 30 days. The lamb feeding studies were also performed for 30 days using one- to two-year-old male and female lambs. The parameters for the feeding studies using both the DOCs and lambs were total bodyweight, mortality, feed conversion ratio, haematological response, liver and kidney functions and feed digestibility. The results of both studies indicated that no significant changes were observed in all parameters tested. They showed that transgenic rice by-products are safe to be used as sources of proteins and fibre for animal feed.

Conclusion

There have been several attempts to develop stem borer resistance by expressing *cry* genes in rice. Some transgenic rice events have undergone tests for food, feed and environmental safety as per Indonesian government regulations, pending submission of regulatory data dossier for assessment by the competent authorities and follow-up of the procedure for release, which is currently being established by the Ministry of Agriculture.

Acknowledgements

Dr Paul Christou (John Innes Center, UK) provided the *cry1Ab* fused to the maize *ubiquitin* promoter *GNA* (*Galanthus nivalis* agglutinin or snowdrop lectin) gene constructs. Dr Immanuel Guiderdoni and Dr Roger Frutos (CIRAD, France) provided the wound-inducible promoter *mpi::cry1B* and *cry1B::cry1Aa* fusion genes construct.

Endnotes

1. https://www.bps.go.id/statictable/2014/09/08/1043/impor-beras-menurut-negara-asal-utama-2000-2017.html
2. https://www.bps.go.id/statictable/ 2014/ 09/08/1043/impor-beras-menurut-negara-asal-utama-2000-2017.html
3. http://www.sciencedev.net/Docs/Iran%20Releases%20World.htm

References

Anderson P, Sharma SR, Sresty TVS, Baisaria S, Kumar PS, Parthasaradhi P, Frutos R, Altosaar I, Ananda Kumar P (2005) Transgenic cabbage (*Brassica oleracea* var. *capitata*) resistant to diamondback moth (*Plutella xylostella*). *Indian Journal of Biotechnology* **4**, 72–77.

Baehaki SE (2013) Hama penggerek batang padi dan teknologi pengendalian. *Iptek Tanaman Pangan* **8**, 1–14.

Baehaki SE, Rifki A, Yahya AS (1991) Penentuan biotipe wereng coklat di daerah sentra produksi padi. *Media Penelitian Sukamandi* **9**, 26–30.

Bennett J, Cohen MB, Katiyar SK, Ghareyazie B, Khush GS (1997) Enhancing insect resistance in rice through biotechnology. In *Advances in Insect Control: The Role of Transgenic Plants.* (Eds N Carrozi and M Koziel) pp. 75–93. Taylor and Francis, London.

Bohorova N, Frutos R, Royer M, Estanol P, Pacheco P, Rascon Q, McLean S, Hoisington D (2001) Novel synthetic *Bacillus thuringiensis cry1B* gene and the *cry1B-cry1Ab* translational fusion confer resistance to southwestern corn borer, sugarcane borer and fall armyworm in transgenic tropical maize. *Theoretical and Applied Genetics* **103**, 817–826. doi:10.1007/s001220100686

Breitler JC, Marfä V, Royer M, Meynard D, Vassal JM, Vercambre B, Frutos R, Messeguer J, Gabarra R, Guiderdoni E (2000) Expression of a *Bacillus thuringiensis cry1B* synthetic gene protects Mediterranean rice against the striped stem borer. *Plant Cell Reports* **19**, 1195–1202. doi:10.1007/s002990000247

Breitler JC, Vassal JM, Catala MC, Meynard D, Marfa V, Melé E, Royer M, Murillo I, San Segundo B, Guiderdoni E, Messeguer J (2004) *Bt* rice harbouring *cry* genes controlled by a constitutive or wound-inducible promoter: protection and transgene expression under Mediterranean field conditions. *Plant Biotechnology Journal* **2**, 417–430. doi:10.1111/j.1467-7652.2004.00086.x

CADIS (2017) *Statistics of Climate, Crop Pest and Climate Change impact 2014–2017.* Center for Agricultural Data and Information System, Ministry of Agriculture, Indonesia. <http://epublikasi.setjen.pertanian.go.id/arsip-perstatistikan/181-statistik/statistik-iklim-opt-dan-dpi>

Chen H, Xu CG, Tang W, Li XH (2005) Transgenic indica rice plants harboring a synthetic *cry2A* gene of *Bacillus thuringiensis* exhibit enhanced resistance against lepidopteran rice pests. *Theoretical and Applied Genetics* **111**, 1330–1337. doi:10.1007/s00122-005-0062-8

Cheng XY, Sardana R, Kaplan H, Altossar I (1998) *Agrobacterium*-transformed rice plants expressing synthetic *cryIA(b)* and *cryIA(c)* gene are highly toxic to striped stem borer and yellow stem borer. *Proceedings of the National Academy of Sciences of the United States of America* **95**, 2767–2772. doi:10.1073/pnas.95.6.2767

Cohen MB, Chen M, Bentur JS, Heong KL, Ye GY (2008) Bt rice in Asia: potential benefits, impact, and sustainability. In *Integration of Insect-resistant Genetically Modified Crops with IPM Systems.* (Eds J Romeis, AM Shelton and GG Kennedy) pp. 223–248. Springer, Dordrecht.

Estiati A, Herman M (2016) Biosafety regulation of genetically modified products in Indonesia. *Analisis Kebijakan Pertanian* **13**, 129–146. doi:10.21082/akp.v13n2.2015.129-146

Estiati A, Rahmawati S, Astuti D, Slamet-Loedin IH (2007) *Agrobacterium*-mediated transformation of javanica rice plants with a *cry1b* gene under the control of wound-inducible gene promoter. *Annales Bogorienses* **11**, 1–5.

Estiati A, Nena A, Nugroho S (2013) Insect bioassay in biosafety containment to select transgenic rice (*Oryza sativa* L.) harboring *cry1B* gene resistant to yellow stem borer (*Scirpophaga incertulas* Walk.). *Annales Bogorienses* **17**, 17–26.

Farias DF, Viana MP, Oliveira GR, Santos VO, Pinto CEM, Viana DA, Vasconcelos IM, Grossi-de-Sa MF, Carvalho AFU (2015) Food safety assessment of Cry8Ka5 mutant protein using Cry1Ac as a control Bt protein. *Food and Chemical Toxicology* **81**, 81–91. doi:10.1016/j.fct.2015.04.008

Goodman RE, Ebisawa M, Ferreira F, Sampson HA, van Ree R, Vieths S, Baumert JL, Bohle B, Lalithambika S, Wise J, Taylor SL (2016) AllergenOnline: a peer-reviewed, curated allergen database to assess novel food proteins for potential cross-reactivity. *Molecular Nutrition & Food Research* **60**, 1183–1198. doi:10.1002/mnfr.201500769

Heinrich EA, Medrano FG, Rapusas HR (1985) *Genetic Evaluation for Insect Resistance in Rice.* International Rice Research Institute, Manila.

Hiei Y, Ohta S, Komari T, Kumashiro T (1994) Efficient transformation of rice (*Oryza sativa* L.) mediated by *Agrobacterium* and sequence analysis of the boundaries of the T-DNA. *Plant Journal* **6**, 271–282. doi:10.1046/j.1365-313X.1994.6020271.x

Ho NH, Baisakh N, Oliva N, Datta K, Frutos R, Datta S (2006) Translational fusion hybrid *Bt* genes confer resistance against yellow stem borer in transgenic elite Vietnamese rice (*Oryza sativa* L.) cultivars. *Crop Science* **46**, 781–789. doi:10.2135/cropsci2005.06-0111

Kalshoven LGE (1981) *The Pests of Crops in Indonesia.* PT Ichtiar Baru-Van Hoeve. Jakarta.

Liu CS, Hallerman E, Peng YF, Li YH (2016) Development of Bt rice and Bt maize in China and their efficacy in target pest control. *International Journal of Molecular Sciences* **17**, 1561–1576. doi:10.3390/ijms17101561

Mulyaningsih ES, Deswina P, Slamet-Loedin IH (2009) Dampak padi transgenik mengekspresikan gen *cry1a(b)* untuk ketahanan terhadap penggerek batang di lapang terbatas terhadap serangga bukan sasaran. *Jurnal Hama dan Penyakit Tumbuhan Tropika* **9**, 85–91.

Perlak FJ, Deaton RW, Armstrong TA, Fuchs RL, Sims SR, Greenplate TJ, Fischoff DA (1990) Insect-resistant cotton plants. *Nature Biotechnology* **8**, 939–943.

Rahmawati S, Slamet-Loedin IH (2006) Introduction of *cryIB-cryIAa* hybrid gene into rice (*Oryza sativa*) genome cv. Rojolele using *Agrobacterium*-mediated transformation. *Hayati* **13**, 19–25. doi:10.1016/S1978-3019(16)30374-6

Rao KV, Rathore KS, Hodges TK, Fu X, Stoger E, Sudhakar D, Williams S, Christou P, Bharathi M, Bown DP, Powell KS, Spence J, Gatehouse AMR, Gatehouse JA (1998) Expression of snowdrop lectin (GNA) in transgenic rice plants confers resistance to rice brown plant hopper. *Plant Journal* **15**, 469–477. doi:10.1046/j.1365-313X.1998.00226.x

Roberts CS, Rajagopal S, Smith LA, Nguyen TA, Yang W, Nugroho S, Ravi KS, Cao ML, Vijayachandra K, Patel V, Harcourt RL, Dransfield L, Desamero N, Slamet I, Keese P, Killian A, Jefferson RA (1997) A comprehensive set of modular vectors for advanced manipulations and efficient transformation of plants by both *Agrobacterium* and direct DNA uptake methods. In *Proceedings of Rockefeller Foundation Meeting of the International Program on Rice Biotechnology*, Malacca, 15–20 September.

Satoto (2003) The Stability, Inheritance and Effectiveness of *GNA* and *cry1Ab* Genes to Brown Plant Hopper and Yellow Stem Borer in Rojolele Rice Transgenic. Program Pasca Sarjana, Institut Pertanian Bogor. <https://repository.ipb.ac.id/jspui/bitstream /123456789/7730/1/2003sat_abstract.pdf>

Satoto, Sulistyowati Y, Hartana A, Slamet-Loedin IH (2008) The segregation pattern of insect resistance genes in the progenies and crosses of transgenic Rojolele rice. *Indonesian Journal of Agricultural Science* **9**, 35–43. doi:10.21082/ijas.v9n2.2008.p35-43

Shelton AM, Hossain MJ, Paranjape V, Azad AK, Rahman ML, Khan ASMMR, Prodhan MZH, Rashid MA, Majumder R, Hossain MA, Hussain SS, Huesing JE, McCandless L (2018) Bt eggplant project in Bangladesh: history, present status, and future direction. *Frontiers in Bioengineering and Biotechnology* **6**, 106. doi:10.3389/fbioe.2018.00106

Shu QY, Ye QY, Cui HR, Cheng XY, Xiang YB, Wu DX, Gao MW, Xia YW, Hu C, Sardana R, Altosaar I (2000) Transgenic rice plants with a synthetic *cry1Ab* gene from *Bacillus thuringiensis* were highly resistant to eight lepidopteran rice pest species. *Molecular Breeding* **6**, 433–439. doi:10.1023/A:1009658024114

Slamet-Loedin IH, Novalina, Satoto, Damayanti D, Sutrisno, Mulyaningsih ES, Christou P, Aswidinoor H (2003) Inheritance of *cry1A(b)* and snowdrop lectin *GNA* genes in transgenic Javanica rice progenies and bioassays for resistance to brown planthopper and yellow stem borer. In *Advances in Rice Genetics.* (Eds GS Khush, DS Brar and B Hardy) pp. 565–566. International Rice Research Institute, Manila.

Somantri IH, Ambarwati AD, Apriana A, Listanto E, Dewi IS, Santoso T, Damayanti D, Altosaar I (2008) Rice transformation for resistance to stem borer. In *Advance in Rice Genetics* (Eds GS Khush, DS Brar and B Hardy), pp. 555–557, International Rice Research Institute, Los Banos, Manila.

Sulistyowati Y, Rachmat A, Zahra F, Rahmawati S, Nugroho S (2011) Genetic transformation of rice cv Ciherang using double T-DNA vector harboring *cry1Ab* gene. *Annales Bogorienses* **15**, 27–32.

Tang W, Chen H, Xu CG, Li XH, Lin YJ, Zhang QF (2006) Development of insect-resistant transgenic indica rice with a synthetic *cry1C** gene. *Molecular Breeding* **18**, 1–10. doi:10.1007/s11032-006-9002-9

Tu J, Datta K, Alam MF, Fan YL, Kush GS, Datta SK (1998) Expression and function of a hybrid Bt toxin gene in transgenic rice conferring resistance to insect pest. *Plant Biotechnology (Sheffield, England)* **15**, 195–203. doi:10.5511/plantbiotechnology.15.195

Tu J, Zhang G, Datta K, Xu C, He Y, Zhang Q, Kush GS, Datta SK (2000) Field performance of transgenic elite commercial hybrid rice expressing *Bacillus thuringiensis* ɗ-endotoxin. *Nature Biotechnology* **18**, 1101–1104. doi:10.1038/80310

Usyati N, Buchori D, Manuwoto S, Hidayat P, Slamet-Loedin IH (2009) Keefektivan Padi Transgenik terhadap Hama Penggerek Batang Padi Kuning *Scirpophaga incertulas* (Walker) (Lepidoptera: Crambidae). *Jurnal Entomologi Indonesia* **6**, 30–41. doi:10.5994/jei.6.1.30

Usyati N, Buchori D, Manuwoto S, Hidayat P, Slamet-Loedin IH (2015) Larvae development and survival of predatoral insect *Verania lineata* on the transgenic Rojolele rice variety. *Penelitian Pertanian Tanaman Pangan* **34**, 187–194. doi:10.21082/jpptp.v34n3.2015.p187-194

Wang Y, Ke K, Li Y, Han L, Liu Y, Hua H, Peng Y (2016) Comparison of three transgenic Bt rice lines for insecticidal protein expression and resistance against a target pest, *Chilo suppressalis* (Lepidoptera: Crambidae). *Insect Science* **23**, 78–87. doi:10.1111/1744-7917.12178

Wunn J, Kloti A, Burkhardt P, Ghosh Biswas G, Launis K, Iglesias VA, Potrykus I (1996) Transgenic indica rice breeding line IR58 expressing a synthetic *cry1A(b)* gene from *Bacillus thuringiensis* provides effective insect pest control. *Biotechnology* **14**, 171–176.

Zhao H, Zhang Y, Zhao K, Peng Y, Guo Y (2004) Expression of Cry1Ac protein in *cry1Ac/CpTI* transgenic rice and its resistance in different stages to *Chilo suppressalis. Journal of Agricultural Biotechnology* **12**, 76–79.

17

Experience in developing genetically engineered potato resistant to late blight disease

Alberta Dinar Ambarwati, Tri Joko Santoso, Kusmana and Muhammad Herman

Introduction

Potato (*Solanum tuberosum* L.) is one of the leading crops with potential in food diversification. The high carbohydrate content in potato makes its tubers a substitute for rice. Potato production in Indonesia over the past few years increased by 3.47%, from ~1.12 million tons in 2013 to ~1.18 million tons in 2018. In Indonesia, potatoes are widely cultivated. However, farmers face several constraints including pests and diseases. The main pests and diseases are potato tuber moth (*Phthorimaea operculella*), potato leafminer (*Liriomyza huidobrensis*), root knot nematode (*Meloidogyne* sp.), golden cyst nematode (*Globodera rostochiensis*), early blight (*Alternaria solani*), late blight (*Phytophthora infestans*), fusarium wilt (*Fusarium oxysporum*), bacterial wilt (*Ralstonia solanacearum*), potato X and potato Y virus. Late blight caused by fungus *P. infestans* is one of the most devastating plant diseases for potato. It can cause significant yield loss in susceptible varieties such as Granola and Atlantic. In Indonesia, loss of yield due to attack of *P. infestans* ranged from 47% to 100%. If weather conditions are very conducive to development of the disease, especially in the rainy season, the yield losses can reach 100% (Hariyadi and Koentjoro 1996; Kusmana 2003). Besides, the high genetic diversity of *P. infestans* causes the resistant potato varieties to break down quickly (Cooke *et al.* 2003).

Until now, chemical control of late blight disease using fungicides was the main choice of potato farmers. Potato farmers in Pangalengan spray fungicides 20–30 times in one growing season to control the disease (Adiyoga 2009). The excessive spraying of fungicides can harm the environment and human health (Schepers and Spits 2006; Forbes 2009). In addition, farmers incur more production costs.

The use of resistant varieties is the best way to control late blight. Several sources of resistance to *P. infestans* were found in wild potato species, such as *S. demissum*, *S. bulbocastanum*, *S. stoloniferum* and *S. microdontum* (Hawkes 1994; Helgeson *et al.* 1998). The utilisation of wild (diploid) species in potato breeding is often difficult because these are genetically incompatible with commercial (tetraploid) potato (*S. tuberosum*) cultivated during the 1990s. Eleven resistance genes (R) originating from *S. demissum* were introduced into various types of potato cultivars (Umaerus and Umaerus 1994). However, these provided race-specific hypersensitive short-term resistance under field conditions (Wastie 1991).

Durable resistance to a broad spectrum of known *P. infestans* races was found in diploid wild potato species *S. bulbocastanum* (Helgeson *et al.* 1998; Naess *et al.* 2000) where the resistance loci of the *RB* gene was mapped to a specific location on chromosome 8 (Song *et al.* 2003; van der Vossen *et al.* 2003). Blight resistant potato plants were developed by introgressing the *RB* gene through genetic engineering (GE) into four popular varieties – Katahdin, Superior, Dark Red Norland and Russet Burbank – in the USA (Song *et al.* 2003; Halterman *et al.* 2008; Jo *et al.* 2014). GE potato plants of Katahdin expressing the *RB* gene were tested in the field at four different locations in PICTIPAPA (Mexico), Washington, Wisconsin and Minnesota, and showed resistance against leaf blight.

Research collaboration between Agricultural Biotechnology Support Project Phase 2 (ABSP II) and the Indonesian Center for Agricultural Biotechnology and Genetic Resources Research and Development (ICABIOGRAD) led to signing of a Material Transfer Agreement. Through the Agreement, GE Katahdin event SP951 expressing the *RB* gene resistant to *P. infestans* was donated to the ICABIOGRAD to be used as a source material in the research program under the ABSP II initiative funded by USAID and managed by Cornell University, USA.

GE Katahdin plants can also be used as a source of resistance to late blight for breeding with local varieties like Granola and Atlantic. Granola, of short duration and extensively adapted, is the main choice of farmers, dominating almost 90% of the planting area in Indonesia (Simatupang *et al.* 1996). Atlantic has good taste and high quality for processing. However, both Granola and Atlantic are susceptible to late blight, which can be controlled by extensive spraying of fungicides (Suryadi *et al.* 1995; Dwiastuti and Djoema'ijah 2000; Kusmana 2003, 2004; Kusmana and Basuki 2004; Basuki *et al.* 2005).

GE Katahdin event SP951 was crossed with Atlantic and Granola. The hybrid clones from crosses with *RB* gene showed blight resistance up to 45–50 days after planting without fungicide application in the confined field trial (CFT) (Ambarwati *et al.* 2011, 2012, 2015). The crop was protected against late blight until harvest with a very few need-based fungicide sprays and reduced cost of fungicide sprays, by as much as 50–80% (Adiyoga 2009).

Despite the benefits of GE crops in agriculture, there are concerns related to their possible adverse impact on the environment, and human and animal health. GE plants intended to be grown in Indonesia to be used as food and feed must meet the safety requirements for environment, food and feed. Testing of environmental safety can be carried out both in a biosafety containment facility (BCF) and in a CFT as per specifications. In Indonesia, the need for biosafety regulation is well recognised (Herman 2010). It was initiated as early as September 1997 and led to the creation of infrastructure along with necessary legal upgrades. GE Katahdin SP951 potato has a food safety certificate through decree of the Director General of the National Agency of Food and Drug Control no. HK.04.01.1.52.12.16.4300 in December 2016 and an environment safety certificate from the Minister of Environment and Forestry no. S.240/MENLHK/KSDAE/KSA.2/7/2018 as at July 2018. With biosafety certificates for food and environmental safety, GE potato became ready for commercialisation after meeting the regulations on variety release or registration.

Research strategy and product development through hybridisation

Development of potato resistant to late blight disease through GE is an excellent method of overcoming the barrier in sexual hybridisation from wild resistant relatives. Eleven R genes originating from *S. demissum* have been introgressed into various types of potato cultivars (Umaerus and Umaerus 1994). Resistant genes against *P. infestans* have been isolated, namely *R1* (El-Kharbotly *et al.* 1994; Ballvora *et al.* 2002) and *R3* (El-Kharbotly *et al.* 1994), *Rpi-blb1* (van der Vossen *et al.* 2003), *RB* (Song *et al.* 2003) and *Rpi-blb2* (van der Vossen *et al.* 2005). These resistant genes have been successfully transformed into susceptible cultivars to develop resistance to *P. infestans*. The *RB* gene for resistance to late blight derived from diploid potato *S. bulbocastanum* is introgressed into the susceptible variety Granola through *Agrobacterium tumefaciens* (Listanto *et al.* 2009). Molecular analysis of *RB* gene and bioefficacy was conducted via CFT in Pasir Sarongge (2009–10), Lembang (2009–10) and Pangalengan (2010–11) (Herman 2011). An alternate strategy is to cross the GE potato expressing *RB* gene with the local potato varieties Atlantic and Granola in Indonesia. The GE Katahdin event SP904 and event SP951 expressing *RB* gene were developed through *A. tumefaciens* transformation (Song *et al.* 2003). Katahdin is a commercial potato cultivar in the USA, but it is susceptible to late blight. The GE Katahdin SP904 and SP951 events showed broad-spectrum resistance to *P. infestans* in greenhouse and in CFTs in Mexico, Washington, Wisconsin and Minnesota (Song *et al.* 2003; Staples 2004; Kuhl *et al.* 2007; Bradeen *et al.* 2009). GE Katahdin SP904 and SP951 also showed resistance to *P. infestans* in CFT in Lembang, West Java in 2007 (Herman 2007). Katahdin SP904 and SP951 were used as resistant parents in crossing with susceptible local parents Atlantic and Granola, and their progenies were characterised for the integration and segregation of *RB* gene and evaluated for efficacy against late blight.

Crossing between GE potato and non-GE potato

Potato resistant to late blight was developed through conventional crossing between Granola or Atlantic and GE Katahdin event SP951 or event SP904. Granola is a high-yielding variety with many good agronomic characters and is therefore a leading variety in ~90% of potato-growing areas in Indonesia. Granola is tolerant to bacterial wilt, potato virus A (PVA) and potato virus Y (PVY) (Simatupang *et al.* 1996). Atlantic has good tuber quality and is used for making chips (Basuki *et al.* 2005). It is also claimed to be highly resistant to race A of golden nematode, potato virus X and tuber net necrosis caused by potato leaf roll virus. However, both Atlantic and Granola are susceptible to late blight.

Atlantic and Granola were used as female parents in crossing with GE Katahdin event SP951. There are four crossing combinations – Atlantic and Katahdin event SP904, Atlantic and Katahdin event SP951, Granola and Katahdin event SP904, and Granola and Katahdin event SP951. A total number of 22 766 seeds were produced from all crossings. A total of 554 F1 progenies were sown and leaves were taken for PCR analysis to detect the presence of the *RB* gene and their segregation.

Bioefficacy of potato hybrid clones to late blight *P. infestans*

Katahdin SP951 showed higher resistance to *P. infestans* than Katahdin SP904 in CFT in Indonesian Vegetable Research Institute (IVEGRI), Lembang (Herman 2007). In further experiments, only the hybrid clones from crossing Atlantic or Granola with Katahdin SP951 were evaluated for their resistance to late blight *P. infestans*.

The bioefficacy of 43 hybrid clones expressing *RB* gene from crosses between Atlantic or Granola and Katahdin SP951 against late blight was tested in CFT in Pasir Sarongge, West Java from September 2008 to January 2009, in Lembang in West Java from December 2009 to April 2010, in Pangalengan in West Java from December 2010 to May 2011 and in Banjarnegara in Central Java from December 2011 to April 2012.

Sweetcorn and spreader rows of susceptible plants were planted around the plot of GE plants as a late blight inoculum source. The GE plants were thus exposed to natural infection, and no artificial inoculation was applied during the experiments. The late blight resistance score was determined visually based on the percentage of infected leaf tissue (Halterman *et al.* 2008). Plants were scored as resistant if the resistance score was ≥7.0 (≤25% infection) (Colton *et al.* 2006; Kramer *et al.* 2009). Plants that scored ≥6–<7 were categorised as moderately resistant and <6 as susceptible. Area under disease progress curve (AUDPC) was calculated using the formula of Campbell and Madden (1990). The higher the AUDPC, the more susceptible the clone.

Bioefficacy studies at four CFT locations led to shortlisting of 12 hybrid clones based on their resistance to late blight (Figure 17.1), i.e. five hybrid clones of crossing between Atlantic and Katahdin SP951 and seven hybrid clones of crossing between Granola and Katahdin SP951. Atlantic, Granola and Katahdin were used as susceptible control, and GE Katahdin SP951 was used as resistance control.

Twelve hybrid clones showed resistance to late blight *P. infestans* in all locations. The hybrid clones had a resistance score varying from 7.65 to 8.23 and were significantly different from the susceptible parents Atlantic and Granola with a resistance score of 3.6 and 3.45. These hybrid clones showed resistance to *P. infestans* for 14–18 days after infection or 40–45 days after planting, in the absence of fungicide spraying (Table 17.1) (Ambarwati *et al.* 2015).

Bioefficacy testing of 12 selected hybrid clones was also conducted in three CFT locations, along with horticultural characterisation. The CFT locations were in Lembang in West Java, 1250 m above sea level (ASL) from October 2011 to March 2012, in Pangalengan in West Java, 1400 m ASL from December 2011 to April 2012 and in Banjarnegara in Central Java, 1800 m ASL from March to September 2012.

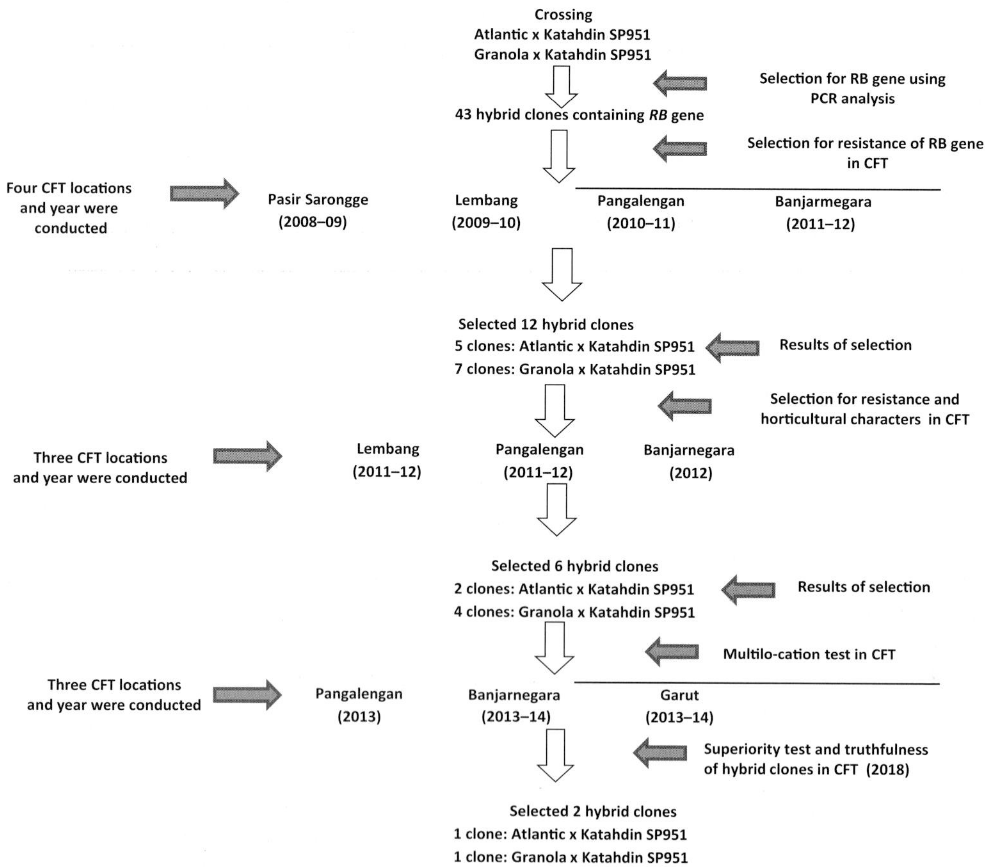

Figure 17.1 The development of selected hybrid clones from crossing Atlantic and Granola with Katahdin SP951 resistant to potato late blight *P. infestans*.

Table 17.1. Selected potato hybrid clones resistant to *P. infestans* without fungicide spraying in four CFTs, Pasir Sarongge, Lembang, Pangalengan and Banjarnegara, 2008–12

No. of genotype	Genotype	Resistance score at 40–45 days after planting	Notes
20	B35	8.13 a	Hybrid clone of Atlantic × Katahdin SP951
26	B169	8.23 a	Hybrid clone of Atlantic × Katahdin SP951
27	B163	7.65 a	Hybrid clone of Atlantic × Katahdin SP951
28	B11	7.93 a	Hybrid clone of Atlantic × Katahdin SP951
29	B162	7.68 a	Hybrid clone of Atlantic × Katahdin SP951
60	D76	7.83 a	Hybrid clone of Granola × Katahdin SP951
62	D12	8.03 a	Hybrid clone of Granola × Katahdin SP951
63	D25	7.93 a	Hybrid clone of Granola × Katahdin SP951
65	D48	7.83 a	Hybrid clone of Granola × Katahdin SP951
66	D38	8.03 a	Hybrid clone of Granola × Katahdin SP951
69	D37	8.18 a	Hybrid clone of Granola × Katahdin SP951
70	D15	7.98 a	Hybrid clone of Granola × Katahdin SP951
74	Atlantic	3.60 b	Susceptible parent
75	Katahdin	4.30 b	Susceptible control
76	Granola	3.45 b	Susceptible parent
79	Katahdin SP951	8.33 a	Resistant parent
	CV (%)	14.03	

Source: Ambarwati *et al.* (2015).

Hybrid clones showed similar response to the results of the previous CFTs in Pasir Sarongge, Lembang, Pangalengan and Banjarnegara. Hybrid clones from crosses of Atlantic × Katahdin SP951 and Granola × Katahdin SP951 were significantly more resistant than the susceptible parents Atlantic and Granola. Six hybrid clones were selected, based on resistance to late blight and horticultural characters, i.e. two hybrid clones from Atlantic × Katahdin SP951 and four hybrid clones from Granola × Katahdin SP951.

The AUDPC value showed a similar pattern to the previous results in four CFT locations (Pasir Sarongge, Lembang, Pangalengan and Banjarnegara). Hybrid clones containing *RB* gene as well as Katahdin SP951 showed lower AUDPC than Katahdin without *RB* gene, in all three CFTs (Lembang, Pangalengan, Banjarnegara), indicating that resistance was mediated by *RB* gene. Hybrid clones showed resistance to *P. infestans* until 51 days after planting (± 22 days after infestation) and were significantly different from Atlantic, Granola and Katahdin, which exhibited a high level of susceptibility (Plate 10).

Six hybrid clones were tested again for their resistance to *P. infestans* and horticultural characters in three CFT locations in Pangalengan, Banjarnegara and Garut during 2013–14 in a multi-location trial. Hybrid clones expressing *RB* gene as well as GE Katahdin SP951 showed lower AUDPC than susceptible varieties without *RB* gene in all three CFTs. Millett *et al.* (2009) reported that Dark Red Norland lines containing *RB* gene had an AUDPC value lower than

Dark Red Norland without *RB* gene. Consistent with this observation, the six hybrid clones showed resistance in all previous CFTs. Plate 11 shows the responses of a hybrid clone from GE Katahdin SP951 and Granola, and non-GE Granola, to blight tested at CFT Pangalengan.

After molecular analysis and assessment of resistance to *P. infestans* and horticultural characters in three CFT locations, only two hybrid clones (one each from an Atlantic × Katahdin SP951 cross and a Granola × Katahdin SP951 cross) were then used for superiority test and truthfulness in the 2018 CFT in Pangalengan, required for variety registration.

Stability analysis of gene for resistance to late blight

Stability analysis of the *RB* gene was conducted in six selected hybrid lines. Plant materials were clonal generation of G0, G1, G2, and G3, the negative controls of non-GE Atlantic, Granola and Katahdin and the positive control GE Katahdin SP951. PCR analysis of the *RB* gene in four clonal generations (G0, G1, G2 and G3) showed stable inheritance of the *RB* gene in the genome of the six hybrid clones, as indicated by the consistent appearance of the 619 bp PCR product (Listanto *et al.* 2015).

Regulation on biosafety

Since 1997, Indonesia has had a regulatory system on biosafety through Decree of the Minister of Agriculture no. 856/Kpts/HK.330/9/1997, Provisions on Biosafety of Genetically Engineered Agricultural Biotechnology Products (GEABP). The decree is intended to regulate and supervise the utilisation of GEABP. Food safety aspects of GEABP was not covered in that decree; therefore, in 1999 it was replaced with Joint Decree of the Minister of Agriculture, Minister of Forestry and Estate Crops, Minister of Health, and State Minister of Food and Horticulture no. 998.1/Kpts/OT.210/9/99; 790.a/Kpts-IX/1999; 1145A/MENKES/SKB/ IX/99; 015A/Nmeneg PHOR/09/1999, Biosafety and Food Safety of Genetically Engineered Agricultural Products. To upgrade the level of biosafety regulation, in 2005 government Regulation no. 21 on Biosafety of Genetically Engineered Products was signed by the President of the Republic of Indonesia (Herman 2010). Since then, the government has developed other regulations to implement biosafety.

Other laws and regulations on biosafety are Law no. 21–2004 concerning Cartagena Protocol ratification; Law no. 32–2009 concerning protection and management of the environment; Law no. 18–2012 concerning food; Decree of the Minister of Environment no. 25–2012 concerning document preparation for risk analysis to the environment of GE products; Decree of the Minister of Environment and Forestry no. P 69/Menlhk/Setjen/Kum.1/8/2016 concerning procedures for testing safety of the environment of GE plants in CFTs; Regulation of the Head of National Agency of Food Drug Control no. 6–2018 concerning supervision of GE food; and Decree of the Minister of Agriculture no. 36/Permentan/LB.070/8/2016 concerning assessment of feed safety of GE products. It is mandatory for all GE product developers to abide by these policies.

Regulation no. 21–2005 is based on a precautionary approach to environmental, food and/or feed safety by considering religious, ethical, sociocultural and aesthetic principles and conservation. Article 8 states that any person conducting research and development of GE products shall prevent and/or eradicate any negative impact on humans and the environment as a result of their activity. Article 9 specifies that testing during the process of research and development must be conducted in a laboratory, biosafety containment facility and/or CFT. A biosafety containment facility used for research on GE plants must ensure that plant material and inserted genes remain inside the facility. It consists of a head house, greenhouse and screen house. Likewise, plant material and inserted genes must be contained within the area used for CFT. Article 10 states that GE products that result from a research and development activity must be tested for efficacy and fulfil biosafety requirements before they are proposed for release and/or distribution (Estiati and Herman 2015).

Regulatory bodies

To implement Regulation no. 21–2005 for biosafety assessment, regulatory bodies such as the Biosafety Committee and Biosafety Technical Team were formed. Article 28 of Regulation no. 21–2005 states that the Biosafety Committee shall assist the Minister of Environment, Minister of Agriculture and Head of Non Department Government Institute (in this case the Head of the National Agency of Food and Drug Control, NAFDC) in giving recommendations about biosafety and conducting controls of importation and utilisation of GE products, as well as inspection and proof of the correctness of reports about negative impacts of GE products. The Committee was formed through Presidential Decree no. 39–2010, Biosafety Committee of Genetically Engineered Products. Its tenure was prolonged through Presidential Decree no. 181–2014, Biosafety Committee of Genetically Engineered Products 2018 and it was newly formed through Presidential Decree no. 50–2018, Biosafety Committee of Genetically Engineered Products. Article 32 of Regulation no. 21–2005 states that the Technical Team shall assist the Committee in conducting technical assessments on biosafety by conducting monitoring during CFTs. The Technical Team was newly formed through Decree of the Chair of Biosafety Committee no. 02–2019.

Guidelines for assessment of biosafety of GE products

In Regulation no. 21–2005, Article 6 para. (1), Article 7 and Article 14 para. (1) state that GE products that are developed both within Indonesia and abroad must meet the requirements for environment, food and/or feed safety before being released and distributed in Indonesia. Paragraph (4) of Article 20 states that the guidelines for the assessment of biosafety are stipulated by the Minister of Environment for environmental safety, Head of NAFDC for Food Safety, and Minister of Agriculture for Feed Safety (Herman 2010).

To conduct risk assessment for biosafety, in 2012 the Minister of Environment prepared Decree no. 25–2012, Guidelines on Documents Preparation of Risk Analysis of Environment of GE Products and Decree of the Minister of Environment and Forestry no. P 69/Menlhk/Setjen/Kum.1/8/2016, Procedure on Testing of Environment Safety of GE Plants in CFT. The Head of NAFDC prepared Regulation no. HK.00.05.23.3541–200, Guidelines of Assessment of Food Safety of GE Products. In 2012, that was revised to Regulation no. HK.03.1.23.03.12.1563 and in 2016 it was revised to Regulation no. 19, Amendment to the Regulation of the Head of NAFDC no. HK.03.1.23.03.12.1563–2012. In 2018, the Regulation of the Head of NAFDC no. 19–2016 was revised again to become Regulation of the Head of NAFDC no. 6–2018, Supervision of GE products. This Regulation specifies the requirements which relate to assessment of food safety of GE products, and food labelling. It superseded Regulation No HK.03.1.23.03.12.1563–2012, Guidelines of Assessment of Food Safety of GE Products and Regulation HK.03.1.23.03.12.1564–2012, Supervision of Food Labelling of GE Products. In 2016, the Minister of Agriculture prepared and launched Decree no. 36/Permentan/LB.070/8/2016, Assessment of Feed Safety of GE Products.

These guidelines were issued to serve as a reference for applicants regarding assessment of environmental, food and/or feed safety requirements of GE products, and to aid regulatory bodies such as the Biosafety Committee and Technical Team in carrying out risk assessment for the environment. They ensure that a precautionary approach has been applied.

Biosafety risk assessment

In Regulation no. 21–2005, Article 1 no. 1 deals with environment, food and/or feed safety of GE products. Article 1 nos 21 and 22 relate to risk assessment of GE products for possible adverse effects on the environment, human health and animal health using scientific methods and relevant documents. In relation to that requirement, a socioeconomic study conducted by IVEGRI in 2009 found indiscriminate excessive use of fungicides in a planting season, and possible cost savings of IDR4.1–6.6 million due to reduced spraying needs (Adiyoga 2009) if farmers planted GE potato resistant to *P. infestans*.

Biosafety assessment of GE products must be carried out before release and distribution. The assessment is based on a written application to the Minister of Agriculture or Head of NAFDC. The Minister of Agriculture or Head of the NAFDC submits a request for GE product biosafety recommendations to the Minister of Environment and Forestry or Chairperson of the Biosafety Committee. The Committee is then tasked with conducting an assessment and, if technical evaluation is involved, the Committee tasks the Technical Team to conduct a technical review of documents and follow-up tests if necessary.

The Biosafety Clearing House (BCH) announces summary results of these biosafety assessments for public notification for 60 days, to allow the community to submit responses. The BCH, as a part of the Biosafety Committee, functions as a means of communication between the Committee and stakeholders.

Environment safety assessment

Decree of the Minister of Environment no. 25–2012 and Decree of the Minister of Environment and Forestry no. P 69/Menlhk/Setjen/Kum.1/8/2016, Procedure on Testing of Environment Safety of GE Plants in CFT, are used by regulatory bodies to conduct environmental safety assessment. The applicant must provide data and information on several aspects, such as genetic and agronomic traits of the GE plant, impact potential on the environment, risk management and monitoring, and communication. Genetic information includes genetic element, source of inserted gene, transformation method and genetic stability. Impact potential on the environment includes impact on non-target organisms, weediness and potential for gene flow.

Risk communication was conducted for the late blight resistant potato by the Indonesian Biotechnology Information Center. Environmental risk communication of GE potato was disseminated through workshops on the role of biotechnology in potato breeding and regulations in Indonesia, and the development of late blight resistant GE potatoes through breeding and biotechnology, and biosafety assessment. Dissemination was carried out in 15 locations such as Wonosobo in Central Java and Lembang and Bandung in West Java from 2011 to 2014. Workshop participants included the chair and members of the Potato Farmers Group, seed growers, potato breeders, research scientists from university and research institutes, extension personnel and staff of the District Agricultural Office. The workshops comprised presentations and discussions, followed by visits to CFTs to observe the performance of GE potato resistant to late blight disease.

Food safety and feed safety assessment

Regulatory bodies implement Regulation no. 6–2018, Supervision of GE products regarding food safety assessment, and Decree of Minister of Agriculture no. 36–2016 regarding feed safety assessment of GE products.

Regulation on plant variety release and registration

After obtaining environmental, food and feed safety certificates, GE plants can be registered for varietal release. After obtaining the release or registration variety certificate from the Minister of Agriculture, GE crops can then be commercialised and utilised (Herman 2010).

Provisions in Article 23 of Regulation no. 21–2005 concerning biosafety of GE products state that the competent authority, which is the Minister of Agriculture, grants varietal release or registration of agricultural GE crops that have obtained biosafety recommendations, in accordance with current laws and regulations. The relevant laws and regulations are Law no. 12–1992, Plant Cultivation System; Law no. 13–2010, Horticulture; Regulation no. 44–1995, Plant Seeds; Regulation no. 21–2005, Biosafety of GE Products; Decree of the Minister of Agriculture no. 38/2019, Plant Variety Release; and Decree of the Minister of Agriculture no. 38/Permentan/OT.140/7/2011, Registration of Horticulture Crops Variety.

Varietal release of food and estate crops

The Variety Assessment and Release Team is the regulatory body tasked with implementing Decree of the Minister of Agriculture no. 38 of 2019 for release of GE food crops which include rice, maize, soybean, sweet potato and cassava, and GE estate crops such as sugarcane. An applicant must conduct adaptation and observation tests through a breeding institution or other institution. An adaptation test is a field test activity in several agroecological areas for annual plants, to determine the superiority and interaction of varieties in the environment; an observation test is a field test activity in several agroecologies for perennial plants, to determine the superior characters and adaptability of varieties to the environment, or for annual plants that are already local varieties becoming common or regular varieties. Adaptation or observation tests are carried out in several development locations and/or laboratories with the number of testing units adjusted to the type of plant. An adaptation or observation test can be harmonised with tests for the benefit of Plant Variety Protection such as distinctness, uniformity and stability (DUS) testing.

Variety registration for horticulture crops

The Horticultural Variety Assessment and Registration Team is the regulatory body tasked with implementing Decree of the Minister of Agriculture no. 38 of 2011 for evaluation and registration of GE horticulture crops such as potato, tomato, cabbage, carrots and hot pepper. An applicant must conduct superiority and truthfulness tests through a government or private testing institution. A superiority test can be carried out in one or more locations (multi-location trial).

Supervision and control of GE products

GE plants that have been granted an environmental safety approval and released by the Minister of Agriculture will be monitored and controlled by the Minister of Agriculture and Minister of Environment and Forestry. If the GE crop has been granted a food and feed safety approval for use as fresh or processed food and forage feed, supervision and control will be carried out by the Minister of Agriculture, Minister of Environment and Forestry and the NADFC. Supervision and control are carried out in accordance with current laws and regulations.

When Bt cotton was limitedly released in South Sulawesi in 2001–03, a GE Cotton Monitoring Team was established through Decree of Minister of Agriculture no. 305/Kpts/Kp.150/5/2001 on 16 May 2001 (Herman 2009). Members of the Monitoring Team were drawn from the offices of the State Minister of Environment/Central Agency Environmental Impact Management, Ministry of Agriculture, Government of South Sulawesi Province, Evaluation and Variety Release Team, Biosafety and Food Safety Committee, and Biotechnology, Environmental, Socio-economic and Breeding Expert Groups. The Monitoring Team has wide responsibilities including the Production and Development Sector; the Assessment Sector, which is divided into the Production Sub-Sector, Environment Risk Analysis Sub-Sector and Socio-Economic Sub-Sector; and Monitoring and Supervision.

Provisions regarding supervision and control are regulated in Chapter VI, Monitoring, Supervision and Reporting of the Joint Decree of Four Ministers concerning Biosafety and

Food Safety of GEAP; in Chapter VI, Supervision and Control of GE Products of Regulation no. 21–2005 concerning Biosafety of GE Products; and in Chapter VI, Withdrawal of Varieties of Decree of Minister of Agriculture no. 61–2011, Testing, Evaluation, Release and Withdrawal of Varieties. According to the provisions in Articles 43–45 of the Rights and Obligations Chapter, and the Chapter on Monitoring, Supervision and Reporting of Joint Decree of 1999 concerning Biosafety and Food Safety of GEAP, monitoring and supervision of GE plant utilisation is carried out by the Minister of Agriculture, assisted by Observers of Pests and Plant Diseases, and Seed Supervisors. Evaluation of biosafety and food safety due to the use of GE plants is carried out by the Biosafety and Food Safety Committee, which was assisted by the Biosafety and Food Safety Technical Team. Every person or legal entity that obtains approval of plant utilisation must make a report to the Minister of Agriculture periodically, i.e. annually or at any time it is necessary if GE plants cause a loss of biosafety and food safety. In addition, every person or legal entity that has obtained approval for the use of GE plants is responsible for and must participate in taking control and mitigation actions if the GE plant turns out to cause loss of biosafety and food safety.

Provisions for monitoring and controlling GE plants according to Regulation no. 21–2005, Articles 25–27 are as follows: if a GE plant that has been released, distributed and/or utilised or imported from abroad has a negative impact on the environment, human health and/or animal health, then the developer who produces it or imports it, or consumers and the public who know that there is a negative impact, must report to the Minister of Environment, and Minister of Agriculture or Head of NAFDC. After receiving a report regarding a negative impact of GE plants on the environment, human health and/or animal health, those bodies assign the Biosafety Committee to examine and verify the truth of the report. If the examination proves that the reported GE plant has a negative impact on the environment, human health and/or animal health, the Minister of Environment proposes to the Minister of Agriculture or Head of NAFDC to withdraw the certificate of release or distribution of the GE plant. Then the Minister of Agriculture or the Head of NAFDC withdraws the certificate of release or distribution of the GE plant. The person in charge of activities must carry out control and mitigation actions, as well as withdraw GE plants from distribution. Guidelines for Supervision and Control of Agricultural GE Plants are in the process of being drafted by the IAARD, taking into account some inputs from the Biosafety Committee.

Environmental biosafety assessment

Effects on non-target organisms

Non-target organisms are defined as all those which are directly and/or indirectly exposed to the GE plant and which are not the targets of the newly expressed gene or gene products in those plants. Results from CFT studies showed that, in general, the cultivation of GE potato Katahdin SP951 has no effect on non-target organisms, such as *Alternaria solani* and *Liriomyza* as well as other insect pests namely potato tuber moth, aphid and mites, thrip and tobacco caterpillar (Riyanti *et al.* 2014). The resistance gene in GE potato Katahdin SP951 is very specific, limited to species or race in causing initiation of a resistance response. Due to the specificity of the response reaction, no effects on organisms other than *P. infestans* are expected bar those that also apply to the interaction of non-GE potato with non-target organisms under conventional agricultural practice.

In addition, the *RB* gene in Katahdin SP951 is constitutively expressed and RB proteins are produced in plants at very low levels. Such proteins are described as intractable proteins (Bushey *et al.* 2014). RB proteins in very small quantity relate to the nature of the resistance to leaf blight disease.

Effects on beneficial soil microorganisms

The potential impact of GE plants on the environment should be investigated. Studies should include unintentional changes in the chemical compositions of root exudates, and direct effects on non-target species of soil microorganisms. Most studies to date suggest that GE plants cause minor changes in microbial community structures that are often transient in duration.

Four studies of the effects of GE potato on soil beneficial microbes conducted in four CFT locations in Pangalengan, Lembang, Garut and Pangalengan from 2011 to 2013 included monitoring of the beneficial microbial population such as nitrogen-fixing bacteria and phosphorus-solubiliser bacteria using plate culture in selective media. GE potato did not show any adverse effect on the beneficial soil microorganisms. Populations of P-solubiliser bacteria had a similar pattern as N-fixing bacteria. There was no difference between non-GE and GE plots. However, there was a significant different in bacterial population between treatments from the soil sample taken before planting, due to the soil fertility.

Populations of soil fungus from the rhizosphere from CFT Pangalengan were investigated using plating method on Potato Dextrose Agar (PDA) medium. The population of soil fungus was $\sim 10^4$ viable cells/g soil; 1.5 months after planting, the population was $\sim 10^5$ viable cells/g soil. There was no significant difference in fungus populations between samples taken from the GE and the non-GE rhizospheres (Riyanti *et al.* 2014).

Carbon/nitrogen ratio in the soil

The C/N ratios for the soils from CFT trials in Lembang (2011–12), Pangalengan (2012), Garut (2013) and Pangalengan (2013) were investigated to know the impact on soil fertility. The results showed that C/N values (most within a range of 8–12) were not significantly different between soil samples taken from the GE and non-GE plots from all CFTs, meaning soil fertility was not affected in the GE potato plantation (Riyanti *et al.* 2014).

Certificate of environment safety

The application for environment safety and all relevant documents for GE potato Katahdin SP951 resistant to late blight disease were submitted to the regulatory bodies. After all documents were assessed and evaluated, the Biosafety Committee recommended that the Ministry of Environment and Forestry grant the environmental safety certificate for GE potato Katahdin SP951. This was done through Letter no. S.240/MENLHK/KSDAE/KSA.2/ 7/2018, Recommendation of Environment Safety of GE Katahdin Event SP951 Resistant to Late Blight Disease.

Food safety assessment

The GE potato differs from the conventional potato varieties in its resistance to *P. infestans* conferred by the *RB* gene. Allergenicity and toxicity are important factors in the food safety of GE potato. As RB protein is very low and is classified as an intractable protein, it is difficult to express in heterologous systems (Bushey *et al.* 2014). Therefore, instead of RB protein, studies use total protein extracted from the GE potato tuber.

Studies on oral toxicity

Analysis of oral acute toxicity was performed on GE hybrid clones of potato tubers compared with a control. The study consisted of one control and five treatment groups, each containing 10 male and 10 female mice. Diet as pellet R-91 was administered *ad libitum* at 1–2 g/day/animal. After administration of potato juice or powder at the highest dose of 21 g/kg bw, there was no mortality in all treatments. This indicated that potato juice and powder from GE potato Katahdin SP951 was safe to mice. Similar results were obtained with potato juice and powder of hybrid clones. In addition, bodyweights of mice fed on GE potato Katahdin SP951 did not differ significantly from those of the controls at day 8 in all treatments. There were no detectable differing clinical symptoms in mice fed separately on GE Katahdin SP951 and on non-GE potato (control) after 14 days testing. Assessed clinical signs included no red and watery eyes, no bloody nose, no body tremors, no weakness, no breathing problem and no diarrhoea. The purpose of the oral acute toxicity testing is aimed at hazard identification and risk management in the context of production, handling and use of the proteins. Most dietary proteins do not cause adverse effects on humans because they are metabolised into amino acids and small peptides in the gut that are readily absorbed as nutrients (Delaney *et al.* 2008). GE potato protein toxins have been evaluated in mice or other laboratory animals. Molecular analysis of GE potato did not show the presence of any other proteins encoded by *RB* gene (Chassy 2004, unpublished).

GE potato did not show acute toxicity to either sex in mice. Similarly, the administration of potato juice and powder from hybrid clones had no effect on the survival of mice (data not shown). This indicated that the GE potato was safe to mice.

The food safety of GE crops has drawn much attention as the use of GE crops spreads across the world. A food safety evaluation of GE crops has been conducted using oral toxicity studies performed in rats (Bravo-Almonacid *et al.* 2012).

Analysis of total glycoalkaloids and anti-nutrients

The total glycoalkaloid analysis of GE Katahdin SP951 and non-GE Katahdin tubers was conducted at EPL Bio Analytical Services in the USA. The glycoalkaloids (α-chaconine and α-solanine) and anti-nutrients (trypsin inhibitor and lectin) components were analysed as specified by the Organization for Economic Co-operation and Development.[1] The studied glycoalkaloids and anti-nutrients of GE potato were compositionally and nutritionally equivalent to those of conventional non-GE potato currently on the market.

Analysis of hybrid clones showed variations at glycoalkaloid and anti-nutrients levels, but well within the 99% tolerance intervals for reference potato (Katahdin, Granola and Atlantic). Therefore, it was concluded that hybrid clones of GE potato Katahdin SP951 were compositionally and nutritionally equivalent to those of conventional potato.

Allergenicity analysis

The parameters of food safety testing include allergenicity testing and protein stability test with simulated gastric fluid (SGF) and simulated intestinal fluid (SIF) with bioinformatics. Bioinformatics analysis was performed using the program FASTA and BLASTP Alignment through the data of protein sequence comparator available on Allergen Online and NCBI database. Studies on SGF analysis used pepsin, and SIF used trypsin enzyme. The bioinformatics analysis did not show any amino acid sequence homology with the known allergens in the database, confirming that GE potato is not likely to be allergenic. Analysis of SGF and SIF testing of allergenicity for GE potato Katahdin SP951 and hybrid lines indicated that the total protein of GE potato Katahdin SP951 and hybrid lines degraded in less than 5 min of incubation after treatment with pepsin or trypsin, thus showing that RB protein was not stable.

Compositional analysis

Compositional analysis was performed to determine the content of dry matter, protein, starch, total fat, fibre, sugar, vitamin C (ascorbic and dehydroascorbic acid), minerals and amino acids of potato tubers of Katahdin SP951 (GE), Katahdin (non-GE), Atlantic, Granola and six hybrid clones. Evaluation of the GE potato tuber composition was based on OECD guidelines.[2]

Analysis of dry matter and protein revealed that Katahdin SP951 (GE), Katahdin (non-GE), Atlantic, Granola and six hybrid clones in CFT Banjarnegara in 2012 and Pangalengan in 2013 fell within the value range of OECD documents. Similar results were observed in other parameters, i.e. starch, crude fibre, sugar, vitamin C (ascorbic and dehydroascorbic acid), minerals and amino acids. Meanwhile, the values of fat and dietary fibres were higher than the OECD range. However,

Katahdin SP951 (GE) and Katahdin (non-GE) had similar values, suggesting that the higher values of fat and dietary fibre of hybrid clones probably originated from the parents (Katahdin SP951, Granola and Atlantic). Mineral K and amino acid alanine in Katahdin SP951 (GE), Katahdin (non-GE), Atlantic, Granola and six hybrid clones were no different.

The results of composition study were consistent across CFT Banjarnegara and Pangalengan. It was concluded that the compositional and nutritional values of Katahdin SP951 (GE) and Katahdin (non-GE) as well six hybrid clones in both CFT Banjarnegara and Pangalengan were substantially equivalent.

Certificate for food safety

The application for food safety and all documents related to GE potato Katahdin SP951 resistant to late blight disease (*P. infestans*) were submitted to regulatory bodies. After all assessment and evaluation, the Biosafety Committee recommended that the Head of NAFDC grant the food safety certificate for GE potato Katahdin SP951. This was done via Decree of NAFDC no. HK.04.01.1.52.12.16.4300 2016, Permit for Food Distribution of Genetically Engineered Potato Katahdin Event SP951.

Tests for variety registration

Superiority of potato varieties

The superiority test is similar to multi-location test or adaptation test. The legal basis for registration of horticultural plant varieties is Law no. 13–2010, Article 58 para. 3 and Article 60 para. 3, and Decree of the Minister of Agriculture no. 38/Permentan/OT.140/7/2011, Registration of Horticulture Crops Variety. Registration must be obtained by the variety developer before seeds can be commercialised.

The variety superiority test can be carried out in three district locations at the same altitude in one planting season. Alternatively, it can be carried out at site-specific locations in one test location with one planting time. The number of locations determines the recommendation for distribution of seeds. If planted in three different locations with different regency, the distribution of seeds is permitted nationally. If it is tested only in one location, the distribution of seeds is permitted only in the testing district. Before testing, the proponent must report to the Horticultural Variety Assessment Registration Team in Jakarta, which often carries out the field monitoring or supervision.

The methodology involves a randomised complete block design with three replications if the test is carried out in three locations or four replications if the test is conducted only in one location. The minimum population is 50 and 75 plants, respectively. Plant observations are carried out for qualitative and quantitative characters and consumer preferences. Plant cultivation includes all recommended agronomic practices. Insect control is performed weekly, with profenophos. To control plant diseases, fungicides are sprayed twice a week. Contact and systemic fungicides are used alternately with total spraying during the planting period up to 20 times.

The traits to be observed, both qualitative and quantitative, are described for each potato variety in the 2016 guidebook on the preparation of horticulture variety descriptions (Direktorat Perbenihan Hortikultura, Direktorat Jenderal Hortikultura, Kementerian Pertanian 2016).[3]

Preference data and analysis

Preference tests involving farmers, seed growers, consumers and traders compared GE potato chips with chips currently sold in the supermarkets. Assessment was based on a range of 1–5, where 1 = not liked and 5 = very liked. The observed characters included the taste, appearance, colour and crispness of the chips. Respondents were also asked to choose their most preferred processed chip genotype, and their reasons.

Quantitative data were analysed by the F test at 5% level of significance. Quantitative data were analysed by PKBT STAT 2.01 software and different tests using the 5% Tukey Test. To verify the results of the multi-location test, a variety of truthfulness tests was conducted at one location of the multi-location test. Data from the multi-location trial must be verified by Local Seed Office Certification or the Office of Head of the Center for Commodity.

Correctness of varieties

The following steps are followed in testing for correctness:

- a proposal to conduct varieties' truthfulness tests is made to the Local Seed Office Certification or the Office of Head of the Center for Commodity;
- the authority appoints a breed officer to check all the characters in the description of the candidate varieties by verifying the data that have been observed by the proponent/ applicant;
- verification is made of the unique characters submitted by the proponent/applicant;
- the results are ratified and signed by the Head of the Office and the truthfulness test officer;
- the truthfulness test document is attached to the proposed variety registration document.

CFTs of late blight resistant GE potato for variety registration in 2013–14

In order to commercialise GE plants, a developer should conduct a multi-location trial. ICABIOGRAD, IAARD as the developer of GE plant in this case carried out three multi-location CFTs of GE/non-GE potato in Pangalengan and Garut (West Java) and Banjarnegara (Central Java) from 2013–2014.

Observations of late blight infection were conducted at 28 and 63 days after planting when the late blight symptoms were detected. At 35 days after planting, Granola showed susceptibility to *P. infestans* and required fungicide spraying, three hybrid clones (B163, D48, D38) showed no decline in symptoms and two hybrid clones (B35 and D37) were moderately resistant until 63 days after planting. Non-GE Atlantic and Granola were susceptible to *P. infestans* even though the plants were sprayed with fungicides (10 and 20 times). All the hybrid clones maintained resistance until 63 days after planting.

The yields of hybrid clones were 10.97–16.59 ton/ha, significantly higher than the control non-GE Atlantic (3.46 ton/ha) and Granola (5.39 ton/ha). The B35 and D38 hybrid clones showed higher yield among the clones of crosses from Atlantic and Katahdin SP951. Tuber yield from control plots without spray was significantly lower than from plots sprayed with fungicides. However, there was no significant difference in yield from plots sprayed 10 times and 20 times. This indicates that late blight resistant potato could reduce the number of fungicide spraying events by 50%. However, the data from the three CFT locations could not be used for varietal registration, because they were quite old.

A visit by the Horticultural Variety Assessment and Registration Team was conducted when the plant was one month old (2 August 2018). Potato plants were next inspected by the Truthfulness Test Team led by Dr Redy Gaswanato and Ms Juniarti Sahat on 3 September 2018, with guests from Michigan State University.

Harvesting was carried out in the CFT Pangalengan plot when the plants reached 100 days old (4 December 2018). The genotypes harvested were three hybrid potato clones derived from non-GE Granola × GE Katahdin SP951 (clones D12, D37 and D48) and one hybrid clone derived from non-GE Atlantic × GE Katahdin SP951 (clone B35). Observations were carried out for quantitative data, qualitative data and consumer preferences. Hybrid clone D48 was selected because of its desirable traits and yield (Table 17.2). It is proposed to be called BioGranola Agrihorti.

Bioefficacy tests conducted at IVEGRI using an isolate of *P. infestans* from Pangalengan showed late blight symptoms after inoculation. These symptoms were observed at four, eight,

Table 17.2. Some important agronomic and late blight resistance traits of hybrid clones of potato

Genotype	Late blight incidence*	Tuber no./ plant (#)	Tuber yield/ plant (g)	Tuber yield (ton/ha)	Sg (m^3)
D12	34.7	9.90 ab	855 ab	21.48 bc	1070
B35	271.8	6.30 b	1095 ab	21.56 bc	1090
D37	0	7.80 b	838 ab	21.75 bc	1079
D48 (BioGranola Agrihorti)	0	8.75 b	928 ab	25.81 ab	1078
Medians	—	13.75 a	1.046 ab	29.41 a	1089
Granola	864.8	9.05 b	1.153 a	27.69 ab	1065
Atlantic	781.9	6.60 b	748 ab	19.43 c	1082
Katahdin	0	8.75 b	676 b	17.41 c	1073
CV (%)	—	19.72	21.34	11.24	

B35 = clone from crossing non-GE Atlantic × GE Katahdin SP951; D12, D37, D38, D48 = clones from crossing non-GE Granola × GE Katahdin SP951.
* Incidence in terms of area under disease progress curve (AUDPC) over period of 16 days after inoculation in glasshouse trial. Figures followed by the same letter in the column do not show significant difference at 5% level.

12 and 16 days after inoculation. Cumulative late blight infection was calculated using AUDPC as per Campbell and Madden (1990). BioGranola Agrihorti showed resistance to late blight as compared to non-GE Granola and non-GE Atlantic (Table 17.2) and was preferred by 22% respondents in a limited survey.

Conclusion: prospects

BioGranola Agrihorti expresses only one gene for late blight resistance. Hence, efforts to increase durable broad-spectrum plant resistance to *P. infestans* are to be pursued by combining several resistance genes (stacked genes) (Halpin 2005). The development of potato plants using a stacked gene with three resistance genes – *Rpi-sto1* gene (*S. stoloniferum*), *Rpi-vnt1.1* (*S. venturii*) and Rpi-blb3 (*S. bulbocastanum*) – was successfully conducted in susceptible Desiree and Victoria varieties. All events are more resistant than control plants or non-GE plants and showed resistance until the end of the growing season in the Netherlands and Belgium, for two consecutive seasons. GE potato events containing a single gene showed lower resistance than events containing three resistance genes (Haesaert *et al.* 2015). In 2016, resistance tests in greenhouses and CFTs in Uganda showed that GE potato varieties of Victoria and Desiree containing *RB*, *Rpi-blb2* (van der Vossen *et al.* 2005) and *Rpi-vnt1.1* (Foster *et al.* 2009; Pel *et al.* 2009) genes were resistant to *P. infestans* until harvesting without spraying of fungicides (Ghislain *et al.* 2019). The successful development of potato with durable resistance to *P. infestans* is expected to facilitate potato production in Indonesia in a sustainable manner.

The development of GE potato using three resistance genes *Rpi-mcq1* (Smilde *et al.* 2005), *Rpi-blb2* and *Rpi-vnt1* is being conducted in the USA on GE Diamant and Granola varieties. In the initial stages, stacked-gene GE potatoes will be tested for their resistance to *P. infestans* in a CFT.

GE stacked-genes Diamant and Granola containing three genes (*Rpi-mcq1, Rpi-blb2, Rpi-vnt1*) will be donated to ICABIOGRAD, IAARD through Feed the Future Biotechnology Potato Partnership Project managed by Michigan State University and funded by USAID. A Material Transfer Agreement for GE stacked-genes Diamant for research purposes has been signed by both parties. The application for a research permit on GE stacked-genes potato resistant to *P. infestans* in CFT is in process, and stacked-genes potato Granola resistant to *P. infestans* are expected to be imported into Indonesia from Michigan State University in 2020. These efforts continue the research and development of stacked-gene late blight resistant potato that is suitable for Indonesia.

Endnotes

1. https://www.oecd.org/env/ehs/biotrack/46815167.pdf
2. https://www.oecd.org/env/ehs/biotrack/46815167.pdf
3. http://hortikultura.pertanian.go.id/

References

Adiyoga W (2009) Costs and benefits of transgenic late blight resistant potatoes in Indonesia. In *Projected Impacts of Agricultural Biotechnologies for Fruits and Vegetables in the Philippines and Indonesia.* (Eds GW Norton and MH Desiree) pp. 86–104. ISAAA *SEAsia* Center, Los Banos.

Ambarwati AD, Herman M, Purwito A, Sumaraw SM, Aswidinnoor H (2011) Resistance evaluation on populations of crosses between transgenic potato Katahdin RB and non-transgenic Atlantic and Granola to late blight (*Phytophthora infestans*) in confined field trial. *Indonesian Journal of Agricultural Science* **12**, 33–39. doi:10.21082/ijas.v12n1.2011.p33-39

Ambarwati AD, Herman M, Listanto E, Suryaningsih E, Sofiari E (2012) Pengujian ketahanan klon-klon hasil silangan tanaman kentang transgenik dengan non transgenik terhadap penyakit hawar daun *Phytophthora infestans* di lapangan uji terbatas. *Jurnal Hortikultura* **22**, 187–196. doi:10.21082/jhort.v22n2.2012.p187-196

Ambarwati AD, Kusmana, Listanto E (2015) Klon-klon kentang transgenik hasil silangan terseleksi tahan terhadap penyakit hawar daun *Phytophthora infestans* tanpa penyemprotan fungisida di empat lapangan uji terbatas. *Jurnal Biologi Indonesia* **11**, 177–186.

Ballvora A, Ercolano MR, Weiss J, Meksem K, Bormann CA, Oberhagemann P, Salamini F, Gebhardt C (2002) The *R1* gene for potato resistance to late blight (*Phytophthora infestans*) belongs to the leucine zipper/NBS/LRR class of plant resistance genes. *Plant Journal* **30**, 361–371. doi:10.1046/j.1365-313X.2001.01292.x

Basuki RS, Kusmana, Dimyati A (2005) Analisis daya hasil, mutu, dan respon pengguna terhadap klon 380584.3, TS-2, FBA-4, I-1085, dan MF-II sebagai bahan baku keripik kentang. *Jurnal Hortikultura* **15**, 160–170.

Bradeen JM, Iorizzo M, Mollov DS, Raasch J, Kramer LC, Millett BP, Phillips SA, Jiang J, Carputo D (2009) Higher copy numbers of the potato rb transgene correspond to enhanced transcript and late blight resistance levels. *Molecular Plant-Microbe Interactions* **22**, 437–446. doi:10.1094/MPMI-22-4-0437

Bravo-Almonacid F, Rudoy V, Welin B, Segretin ME, Bedogni MC, Stolowicz F, Criscuolo M, Foti M, Gomez M, Lopez M, Serino G, Cabral S, Dos Santos C, Huarte M, Mentaberry A (2012) Field testing, gene flow assessment and pre-commercial studies on transgenic *Solanum tuberosum* spp. *tuberosum* (cv. Spunta) selected for PVY resistance in Argentina. *Transgenic Research* **21**, 967–982. doi:10.1007/s11248-011-9584-9

Bushey DF, Bannon GA, Delaney BF, Graser G, Hefford M, Jiang X, Lee TC, Madduri KM, Pariza M, Privalle LS, Ranjan R, Saab-Rincon B, Schafer BW, Thelen JJ, Zhang JXQ, Harper MS (2014) Characteristics and safety assessment of intractable proteins in genetically modified crops. *Regulatory Toxicology and Pharmacology* **69**, 154–170. doi:10.1016/j.yrtph.2014.03.003

Campbell CL, Madden LV (1990) *Introduction to Plant Disease Epidemiology.* Wiley, New York.

Colton LM, Groza HI, Wielgus SM, Jiang J (2006) Marker-assisted selection for the broad-spectrum potato late blight resistance conferred by gene *RB* derived from a wild potato species. *Crop Science* **46**, 589–594. doi:10.2135/cropsci2005.0112

Cooke DEL, Young V, Birch PRJ, Toht R, Gourlay F, Day JP, Carnegie SF, Duncan JM (2003) Phenotypic and genotypic diversity of *Phytophthora infestans* populations in Scotland (1995–97). *Plant Pathology* **52**, 181–192. doi:10.1046/j.1365-3059.2003.00817.x

Delaney B, Astwood JD, Cunny H, Conn RE, Herouet-Guicheney C, MacIntosh S, Meyer LS, Privalle L, Gao Y, Mattsson J, Levine M (2008) Evaluation of protein safety in the context of agricultural biotechnology. *Food and Chemical Toxicology* **46**(Suppl 2), S71–S97. doi:10.1016/j.fct.2008.01.045

Direktorat Perbenihan Hortikultura, Direktorat Jenderal Hortikultura, Kementerian Pertanian (2016) Pedoman pendaftaran varietas.

Dwiastuti ME, Djoema'ijah (2000) Ketahanan beberapa klon kentang terhadap *Phytophthora infestans* Mont.D.By. di Sumber Brantas. *Jurnal Hortikultura* **10**, 24–29.

El-Kharbotly A, Leonards-Schippers C, Huigen DJ, Jacobsen E, Pereira A, Stiekema WJ, Salamini F, Gebhardt C (1994) Segregation analysis and RFLP mapping of the *R1* and *R3* alleles conferring race-specific resistance to *Phytophthora infestans* in progeny of dihaploid potato parents. *Molecular & General Genetics* **242**, 749–754. doi:10.1007/BF00283432

Estiati A, Herman M (2015) Regulasi keamanan hayati produk rekayasa genetik di Indonesia. *Analis Kebijakan Pertanian* **13**, 129–146. doi:10.21082/akp.v13n2.2015.129-146

Forbes GA (2009) Late blight in developing countries and the role of the global initiative on late blight. In *GILB Proceedings of the 11th Euroblight Workshop*, Wageningen, pp. 37–43.

Foster SJ, Park TH, Pel M, Brigneti G, Sliwka J, Jagger L, van der Vossen E, Jones JDG (2009) *Rpi-vnt1.1*, a *Tm-2*^{2} Homolog from *Solanum venturii*, confers resistance to potato late blight. *Molecular Plant-Microbe Interactions* **22**, 589–600. doi:10.1094/MPMI-22-5-0589

Ghislain M, Byarugaba AA, Magembe E, Njoroge A, Rivera C, Roman ML, Tover JC, Gamboa S, Forbes GA, Kreuze JF, Barekye A, Kiggundu A (2019) Stacking three late blight resistance genes from wild species directly into African highland potato varieties confers complete field resistance to local blight races. *Plant Biotechnology Journal* **17**, 1119–1129. doi:10.1111/pbi.13042

Haesaert G, Vossen JH, Custers R, Loose MD, Haverkort A, Heremans B, Hutten R, Kessel G, Landschoot S, Droogenbroeck BV, Visser RGF, Gheysen G (2015) Transformation of the potato variety desiree with single or multiple resistance genes increases resistance to late blight under field conditions. *Crop Protection* **77**, 163–175. doi:10.1016/j.cropro.2015.07.018

Halpin C (2005) Gene stacking in transgenic plants-the challenge for 21st century plant biotechnology. *Plant Biotechnology Journal* **3**, 141–155. doi:10.1111/j.1467-7652.2004.00113.x

Halterman DA, Kramer LC, Wielgus S, Jiang J (2008) Performance of transgenic potato containing the late blight resistance gene *RB*. *Plant Disease* **92**, 339–343. doi:10.1094/PDIS-92-3-0339

Hariyadi, Koentjoro Y (1996) Penampakan galur-galur kentang (*Solanum tuberosum*. L) resisten terhadap penyakit hawar daun (*Phytophthora infestans* Mont.d.By). In *Prosiding Simposium Pemuliaan Tanaman IV*, UPN Jatim, 24–25 Mei 1996, pp. 241–248.

Hawkes JG (1994) Origins of cultivated potatoes and species relationships. In *Potato Genetics*. (Eds JE Bradshaw and GR Mackay) pp. 3–42. CAB International, Wallingford.

Helgeson JP, Pohlman JD, Austin S, Haberlach GT, Wielgus SM, Ronis D, Zambolim L, Tooley P, McGrath JM, James RV, Stevenson WR (1998) Somatic hybrids between *Solanum bulbocastanum* and potato: a new source of resistance to late blight. *Theoretical and Applied Genetics* **96**, 738–742. doi:10.1007/s001220050796

Herman M (2007) *ABSP II Technical Report of Confined Field Trial of Late Blight Resistant (LBR) Potato Activities in Indonesia*. USDA/ABSP II, Cornell University, Ithaca, NY, USA.

Herman M (2009) Pelepasan, pemanfaatan, pengawasan, dan pengendalian tanaman PRG. In *Tanaman Produk Rekayasa Genetik dan Kebijakan Pengembangannya: Status Global Tanaman Produk Rekayasa Genetik dan Regulasinya*. (Eds B Purwantara dan M Thohari) Balai Besar Penelitian dan Pengembangan Bioteknologi dan Sumber Daya Genetik Pertanian. Badan Penelitian dan Pengembangan Pertanian, Departemen Pertanian, Bogor.

Herman M (2010) Empat belas tahun perkembangan peraturan keamanan hayati dan keamanan pangan produk rekayasa genetik dan implementasinya di Indonesia. *Jurnal AgroBiogen* **6**, 113–125. doi:10.21082/jbio.v6n2.2010.p113-125

Herman M (2011) *Current Status and Progress of ABSPII Activities in Indonesia: Research on Late Blight Resistant Potato to Phytophthora infestans and Outreach Communication Activities*. Annual Advisory ABSPII Board Meeting. Cox's Bazaar, Bangladesh, 31 January–1 February.

Jo KR, Kim CJ, Kim SJ, Kim TY, Bergervoet M, Jongsma MA, Visser RGF, Jacobsen E, Vossen JH (2014) Development of late blight resistant potatoes by cisgene stacking. *BMC Biotechnology* **14**, 50. doi:10.1186/1472-6750-14-50

Kramer LC, Choudoir MJ, Wielgus SM, Bhaskar PB, Jiang J (2009) Correlation between transcript abundance of the *RB* gene and the level of the *RB*-mediated late blight resistance in potato. *Molecular Plant–Microbe Interactions* **22**, 447–455. doi:10.1094/MPMI-22-4-0447

Kuhl JC, Zarka K, Coombs J, Kirk WW, Douches DS (2007) Late blight resistance of RB transgenic potato lines. *Journal of the American Society for Horticultural Science* **132**, 783–789. doi:10.21273/JASHS.132.6.783

Kusmana (2003) Evaluasi beberapa klon kentang asal stek batang untuk uji ketahanan terhadap *Phytophthora infestans*. *Jurnal Hortikultura* **13**, 220–228.

Kusmana (2004) Evaluasi resistensi 26 genotip kentang terhadap penyakit busuk daun di Cibodas, Lembang. *Jurnal Hortikultura* **14**, 15–24.

Kusmana, Basuki R (2004b) Produksi dan mutu umbi klon kentang dan kesesuainnya sebagai bahan baku kentang goreng dan keripik kentang. *Jurnal Hortikultura* **14**, 246–252.

Listanto E, Wattimena GA, Armini NM, Sinaga MS, Sofiari E, Herman M (2009) Regenerasi beberapa kultivar kentang dan transformasi kentang dengan gen RB melalui *Agrobacterium tumefaciens*. *Jurnal Hortikultura* **19**, 137–147.

Listanto E, Riyanti EI, Santoso TJ, Hadiarto T, Ambarwati AD (2015) Genetic stability analysis of RB gene in genetically modified potato lines tolerant to *Phytophthora infestans*. *Indonesian Journal of Agricultural Science* **16**, 51–58. doi:10.21082/ijas.v16n2.2015.pp.51-58

Millett BP, Mollov DS, Iorizzo M, Carputo D, Bradeen JM (2009) Changes in disease resistance phenotypes associated with plant physiological age are not caused by variation in *R* gene transcript abundance. *Molecular Plant–Microbe Interactions* **22**, 362–368. doi:10.1094/MPMI-22-3-0362

Naess SK, Bradeen JM, Wielgus SM, Haberlach GT, Mc Grath JM, Helgeson JP (2000) Resistance to late blight in *Solanum bulbocastanum* is mapped to chromosome 8. *Theoretical and Applied Genetics* **101**, 697–704. doi:10.1007/s001220051533

Pel M, Foster SJ, Park TH, Rietman H, van Arkel G, Jones JDG, Jacobsen E, Visser R, van der Vossen E (2009) Mapping and cloning of late blight resistance genes from *Solanum venturii* using an interspecific candidate gene approach. *Molecular Plant–Microbe Interactions* **22**, 601–615. doi:10.1094/MPMI-22-5-0601

Riyanti EI, Listanto E, Ambarwati AD (2014) Effects of late blight resistant potato containing *RB* gene on the soil microbes, pests and diseases. *Indonesian Journal of Agricultural Science* **15**, 47–54.

Schepers HTAM, Spits HG (2006) The development and control of *Phytophthora infestans* in Europe in 2004–2005. In *Proceedings of the 9th Workshop of a European Network for Development of an Integrated Control Strategy of Potato Late Blight*. (Eds CE Westerdijk and HTAM Schepers), PPO Special Report no. 11, pp. 11–22.

Simatupang S, Hutagalung L, Sembiring T, Bahar FA (1996) Adaptasi varietas kentang di dataran medium Kabupaten Simalungun Sumatra Utara. *Jurnal Hortikultuta* **6**, 249–254.

Smilde WD, Brigneti G, Jagger L, Perkins S, Jones JDG (2005) Solanum mochiquense chromosome IX carries a novel late blight resistance gene *Rpi-moc1*. *Theoretical and Applied Genetics* **110**, 252–258. doi:10.1007/s00122-004-1820-8

Song J, Bradeen JM, Naess SK, Raasch JA, Wielgus SW, Haberlach JT, Liu J, Kuang H, Austin-Phillips S, Buell CR, Helgeson JP, Jiang J (2003) Gene *RB* cloned from *Solanum bulbocastanum* confers broad spectrum resistance to potato late blight. *Proceedings of the National Academy of Sciences of the United States of America* **100**, 9128–9133. doi:10.1073/pnas.1533501100

Staples RC (2004) Race non-specific resistance for potato late blight. *Trends in Plant Science* **9**, 5–6. doi:10.1016/j.tplants.2003.10.009

Suryadi S, Sahat, Gaos MAH (1995) Pengujian resistensi beberapa varietas/klon kentang (*Solanum tuberosum* L.) terhadap penyakit busuk daun (*P. infestans*). *Bulletin Penelitian Hortikultura* **22**, 57–60.

Umaerus V, Umaerus M (1994) Inheritance of resistance to late blight. In *Potato Genetics*. (Eds JE Bradshow and GR Mackay) pp. 365–401. CAB International, Wallingford.

van der Vossen E, Sikkema A, Hekkert Baste L, Gros J, Stevens P, Muskens M, Wouters D, Pereira A, Stiekema W, Allefs S (2003) An ancient *R* gene from the wild potato species *Solanum bulbocastanum* confers broad-spectrum resistance to *Phytophthora infestans* in cultivated potato and tomato. *Plant Journal* **36**, 867–882. doi:10.1046/j.1365-313X.2003.01934.x

van der Vossen EA, Gros J, Sikkema A, Muskens M, Wolters DP, Pereira A, Allefs S (2005) The *Rpi-blb2* gene from *Solanum bulbocastanum* is an *Mi-1* gene homolog conferring broad-spectrum late blight resistance in potato. *Plant Journal* **44**, 208–222. doi:10.1111/j.1365-313X.2005.02527.x

Wastie RL (1991) Breeding for resistance. In *Advances in Plant Pathology: Phytophthora infestans – The Cause of Late Blight in Potato*. 7 (Eds DS Ingram and PH Williams) pp. 193–224. Academic Press, London.

18

Genetic modifications for disease resistance in crops

SV Ramesh, Satendra K Mangruthia and Shelly Praveen

Introduction

It is estimated that the global human population will rise from the current 7.7 billion to ~8.5 billion by 2030 and 9.7 billion by 2050. According to a UN Food and Agricultural Organization report (FAO 2018), over 821 million people are undernourished in the world; ~200 million of them are in India alone. Hence, eradication of hunger (zero hunger) is one of the Sustainable Development Goals that the UN has set to achieve by 2030. To achieve the goal of zero hunger, we need to increase food production against the background of resource and climate constraints that the developing world is facing. Around 20–30% of the world's food grain production is lost due to damages caused by plant diseases and pests (Savary *et al.* 2012, 2019). In India, the importance of plant diseases was realised in the first half of the 19th century. In 1937, Mitra reviewed crop losses due to pests and diseases, and found that different rusts of wheat caused as much as UK£40 million of losses worldwide, UK£4 million of it in India (Mitra 1937). In 1942, brown spot of paddy caused by fungus was partially responsible for failure of the rice crop in Bengal, with 40–90% yield reduction leading to the starvation death of ~2 million people during the great Bengal famine of 1943. Plant diseases cause ~30–50% of total crop losses due to all kinds of biotic stresses. These crop losses are aggravated by climate change, intensive cropping, introduction of new pathogens and strains and change in their virulence status, facilitated with global trade.

Crop disease resistance: management practices

Management of crop diseases generally involves four main approaches: host plant resistance, chemical control, biological control and crop husbandry practices. Even though cultural practices such as managing the planting date, crop rotation and ecological engineering via multi-species cropping systems have yielded desired results in the management of diseases, at times these measures are inadequate or practically not feasible. Biological control has not yielded desired results in many destructive and rapidly spreading crop diseases. Over-dependence on pesticides has caused pollution and ecological imbalance in the agroecosystem. However, growing awareness about environmental pollution, pesticide residues in food products and the evolution of resistance in plant pathogens has led to the implementation of integrated disease management (IDM) strategies. Hence, IDM complements all tactics like chemical, cultural, crop resistance, biological and even transgenics in the best possible manner to achieve efficient and economic control of diseases in the environmental context.

Host plant resistance is an ecologically safe way of managing diseases. It continues to contribute to crop production in a significant manner. Conventional breeding techniques used by traditional breeders involve selection of parents, hybridisation and effective selection for desired resistance traits among the progenies for multiplication, which requires enormous amounts of time in screening the plant germplasm. Our initial understanding of genetic resistance was based on the concept of 'gene-for-gene' which requires a resistance gene (*R*) in the host plant and avirulence gene (*Avr*) in the pathogen (Flor 1955). However, it took three decades to clone the first *Avr* gene (Staskawicz *et al.* 1984) and still longer to clone the single resistance-conferring loci (*R*) gene (Martin *et al.* 1993). Thus, the *R* gene from tomato *Pto* conferred resistance to the *Avr* genes *AvrPto* and *AvrPtoB* carried by the pathogen *Pseudomonas syringae* causing bacterial speck disease of tomato (Scofield *et al.*1996; Tang *et al.* 1996; Kim *et al.* 2002). Simultaneously, nucleotide-binding site leucine-rich repeat (*NBS-LRR*) kind of *R* genes were characterised (Bent *et al.* 1994). Breeders have developed strategies to select and multiply the specific genotypes that harbour *R* gene so as to combat the crop diseases. DNA-based molecular markers linked to the *R* gene of interest are identified to expedite the varietal development, in the process called marker-assisted selection (MAS). MAS has greatly aided conventional breeding techniques including backcross breeding, widely used to develop resistant cultivars, by reducing the time involved developing a variety in an efficient manner (Farokhzadeh and Ali Fakeheri 2014). Rice breeding has witnessed important advancements as several genes conferring resistance to blast, bacterial blight, tungro and sheath blight diseases have been identified, characterised and used in resistance breeding programs (Hittalmani *et al.* 2000; Zhu *et al.* 2016). In India, ICAR-Indian Institute of Rice Research (formerly Directorate of Rice Research) Hyderabad, ICAR-National Rice Research Institute, Cuttack, along with IARI, New Delhi and several other ICAR institutes and agricultural universities have made significant progress in achieving disease resistance in rice crop through molecular breeding, and have developed commercial cultivars resistant to blast and bacterial blight diseases. All India Coordinated Rice Improvement Programme (AICRIP) has been instrumental in rice resistance breeding. Genetic studies of Indian rice land races not only identified molecular markers linked to resistance traits but also identified novel resistant sources (Yadav *et al.* 2019). Similarly, resistance to the fungal pathogen *Fusarium oxysporum* f. sp. *ciceris* (Foc) that causes Fusarium wilt in chickpea was effectively complemented by the identification of suitable molecular markers linked to the disease resistance (Jendoubi *et al.* 2017). The tomato genome sequencing project has yielded the complete repertoire of 365 *NBS-LRR* genes of tomato (Sato *et al.* 2012; Jupe *et al.* 2013; Andolfo *et al.* 2014). This has greatly helped in mapping and cloning of disease *R* genes by combining genetics and crop genomics. At the field level, deploying single *R* gene has not been very effective as resistance breakdown occurs soon and the genetic mutation in pathovars switches them to novel pathovars (Sprague *et al.* 2006; Bousset *et al.* 2018). In *Brassica napus*, *RLM*-1 major gene that confers resistance to the pathogen *Leptosphaeria maculans* was overcome by the emergence of a virulent pathogen population in only five years (Sprague *et al.* 2006). Another approach to avoid or impede the occurrence of resistance breakdown was to grow multi-lines wherein a mixture of various pure lines, each incorporated with a different *R* gene, is cultivated. This approach provided resistance to disease to some extent compared to monoculturing of a single genotype. In cultivated potato, resistance to late blight disease was achieved through incorporation of major (*R*) genes from the wild relative *Solanum demissum*. Various *R* genes introgressed into potato provided control of late blight causing *Phytophthora infestans* populations. Hence, to avoid the breakdown of resistance, gene pyramiding or stacking has been the method of choice wherein two or more *R*-genes have been introgressed into a single genotype (Kim *et al.* 2012; Jo *et al.* 2014).

Transgenic strategies for disease resistance

To overcome the shortcomings of conventional and molecular breeding techniques, genetic transformation has been exploited to develop disease-resistant transgenic plants. Transgenic crops circumvent problems such as linkage drag, sexual incompatibility or cross-incompatibility, and repulsion-linked genes (Rodriguez-Moreno *et al.* 2017). The concept of gene cassettes, wherein clusters of various classes of *R* genes are cloned in a single vector and elite plant varieties are transformed to engineer durable resistance, has been proposed (Wulff and Moscou 2014). Ectopic expression of proteins to develop resistance status, suppression of RNA expression though RNA-based gene silencing mechanisms utilising antisense RNAs, small interfering RNAs (siRNAs) and artificial microRNAs (amiRNAs) have also been employed (Praveen *et al.* 2005, 2010; Ramesh *et al.* 2007; Koundal and Praveen 2010). Another promising approach is host-induced gene silencing (HIGS), aimed at down-regulating the key genes of pathogen origin, especially in plant–fungal interactions (Nunes and Dean 2012).

Protein expression-based crop disease resistance

Genetic engineering of crop plants originally referred to the expression of alien genes to incorporate a desired trait which the superior or elite genotype lacks. Genetic engineering approaches have allowed for expression of resistant genes (*R* genes) from wild or cross-incompatible species. Identification and expression of *R* genes from the wild relative of a cultivated plant has yielded disease-resistant phenotypes. There are numerous examples of successful utilisation of *R* gene-mediated resistance in field crops through transgenic technology (van der Vossen *et al.* 2005; Zhao *et al.* 2005; Yang *et al.* 2008). The strategy of *R* gene-mediated resistance has successfully been exploited against specialised, obligate (biotrophic) pathogens. However opportunistic, non-obligate (necrotrophic) pathogens produce many toxins and cell wall-degrading enzymes (polygalacturonases) as pathogenicity factors. Hence, transgenic expression of polygalacturonase-inhibitory proteins has been exploited to control the nectrophic pathogens (Manfredini *et al.* 2005; Joubert *et al.* 2007; Janni *et al.* 2008). Similarly, oxalic acid-degrading enzymes such as oxalate oxidase and oxalate decarboxylase are expressed in transgenic plants to confer resistance against *Botrytis cinerea* infecting tomato (Walz *et al.* 2008). Field trials of transgenic Cavendish banana against the infection of a virulent fungus, *F. oxysporum* f. sp. *cubense* tropical race 4 (TR4) showed promising results. In the field trial, two different Cavendish lines, one transformed with *RGA2* (*R* gene) isolated from the resistant diploid relative and another transformed with *Ced* 9, a nematode-derived gene that negatively regulates programmed cell death, were found to be resistant (Dale *et al.* 2017). Interestingly, the endogenous or *RGA2* homologues of Cavendish also showed expression but 10-fold lower than in transgenic lines. This led the authors to propose the utilisation of gene editing technologies to engineer homologues of RGA2 to obtain non-GM banana that is resistant to this fungal disease.

Non-host resistance and crop genetic modification

The virulence of pathogen strains and emergence and re-emergence of pathogens has led to the exploration of plant resistance genes from non-host organisms. Hence, non-host resistance (NHR) is defined as the resistance exhibited by the plant species towards all the variants of a non-adapted pathogen (Uma *et al.* 2011). In contrast, host resistance (HR) is based on the concept of gene-for-gene disease resistance which is race- or cultivar-specific (Flor 1955). NHR can be either passive or active wherein physical or chemical barriers in the plants cause incompatible reactions. If the NHR occurs in the absence of the pathogen it is called passive NHR (Heath 2000); where barriers are formed in the presence of the pathogen it is termed active NHR (Fonseca and Mysore 2019).

NHR has achieved much significance recently in the development of resistant crop phenotypes owing to the broad-spectrum and durable nature of the resistance. The broad-spectrum resistance of NHR can be attributed to its continuum of layered defences. Notably, most of the NHR genes identified to date are limited to very few plants such as *Arabidopsis*, *Nicotiana benthamiana* and *Oryza sativa* (Sumit *et al.* 2012; Lee *et al.* 2013; Senthil-Kumar and Mysore 2013, Gill *et al.* 2015; Fonseca and Mysore 2019). The major plant defence mechanisms include pattern recognition receptors (PRRs) that effectively recognise the pathogen-associated molecular patterns (PAMP), and effector-triggered immunity (ETI), which is otherwise called *R* gene-mediated resistance (Moffett 2009). The PRRs form the basis of NHR wherein the growth of pathogen is restricted due to the limiting nutrient condition, formation of papilla, deposition of lignin, callose or through peroxisome-based biosynthesis (Senthil-Kumar and Mysore 2013). Most of the NHR genes identified have been demonstrated to have multi-functional roles including modulation of the plant developmental activities. For instance, enzymes involved in ROS-mediated signal transduction pathways have a role in NHR against bacterial pathogens (Rojas *et al.* 2012; Senthil-Kumar and Mysore 2013). Heterotrimeric G-proteins that are well conserved in eukaryotes confer NHR against fungal and bacterial pathogens (Liang *et al.* 2016). The potential application of NHR-conferring genes in disease resistance development has been demonstrated (Lacombe *et al.* 2010; Jiwan *et al.* 2013; Lee *et al.* 2016). Transgenic expression of *Arabidopsis NPR*1 (*non-expresser of PR* genes 1) in strawberry confers broad-spectrum resistance against a variety of pathogens such as *Colletotrichum* sp. causing anthracnose, *Podosphaera aphanis* causing powdery mildew, and *Xanthomonas fragariae*-induced angular leaf spot (Silva *et al.* 2015). Since it was deciphered that loss-of-function mutant alleles of barley mildew resistance locus O (*MLO*) provided resistance to powdery mildew, RNA-mediated down-regulation of peach *MLO* gene conferred cross-species resistance to strawberry powdery mildew (Jiwan *et al.* 2013). NHR has been successfully demonstrated to develop fungal and bacterial diseases in a variety of crops (Tai *et al.* 1999; Jafary *et al.* 2008; Jeuken *et al.* 2008; Qin *et al.* 2008; Lacombe *et al.* 2010; Silva *et al.* 2015). NHR-based genetically engineered disease protection was successfully demonstrated in field-grown soybean (Wang *et al.* 2018a) and a few other crop species (Fonseca and Mysore 2019).

Pathogen-derived resistance and RNA interference to combat viral infections

Pathogen-derived resistance (PDR) refers to the transgenic expression of genetic elements of phytopathogenic viruses to confer disease resistance. The technique has been widely employed in the development of virus-resistant transgenic plants (VRTPs). The nature and inheritance of the resistance have varied, depending on the point of insertion of virus-derived gene in the genomic loci of host plants. Generally, virus-derived coat protein (CP), replicase protein (Rep) and movement protein (MP) are expressed in the susceptible genotypes to induce virus resistance phenotype (Plate 12). The mechanism of the resistance has been identified and the level of resistance has been at the protein and the RNA levels (Baulcombe 1996; Beachy 1997).

RNA-based gene silencing approaches have been the mainstay for controlling plant diseases caused by phytopathogenic viruses. RNA interference (RNAi) phenomenon involves expression of double-stranded RNA leading to the production of a variety of small non-coding RNAs (sncRNAs) such as siRNAs and miRNAs that act as effectors of gene regulation and silence the expression of complementary mRNA (Plate 13). This phenomenon has been widely employed in disease resistance against the infection of viral pathogens. Considering the relatively small genome size of plant viruses and the ability to manipulate viral genes for RNA-based gene silencing, many crops have been developed and commercialised for cultivation based on sncRNAs (Table 18.1).

Besides these interventions, many successful instances of genetic modifications to control the diseases of bacterial, fungal and oomycete pathogens are available worldwide (van Esse *et al.* 2020). RNAi-based resistance has been achieved against *Rhizoctonia solani* causing sheath blight of rice by targeting the PATHOGENICITY MAP KINASE 1 (PMK1) homologues (RPMK1-1 and RPMK1-2) and polygalacturonase-encoding genes of the pathogen (Tiwari *et al.* 2017; Rao *et al.* 2019). Resistance development against this fungal pathogen of rice is a significant achievement as no natural resistance source is recorded in available rice germplasm. Notable examples of induced fungal resistance include introgression of late blight resistance gene (*Rpi-Vnt1*) derived from *Solanum venturii* in potato (Foster *et al.* 2009). Silencing of effector target protein VASCULAR HIGHWAY1-interacting kinase (StVIK) conferred resistance to *P. infestans* (Murphy *et al.* 2018). Down-regulation of host transcription factor (IPA1/OsSPL14) not only yielded a blast-resistant rice genotype but increased the economic yield (Wang *et al.* 2018b).

We have demonstrated the application of transgenic technology in developing disease-resistant field crops, especially for the control of tomato leaf curl disease afflicting tomato. The genetic diversity of begomoviruses inflicting the leaf curl disease in tomato, coupled with the limited genetic resistance sources for the disease, prompted us to develop transgenic technology utilising the viral nucleic acids and the phenomenon of pathogen-derived resistance. Antisense RNA technology has been successfully demonstrated to suppress the virus establishment in tomato by expressing the antisense *rep* gene (coding for replication associated protein) derived from Tomato leaf curl New Delhi virus (ToLCNDV) (Praveen *et al.* 2005). Genetically engineered tomato with *rep* gene was found to be safe following the toxicity and allergenicity tests (Singh *et al.* 2009). Later, RNAi-based suppression of ToLCNDV was achieved by the constitutive expression of siRNAs targeting viral genes. The siRNAs were derived from the conserved region of the *rep* gene with overlapping sequence from AC4 gene (Ramesh *et al.* 2007). These strategies imply that ToLCV *rep*-driven RNAi, targeting AC4 and conserved viral sequences, provides a promising approach to suppress a wide-spectrum ToLCV infection in the tomato (Ramesh *et al.* 2007). The strategy was further improved and artificial microRNA (amiRNA) has been utilised to express *rep* gene-derived sequences to counteract the expression of ToLCV (Koundal and Praveen 2010).

Indian research and development efforts to protect crops from virus infection through transgenic approaches have gained momentum in the last decade (Table 18.2). Diverse crop species including potato (resistant to Potato virus Y), rice (resistant to Rice tungro virus), cotton (resistant to Cotton leaf curl virus), papaya (resistant to Papaya ring spot virus), cucumber (resistant to Cucumber mosaic virus), soybean (resistant to Mungbean yellow mosaic India virus) etc. were genetically modified. The transgenic strategies varied from sense expression of protein and antisense RNA expression, to RNA interference wherein expression of hairpin RNA and siRNAs targeting viral genomic regions successfully yielded trait-stable transgenics. Short hairpin RNA (shRNA) expressing siRNAs targeted against the MYMIV encoded viral suppressor of RNA silencing (AC2) showed promise for the development of transgenic soybean (Ramesh *et al.* 2019). Recently, Malathi *et al.* (2019) developed transgenic RNAi lines of rice using *CP3* gene of Rice tungro spherical virus (RTSV) to develop tungro disease resistance. The transgenic lines showed inability to transmit the virus complex and were evaluated for yield and other morphological traits. RNAi-based virus resistance against Rice tungro bacilliform virus has been further introgressed into five popular rice cultivars using marker-assisted backcross methodology (Kumar *et al.* 2019). Thus, application of transgenic technology and molecular breeding has been effectively utilised to develop virus-resistant phenotypes in crops of

Table 18.1. Summary of approved GM crops based on protein and RNA-mediated crop disease resistance

Crop	Event or trade name	Developer	Country and year of approval
Common bean (*Phaseolus vulgaris*) Resistant to Bean golden mosaic virus (BGMV)	EMBRAPA 5.1	EMBRAPA	Brazil 2011
Papaya (*Carica papaya*) Resistant to Papaya ring spot virus (PRSV)	Rainbow, SunUp	Cornell University and University of Hawaii	Canada 2003; Japan 2011; USA 1996
	Event 63–1	Cornell University and University of Hawaii	USA 1996
	Huanong no. 1	South China Agricultural University	China 2006
	X17–2	University of Florida	USA 2008
Plum (*Prunus domestica*) Resistant to Plum pox virus (PPV)	C-5	US Department of Agriculture	USA 2007
Potato Resistant to Potato virus Y (PVY)	Hi-Lite New Leaf™ Y potato (event HLMT15–15, HLMT15–3 HLMT15–46)	Monsanto	USA 1998
	New Leaf™ Y Russet Burbank potato (event RBMT15–101)	Monsanto	Australia 2001; Canada 1999; Japan 2003; Mexico 2001; New Zealand 2001; Philippines 2003; South Korea 2004; USA 1997
Potato Resistant to Potato leaf roll virus (PLRV)	New Leaf™ plus Russet Burbank potato (event RBMT21–129, RBMT21–350, RBMT22–082)	Monsanto	Australia 2001; Canada 1999; Japan 2001; Mexico 2001; New Zealand 2001; Philippines 2004; South Korea 2004; USA 1997
	New Leaf™ plus Russet Burbank potato (event RBMT21–152, RBMT22-186, RBMT22–238, RBMT22–262)	Monsanto	USA 1998
Potato Resistant to Potato virus Y (PVY)	Shepody New Leaf™ Y potato (event SEMT15–02, SEMT15–15)	Monsanto	Australia 2001; Canada 1999; Japan 2003; Mexico 2001; New Zealand 2001; Philippines 2003; South Korea 2004; USA 1997
	Shepody New Leaf™ Y potato (event SEMT15–07)	Monsanto	USA 1998

Crop	Event or trade name	Developer	Country and year of approval
Squash (*Cucurbito pepo*) Broad-spectrum virus resistance against Cucumber mosaic cucumovirus (CMV), Zucchini yellow mosaic potyvirus (ZYMV), Watermelon mosaic potyvirus 2 (WMV2)	Event: CZW3	Seminis Vegetable Seeds (Canada) and Monsanto Co. (Asgrow)	Canada 1998; USA 1994
Squash (*Cucurbito pepo*) Resistant to Zucchini yellow mosaic potyvirus (ZYMV), Watermelon mosaic potyvirus 2 (WMV2)	Event: ZW20	Seminis Vegetable Seeds (Canada) and Monsanto Co. (Asgrow)	USA 1994
Sweet pepper (*Capsicum annuum*) Resistant to Cucumber mosaic cucumovirus (CMV)	PK-SP01	Beijing University	China 1998
Tomato (*Lycopersicon esculentum*) Resistant to Cucumber mosaic cucumovirus (CMV)	Event: PK-TM8805R (8805R)	Beijing University	China 1999

Source: compiled from International Service for the Acquisition of Agri-biotech Applications (ISAAA), GM Approval Database (2019) <http://www.isaaa.org/gmapprovaldatabase/default.asp>

economic importance. Further, use of genome editing technologies to develop begomovirus resistance in a commercially important crop like cotton against cotton leaf curl disease has been initiated (Uniyal *et al.* 2019).

Application of gene-editing tools for disease resistance

Classical breeding methods are still being used to incorporate resistance trait to an otherwise superior variety. However, the process requires several rounds of genetic recombination and rigorous screening to obtain the natural or induced variants with the resistance-conferring trait. Another feasible approach is to generate desired variations through random mutagenesis. However, selection of the desired variant again involves screening of large populations. In this context, site-directed mutagenesis using gene-editing technologies is a pragmatic technique that can provide the desired resistance in the shortest possible time without compromising the original traits of the plant genotype. Gene-editing technologies function on the principle of creating double-stranded DNA breaks using specific nucleases followed by the action of DNA repair mechanisms, by both non-homologous end joining and homologous recombination, to effect small deletions, insertions and modifications (Plate 14). The nucleases such as zinc finger nucleases (ZFN) and transcription activator-like effector nucleases (TALENs) were initially used for the purpose (Li *et al.* 2012; Bogdanove and Booher 2016). The RNA-guided nucleases-based CRISPR (clustered regularly interspaced short palindromic repeats)/Cas9 (CRISPR-associated protein-9nuclease) is a preferred mode of gene editing owing to its versatility (Jinek

Table 18.2. Salient achievements of transgenic virus resistance in major crops grown in India

Crop	Target virus	Viral gene modified	GM technology	Developer	Institutional type
Banana	Banana bunchy top virus (BBTV)	Replicase	RNAi	NRCB, Trichy	Public
Citrus	Citrus tristeza virus (CTV)	Coat protein (CP)	Sense	IARI, New Delhi and Delhi University	Public
Cotton	Cotton leaf curl virus (CLCuV)	AV_2	Antisense	IISc. Bangalore and Mahyco, Jalna	Public and private
	CLCuV	AC_1	Antisense	IARI, New Delhi and CICR, Nagpur	Public
	CLCuV	—	Genome editing	Central University of Punjab and CIAB, Mohali	Public
	Tobacco streak virus (TSV)	CP	-	Mahyco, Jalna	Private
Cucumber	Cucumber mosaic virus (CMV)	CP	Sense, RNAi	IARI	Public
Groundnut	Groundnut bud necrosis virus (GBNV) + TSV	Np + CP	Sense	IARI, New Delhi and DGR Junagarh	Public
Papaya	Papaya ring spot virus (PRSV)	CP	Sense, RNAi	IARI and CISH, Lucknow	Public
		HcPro	RNAi	IARI and CISH, Lucknow	Public
Potato	Potato virus Y (PVY)	CP	Sense	IARI and CPRI Shimla	Public
	PVY	CP	Sense	IARI	Public
	Potato leaf roll virus (PLRV)	CP	Sense	IARI and CPRI Shimla	Public
	Tomato leaf curl New Delhi virus (ToLCNDV)	AC_1	RNAi	CPRI Shimla	Public
Rice	Rice tungro bacilliform virus (RTBV)	ORF IV	RNAi	Delhi University, IIRR, Hyderabad, TNAU, Coimbatore, BCKV, Kalyani	Public
	Rice tungro spherical virus (RTSV)	-	Sense, antisense	Delhi University, IIRR, Hyderabad	Public

Soybean	Mungbean yellow mosaic India virus (MYMIV)	AC_1, AC_2	Antisense, RNAi	IARI, New Delhi and IISR, Indore	Public
Sunflower	TSV	CP	Sense	IARI, New Delhi, J.K. Seeds and IIOR, Hyderabad	Public and Private
Tomato	ToLCV	AV_2	Antisense	IISc. Bangalore and Rallis India	Public and private
	ToLCV	CP (coat protein)	Sense	NBRI, Lucknow	Public
	ToLCNDV	AC_1 (Replicase)	Antisense	IARI, New Delhi	Public
		AC_1 (Replicase) AC_4 (RNAi Suppressor)	RNAi	IARI, New Delhi	Public
			RNAi	IARI and Bejo Sheetal Pvt. Ltd, Jalna	Public and private
	Cucumber mosaic virus (CMV)	CP	Sense, antisense, RNAi	IARI, New Delhi	Public
	Groundnut bud necrosis virus (GBNV)	Np (nucleo-capsid gene)	Sense	IARI, New Delhi and IISc, Bangalore	Public
		NSs (RNAi suppressor)	RNAi	IARI, New Delhi	Public
	ToLCNDV + GBNV	AC_1 (truncated + NSs)	RNAi	IARI, New Delhi and Advanta India Ltd, Hyderabad	Public and private

Source: Adapted from Praveen *et al.* (2017).

et al. 2012; Aglawe *et al.* 2018). CRISPR/Cas9 and homologous repairing-based gene editing has greater application in engineering novel alleles of an endogenous gene. It also has the potential to introgress multiple transgenes in a single locus or perform concurrent editing at multiple loci across the genome (Cong *et al.* 2013; Wang *et al.* 2014; Biswal *et al.* 2019).

Editing of host genes

Various players in plant–pathogen interactions, especially host susceptibility genes (*S*) and transcription or translation factors, have been used in gene editing to create resistant alleles to impart disease resistance. Heritable resistance in wheat against powdery mildew disease was engineered by manipulating three mildew-resistance locus O (*MLO*) homeoalleles (Wang *et al.* 2014). A similar knock-out of *SlMLO* locus in tomato confers resistance to fungi *Oidium neolycopersici* causing powdery mildew (Nekrasov *et al.* 2017). In rice, gene editing has been used to boost resistance against pathogens causing bacterial blight, blast and tungro diseases by editing Sweet genes (*SWEET11, SWEET13* and *SWEET14*), ERF transcription factor gene (*OsERF922*) and translation initiation factor 4 gamma gene (*eIF4G*), respectively (Liu *et al.* 2012; Wang *et al.* 2016; Macovei *et al.* 2018; Kim *et al.* 2019; Oliva *et al.* 2019). CRISPR/Cas9-based targeted editing of the susceptibility gene *CsLOB1* promoter in citrus conferred citrus canker resistance (Peng *et al.* 2017). The host gene required for virus establishment, for instance eIF4E (eukaryotic translation initiation factor 4E), has been modulated by gene editing to confer broad-spectrum virus resistance against Cucumber vein yellowing virus, Zucchini yellow mosaic virus and Papaya ring spot mosaic virus type-W (Chandrasekaran *et al.* 2016). Similarly, the host gene involved in the survival of the viral pathogen has been edited to develop potyvirus resistance (Chandrasekaran *et al.* 2016; Pyott *et al.* 2016).

Editing of pathogen genes

The pathogen-derived gene that activates immunity (*Avr4/6*) has been modulated in a destructive oomycete pathogen such as *Phytophthora sojae* to incorporate resistance (Fang and Tyler 2016). DNA-virus resistance in plants employing the CRISPR-Cas system was demonstrated in model species such as *Arabidopsis* and *N. benthamiana* against beet severe curly top virus (Ji *et al.* 2015). Later, the utility of the technique in conferring virus resistance against the infection of Tomato yellow leaf curl virus, Merremia mosaic virus (Ali *et al.* 2015) and Bean yellow dwarf virus (Baltes *et al.* 2015) were shown in *N. benthamiana*.

RNAi *v.* gene editing

While RNAi is very much established technology to develop disease resistance in crops, gene editing has shown great potential to address the menace of plant diseases or pathogens in a very short time. Both the technologies are important and relevant in certain contexts, and complement each other's limitations (Table 18.3). RNAi is primarily used to target the actively transcribing genetic regions while gene editing can act on regulatory sequences as well as protein or RNA coding units. Gene editing is more suitable technology for developing disease resistance through modification of host genes, while RNAi is more appropriate when pathogen genes are targeted. Targeting of pathogen genes to impart disease resistance through gene editing would require continuous production of Cas protein in host genome, which may have biosafety concerns. RNAi technology can be used to knock down the genes/genetic regions where a suitable PAM (protospacer adjacent motif) site is unavailable for editing through nucleases. Positive effects of RNAi have been demonstrated by exogenous application of RNA molecules in plants (Dalakouras *et al.* 2019), hence can be used as eco-friendly pesticide to control the diseases without altering the plant genome, while gene editing necessitates modification of the genome to attain the phenotype. Gene editing is biologically more

Table 18.3. Comparative analysis of RNAi and gene editing technologies to deliver disease-resistant crop phenotypes

Sl. no	Features	RNAi	Gene editing
1	Principle	Down-regulation of gene expression by targeting RNA (gene knockdown)	Selective modification of gene or its regulatory regions to completely eliminate its expression (gene knockout)
2	Targets	Transcribing region including non-coding RNAs	Regulatory sequences, RNA coding units
3	Design	Sequence of target RNA region is sufficient	Sequence information of target genomic loci along with the presence of PAM sequence
4	Applicability	Widely employed to silence the genes of pathogen origin	Suitable for modification of host genes
5	Specificity	Relatively high off-targets	Minimal off-target effects (chemically modified sgRNAs avoid off-target effects)
6	Effect of viral suppressor proteins	Viral suppressors alter the efficacy of RNAi	No such interventions
7	Continuous production of effector biomolecules	Continuous expression of RNAi effector is required	Expression of Cas protein/nucleases not required once intended genetic changes were made
8	Alternative application	Feasible: application of dsRNA-based biopesticides has been proven	Genetic modification is mandatory

economic than RNAi, as expression of Cas protein/nucleases and guide RNA is not required after the intended changes have been made in the plant genome. In contrast, RNAi requires continuous production of small RNAs to get the desired phenotype. Gene editing can be used to alter a specific domain/motif/nucleotide of a gene without altering any other sequence; knockdown of a specific domain or motif of a gene is not possible through RNAi. The positive effects of RNAi can be suppressed by viral-encoded RNAi suppressor proteins (Praveen and Mangrauthia 2006; Mangrauthia *et al.* 2010), but the phenotype developed through editing cannot be influenced by external proteins/factors. Both the technologies have applications in developing disease-resistant crops, and the selection of appropriate technology has to be made on a case-by-case basis.

Transgenic approaches are fraught with random insertions in plant genomes producing random phenotypes, though not always. In contrast, gene-editing technologies yield defined genotypes due to site-directed mutagenesis. Hence, it is the preferred method of genetic modification for studies in functional genomics and in the development of an improved crop phenotype. Also, as genome-edited crops generally invite lesser or no regulatory norms they are accepted relatively easily and without many complicated regulatory procedures compared to transgenic crops based on protein or RNA expression (Waltz 2018). CRISPR-Cas has been an efficient and precise replacement for RNAi owing to its versatility. The limitations of RNA-silencing technologies, such as incomplete loss-of-function and off-target effects, were overcome by the efficient CRISPR-Cas system. Another important advantage of CRISPR-Cas is its ability to perform multiplex gene knockouts which is important in reducing the genetic redundancy associated with paralogous sequences. Generally, CRISPR-Cas9 DNA constructs are delivered via *Agrobacterium*-mediated infiltration which causes the inadvertent insertion of foreign DNA such as selectable marker gene, leading to off-target effects, mosaicism or reduction of the efficiency of the gene-editing process (Mao *et al.* 2013). Hence, to alleviate the disadvantages of plasmid-based delivery, DNA-free genome editing is performed utilising Cas9-ribonucleoproteins (RNPs) in isolated protoplasts (Zhang *et al.* 2017). It is pertinent to have plant regeneration from protoplasts, which is not available for most monocot crops. Hence, embryonic cells were delivered with DNA-free RNPs (Svitashev *et al.* 2015).

Regulatory set-up

In Canada, plant phenotypes developed through conventional breeding techniques, mutation breeding, transgenic development and genome editing techniques are subjected to identical regulatory approval depending on the final 'product'. In the EU, crops developed as a result of genetic modification are subjected to designated regulatory approvals whereas crops developed through mutagenesis are excluded. Hence, many years are added to the timelines for development and release of GM crops within Europe or for the European market. The novel gene-editing technologies using ZFN, TALENS and CRISPR-Cas have not been brought under the GMO-related regulatory regimes of many developed countries including the USA. The USDA has eased GMO regulatory rules against CRISPR-Cas edited crops, which can be commercially sold without any regulatory monitoring (Waltz 2018). In India, the Department of Biotechnology has constituted a taskforce on genome engineering technologies and their applications with a vision to harvest the benefits of gene-editing technologies. Currently, in the absence of any fresh guidelines, the Rules 1989 notified by Ministry of Environment, Forest and Climate Change encompass gene-editing technologies (Chimata and Bharti 2019). Appropriate guidelines and regulatory procedures are yet to be drafted and published (Friedrichs *et al.* 2019). In the meantime, the Department of Biotechnology has proposed a draft regulatory framework and guidelines for genome-edited organisms in Indian context (DBT 2020). The draft guidelines aim to follow a tiered approach for risk assessment of genetically edited (GEd) products/organisms based on the different processes followed to develop the GEd organisms and their resultant characteristic features. GEd organisms or products with single or few base pair edits or small deletions are placed in GEd Group I, GEd cells/organisms harbouring targeted few/several base pair edits are in Group II, whereas GEd cells/organisms harbouring targeted edit(s) synthetic/foreign DNA are categorised as Group III. The natural history of safe use of products derived from Group I organisms is recognised, whereas GEd plants in Group III have to undergo the safety guidelines procedure as laid out in Guidelines for the Environmental Risk Assessment of Genetically Engineered Plants (2016) and the Guidelines for the Safety Assessment of Foods Derived from Genetically Engineered Plants (2008).

Conclusion

Tremendous advancements have been made in the field of disease-resistant crop phenotypes since development of the first GM crop resistant to disease, tobacco expressing viral coat protein (Abel *et al.* 1986). Among the field-grown disease resistance crops obtained through GM technology, antiviral resistance holds prime position. Despite the improvements made in developing disease resistance in important cultivated crops, the notable exception is wheat, a widely cultivated and consumed food crop which lacks genetic modification for disease resistance. Although molecular and conventional breeding have been shielding wheat cultivars from diseases, the emergence and spread of new threats such as wheat blast cannot be ignored

(Islam *et al.* 2016). Wulff and Dhugga (2018) suggest the need to apply transgenic development in wheat, to introgress blast resistance-conferring genes from a wild progenitor into the cultivated variety to halt the spread of this devastating disease in south and south-east Asia. Development of disease resistance is one of the most exciting applications of gene-editing technology in crop improvement. During the last few years, there has been tremendous progress in pathogenomics, host genomics and the emergence of more precise and effective nucleases suitable for gene editing. In addition, base editing and RNA editing using CRISPR technology offer more advanced applications to alter plant phenotypes (Molla and Yang 2019). However, challenges remain in identifying the ideal target genes for editing and the availability of standardised gene transformation and regeneration protocols in the target crop. More extensive studies on functional characterisation of host and pathogen genes will facilitate the appropriate use of gene editing in disease resistance. While some of the pathosystems are characterised at molecular level, the majority are still not well understood. Gene editing can also be helpful in molecular characterisation of pathosystems and functional genomics of host and pathogens.

Acknowledgements

The authors gratefully acknowledge funding from various organisations including the Department of Biotechnology, Department of Science and Technology, Indian Council of Agricultural Research, Government of India for facilitating research in the field of virus-resistant transgenic plants.

References

Abel PP, Nelson RS, De B, Hoffmann N, Rogers SG, Fraley RT, Beachy RN (1986) Delay of disease development in transgenic plants that express the tobacco mosaic virus coat protein gene. *Science* **232**, 738–743. doi:10.1126/science.3457472

Aglawe S, Barbadikar K, Mangrauthia S, Madhav M (2018) New breeding technique 'genome editing' for crop improvement: applications, potentials and challenges. *3Biotech* **8**, 336. doi:10.1007/s13205-018-1355-3

Ali Z, Abulfaraj A, Idris A, Ali S, Tashkandi M, Mahfouz MM (2015) CRISPR/Cas9-mediated viral interference in plants. *Genome Biology* **16**, 238. doi:10.1186/s13059-015-0799-6

Andolfo G, Jupe F, Witek K, Etherington GJ, Ercolano MR, Jones JD (2014) Defining the full tomato NB-LRR resistance gene repertoire using genomic and cDNA RenSeq. *BMC Plant Biology* **14**, 120. doi:10.1186/1471-2229-14-120

Baltes NJ, Hummel AW, Konecna E, Cegan R, Bruns AN, Bisaro DM, Voytas DF (2015) Conferring resistance to geminiviruses with the CRISPR–Cas prokaryotic immune system. *Nature Plants* **1**, 15145. doi:10.1038/nplants.2015.145

Baulcombe DC (1996) Mechanisms of pathogen-derived resistance to viruses in transgenic plants. *Plant Cell* **8**, 1833–1844. doi:10.2307/3870233

Beachy RN (1997) Mechanisms and applications of pathogen derived resistance in transgenic plants. *Current Opinion in Biotechnology* **8**, 215–220. doi:10.1016/S0958-1669(97)80105-X

Bent AF, Kunkel BN, Dahlbeck D, Brown KL, Schmidt R, Giraudat J, Leung J, Staskawicz BJ (1994) RPS2 of *Arabidopsis thaliana*: a leucine-rich repeat class of plant disease resistance genes. *Science* **265**, 1856–1860. doi:10.1126/science.8091210

Biswal AK, Mangrauthia SK, Reddy MR, Yugandhar P (2019) CRISPR mediated genome engineering to develop climate smart rice: challenges and opportunities. *Seminars in Cell & Developmental Biology* doi:10.1016/j.semcdb.2019.04.005.

Bogdanove AJ, Booher NJ (2016) Online tools for TALEN design. In *TALENs: Methods and Protocols.* (Eds R Kühn, W Wurst and B Wefers) pp. 43–47. Humana Press (Springer), New York.

Bousset L, Sprague SJ, Thrall PH, Barrett LG (2018) Spatio-temporal connectivity and host resistance influence evolutionary and epidemiological dynamics of the canola pathogen *Leptosphaeria maculans*. *Evolutionary Applications* **11**, 1354–1370. doi:10.1111/eva.12630

Chandrasekaran J, Brumin M, Wolf D, Leibman D, Klap C, Pearlsman M, Sherman A, Arazi T, Gal-On A (2016) Development of broad virus resistance in non-transgenic cucumber using CRISPR/Cas9 technology. *Molecular Plant Pathology* **17**, 1140–1153. doi:10.1111/mpp.12375

Chimata MK, Bharti G (2019) Regulation of genome edited technologies in India. *Transgenic Research* **28**, 175–181. doi:10.1007/s11248-019-00148-z

Cong L, Ran FA, Cox D, Lin S, Barretto R, Habib N, Hsu PD, Wu X, Jiang W, Marraffini LA, Zhang F (2013) Multiplex genome engineering using CRISPR/Cas systems. *Science* **339**, 819–823. doi:10.1126/science.1231143

Dalakouras A, Lauter A, Bassler A, Krczal G, Wassenegger M (2019) Transient expression of intron-containing transgenes generates non-spliced aberrant pre-mRNAs that are processed into siRNAs. *Planta* **249**, 457–468. doi:10.1007/s00425-018-3015-6

Dale J, James A, Paul JY, Khanna H, Smith M, Peraza-Echeverria S, Garcia-Bastidas F, Kema G, Waterhouse P, Mengersen K, Harding R (2017) Transgenic Cavendish bananas with resistance to *Fusarium* wilt tropical race 4. *Nature Communications* **8**, 1496. doi:10.1038/s41467-017-01670-6

DBT (2020) *Draft Document on Genome Edited Organisms: Regulatory Framework and Guidelines for Risk Assessment.* <http://dbtindia.gov.in/sites/default/files/Draft_Regulatory_Framework_Genome_Editing-9jan2020.pdf>

Fang Y, Tyler BM (2016) Efficient disruption and replacement of an effector gene in the oomycete *Phytophthora sojae* using CRISPR/Cas9. *Molecular Plant Pathology* **17**, 127–139. doi:10.1111/mpp.12318

FAO (2018) *World Food and Agriculture Statistical Pocketbook 2018.* Food and Agriculture Organization, Rome. <http://www.fao.org/3/CA1796EN/ca1796en.pdf>

Farokhzadeh S, Ali Fakeheri B (2014) Marker-assisted selection for disease resistance: applications in breeding. *International Journal of Agriculture and Crop Sciences* **7**, 1392–1405.

Flor HH (1955) Host–parasite interaction in flax rust: its genetics and other implications. *Phytopathology* **45**, 680–685.

Fonseca JP, Mysore KS (2019) Genes involved in non-host disease resistance as a key to engineer durable resistance in crops. *Plant Science* **279**, 108–116. doi:10.1016/j.plantsci.2018.07.002

Foster SJ, Park TH, Pel M, Brigneti G, Sliwka J, Jagger L, van der Vossen E, Jones JDG (2009) *Rpi-vnt1.1*, a Tm-22 homolog from *Solanum venturii*, confers resistance to potato late blight. *Molecular Plant-Microbe Interactions* **22**, 589–600. doi:10.1094/MPMI-22-5-0589

Friedrichs S, Takasu Y, Kearns P, Dagallier B, Oshima R, Schofield J, Moreddu C (2019) Meeting report of the OECD conference on Genome Editing: Applications in Agriculture – Implications for Health, Environment and Regulation. *Transgenic Research* **28**, 419–463. doi:10.1007/s11248-019-00154-1

Gill US, Lee S, Mysore KS (2015) Host versus non-host resistance: distinct wars with similar arsenals. *Phytopathology* **105**, 580–587. doi:10.1094/PHYTO-11-14-0298-RVW

Heath MC (2000) Non-host resistance and nonspecific plant defenses. *Current Opinion in Plant Biology* **3**(4), 315–319. doi:10.1016/S1369-5266(00)00087-X

Hittalmani S, Parco A, Mew TV, Zeigler RS, Huang N (2000) Fine mapping and DNA marker-assisted pyramiding of the three major genes for blast resistance in rice. *Theoretical and Applied Genetics* **100**, 1121–1128. doi:10.1007/s001220051395

Islam MT, Croll D, Gladieux P, Soanes DM, Persoons A, Bhattacharjee P, Hossain MS, Gupta DR, Rahman MM, Mahboob MG, Cook N (2016) Emergence of wheat blast in Bangladesh was caused by a South American lineage of *Magnaporthe oryzae*. *BMC Biology* **14**, 84. doi:10.1186/s12915-016-0309-7

Jafary H, Albertazzi G, Marcel TC, Niks RE (2008) High diversity of genes for nonhost resistance of barley to heterologous rust fungi. *Genetics* **178**, 2327–2339. doi:10.1534/genetics.107.077552

Janni M, Sella L, Favaron F, Blechl AE, De Lorenzo G, D'Ovidio R (2008) The expression of a bean PGIP in transgenic wheat confers increased resistance to the fungal pathogen *Bipolaris sorokiniana*. *Molecular Plant–Microbe Interactions* **21**, 171–177. doi:10.1094/MPMI-21-2-0171

Jendoubi W, Bouhadida M, Boukteb A, Béji M, Kharrat M (2017) Fusarium wilt affecting chickpea crop. *Agriculture* **7**(3), 23. doi:10.3390/agriculture7030023

Jeuken MJ, Pelgrom K, Stam P, Lindhout P (2008) Efficient QTL detection for non-host resistance in wild lettuce: backcross inbred lines versus F (2) population. *Theoretical and Applied Genetics* **116**, 845–857. doi:10.1007/s00122-008-0718-2

Ji X, Zhang H, Zhang Y, Wang Y, Gao C (2015) Establishing a CRISPR-Cas-like immune system conferring DNA virus resistance in plants. *Nature Plants* **1**, 15144. doi:10.1038/nplants.2015.144

Jinek M, Chylinski K, Fonfara I, Hauer M, Doudna JA, Charpentier E (2012) A programmable dual-RNA-guided DNA endonuclease in adaptive bacterial immunity. *Science* **337**, 816–821. doi:10.1126/science.1225829

Jiwan D, Roalson EH, Main D, Dhingra A (2013) Antisense expression of peach mildew resistance locus O (*PpMlo1*) gene confers cross-species resistance to powdery mildew in *Fragaria xananassa*. *Transgenic Research* **22**, 1119–1131. doi:10.1007/s11248-013-9715-6

Jo KR, Kim CJ, Kim SJ, Kim TY, Bergervoet M, Jongsma MA, Visser RG, Jacobsen E, Vossen JH (2014) Development of late blight resistant potatoes by cis gene stacking. *BMC Biotechnology* **14**, 50. doi:10.1186/1472-6750-14-50

Joubert DA, Kars I, Wagemakers L, Bergmann C, Kemp G, Vivier MA, van Kan JA (2007) A polygalacturonase-inhibiting protein from grapevine reduces the symptoms of the endo polygalacturonase BcPG2 from *Botrytis cinerea* in *Nicotiana benthamiana* leaves without any evidence for in vitro interaction. *Molecular Plant–Microbe Interactions* **20**, 392–402. doi:10.1094/MPMI-20-4-0392

Jupe F, Witek K, Verweij W, Śliwka J, Pritchard L, Etherington GJ, Maclean D, Cock PJ, Leggett RM, Bryan GJ, Cardle L, Hein I, Jones JD (2013) Resistance gene enrichment sequencing (RenSeq) enables reannotation of the *NB-LRR* gene family from sequenced plant genomes and rapid mapping of resistance loci in segregating populations. *Plant Journal* **76**, 530–544. doi:10.1111/tpj.12307

Kim YJ, Lin N-C, Martin GB (2002) Two distinct *Pseudomonas* effector proteins interact with the Pto kinase and activate plant immunity. *Cell* **109**, 589–598. doi:10.1016/S0092-8674(02)00743-2

Kim HJ, Lee HR, Jo KR, Mortazavian SM, Huigen DJ, Evenhuis B, Kessel G, Visser RG, Jacobsen E, Vossen JH (2012) Broad spectrum late blight resistance in potato differential set plants MaR8 and MaR9 is conferred by multiple stacked *R* genes. *Theoretical and Applied Genetics* **124**, 923–935. doi:10.1007/s00122-011-1757-7

Kim D, Luk K, Wolfe SA, Kim JS (2019) Evaluating and enhancing target specificity of gene-editing nucleases and deaminases. *Annual Review of Biochemistry* **88**, 191–220. doi:10.1146/annurev-biochem-013118-111730

Koundal V, Praveen S (2010) MicroRNA-based RNA interference vector for gene silencing in plants. *Journal of Plant Biochemistry and Biotechnology* **19**, 79–82. doi:10.1007/BF03323439

Kumar G, Jyothsna M, Valarmathi P, Roy S, Banerjee A, Tarafdar J, Senapati BK, Robin S, Manonmani S, Rabindran R, Dasgupta I (2019) Assessment of resistance to rice tungro disease in popular rice varieties in India by introgression of a transgene against Rice tungro bacilliform virus. *Archives of Virology* **164**, 1005–1013. doi:10.1007/s00705-019-04159-3

Lacombe S, Rougon-Cardoso A, Sherwood E, Peeters N, Dahlbeck D, Van Esse HP, Smoker M, Rallapalli G, Thomma BP, Staskawicz B, Jones JD (2010) Interfamily transfer of a plant pattern-recognition receptor confers broad-spectrum bacterial resistance. *Nature Biotechnology* **28**, 365–369. doi:10.1038/nbt.1613

Lee S, Rojas CM, Ishiga Y, Pandey S, Mysore KS (2013) Arabidopsis heterotrimeric G-proteins play a critical role in host and non-host resistance against *Pseudomonas syringae* pathogens. *PLoS One* **8**, e82445. doi:10.1371/journal.pone.0082445

Lee S, Whitaker VM, Hutton SF (2016) Mini review: potential applications of non-host resistance for crop improvement. *Frontiers of Plant Science* **7**, 997. doi:10.3389/fpls.2016.00997

Li T, Liu B, Spalding MH, Weeks DP, Yang B (2012) High-efficiency TALEN-based gene editing produces disease-resistant rice. *Nature Biotechnology* **30**, 390–392. doi:10.1038/nbt.2199

Liang X, Ding P, Lian K, Wang J, Ma M, Li L, Li L, Li M, Zhang X, Chen S, Zhang Y (2016) *Arabidopsis* heterotrimeric G proteins regulate immunity by directly coupling to the FLS2 receptor. *eLife* **5**, e13568. doi:10.7554/eLife.13568

Liu D, Chen X, Liu J, Ye J, Guo Z (2012) The rice ERF transcription factor OsERF922 negatively regulates resistance to *Magnaporthe oryzae* and salt tolerance. *Journal of Experimental Botany* **63**, 3899–3911. doi:10.1093/jxb/ers079

Macovei A, Sevilla NR, Cantos C, Jonson GB, Slamet-Loedin I, Čermák T, Voytas DF, Choi IR, Chadha-Mohanty P (2018) Novel alleles of rice eIF4G generated by CRISPR/Cas9-targeted mutagenesis confer resistance to *Rice tungro spherical virus*. *Plant Biotechnology Journal* **16**, 1918–1927. doi:10.1111/pbi.12927

Malathi P, Muzammil SA, Krishnaveni D, Balachandran SM, Mangrauthia SK (2019) Coat protein 3 of *Rice tungro spherical virus* is the key target gene for development of RNAi mediated tungro disease resistance in rice. *Agri Gene* **12**, 100084. doi:10.1016/j.aggene.2019.100084

Manfredini C, Sicilia F, Ferrari S, Pontiggia D, Salvi G, Caprari C, Lorito M, De Lorenzo G (2005) Polygalacturonase-inhibiting protein 2 of *Phaseolus vulgaris* inhibits BcPG1, a polygalacturonase of *Botrytis cinerea* important for pathogenicity, and protects transgenic plants from infection. *Physiological and Molecular Plant Pathology* **67**, 108–115. doi:10.1016/j.pmpp.2005.10.002

Mangrauthia SK, Singh P, Praveen S (2010) Genomics of helper component proteinase reveals effective strategy for *Papaya ringspot virus* resistance. *Molecular Biotechnology* **44**, 22. doi:10.1007/s12033-009-9205-5

Mao Y, Zhang H, Xu N, Zhang B, Gou F, Zhu JK (2013) Application of the CRISPR–Cas system for efficient genome engineering in plants. *Molecular Plant* **6**, 2008–2011. doi:10.1093/mp/sst121

Martin GB, Brommonschenkel SH, Chunwongse J, Frary A, Ganal MW, Spivey R, Wu T, Earle ED, Tanksley SD (1993) Map-based cloning of a protein kinase gene conferring disease resistance in tomato. *Science* **262**, 1432–1436. doi:10.1126/science.7902614

Mitra A (1937) Some plant diseases and pests of India and their control. *Science & Culture (Calcutta)* **2**(8), 392–395; **2**(9), 417–423.

Moffett P (2009) Mechanisms of recognition in dominant R gene mediated resistance. *Advances in Virus Research* **75**, 1–33. doi:10.1016/S0065-3527(09)07501-0

Molla KA, Yang Y (2019) CRISPR/Cas-mediated base editing: technical considerations and practical applications. *Trends in Biotechnology* **37**, 1121–1142. doi:10.1016/j.tibtech.2019.03.008

Murphy F, He Q, Armstrong M, Giuliani LM, Boevink PC, Zhang W, Tian Z, Birch PRJ, Gilroy EM (2018) The potato MAP3K StVIK is required for the *Phytophthora infestans* RXLR effector Pi17316 to promote disease. *Plant Physiology* **177**, 398–410. doi:10.1104/pp.18.00028

Nekrasov V, Wang C, Win J, Lanz C, Weigel D, Kamoun S (2017) Rapid generation of a transgene-free powdery mildew resistant tomato by genome deletion. *Scientific Reports* **7**(1), 482. doi:10.1038/s41598-017-00578-x

Nunes CC, Dean RA (2012) Host-induced gene silencing: a tool for understanding fungal host interaction and for developing novel disease control strategies. *Molecular Plant Pathology* **13**, 519–529. doi:10.1111/j.1364-3703.2011.00766.x

Oliva R, Ji C, Atienza-Grande G, Huguet-Tapia JC, Perez-Quintero A, Li T, Eom JS, Li C, Nguyen H, Liu B, Auguy F (2019) Broad-spectrum resistance to bacterial blight in rice using genome editing. *Nature Biotechnology* **37**, 1344–1350. doi:10.1038/s41587-019-0267-z

Peng A, Chen S, Lei T, Xu L, He Y, Wu L, Yao L, Zou X (2017) Engineering canker-resistant plants through CRISPR/Cas9-targeted editing of the susceptibility gene CsLOB1 promoter in citrus. *Plant Biotechnology Journal* **15**, 1509–1519. doi:10.1111/pbi.12733

Praveen S, Mangrauthia SK (2006) Viral suppressors: small RNAs regulators. *Indian Journal of Virology* **17**, 62–77.

Praveen S, Mishra AK, Dasgupta A (2005) Antisense suppression of replicase gene expression recovers tomato plants from leaf curl infection. *Plant Science* **168**, 1011–1014. doi:10.1016/j.plantsci.2004.12.008

Praveen S, Ramesh SV, Mishra AK, Koundal V, Palukaitis P (2010) Silencing potential of viral derived RNAi constructs in tomato leaf curl virus-AC4 gene suppression in tomato. *Transgenic Research* **19**, 45–55. doi:10.1007/s11248-009-9291-y

Praveen S, Ramesh SV, Mangrauthia SK (2017) Transgenic approaches to combat plant viruses occurring in India. In *A Century of Plant Virology in India*. (Eds B Mandal, GP Rao, VK Baranwal and RK Jain) pp. 783–805, Springer, Singapore.

Pyott DE, Sheehan E, Molnar A (2016) Engineering of CRISPR/Cas9-mediated potyvirus resistance in transgene-free *Arabidopsis* plants. *Molecular Plant Pathology* **17**, 1276–1288. doi:10.1111/mpp.12417

Qin Y, Teixeira Da Silva JA, Zhang L, Zhang S (2008) Transgenic strawberry: state of the art for improved traits. *Biotechnology Advances* **26**, 219–232. doi:10.1016/j.biotechadv.2007.12.004

Ramesh SV, Mishra AK, Praveen S (2007) Hairpin RNA-mediated strategies for silencing of *Tomato leaf curl virus* AC1 and AC4 genes for effective resistance in plants. *Oligonucleotides* **17**, 251–257. doi:10.1089/oli.2006.0063

Ramesh SV, Shivakumar M, Praveen S, Chouhan BS, Chand S (2019) Expression of short hairpin RNA (shRNA) targeting AC2 gene of *Mungbean yellow mosaic India virus* (MYMIV) reduces the viral titre in soybean. *3 Biotech* **9**, 334. doi:10.1007/s13205-019-1865-7

Rao TB, Chopperla R, Methre R, Punniakotti E, Venkatesh V, Sailaja B, Reddy MR, Yugander A, Laha GS, Madhav MS, Sundaram RM, Ladhalakshmi D, Balachandran SM, Mangrauthia SK (2019) Pectin induced transcriptome of a *Rhizoctonia solani* strain causing sheath blight disease in rice reveals insights on key genes and RNAi machinery for development of pathogen-derived resistance. *Plant Molecular Biology* **100**, 59–71. doi:10.1007/s11103-019-00843-9

Rodriguez-Moreno L, Song Y, Thomma BP (2017) Transfer and engineering of immune receptors to improve recognition capacities in crops. *Current Opinion in Plant Biology* **38**, 42–49. doi:10.1016/j.pbi.2017.04.010

Rojas CM, Senthil-Kumar M, Wang K, Ryu CM, Kaundal A, Mysore KS (2012) Glycolate oxidase modulates reactive oxygen species-mediated signal transduction during non-host resistance in *Nicotiana benthamiana* and *Arabidopsis*. *Plant Cell* **24**, 336–352. doi:10.1105/tpc.111.093245

Sato S, Tabata S, Hirakawa H, Asamizu E, Shirasawa K, Isobe S, Kaneko T, Nakamura Y, Shibata D, Aoki K (2012) The tomato genome sequence provides insights into fleshy fruit evolution. *Nature* **485**, 635–641. doi:10.1038/nature11119

Savary S, Ficke A, Aubertot J-N, Hollier C (2012) Crop losses due to diseases and their implications for global food production losses and food security. *Food Security* **4**, 519–537. doi:10.1007/s12571-012-0200-5

Savary S, Willocquet L, Pethybridge SJ, Esker P, McRoberts N, Nelson A (2019) The global burden of pathogens and pests on major food crops. *Nature Ecology & Evolution* **3**, 430–439. doi:10.1038/s41559-018-0793-y

Scofield S, Tobias C, Rathjen J, Chang J, Lavelle D, Michelmore R, Staskawicz BJ (1996) Molecular basis of gene-for-gene specificity in bacterial speck disease of tomato. *Science* **274**, 2063–2065. doi:10.1126/science.274.5295.2063

Senthil-Kumar M, Mysore KS (2013) Non-host resistance against bacterial pathogens: retrospectives and prospects. *Annual Review of Phytopathology* **51**, 407–427. doi:10.1146/annurev-phyto-082712-102319

Silva KJ, Brunings A, Peres NA, Mou Z, Folta KM (2015) The Arabidopsis NPR1 gene confers broad-spectrum disease resistance in strawberry. *Transgenic Research* **24**, 693–704. doi:10.1007/s11248-015-9869-5

Singh AK, Praveen S, Singh BP, Varma A, Arora N (2009) Safety assessment of leaf curl virus resistant tomato developed using viral derived sequences. *Transgenic Research* **18**, 877–887. doi:10.1007/s11248-009-9274-z

Sprague SJ, Balesdent MH, Brun H, Hayden HL, Marcroft SJ, Pinochet X, Rouxel T, Howlett BJ (2006) Major gene resistance in *Brassica napus* (oilseed rape) is overcome by changes in virulence of populations of *Leptosphaeria maculans* in France and Australia. *European Journal of Plant Pathology* **114**, 33–40. doi:10.1007/s10658-005-3683-5

Staskawicz BJ, Dahlbeck D, Keen NT (1984) Cloned avirulence gene of *Pseudomonas syringae* pv.*glycinea* determines race-specific incompatibility on *Glycine max* (L.) Merr. *Proceedings of the National Academy of Sciences of the United States of America* **81**, 6024–6028. doi:10.1073/pnas.81.19.6024

Sumit R, Sahu BB, Xu M, Sandhu D, Bhattacharyya MK (2012) *Arabidopsis* non-host resistance gene *PSS1* confers immunity against an oomycete and a fungal pathogen but not a bacterial pathogen that cause diseases in soybean. *BMC Plant Biology* **12**, 87. doi:10.1186/1471-2229-12-87

Svitashev S, Young JK, Schwartz C, Gao H, Falco SC, Cigan AM (2015) Targeted mutagenesis, precise gene editing, and site-specific gene insertion in maize using Cas9 and guide RNA. *Plant Physiology* **169**, 931–945. doi:10.1104/pp.15.00793

Tai TH, Dahlbeck D, Clark ET, Gajiwala P, Pasion R, Whalen MC, Stall RE, Staskawicz BJ (1999) Expression of the *Bs2* pepper gene confers resistance to bacterial spot disease in tomato. *Proceedings of the National Academy of Sciences of the United States of America* **96**, 14153–14158. doi:10.1073/pnas.96.24.14153

Tang X, Frederick R, Zhou J, Halterman D, Jia Y, Martin G (1996) Initiation of plant disease resistance by physical interaction of Avr Pto and Pto kinase. *Science* **274**, 2060–2063. doi:10.1126/science.274.5295.2060

Tiwari IM, Jesuraj A, Kamboj R, Devanna BN, Botella JR, Sharma TR (2017) Host delivered RNAi: an efficient approach to increase rice resistance to sheath blight pathogen (*Rhizoctonia solani*). *Scientific Reports* **7**, 7521. doi:10.1038/s41598-017-07749-w

Uma B, Rani TS, Podile AR (2011) Warriors at the gate that never sleep: non-host resistance in plants. *Journal of Plant Physiology* **168**, 2141–2152. doi:10.1016/j.jplph.2011.09.005

Uniyal AP, Yadav SK, Kumar V (2019) The CRISPR–Cas9, genome editing approach: a promising tool for drafting defense strategy against begomoviruses including cotton leaf curl viruses. *Journal of Plant Biochemistry and Biotechnology* **28**, 121–132. doi:10.1007/s13562-019-00491-6

van der Vossen EA, Gros J, Sikkema A, Muskens M, Wouters D, Wolters P, Pereira A, Allefs S (2005) The *Rpi-blb2* gene from *Solanum bulbocastanum* is an *Mi-1* gene homolog conferring broad-spectrum late blight resistance in potato. *Plant Journal* **44**, 208–222. doi:10.1111/j.1365-313X.2005.02527.x

van Esse HP, Reuber TL, van der Does D (2020) Genetic modification to improve disease resistance in crops. *New Phytologist* **225**, 70–86. doi:10.1111/nph.15967

Waltz E (2018) With a free pass, CRISPR-edited plants reach market in record time. *Nature Biotechnology* **36**, 6–7. doi:10.1038/nbt0118-6b

Walz A, Zingen-Sell I, Loeffler M, Sauer M (2008) Expression of an oxalate oxidase gene in tomato and severity of disease caused by *Botrytis cinerea* and *Sclerotinia sclerotiorum*. *Plant Pathology* **57**, 453–458. doi:10.1111/j.1365-3059.2007.01815.x

Wang Y, Cheng X, Shan Q, Zhang Y, Liu J, Gao C, Qiu JL (2014) Simultaneous editing of three homoeoalleles in hexaploid bread wheat confers heritable resistance to powdery mildew. *Nature Biotechnology* **32**, 947–952. doi:10.1038/nbt.2969

Wang F, Wang C, Liu P, Lei C, Hao W, Gao Y, Liu YG, Zhao K (2016) Enhanced rice blast resistance by CRISPR/Cas9-targeted mutagenesis of the ERF transcription factor gene *OsERF922*. *PLoS ONE* **11**, e0154027. doi:10.1371/journal.pone.0154027

Wang B, Sumit R, Sahu BB, Ngaki MN, Srivastava SK, Yang Y, Swaminathan S, Bhattacharyya MK (2018a) *Arabidopsis* novel glycine-rich plasma membrane PSS1 protein enhances disease resistance in transgenic soybean plants. *Plant Physiology* **176**, 865–878. doi:10.1104/pp.16.01982

Wang J, Zhou L, Shi H, Chern M, Yu H, Yi H, He M, Yin J, Zhu X, Li Y, Li W (2018b) A single transcription factor promotes both yield and immunity in rice. *Science* **361**, 1026–1028. doi:10.1126/science.aat7675

Wulff BB, Dhugga KS (2018) Wheat: the cereal abandoned by GM. *Science* **361**, 451–452.

Wulff BB, Moscou MJ (2014) Strategies for transferring resistance into wheat: from wide crosses to GM cassettes. *Frontiers of Plant Science* **5**, 692. doi:10.3389/fpls.2014.00692

Yadav MK, Aravindan S, Ngangkham U, Raghu S, Prabhukarthikeyan SR, Keerthana U, Marndi BC, Adak T, Munda S, Deshmukh R, Pramesh D, Samantaray S, Rath PC (2019) Blast resistance in Indian rice landraces: genetic dissection by gene-specific markers. *PLoS One* **14**(3), e0213566. doi:10.1371/journal.pone.0213566

Yang S, Gao M, Xu C, Gao J, Deshpande S, Lin S, Roe BA, Zhu H (2008) Alfalfa benefits from *Medicago truncatula*: the *RCT1* gene from *M. truncatula* confers broad-spectrum resistance to anthracnose in alfalfa. *Proceedings of the National Academy of Sciences of the United States of America* **105**, 12164–12169. doi:10.1073/pnas.0802518105

Zhang K, Raboanatahiry N, Zhu B, Li M (2017) Progress in genome editing technology and its application in plants. *Frontiers of Plant Science* **8**, 177. doi:10.3389/fpls.2017.00177

Zhao BY, Lin XH, Poland J, Trick H, Leach J, Hulbert S (2005) A maize resistance gene functions against bacterial streak disease in rice. *Proceedings of the National Academy of Sciences of the United States of America* **102**, 15383–15388. doi:10.1073/pnas.0503023102

Zhu D, Kang H, Li Z, Liu M, Zhu X, Wang Y, Wang D, Wang Z, Liu W, Wang GL (2016) A genome-wide association study of field resistance to *Magnaporthe oryzae* in rice. *Rice* **9**(1), 44. doi:10.1186/s12284-016-0116-3

19

Nanobiotechnology for plant genome engineering and crop protection

Poonam Sashidhar, Sagar S Arya, Ratul K Das, Mukul K Dubey and Sangram K Lenka

Introduction

'There's plenty of room at the bottom', a lecture by Richard Feynman, came to be looked upon as the conceptual beginning of nanotechnology (Feynman 1960). This scientific discipline deals with processes that take place on the nanometre scale (1–100 nm) at least in one dimension. In the last 50 years, nanotechnology has gained significant recognition in terms of its manufacturing and applications to newer scientific disciplines. One of the nanotechnology applications that has recently emerged is plant nanobiotechnology. Plant nanobiotechnology aims to improve crop productivity and crop protection. Extensive studies have been carried out in various aspects of plant nanobiotechnology including uptake of nanomaterials by plants, toxicity, diagnostics, treatment, translocation and accumulation of nanomaterials (Kole *et al.* 2016). One of the thrust areas in plant science is to design an efficient genetic transformation system for application in plant genetic engineering and genome editing (Ramkumar *et al.* 2020). Nanobiotechnology promises to revolutionise the development of efficient carrier molecules in delivering biomolecules. Different nanoparticle-based delivery systems have been reported for successful delivery of biomolecules. These delivery systems have a unique ability to traverse the plant cell wall barrier, which is technically challenging and resource consuming by conventional delivery methods. Nanoparticle-based delivery systems have the added advantage of broad host range in plant genetic engineering (Cunningham *et al.* 2018). Advanced genome editing technology such as bacterially derived CRISPR/Cas9 system (clustered regularly interspaced short palindromic repeat) has been used for precise targeting of a gene of interest to develop the desirable genetic traits (Jaganathan *et al.* 2018).

Another promising application of nanotechnology in plant science is for crop protection. Crops suffer from damages caused by different biotic and abiotic factors and thus long-term sustainability is a challenge. Designing nanoformulations of agrochemicals has emerged as a new concept under the domain of 'minimal usage with maximum effect' for agricultural applications (Tripathi *et al.* 2017). However, some metallic nanoformulations are phytotoxic (Ma *et al.* 2015). Overall, plant nanobiotechnology strongly suggests three research areas; namely, crop production, diagnostics and crop protection. As plant nanobiotechnology is in its infancy, it is necessary to review all the scattered vital information to serve as a basis for future studies. This chapter discusses the fundamentals, application scope, recent progress and challenges of plant nanobiotechnology.

Nanomaterials as delivery vehicles of biomolecules for genetic transformation

Mesoporous silica

Silica nanoparticles have been established as an important material in biomedical research for application in drug delivery and therapeutics. Mesoporous silica nanoparticles (MSNs) synthesised using polymerisation of silica in the presence of surfactants have various attributes that aid in the intracellular delivery of compounds. MSNs have tuneable particle size (50–300 nm), high porosity and surface area (>900 m^2/g), uniformly dispersed pores (2–6 nm) and well-defined structural frameworks (Lodha *et al.* 2012). Nanoparticles formed by well-ordered pore matrix, even of larger sizes around 30 nm, confer a high loading capacity for molecules like DNA and proteins (Popat *et al.* 2011).

MSNs engineered for animal systems cannot be used for plants due to the existence of a cell wall in the latter. However, Torney *et al.* (2007) demonstrated the ability of modified MSNs to deliver DNA into tobacco plant cells and leaves. They used a plasmid containing a green fluorescent protein (GFP) and its inducible promoter along with a chemical inducer, β-oestradiol. The construct was bound to MSNs capped with gold nanoparticles to prevent leaching of molecules. After bombardment, the GFP showed induced expression in tobacco cotyledon and revealed the possible use of MSNs and this construct for gene delivery. In contrast to gold nanoparticle-based gene delivery where the genetic material is surface coated, MSNs offers more loading capacity due to their mesoporous structure as well as large surface area.

In recent years, MSNs were explored for the delivery of proteins for developing non-transgenic lines (Martin-Ortigosa *et al.* 2014). Gold-coated MSNs were used as carriers for Cre-recombinase protein into plant cells harbouring loxP sites on either side of a selection gene and a reporter gene. The MSN-mediated Cre protein delivery independent of target tissue or organism exhibited a recombination efficiency of 20%. In another study, different positively functionalised MSNs (~50 nm) doped with fluorescein (F-MSNs) were used to carry plasmid DNAs into different root tissues with a simple co-culture method (Chang *et al.* 2013). These results showed the presence of MSNs in the endodermal tissues of *Arabidopsis thaliana* root and the inner vascular bundle. This co-culture method was found superior to other techniques such as particle bombardment, which is not efficient to deliver the DNA into deeper plant tissues. Investigations on the uptake efficiency of functionalised MSNs showed that trimethoxy[3-(methlyamino) propyl]silane (TMAPS) functionalised MSNs had higher uptake efficiency in the cells as compared to other functionalised MSNs. The higher uptake efficiency of TMAPS/F-MSNs might be due to higher zeta potential in comparison to others (Chang *et al.* 2013). The study also revealed that cell-wall pore diameter (5–20 nm) does not interfere with the uptake of MSNs due to the relaxation of micro-fibrilar network of cell walls and its uptake by endocytosis.

The quest for delivering plasmid DNA-loaded MSNs into deeper tissues of dicot plants led to the identification of an efficient method for this application. Injecting the DNA-MSNs through the abaxial surface of leaves was found to be better than spraying on leaves and shoot injection. The method confirmed the movement of MSNs through the stomatal pore and through the plasmodesmata. This method can be time- and energy-saving and does not necessitate additional mechanical or chemical treatment (Plate 16) (Hajiahmadi *et al.* 2019).

To further confirm the efficiency of the delivery system, MSNs carrying *cry1Ab* gene were delivered to tomato. This was the first study to report MSN-mediated transient gene transformation performed successfully under *in vivo* condition. Transformed tomato were resistant to the South American tomato pinworm (also called tomato leafminer) *Tuta absoluta*, thus suggesting a new tool for crop genetic engineering (Hajiahmadi *et al.* 2019).

Polymeric nanoparticles

Polymeric nanoparticles (PNPs) as a delivery vehicle have many attributes like biocompatibility, multifaceted mode of action, scalability and biodegradability (Athanassiou *et al.* 2018). They are typically composed of a core-shell structure enclosed in a polymeric matrix. The core-shell structure of PNP acts as a reservoir for the drug and the matrix allows diffusion-based controlled release depending upon the permeability and degradation properties of the matrix used.

Cationic polymers are predominantly used for the delivery of nucleic acids, allowing their internalisation into the cells and protection against the nucleases (Gary *et al.* 2007). The stable complexation of the cationic polymer and nucleic acids plays a key role in its successful delivery. The ratio of the Protonatable amine groups of polymers to the phosphate group of nucleic acids, pH and ionic strength of the medium also affect the electrostatic interactions between a cationic polymer and nucleic acids (Cherng *et al.* 1999). Since the condensation and assembly of DNA and RNA is regulated by electrostatic attractions, it is understood to favour the use of polycationic polymers. Polycations like poly (L-lysine) and polyethylene amine (PEI) have been extensively used for *in vitro* delivery of DNA into mammalian cell lines. However, the *in vivo* transfection efficiency reported for polycationic polymers is fairly low (Abdallah *et al.* 1996). The low efficiency has been reported to be due to the strong electrostatic binding between the DNA and polycations impeding the release of DNA at the appropriate stage and thus resulting in low level of gene expression and blocking transcription. To overcome this, core-shell structures with block co-polymers have been proposed. The co-polymer consisting of polyethylene glycol (PEG) and polycations is mixed with DNA (Davis 2002). The cationic segments bind to DNA through electrostatic interaction (polyplex structure), whereas a PEG layer forms a protective shield to the polyplex core stabilising the steric hindrance against undesirable interactions in the surrounding environment.

To investigate the functioning of PNPs for delivering gene vectors into plants, Jiang *et al.* (2014) explored the possibility of using perylene 3,4,9,10-tetracaboxyamide chromophore with cationic groups attached to the periphery. The central chromophore allowed tracking of nanoparticles inside the plants that aided in demonstrating the ability of cationic fluorescent nanoparticles to deliver dsRNA and DNA in *Arabidopsis*. The use of a dsRNA-fluorescent nanoparticle mixture led to systematic gene silencing of important developmental genes, resulting in distinct phenotypes. A study based on the post-transcriptional gene silencing mechanism used fluorescent cationic polymers to deliver siRNA and knock down targeted genes in BY-2 protoplast. These cationic polymers have a potential to self-track during RNA interference experiments in addition to protecting siRNA from degradation with RNAse. The microscopic studies revealed that the penetration of nanoparticles bound siRNA into the protoplast within two hours of delivery and showed no potential toxicity. The siRNA knocked down the intended genes of cell-wall synthesis during the early stages of protoplast synthesis, proving the potential of using PNP as a non-viral mode of gene delivery (Silva *et al.* 2010).

Carbon-based nanomaterials

Current nanotechnological research is focused on creating a plethora of novel materials to be used in drug delivery and biomedical applications (Vance *et al.* 2015). In addition to the nanoparticles mentioned above, the discovery of carbon nanotubes (CNTs) has opened up novel applications in drug delivery, therapeutic and molecular transporters in plants (Liu *et al.* 2009).

CNTs inherently consist of carbon arranged repetitively in a hexagonal pattern to form a hollow cylindrical structure (Bianco 2004). They offer high surface area of ~2600 m^2/g. The wall of single-walled carbon nanotubes (SWCNTs) is extremely hydrophobic, so functionalisation has been actively carried out to increase its water solubility for utilisation in medicine and biology (Liu *et al.* 2007). CNTs have an extraordinary ability to carry a wide array of molecules like DNA, proteins and other moieties through covalent or non-covalent attachment (Plate 15). While CNTs have been used for the delivery of genetic materials into animal cells, their utilisation in plant cells is still in its infancy (Demirer and Landry 2017). In plants, to understand the mechanism of CNT trafficking and subcellular localisation, Serag *et al.* (2011) carried out a size-dependent study of CNT. Fluorescein isothiocyanate (FITC) labelled long multi-walled carbon nanotubes (MWCNT) (>200 nm) was found inside the subcellular structure, whereas the shorter ones (30–100 nm) were localised in vacuoles, nucleus and plastids. The observed differential penetration was attributed to the passage present in vacuolar, nuclear and plastid membrane that allowed the passage of CNTs in a size range of 30–100 nm. Further, the membrane of other subcellular organelles like endoplasmic reticulum and mitochondria were found intact, which might have hampered the passage of short MWCNTs. This strategy was found promising for use in targeted delivery into nuclear and plastid regions (Serag *et al.* 2011).

A recent report studied the mechanism of entry and trapping of SWCNTs in chloroplast via single particle tracking of intrinsic photoluminescence generated by $ss(AT)_{15}$-SWCNTs (Wong *et al.* 2016). The use of single particle tracking allowed the elucidation of the complete lipid exchange envelope penetration (LEEP) pathway describing various events of transport outside the chloroplast, traversing across the membrane, diffusion within the chloroplast and its kinetic trapping supported by mathematical modelling. The SWCNTs present in the chloroplast were hypothesised to be trapped in the thylakoid membrane, enhancing the photosynthetic efficiency of chloroplast (Giraldo *et al.* 2014). Recently, the applicability of chloroplast transformation using chitosan-SWCNTs complex as delivery vehicles was demonstrated by Kwak *et al.* (2019). The chitosan-based carriers conjugated to SWCNTs were positively charged to allow the penetration of cargo with the release mechanism tuned at pH 8 to mimic the biological environment of chloroplast stroma. As discussed above, the carrier was designed to traverse the chloroplast using the LEEP model to allow maximal penetration. A reporter gene attached to plasmid DNA was used to validate the transgene expression in mature plants like arugula, watercress, spinach and tobacco in addition to isolated *A. thaliana* mesophyll protoplasts (Kwak *et al.* 2019).

This specific and selective transgene delivery system to the chloroplast offers the advantages of simplicity, cost-effectiveness and wide applicability in different species. In addition to the extracellular DNA, its applicability can be expanded to genome editing in chloroplast using the CRISPR-Cas9 system as well as zinc finger nucleases. The toxicity of CNTs in plants is of concern (Lin and Xing 2007). For instance, mesoporous CNTs with a concentration of 150 mg/L decreased the root and shoot length of rice by 70% and 57%, respectively (Hao *et al.* 2019). However, the functionalisation strategies have dramatically reduced its toxicity. Carboxylic acid functionalisation enhanced the dispersibility of surface-modified CNTs in polar solvents and also reduced toxicity (Kim *et al.* 2012).

Functionalisation strategies were also devised for attachment of genetic material through electrostatic interactions. Ammonia-functionalised CNTs were bound to plasmid DNA through electrostatic interaction, which resulted in their penetration of the cell membrane of mammalian cells through their spontaneous behaviour as nano-needles. The functionalised CNTs exhibited low cytotoxicity and efficient delivery of DNA, resulting in 10-fold higher gene expression in comparison to the delivery of DNA alone (Pantarotto *et al.* 2004). Demirer *et al.* (2019) reported the delivery of plasmid DNA and siRNA using a co-culture approach with functionalised CNT for transient expression of GFP in *Eruca sativa* (arugula) leaves and silencing of constitutively expressed GFP in leaves of *Nicotiana benthamiana* (Plate 17).

The study showed the potential of safely delivering and transiently expressing genes using CNTs, thereby showcasing nanoparticle-mediated genetic engineering in model and crop species. In order to use CNTs as a delivery vehicle for DNA, the DNA must be condensed onto the nanotube, directed to the targeted cell, detach intracellularly and gain entry into the nucleus before transgene expression (Singh *et al.* 2005). The same report also studied the attachment of DNA and CNT using SEM, which evidently showed the formation of lattice in a supramolecular arrangement, where parallel groups of nanotubes were arranged with localisation of condensed packets of DNA. To increase the penetration of CNTs, a unique strategy to deliver CNT into plants using cellulase immobilised on their surface was designed. These CNTs were observed in the interior of *A. thaliana* cells, suggesting the local hydrolysis of cell wall by cellulase enzyme (Serag *et al.* 2012).

Inorganic nanoparticles

Inorganic nanomaterials such as gold and iron oxide are highly sought for gene delivery because of their unique electrical and optical attributes (quantum dots), ease of functionalisation strategies and biocompatibility (Erathodiyil and Ying 2011). Gold nanoparticles (AuNPs) in particular offer a dual functional platform for therapeutic as well as diagnostic applications. AuNPs functionalised with amino acids provide scaffolds for effective binding of DNA that can be used for gene delivery application (Ghosh *et al.* 2008). Creating multifunctional monolayers by adding diverse functional groups allows nucleic acids as well as targeting agents to be placed on a single platform (Ryan *et al.* 2007; Bowman *et al.* 2008). The AuNP based gene delivery vehicle is typically constructed via a surface functionalisation strategy using amino acids, cationic peptides and molecules constituting tertiary amine. The assembly of thiol groups to directly conjugate nucleic acids with the AuNPs–thiol bridge, and layer-by-layer assembly of polyethylenimine (PEI), are other approaches to the construction of gold-based gene carriers (Loh *et al.* 2016).

Direct delivery of Cas9 ribonucleoprotein and donor DNA by constructing CRISPR-gold is reported in animal cells *in vivo*. The whole system comprises AuNPs conjugated with donor DNA, Cas9 moiety and an endosomal disruptive polymer composed of poly-aspartic acid (diethylenetriamine). The system could be internalised inside the cell via endocytosis due to the polymer, triggering the endosomal disruption and causing the release of CRISPR-gold into the cytoplasm (Plate 18) (Lee *et al.* 2017).

Although a recent strategy utilising covalent modification is gaining attention for gene delivery in animal cells, its applicability in plants remains a topic for exploration (Cunningham *et al.* 2018).

Zhang *et al.* (2016) demonstrated the delivery of CRISPR/Cas9 DNA into wheat callus using AuNPs. This study did not use selection free medium, in contrast to conventional genome editing experiments where a herbicide or antibiotic was added to the medium to select the transformed cells. The total time required to obtain the edited plants in this study was six to eight weeks, which was two to four weeks less than genome editing protocol reported previously. To investigate the integration of CRISPR/Cas9 DNA into the genome, the T_0 population was checked for the presence of plasmid DNA through PCR (Zhang *et al.* 2016). Another study reported covalent attachment of PBI221 plasmid and FITC molecules on the surface of magnetic AuNPs. In presence of an external magnetic field, these nanoparticles were delivered into the plant cells as observed by confocal imaging of a nanoparticle–FITC complex within the intact cells (Hao *et al.* 2013). As discussed above, AuNPs have also been used in composite with MSNs to deliver biomolecules into plant cells. An iron oxide (Fe_3O_4)–DNA complex was delivered to transgenic cotton through pollen-based magnetofection, offering an efficient platform for genetic transformation (Zhao *et al.* 2017).

Other emerging inorganic nanoparticles like calcium phosphate and barium sulphate have exhibited potential in terms of usage and application in numerous frontiers of material science, particularly bone tissue engineering. Escape of the inorganic nanoparticles from the microbial attack, enhanced storage of cargo, inexpensive preparation methods, targeted delivery and the ease of application to monocots and dicots opens up new avenues in plant genetic engineering. Calcium phosphate nanoparticles were used for efficient introduction of *hmgr* gene (involved in mevalonate and phenylpropanoid pathway) in *Cichorium intybus* and resulted in a transformation efficiency of 9.6%. The transformants possessed high chlorophyll, protein and esculin content in comparison to the non-transformed ones, attributed to higher transgene expression (Rafsanjani *et al.* 2016).

Quantum dots

Quantum dots (QDs) are semiconductor-based, nanometre-sized monodisperse particles. Due to the size-dependent high tenability in their optical and electrical properties (Collier *et al.* 1998), they can be applied to different biochemical specificities. They have a broad absorption range and a narrow emission range, allowing simultaneous excitation and detection (Wang and Wang 2014). The size of QDs has been correlated directly to its uptake and efficiency of gene delivery (Wu and Yan 2013). Several studies concerning gene delivery application using QDs (made up of zinc sulphide, cadmium sulphide and cadmium selenide) have reported the display of high and stable transformation efficiency in animal cells. Transition metal doped QDs like zinc sulphide (ZnS) or zinc selenide (ZnSe) have exhibited low cytotoxicity, making them promising candidates for drug delivery applications (Wu and Yan 2013). Glycopeptide coated manganese doped ZnS QDs were used for controlled delivery of ibuprofen. The ZnS QDs core and glycopeptide shell exhibited low cytotoxicity, high loading capacity and targeted drug delivery (Xu *et al.* 2011). These studies have laid a foundation for the utilisation of QDs in plant genetic modification experiments.

ZnS QDs with a diameter of 3–5 nm modified with poly L-lysine (PLL) were bound to negatively charged plasmid DNA (Fu *et al.* 2012). ZnS-mediated plant genetic transformation was carried out in tobacco leaves using an ultrasound assisted technique. ZnS-PLL nanoparticles protected DNA from ultrasound damage and demonstrated a higher transformation efficiency of 43–53% in comparison to the conventional ultrasound mediated gene delivery (30–40%). Tri-octyl phosphine oxide (TOPO) coated cadmium selenide (CdSe)/ZnS PEGylated QDs were used to introduce linearised DNA in *Arabidopsis* floral buds. PEG was conjugated to the functionalised QDs to protect the cells from cytotoxicity. The resultant transformation was stable, providing evidence of genomic DNA integration in the T1 population of *Arabidopsis* (Samboju *et al.* 2014). Similar types of QDs were used for the delivery of sequence-specific nuclease to introduce double-stranded breaks envisioning stimulated gene targeting in plants (Samuel *et al.* 2015).

Nanoparticles for crop protection

With the advent of nanoparticle-based delivery of genetic material, we are now able to explore these nano-delivery systems for crop protection. This next-generation approach will assist the rapid development of virus-, weed- and parasite-resistant plants. The use of nanotechnology in pesticide delivery is relatively new and is gaining momentum (Nuruzzaman *et al.* 2016).

One technique involves use of nanoclay-mediated delivery of RNAi in human cell lines. This innovative technique was efficiently translated for use in plant systems by Mitter *et al.* (2017). DsRNA was loaded on non-toxic, designer, degradable, layered double hydroxide (LDH) clay nanosheets. This nanoclay with dsRNA, when applied on plants through a foliar route, showed sustained release and was detected even 30 days after application. It was also demonstrated that LDH is degraded and dsRNA was taken up in plant cells, further silencing homologous cucumber mosaic and pepper mild mottle viral RNA (Mitter *et al.* 2017). A polycationic gene vector was employed, N-(dimethyl aminoethyl) methacrylate (DMAEMA) and pentaerythritol to initiate the polymerisation. The gene vector was used to deliver dsRNA conferring resistance to *Agrotis ypsilon* by RNAi (Li *et al.* 2019). In addition to the polymers mentioned above, chitosan has also been used for nucleic acid delivery in plants. Chitosan, being an alkaline biodegradable polymer binding strongly to the negatively charged nucleic acid, was sought to induce resistance to fungal pathogens in potato. Chitosan nanoparticles were complexed with two antifungal thionin protein encoding genes, transformed into potato then the transgenic potato was tested for resistance against *Alternaria alternata* and *Rhizoctonia solani* (Abdel-Razik *et al.* 2017). The fungal bioassay confirmed the antifungal activity of the expressed thionin proteins in transgenic potato. As discussed earlier, MSNs were used to deliver *cry1Ab* gene to tomato, thereby imparting resistance to tomato leaf miner.

Nanocapsules for controlled release of bioactive molecules

Stimuli-responsive nanoparticles can be designed by altering their physical and chemical properties in response to a change in external environment. It leads to alterations in particle properties such as structure, dimension, interaction with other molecules and surface properties (López-Saucedo *et al.* 2019). The concept of stimuli responsiveness has been used in the development of nanoparticles with superparamagnetic properties triggered by changes in pH and light, for biomedical application. A system comprising Fe_3O_4 core nanoparticles with a mesoporous silica hybrid shell has been used to effectively release doxyrubicin hydrochloride (DOX) and rhodamine 123 (Rh123). The release of the model drug molecule was due to change in its structural conformation in response to external UV light and intracellular pH (Moorthy *et al.* 2016).

Magnetic field-responsive polymers appear to be an interesting candidate for plant application as they can be used for encapsulation, permitting a localised release of 'cargo'. Reports on the delivery of genetic material using magnetic nanoparticles are discussed below. However, there is not enough data available regarding the safety of magnetic nanoparticles and the applied magnetic field and some studies show their negative effects. An investigation was carried out on the possible effects on plant tissues by inducing magnetic fluid accumulation. To evaluate the oxidative stress, biochemical assays of photo-assimilatory pigments and nucleic acid contents were performed 12 days post treatment. A decrease in the chlorophyll ratio (chlorophyll a and b) of up to 30% was observed along with a reduction in plant length by 40% when exposed to an electromagnetic field (Răcuciu *et al.* 2009). Hence, a suitable concentration range of magnetic nanoparticles must be taken into consideration when designing a system for magnetic stimuli-responsive release of genetic material on plants.

Among the stimuli used for designing a delivery system, redox- and thiol-responsive nanocarriers containing disulphide bonds have received considerable attention. A glutathione-responsive nanocarrier has been used for intracellular delivery of cargo by taking advantage of the difference in concentration of intracellular and extracellular environments (Li *et al.* 2009). Glutathione (GSH) is a tripeptide widely present not only in humans but also in plants at the concentration of 2–3 mM required to cleave the disulphide linkages (Ren *et al.* 2011). GSH-responsive nanocarriers have been used for delivery of a model herbicide-diuron in two plant species. Nanoparticles of carboxymethyl chitosan had a spherical core-shell structure with a size of 250 nm. The ionisable carboxyl groups provided the nanoparticles with pH sensitivity and the reducible disulphide linkage provided glutathione responsiveness. The release of diuron at 2 mM GSH and no release in absence of GSH showcased the glutathione-responsive behaviour of herbicide (Yu *et al.* 2015). Similarly, GSH-responsive MSNs loaded with salicylic acid (SA) capped with decanethiol gatekeepers through disulphide linkages were used to study the release of SA. The release of SA in *Arabidopsis* was correlated with accumulation of GSH in plant cells and high expression levels of plant pathogen gene (PR-1) (Yi *et al.* 2015). Although such approaches were investigated for agrochemicals, they can also be exploited for the delivery of genetic materials in plants. Nanoparticles functionalised with thiol groups can be used to attach genetic material which can be released effectively in the presence of GSH.

Besides the non-selective stimuli mentioned above, a new class of stimuli-responsive material is found to be susceptible to changes in the molecular assembly in the presence of selective enzymes. The use of enzymes as biological stimuli opens up new avenues of selectivity and requires mild conditions for reaction (pH 5–8). The overall design and assembly of enzyme-responsive polymer was reported (Thornton *et al.* 2005). Polyethylene glycol-acrylamide beads ($PEGA_{800}$) were modified with a protease cleavable tri-peptide sequence consisting of a charged arginine residue. Choosing a peptide linker recognised by the target enzyme results in release of material by subsequent peptide hydrolysis and removal of charged residue. These enzyme-responsive polymers can respond to a target protease in a mixture of other enzymes and thus can allow targeted delivery of genetic material. Stimuli-responsive release of nano-encapsulated bioactive material regulated by plant stress enzymes can be foreseen (Peteu *et al.* 2010).

Thus, different nanoparticles with varying physiochemical characteristics have been explored and are anticipated to make a mark in the field of plant nanobiotechnology. However, not only the nature of the nanoparticle but its delivery into plant tissues/cells is an important factor that will influence the success of plant genetic transformation. Therefore, we attempt to shed light on existing nanoparticle delivery methods and techniques applicable in future.

Modes of nanoparticle-mediated genome engineering

Biolistic method

Gene delivery using biolistics (also called particle bombardment) is a useful method for transfection of DNA and is widely used for plant genetic transfection. There are several systems for the biolistic method of transfection. The Helios gene gun (Bio-Rad, Hercules, CA) is one of the most widely used biolistic systems (O'Brien and Lummis 2011). High-velocity micro-projectiles are used to carry the genetic materials and to pass the cell membrane. Gene transfer through the bombardment of DNA-adsorbed gold particles has been widely used for generation of transgenic lines. The biolistic transformation process is effective regardless of plant species or tissue type, and the simple and rapid procedure and its suitability to transform the organelles make it indispensable in plant transformation (Sanford 1990). Tobacco mosaic virus RNA was the first particle to be delivered into plants to demonstrate the efficiency of particle bombardment. A potential advantage of biolistic delivery over *Agrobacterium*-mediated transformation is the absence of a host plant defence mechanism against the *Agrobacterium*, which can negatively impact the growth of stable transformants (Hansen and Wright 1999). 'Proteolistics' is a specialised protocol to deliver proteins by simply mixing the protein suspension and gold particles, placing the mixture into the gene-gun cartridge and lyophilising to dehydrate the mixture (Martin-Ortigosa and Wang 2014). It is highly acceptable for genome editing applications where DNA-free delivery is a prerequisite (Kanchiswamy *et al.* 2015).

Recently, various types of nanoparticles like gold, iron oxide and chitosan conjugated with biomolecules (bovine serum albumin and RNA) were delivered to *N. benthamiana* leaves using the biolistic and infiltration method (Makhotenko *et al.* 2018). The type of particle used in the biolistic method is one of the most important factors affecting the delivery of biomolecules into the tissue (Zhang *et al.* 2007). The major challenges in nanoparticle-mediated delivery are size and low weight as compared to the conventionally used micro-carriers. The size and surface characteristics of nanoparticles might not be favourable for delivering DNA in all circumstances. Various modifications have been suggested, like co-bombardment of nanoparticles with 0.6 µm gold micro-projectiles and DNA to increase nanoparticle density with gold coating and thereby transformation efficiency (Martin-Ortigosa *et al.* 2012b; Mortazavi and Zohrabi 2018).

Magnetofection

Magnetofection is considered a method of choice for efficient gene transfer. It utilises magnetic force to facilitate the uptake of DNA bound to magnetic nanoparticles into the target cells (Scherer *et al.* 2002; Dobson 2006). Magnetofection with paramagnetic particles as drug carriers, guiding their accumulation in the target cells with a strong and locally applied magnetic field, has been focused mainly in medical and animal research (Zhao *et al.* 2017). Researchers consider that two different mechanisms aid the delivery of the biomolecule and magnetic reagent complex. Some researchers consider the process of endocytosis and pinocytosis of cell membrane to be accelerated in the presence of a magnetic field (Gupta *et al.* 2014; Das *et al.* 2015; Herrero *et al.* 2017) while others suggest the mechanism of magnetofection similar to electroporation.

In the presence of a changing magnetic field, there is an induction of the electric field modulating the transmembrane potential of cell membrane that ultimately creates perforations in it (Jakutavičiūtė *et al.* 2017). In plants, magnetic AuNPs were used for the delivery of FITC and DNA in canola cells with and without cell wall. The fluorescence signals and the expression of β-glucuronidase (GUS) gene inside the cells revealed the successful delivery of FITC molecule and plasmid DNA to canola, suggesting the use of magnetic AuNPs as delivery molecules (Hao *et al.* 2013). To avoid the constraints of *in vitro* tissue culture – regeneration of transformed plants – a group of researchers studied the delivery of DNA using pollen magnetofection in cotton. In this study, positively charged Fe_3O_4 nanoparticles coated with PEI were used to carry the negatively charged DNA. To direct the nanoparticle-DNA complex into the pollen through its apertures, an external magnetic field was applied. Transformed cotton seeds were obtained by antibiotic selection (Zhao *et al.* 2017). This research addresses the major bottleneck in genetic transformation in plants and can be broadly applicable to many other plant systems.

Although magnetofection in animal cells has been shown to yield high transformation efficiency, the thick walls surrounding plant cells, particularly pollen, is a major constraint. The aperture of pollen is a crucial factor. By carefully determining the pollen diameter and tuning the size of nanoparticles, DNA loaded onto the nanoparticles can be easily penetrated into the cells (Ruf and Bock 2017). This has been used for pollen magnetofection in crops like pepper and pumpkin that have high pollen apertures and are difficult to transform (Zhao *et al.* 2017). Transgenic pepper and pumpkin plants were successfully created through this technique and showed stable inheritance of exogenously integrated DNA.

Electroporation

Electroporation based on electric field-mediated membrane penetration is a simple technique for altering the genetic make-up of plant cells. Electroporation influences transient variation in the membrane arising from a high-voltage electrical impulse. The formation of pores in the cell membrane is a result of diffusional exchanges of molecules across the membrane, termed electroosmosis (Dimitrov and Sowers 1990). The relevance of electroporation in plants was demonstrated by Fromm *et al.* (1985). They showed the uptake of DNA in the protoplast of monocots (maize) as well as dicots (carrots and tobacco) (Fromm *et al.* 1985). Electroporation negates the limitation of host range and offers precise transfer of small number of genes to the recipient.

Electric field-induced transformation does not involve chemical, mechanical or viral vector-mediated fusion, thereby avoiding cell toxicity. For efficient gene transfer using electroporation, optimisation of parameters concerning the electric pulse, its duration and field strength need to be worked out properly. For the delivery of DNA into plant cells, it must be easily accessible to the pores created during the process. Unlike cell membranes, electroporation on cell walls is unlikely to occur since it does not constitute a lipid bilayer composition. Hence its applicability is reduced to only protoplasts. One of the major drawbacks is a noteworthy loss of protoplast viability. Taking into consideration the plant species and the voltage applied, a viability loss of up to 50% may occur (Chadegani *et al.* 1989; Fisk and Dandekar 2005). However, a report suggests that direct gene transfer in intact cells is possible through electroporation. This was achieved by plasmolyzing intact cells in the presence of plasmid DNA bound to a reporter gene followed by electroporation for introduction of the plasmid construct into the cell. The plasmolysis could create a temporary protoplast within the intact cell in order to allow a quick transition to a normal walled state (Wu and Feng 1999). Oligonucleotide-modified AuNP delivery using electroporation has been investigated in mammalian models as an alternative to viral-mediated transfection for gene therapy. However, there is no report on use of this technique in plant systems for nanoparticle-mediated gene delivery (Jen *et al.* 2004).

Infiltration

Agro-infiltration began with studying the plant–virus interaction, where exogenous DNA inserted in *Agrobacteria or Agrobacterium* was infiltrated into the intracellular plant tissues to allow the integration of exogenous DNA in plants (Grimsley *et al.* 1986). The most prevalent method of infiltration is syringe infiltration, using insertion of a needleless syringe to introduce *Agrobacterium* into plant cells (Santi *et al.* 2008). A small nick is created in the interstitial spaces of leaves, followed by injection of the *Agrobacterium* harbouring the gene of interest. Syringe infiltration is known to be adapted for a wide range of plant sciences (Wroblewski *et al.* 2005) without need for specialised techniques. It has the flexibility to introduce multiple constructs in different leaf zones, offering multitude assays in a single platform and infiltrating the entire area of leaf with one gene construct (Vaghchhipawala *et al.* 2011). Hence, it is a preferred method for transient expression of genes for plant genetic modification (Ramkumar *et al.* 2020). Nanoparticle infiltration is an effective approach that offers quicker working time and reduced costs as compared to other transient gene transformation methods. Syringe infiltration has been successfully implemented to inject plasmid-carrying MSNs in tomato, offering enhanced resistance to tomato leafminer by expression of the delivered *cryIAb* gene (Hajiahmadi *et al.* 2019).

Advantages of nanoparticle-mediated delivery of genetic material in plants

Efficient gene delivery systems in plants are crucial for genetic improvement of plants. With recombinant DNA technology applications, the major goal is to deliver the DNA into the plant cell and acquire stable integration of the transgene into the genome of regenerative cells (Que *et al.* 2019). During the last 40 years, research investigated many macromolecular delivery methods in plants to generate transgenic lines. The first report of successful genetic transformation of transgene into plants by *Agrobacterium*-mediated delivery was reported 35 years ago (Barton *et al.* 1983). Following the success of *Agrobacterium*-mediated transformation, direct DNA delivery methods like biolistics, electroporation, microinjection, agroinfiltration PEG-mediated protoplast transformation and liposome-mediated delivery were reported (Que *et al.* 2019). Despite advancements in technologies for genetic transformation, the production of transgenic lines through routine methods and regeneration remains expensive and time-consuming (Altpeter *et al.* 2016). Primarily, the entry of gene constructs is hindered by cell walls in *Agrobacterium*-mediated transformation. In direct gene transfer, there is considerable damage to the cells due to larger size of the carrier (gold and tungsten) (Husaini *et al.* 2010).

Application of nanotechnology to plant science, especially the ability of nanoparticles to deliver macromolecules and genetic material, is still in its infancy (Mao *et al.* 2019). As mentioned in Table 19.1, nanoparticles like MSNs, single or multi-walled CNTs, AuNPs, magnetic nanoparticles, PNPs and QDs have been associated with plants for transformation purposes. In comparison to conventional genetic modification, nanoparticle-mediated approaches offer several advantages. Decreasing the size of the particle from micrometre to nanometre for biolistic delivery can decrease the hindrance of the cell wall and minimise damage to the cells (Husaini *et al.* 2010). Nanoparticles can co-deliver more than one biogenic molecule simultaneously to the target cell. For example, Torney *et al.* (2007) used MSNs for the delivery of DNA and its activator that not only delivered it to the target site but also released the encapsulated material in a controlled fashion. Unlike conventional biolistic methods that require detachment of leaves from the plant for targeted delivery, nanoparticles enable translocation of genes to individual cells, organs or tissues (Wang *et al.* 2016). Nanoparticle-mediated delivery enables DNA-free editing of plant genomes in a controlled manner which includes gene silencing and generates accurately modified non-transgenic lines differing from the conventional genetic transformation method. The drawback of *in vitro* tissue culture and regeneration has been addressed by using pollen magnetofection with magnetic nanoparticles (Zhao *et al.* 2017).

Nanoparticles allow intracellular labelling as well as imaging. Nanoparticles conjugated to imaging agents like fluorophores have circumvented the cell wall and plasma membrane barrier in plant cells (Martin-Ortigosa *et al.* 2012a). This could be an excellent molecular tool for tracking the movement of external genes along with observing the expression of the target genes. With significant progress in the application of QDs, they have been used for *in vivo* imaging to understand the uptake and translocation of nanoparticles in intact *Arabidopsis* (Hischemöller *et al.* 2009). Therefore, QDs could be explored for monitoring real-time gene expression in a non-invasive manner.

As mentioned earlier, nanomaterials are potential candidates for intracellular delivery of cargoes/biomolecules owing to their ability to cross the biological membrane. However, the uptake and translocation mechanisms in plants remain largely overlooked. Hence, the next section offers an overview of the uptake and translocation routes and barriers associated with it.

Uptake, entry and barriers for nanoparticles in plants

Foliar uptake of nanoparticles and its associated barrier

Depending on the mode of application, nanoparticles are subjected to foliar and root exposure. Foliar entry of nanoparticles occurs through the stomatal pores or cuticles penetrating the leaf surface. For example, foliar uptake and translocation of ceria nanoparticles (~250 nm) were reported in cucumber, *Cucumis sativus* (Hong *et al.* 2014). Karny *et al.* (2018) loaded nanoscale drug-delivery systems with agricultural nutrients and applied them to the leaves of tomato plants. They showed that the nanoparticles, liposomes composed of plant-derived lipids, penetrate the leaf and translocate in a bidirectional manner, distributing to other leaves and to the roots. The liposomes were then internalised by the plant cells, where they released their active ingredient. Up to 33% of the applied nanoparticles penetrated the leaf, compared to <1%

Table 19.1. Different nanoparticles used for the delivery of biomolecules

Nanoparticles (NPs)	Cargo	Delivery method	Target species/tissue	Remarks	Reference
Mesoporous silica nanoparticles (MSNs)					
MSNs	DNA	Biolistic	Tobacco cells and leaves	Loading capacity Surface area	Torney *et al.* (2007)
MSNs doped with fluorescein (F-MSNs)	DNA	Co-culture	Root	Loading capacity Surface area Penetration	Chang *et al.* (2013)
TMAPS functionalised F-MSNs	DNA	Co-culture	Root	Loading capacity Surface area Penetration Uptake	Chang *et al.* (2013)
Gold-plated MSNs	Protein (Cre-recombinase)	Biolistic	Embryos of *Zea mays*	Loading capacity Surface area DNA-free genome editing	Martin-Ortigosa *et al.* (2014)
MSNs	DNA	Injection	Abaxial surface of tomato leaf	Loading capacity Surface area Uptake Systemic movement	Hajiahmadi *et al.* (2019)
Carbon-based nanomaterials					
Pristine-carbon nanotubes (CNTs)	DNA	Co-culture	*Erucasativa* and *Triticum aestivum* (leaves and protoplasts)	Penetration Efficiency Uptake	Demirer and Landry (2017)
Carbon nanotubes (CNTs)	siRNA	Infiltration	*Nicotiana benthamiana* leaves	Penetration Efficiency Uptake	Demirer and Landry (2017)
Polymeric NPs (PNPs)					
Perylene3,4,9,10-tetracarboxydiimide (PDI) fluorescent NPs	dsRNA, DNA	Root application	*Arabidopsis* root	Gene silencing Uptake Transfection efficiency	Jiang *et al.* (2014)

Fluorescent conjugated polymer NPs	siRNA	Co-culture	BY-2 protoplast	Penetration Gene silencing Time	Silva *et al.* (2010)
Chitosan NPs	DNA	Tissue dip	Potato nodal explants	Penetration Time	Abdel-Razik *et al.* (2017)
Metal NPs (MNPs)					
AuNPs	Ribonucleoprotein (CRISPR/Cas9)		Wheat callus	Time Marker free selection	Zhang *et al.* (2016)
AuNPs	Ribonucleoprotein (CRISPR/Cas9)	Biolistic	Axillary bud and apical meristems of potato plant		Khromov *et al.* (2018)
Iron oxide NPs	DNA	Magnetofection	Cotton pollen	Time Culture free Genotype independent Multi-gene transformation Stable integration and expression	Zhao *et al.* (2017)
Calcium phosphate NPs	DNA	Tissue dip	*Cichoriumintybus* L. leaves	Size Time	Rafsanjani *et al.* (2016)
ZnS NPs	DNA	Ultrasonication	Tobacco cells	Stable integration and expression Cost	Fu *et al.* (2012)
Quantum dots					
TOPO (tri-octyl phosphine oxide)-coated CdSe/ZnS quantum dots and polyethylene glycol	Linearised DNA	Floral dip	*Arabidopsis* floral buds	Linearised DNA Uptake Translocation	Samboju *et al.* (2014)
Luminescent semiconductor nanocrystals (TOPO - CdSe/ZnS core-shell QDs)	Protein (sequence specific nuclease)	Co-culture	Tobacco, carrot, maize, canola, rapeseed, cotton, palm, peanut, soybean, *Oryza* sp., *Arabidopsis* sp., *Ricinus* sp., and sugarcane cells	Uptake	Samuel *et al.* (2015)

of free molecules applied in a similar manner. In this study, tomato plants treated with liposomes loaded with Fe and Mg overcame acute nutrient deficiency which was not treatable using ordinary agricultural nutrients. To address regulatory concerns regarding airborne nanoparticles, the study designed liposomes that were stable only over short spraying distances (<2 m) and disintegrated into safe molecular building blocks (phospholipids) over longer airborne distances. These findings expand the utility of nanotechnology for delivering micronutrients to agricultural crops to increase yield (Karny *et al.* 2018). Foliar application of nanoparticles has displayed translocation through the stomatal cavities to the plant roots and vasculature (Schwab *et al.* 2016). The stomatal pathway can transport hydrophilic substances through the stomatal pores in leaves ranging from<4 nm to >40 nm (Schreiber 2005; Eichert and Goldbach 2008). Eichert *et al.* (2008) demonstrated the internalisation of fluorescent polystyrene nanoparticles and concluded that 43 nm sized nanoparticle could pass through the stomata in fava bean or broad bean leaves, in contrast to 1 μm sized nanoparticles. This experiment suggested that the stomatal pathway can allow particulate size much higher than the previously considered 1 nm limit. However, high variability in stomatal permeability due to unknown reasons renders this pathway unpredictable.

Apart from the stomatal uptake pathway, small lipophilic nanoparticles can be taken up by cuticle and subsequently transported through polar and non-polar pathways (Dietz and Herth 2011). The non-polar or lipophilic path of diffusion is dominated by lipophilic or uncharged molecules, preferentially across cutin or wax domains. In contrast, polar molecules follow the hydrophilic pathway via pores of 0.6–4.8 nm diameter (Schreiber 2010). Nanoparticles <4.8 nm might directly traverse the leaf through the cuticular pathway. Contemporary literature has reported that most reported foliar uptakes of nanoparticles are above 5 nm, with no direct evidence on the pathway of uptake. Hence, nanoparticles >5 nm in size should be investigated for cuticular transport (Lv *et al.* 2019). As the cuticular structure might be subjected to change with climatic condition and plant life-cycle, nanoparticles may be able to bypass the cuticle barrier in some circumstances (Lv *et al.* 2019).

Root uptake of nanoparticles and its associated barrier

As discussed earlier, nanoparticles delivered to the plant can be taken up through the root system. This is affected by multiple factors like particle size, surface chemistry, type of plant, its growth stage and the integrity of root morphology (Lv *et al.* 2019).

One of the most important aspects of root uptake is particle size. Several studies have focused on understanding the role of particle size in root uptake, but the results are inconsistent (Lv *et al.* 2019). Larue *et al.* (2012a) revealed size-dependent uptake and translocation of titanium dioxide (TiO_2) nanoparticles in wheat. With evidence provided by TEM and μ-XRF, it was ascertained that TiO_2 nanoparticles <36 nm accumulated in roots and could translocate throughout the whole plant system. TiO_2 particles in the size range of 36–140 nm accumulated in root cortical region (parenchyma) without passing into the interiors, whereas particles >140 nm did not accumulate in the roots. The size of 140 nm was suggested as a threshold diameter above which no uptake or accumulation is observed (Larue *et al.* 2012a). Carbon-coated magnetic nanoparticles ranging from 5–50 nm were taken up by the roots of sunflower, tomato pea and wheat. The particles spread to the entire plant using its vascular system, differing in distribution pattern depending on species (Cifuentes *et al.* 2010). It is difficult to estimate the size exclusion limits for nanoparticle uptake by the root system as each plant differs in physiological, developmental, morphological, anatomical and architectural traits.

A study examined the cell internalisation and translocation of zinc oxide (ZnO) nanoparticles in ryegrass (Lin and Xing 2008). Zn^{2+} ions were employed to compare the root uptake of ZnO nanoparticles in ryegrass and investigate its phytotoxicity in a hydroponic culture medium. The ZnO nanoparticles were observed onto the root surface with individual particles being located in the apoplast and protoplast of the root epidermis. The translocation of ZnO nanoparticles was considerably lower from root to shoot than the corresponding Zn^{2+} treatments, implying the low translocation ability of ZnO. Similar evaluation of ZnO nanoparticle uptake in *Zea mays* (maize) was investigated (Zhao *et al.* 2012). Confocal microscopy images showed that ZnO nanoparticles applied in sandy loam soil penetrated the root epidermis and cortex by apoplastic pathway. Presence of ZnO aggregates in xylem vessels suggests the translocation of these aggregates through symplastic pathway bypassing the endodermis.

Surface charge is another crucial factor facilitating and controlling the root uptake of nanoparticles. Preliminary studies on root uptake of metal nanoparticles suggest that the uptake is charge-dependent. Positively charged nanoparticles have been found to be present on mainly root surfaces, whereas the uptake and translocation of negatively charged nanoparticle is much more efficient (Zhu *et al.* 2012). The accumulation of positively charged AuNPs in the study mentioned above was explained by the considerable electrostatic attraction between the positively charged nanoparticles and the negatively charged root surface; the negatively charged nanoparticles accumulate least, due to electrostatic repulsion.

The root uptake of nanoparticles also depends on the physiological status of the plant. It is influenced by factors like the presence of root hair and transformation in the rhizosphere due to the presence of root exudates or microbes (Seshadri *et al.* 2015). For instance, Huang *et al.* (2017) suggested that the strong interaction of copper nanoparticles (~40 nm) and synthetically produced root exudate decreased the uptake and bioaccumulation of Cu in *Cucumis sativus* (cucumber). It also influenced the oxidation and transformation of Cu nanoparticles to Cu (I) and Cu (II) (Huang *et al.* 2017). Plants possessing long root hair that increases the surface area for water and nutrient uptake show more nanoparticle localisation in the root irrespective of surface charge (Zhu *et al.* 2012). The changes in these soil factors can positively or negatively affect nanoparticle transformation (Nowack and Bucheli 2007).

Translocation of nanoparticles in plants

Following the entry of nanoparticles into plants through root or foliar exposure, further transport requires their circulation across the xylem or phloem. The extent of nanoparticle translocation in higher plants is determined by their vasculature, as different plants possess different structural architecture of the xylem and phloem (Schwab *et al.* 2016). The transport of nanoparticle in plants takes place via apoplastic and symplastic pathways (Miralles *et al.* 2012). The apoplastic pathway helps in the radial movement of nanoparticles, facilitating the transport to aerial parts of the plants. Once the nanoparticle enters the central cylinder, it follows the transpiration stream to reach the aerial parts (Larue *et al.* 2012b; Wang *et al.* 2012b; Sun *et al.* 2014). In the apoplastic pathway, nanoparticles traverse a range of physiological barriers then travel through cell wall pores and intracellular spaces of the cell without crossing the cell membrane (Sattelmacher 2001; Geisler-Lee *et al.* 2012). The casparian strip forms a layer of impermeable barrier between the apoplast and cortex, preventing the uptake of nanoparticles in the vascular system (Schwab *et al.* 2016). Numerous nanoparticles have been reported to be stopped and accumulated in the casparian strip; however, some are present in the vasculature, which suggests circumvention of the casparian strip (Sun *et al.* 2014; Zhai *et al.* 2014). Circumvention is feasible due to the presence of permeable regions in the casparian strip in endodermis, or entry through the root tip where the casparian strip is not yet formed (Lv *et al.* 2015; Schymura *et al.* 2017).

Nanoparticles can also follow the symplastic route, with the help of sieve tube elements in phloem, and distribute themselves in non-photosynthetic organs and tissues (Wang *et al.* 2012b; Raliya *et al.* 2016). Nanoparticles must cross the cell membrane through the endocytic or non-endocytic pathway which involves movement of the particles from cell to cell through the plasmodesmata, sieve plates and primary pits (Lee and Lu 2011; Zangi and Filella 2012). As the symplastic pathway involves a range of barriers, movement of nanoparticles through the apoplastic pathway (i.e passage outside the cell membrane) is considered the pathway of least resistance (Raven 2003; Sattelmacher and Horst 2007).

Nanoparticles and their mode of delivery have a significant impact on plant cells and tissues. This may lead to adverse events that impede the normal functioning of the plant system. Such adverse events can be predicted by computational modelling which therefore acts as a platform for designing of efficient nanocarriers with minimal or no side effects on plant health. To address this issue, computational models that predict nanostructure–activity relationships have been devised.

Structure–activity relationship

Most of the literature deals with different types of nanoparticles, their treatment regimens and their beneficial and phytotoxic effects. However, no comparison could be drawn to find their uptake efficiency with respect to the structure and strategies used for engineered nanoparticles. Although *in vitro* and *in vivo* studies are recommended, they are time-consuming, labour-intensive and require expertise. Therefore, predicting the structure and activity relationship of engineered nanoparticles remains a significant challenge. A smart way of integrating cellular and molecular responses with that of the nanoparticles used should be devised. Computational chemistry can offer a smart and fastidious approach to analyse the structure–activity-relationship of any nanocarrier.

Harnessing the potential of nanocarriers for genetic engineering

So far, the quantitative structure–activity relationship (QSAR) modelling approach is used to predict the biological response to any chemical. A typical QSAR model first summarises the relationship between different chemical structures to that of their subsequent bioactivities; it can predict the activity of new chemicals based on the previously available datasets. It is a mathematical relationship in the form of an equation between measurable physicochemical properties of a nanocarrier and its biological activity. This system was used to investigate the structure–activity relationship of nanoparticles and animal cells (Fourches *et al.* 2010). Similar to general QSAR strategies, the quantitative nanostructure–activity relationship (QNAR) model incorporates a dataset of manufactured nanoparticles and their measured bioactivities, namely cell viability or cellular uptake. This model can play a vital role in deciphering the biological effects of newly designed or commercially available nanocarriers. Fourches *et al.* (2010) investigated and developed a QNAR platform to predict the bioactivities of different nanostructures. First, 51 diverse magnetic nanoparticles (MNPs) with different metal core and surface functionalisation were tested in *in vitro* cell-based assays. Second, 109 nanoparticles were tested with the same core but different surface functionalisation, for their cellular uptake efficiency. Conventional cheminformatics techniques such as cluster analysis to determine whether MNPs with similar bioactivities have similar structures, and establishment of a QNAR model to correlate the bioactivities of available MNPs, were used to assign and create this dataset. The QNAR predictions for MNPs and their *in vitro* outcomes led to statistically validated models which also quantitatively relate the physical, chemical and geometrical properties of MNPs to their bioactivities. This approach can closely evaluate the activity of a new MNP which is physicochemically similar to the MNPs available in the database.

Numerous reports are available on the physicochemical properties of nanoparticles as well their beneficial and phytotoxic effects. However, little attention has been given to the accountability of using a particular nanoparticle for a specific application. Consolidated studies on different nanoparticles with various sizes, shapes, modifications, bioactivities and cargos are recommended in order to build a computational model similar to QNAR, in order to predict the nature and efficacy of any engineered nanoparticle. Although the studies on effect and use of nanoparticles on plants are discrete, the outcome of such studies can be added into a database. Open access Nano-QSAR tools like NanoBRIDGES can accelerate nanocarrier design and validation procedures computationally (Ambure *et al.* 2015). Quasi-SMILES is a novel computational tool to predict the endpoints of nanomaterials (Toropova *et al.* 2017). Nano-QSAR was used to predict the cytotoxicity of metal oxide nanoparticles using novel descriptors (Pan *et al.* 2016). Nano-QSAR modelling can also be used to design eco-safe nanoparticles with special catalytic properties (Mikolajczyk *et al.* 2018). Similar to the design and nanotoxicology approach used for animal models (Winkler *et al.* 2013), a need-based plant QNAR can accelerate the use of nanoparticles for genetic engineering of plants, and predict the possible outcome.

An analysis on uptake, translocation and accumulation of various gold nanostructures was done in watermelon plants (Raliya *et al.* 2016). The cellular uptake and accumulation measured by inductively coupled plasma-mass spectroscopy (ICPMS) suggested that translocation and accumulation depend on the shape of nanoparticle, application method and type of plant tissue. Nanocarriers coated/loaded with macromolecules like DNA, RNA, protein and ribonucleoproteins for plant genetic engineering can be developed and their outcome can be integrated into a plant-centric Nano-QSAR database.

A biological surface adsorption index and an enthalpy descriptor is proposed as a nano descriptor which includes surface stability/activity, chemical stability, the interaction of nanocarriers with the biological interface etc. (Puzyn *et al.* 2011). Although this approach is time-consuming, it will empower plant molecular biologists to predict and validate nanocarriers even before their use in genetic engineering. Family-specific models were recommended and proposed by Forest *et al.* (2019). The combination of these two simple methods as pre-screening tools offers a compromise that can bridge the gap between case-by-case studies (expensive and time-consuming) and sophisticated nano-QSAR models (not suitable for small datasets) (Forest *et al.* 2019).

Nano-QSAR and regulatory compliance

Use of nanoparticles for plant genetic improvement demands a protocol which is regulatory compliant. Nano-QSAR modelling tools will play a prime role in assessment of nanoparticles and validate the authorisation of nanocarriers for large-scale use. Similarly, using Nano-QSAR in devising a legislative framework for agriculturally useful nanoformulations is under consideration in the European legislative framework (Villaverde *et al.* 2018).

Plant nanotoxicology: toxic and beneficial effects

Application of nanotechnology in plant science has opened new research avenues that will benefit agriculture in terms of higher production and better crop protection. However, data suggest there are negative impacts of nanomaterials on plants. Due to their unique properties, nanomaterials have been found to cause toxicities in plants. This necessitated the introduction of a new scientific discipline called Plant Nanotoxicology, that specifically studies toxicity effects and the mechanisms induced by different nanomaterials used for agricultural purposes or exposed accidentally. Studies have been carried out in the last two decades on the role of different nanomaterials in causing toxicity in plants involving material-specific responses, transport and surface interactions. Different factors related to nanomaterials including size, shape, surface chemistry, dose (concentration × time), permeation behaviour, uptake pathways and uptake rate can influence the overall toxicity. Biomechanical properties of the outer surfaces of plant parts and environmental conditions (e.g. soil type, temperature, humidity etc.) have also been found to play decisive roles in imparting the toxicity of nanomaterials. Nanomaterials that have been explored for toxicity studies include carbon-based such as CNTs (single- and multi-walled) and fullerene; metal and metal oxide nanoparticles (e.g. CuO, TiO_2, ZnO, Ag); rare earth oxide nanoparticles (e.g. CeO_2, La_2O_3, Gd_2O_3, Yb_2O_3); and quantum dots.

It is almost impossible to track down all the beneficial/toxicological studies carried out on plants against different nanomaterials; however, Table 19.2 offers a summary of the beneficial/toxic effects caused by different nanomaterials tested on different plant species. Most of the studies were carried out on commercially important plant species. The beneficial effects exhibited by different nanomaterials are encouraging but not conclusive. Moreover, regulatory aspects were not taken into consideration in dose optimisation. Variations in the selection of plant species, applied doses and methods of molecular-level study make it very difficult to draw broadly accepted conclusions on the specific beneficial/toxic effects of a particular nanomaterial. The lack of field trials in most of the studies is another constraint that delays the process of setting guidelines for safer practices of plant nanotechnology for agricultural application. As shown in Table 19.2, tomato has been tested against the majority of nanomaterials, with contradictory results. This observation can be extrapolated for other plant species used in beneficial/toxic effect studies. Molecular-level studies helped to explore the mechanisms behind the beneficial/toxic effects of different nanomaterials. Generation of reactive oxygen species (ROS) has been found to be associated with the toxic effects exhibited by a few types of nanomaterial. Overall, it can be concluded that plant responses to different nanomaterials are better studied in a case-by-case manner.

Phytotoxicity influencing factors

Extensive studies have been carried out to better understand the issue of toxicity in relation to the physicochemical properties of nanomaterials. However, there is no consensus on the particular aspects of the physicochemical properties of nanomaterials that finally contribute towards their toxicity determination against plants. Environmental factors can also influence phytotoxicity towards plants. Important physicochemical aspects and environmental factors are discussed below.

Surface capping

The interaction of different metallic nanoparticles (MNPs) with plants has been found to be affected by the surface chemistry of the MNPs. Studies have confirmed the role of capping agents in diminishing the toxicity of MNPs towards plants. Surface modifications of MNPs with different capping agents can have different impacts on plant growth. Capping causes changes in the physicochemical properties of MNPS, leading to a different local environment that affects the cellular level. In one study, polymer-coated CuO nanoparticles were found to reduce plant growth and enhance ROS formation in duckweed *Lemna gibba* as compared to the non-effects of uncoated CuO nanoparticles (Perreault *et al.* 2014). In another study (Su *et al.* 2009), ZnS coating was found to enhance the biocompatibility of quantum dots. Hoshino *et al.* (2004) reported changes in the cellular interaction of quantum dots in response to the composition of capping materials. CdSe and CdSe/ZnS-based quantum dots showed variations in plant toxicity with surface modifications (Green and Howman 2005; Kirchner *et al.* 2005)

Dissolution and bioavailability

Very limited studies have been carried out on the mechanism of MNP bioavailability to plants. The bioavailability of MNPs to plants depends on proper transport through the root followed by redistribution via the xylem and phloem. Wang *et al.* (2015) studied the dissolution

Table 19.2. Effect of different nanomaterials on plants

Types of nanomaterials	Plant species	Observed effects	Molecular interpretation	Reference
Single-walled carbon nanotubes	Tomato	Increase in seed germination and growth rate	SWCNTs acted as effective nanotransporters	Srinivasan & Saraswathi (2010)
	Cabbage, carrot, cucumber, lettuce, onion and tomato	Affected root length	Not mentioned	Cañas *et al.* (2008)
	Arabidopsis and rice	Programmed cell death	Production of ROS, chromatins condensation	Shen *et al.* (2010)
	Rice	Delayed the flowering and yield	Natural organic matter elicited the mobility of SWCNTs	Lin *et al.* (2009)
Multi-walled carbon nanotubes	Tomato	Increase in seed germination rate and plant growth	Activation of water channel proteins	Villagarcia *et al.* (2012)
	Maize	Enhanced the germinative growth of seedlings	Improved water delivery	Tiwari *et al.* (2014)
	Barley, soybean and corn	Early seed germination and activation of growth in exposed seedlings	Expression of genes encoding several types of water channel proteins was increased	Lahiani *et al.* (2013)
	Tobacco cell culture	55–64% increase in growth of tobacco cell culture over control	Up-regulation of genes involved in cell division/cell wall formation and water transport	Khodakovskaya *et al.* (2012)
	Wheat	Enhanced root elongation	Faster root growth and higher vegetative biomass No effects on seed germination	Wang *et al.* (2012a)
	Onobrychisarenaria	Stimulated the growth of roots and stems, enhanced the peroxidase activity	Primary uptake, accumulation and translocation of MWCNTs	Smirnova *et al.* (2012)
	Tomato	Produced two times more flowers and fruit compared to control	Increased bacteroidetes	Khodakovskaya *et al.* (2013)
	Rice	Reduced growth	Increased production of ROS	Tan *et al.* (2009)
	Lettuce	Reduced root length	Not mentioned	Lin and Xing (2007)
	Red spinach	Growth inhibition and cell death	ROS production and expression of biomarkers	Begum and Fugetsu (2012)

Continued

Table 19.2. continued

Types of nanomaterials	Plant species	Observed effects	Molecular interpretation	Reference
Fullerene	Bitter melon	Increased biomass and fruit number	Not mentioned	Kole *et al.* (2013)
	Corn and soybean	Reduced biomass	Influenced the accumulation of weathered pesticides	De La Torre-Roche *et al.* (2013)
	Tobacco	Inhibited cell growth	Cell membrane disruption and ROS production	Liu *et al.* (2013)
	Arabidopsis thaliana	Reduced root length and root gravitropism	Abnormalities in cell division, microtubule organisation and mitochondrial activity	Liu *et al.* (2010)
	Onion	Cell death	Apoplastic pathway	Chen *et al.* (2010)
	Arabidopsis thaliana	Up-regulation of 232 stress-responsive genes	Up-regulation of stress-related genes	Landa *et al.* (2012)
TiO_2NPs	*Ulmuselongata*	Reduced photosynthetic rate, defoliation, increased chlorosis and disrupted growth	Synthesis of carbohydrate and lipid compounds as defence mechanism	Gao *et al.* (2013)
	Tomato	Increased root elongation	Not mentioned	Song *et al.* (2013a, b)
	Fennel seeds	Increased germination rate, reduced germination time	Not mentioned	Feizi *et al.* (2013)
CuO NPs	Radish, ryegrass	Strong growth inhibition	DNA damaging	Atha *et al.* (2012)
	Buckwheat	Reduced root growth, alteration of root morphology, genotoxic effects	Altered gene expression levels	Lee *et al.* (2013)
	Wheat	Reduction in root length	Oxidative stress	Dimkpa *et al.* (2012)
ZnO NPs	Buckwheat	Reduced root growth, changes in root morphological feature, genotoxic effects	Different DNA polymorphisms	Lee *et al.* (2013)

Ag NPs	Tomato	Decreased root elongation, lower chlorophyll contents, higher superoxide dismutase activity, decreased fruit productivity	Not mentioned	Song *et al.* (2013a)
	Arabidopsis thaliana	Plant growth inhibition	ROS production	Qian *et al.* (2013)
	Maize and cabbage	Structural changes in primary root cells	Functional impairment	Pokhrel and Dubey (2013)
	Potamogeton crispus	Enhanced senescence	Alteration of sulfhydryl groups of the photosynthetic enzymes	Xu *et al.* (2010)
Au NPs	*Arabidopsis thaliana*	Enhanced defence system, germination rate, antioxidant potential and vegetative growth	Expression of microRNAs (miRs)	Kumar *et al.* (2013)
NiO NPs	Tomato	Induced cell death	ROS generation, apoptosis, necrosis	Faisal *et al.* (2013)
CeO_2, La_2O_3, Gd_2O_3&Yb_2O_3 NPs	Lettuce	Inhibited root elongation and seedling lengths		Ma *et al.* (2010)
Quantum dots (CdSe/ZnS)	*Arabidopsis thaliana*	Oxidative stress	Ratio of GSH to GSSG reduced	Navarro *et al.* (2012)

behaviour of CuO nanoparticles in rice plants, in which fluorescent labelled CuO nanoparticles were located in the meristem zone of the root. Similar studies showed that dissolution and bioavailability of CuO nanoparticles can be confirmed by detection of copper sulfide deposits in leaves and rhizosphere (Dimkpa *et al.* 2012; Peng *et al.* 2015). Dissolution and bioavailability of MNPs have also been found to be influenced by the presence of natural organic matter such as humic acids in soil and secretion of indole-3-acetic acid (IAA) and protons into the rhizosphere by roots (Adams *et al.* 2017).

Doses

Dose determination for beneficial application of MNPs in agriculture is the most challenging and explored domain in plant nanotoxicology. There are numerous reports on the effects of concentration of MNPs and time duration of exposure in determination of phytotoxicity. All those studies concluded that increases in applied concentration cause a sharp increase in phytotoxicity, and that effective doses vary with the type of MNP and plant species (Dietz and Herth 2011; Yang *et al.* 2017).

Soil type

Soil components such as natural organic matter, colloids, heavy metals and nutrients can influence the phytotoxicity of MNPs. Ionic strength, type of cations present in soil and the pH of soil can also play important roles in enhancing or diminishing toxic effects on plants. Soil, being a very complex matrix, influences the colloidal behaviour (homoaggregation, heteroaggregation and kinetic stability) and fate of MNPs, which in turn modifies the mobility of MNPs in soil and finally affects the bioavailability (Kabata-Pendias and Pendias 1993; Hotze *et al.* 2010; El-Temsah and Joner 2012).

Environmental factors

Phytotoxicity of MNPs is shown to be affected by environmental conditions such as temperature and light. Only a few studies have been carried out on this aspect. One important conclusion so far is that an increase in temperature and ageing cause a decrease in the phytotoxicity of MNPs (Jośko and Oleszczuk 2013). The role of light in altering the phytotoxicity of AgNPs was studied in relation to Asian watermeal, *Wolffia globosa*. Light induced AgNP toxicity by altering photosynthetic pigments and protein content of the plant (Zou *et al.* 2016). More research at the molecular level is warranted for exploring the roles of environmental factors in delivering phytotoxicity by MNPs.

Conclusion

Plant genome engineering has allowed scientists and commercial developers to improve plant growth characteristics, improve biotic and abiotic stress tolerance and enhance nutritional values. Also, site-specific genome editing techniques have enhanced the efficiency of genetic engineering compared to conventional gene delivery systems. Nanomaterials will further enhance the efficacy of plant genome engineering. Their ability to carry high concentrations of biomolecules through the cell wall and give efficient uptake into the plant and other cellular compartments showcases them as potential next-generation biomolecule carriers. Nanocarrier systems can reduce the cost and time of genetic manipulation and offer broad-spectrum applicability. However, toxicity issues, the structure–function relationship and the fate of nanocarriers in plant systems still need to be addressed to make them an ideal choice for genetic manipulation. The initial convincing outputs suggest that nanocarriers will enable genome modification to develop superior crop genotypes for a sustainable future.

References

Abdallah B, Hassan A, Benoist C, Goula D, Behr JP, Demeneix BA (1996) A powerful nonviral vector for *in vivo* gene transfer into the adult mammalian brain: polyethylenimine. *Human Gene Therapy* **7**, 1947–1954. doi:10.1089/hum.1996.7.16-1947

Abdel-Razik AB, Hammad IA, Tawfik E (2017) Transformation of Thionin genes using chitosan nanoparticle into potato plant to be resistant to fungal infection. *IOSR Journal of Biotechnology and Biochemistry* **3**, 1–13. doi:10.9790/264X-03030113

Adams J, Wright M, Wagner H, Valiente J, Britt D, Anderson A (2017) Cu from dissolution of CuO nanoparticles signals changes in root morphology. *Plant Physiology and Biochemistry* **110**, 108–117. doi:10.1016/j.plaphy.2016.08.005

Altpeter F, Springer NM, Bartley LE, Blechl A, Brutnell TP, Citovsky V, Conrad L, Gelvin SB, Jackson D, Kausch AP (2016) Advancing crop transformation in the era of genome editing. *The Plant Cell* **28**, 1510–1520.

Ambure P, Aher RB, Gajewicz A, Puzyn T, Roy K (2015) NanoBRIDGES software: open access tools to perform QSAR and nano-QSAR modeling. *Chemometrics and Intelligent Laboratory Systems* **147**, 1–13. doi:10.1016/j.chemolab.2015.07.007

Atha DH, Wang H, Petersen EJ, Cleveland D, Holbrook RD, Jaruga P, Dizdaroglu M, Xing B, Nelson BC (2012) Copper oxide nanoparticle mediated DNA damage in terrestrial plant models. *Environmental Science & Technology* **46**, 1819–1827. doi:10.1021/es202660k

Athanassiou C, Kavallieratos N, Benelli G, Losic D, Rani PU, Desneux N (2018) Nanoparticles for pest control: current status and future perspectives. *Journal of Pest Science* **91**, 1–15. doi:10.1007/s10340-017-0898-0

Barton KA, Binns AN, Matzke AJ, Chilton MD (1983) Regeneration of intact tobacco plants containing full length copies of genetically engineered T-DNA, and transmission of T-DNA to R1 progeny. *Cell* **32**, 1033–1043. doi:10.1016/0092-8674(83)90288-X

Begum P, Fugetsu B (2012) Phytotoxicity of multi-walled carbon nanotubes on red spinach (*Amaranthus tricolor* L) and the role of ascorbic acid as an antioxidant. *Journal of Hazardous Materials* **243**, 212–222. doi:10.1016/j.jhazmat.2012.10.025

Bianco A (2004) Carbon nanotubes for the delivery of therapeutic molecules. *Expert Opinion on Drug Delivery* **1**, 57–65. doi:10.1517/17425247.1.1.57

Bowman MC, Ballard TE, Ackerson CJ, Feldheim DL, Margolis DM, Melander C (2008) Inhibition of HIV fusion with multivalent gold nanoparticles. *Journal of the American Chemical Society* **130**, 6896–6897. doi:10.1021/ja710321g

Cañas JE, Long M, Nations S, Vadan R, Dai L, Luo M, Ambikapathi R, Lee EH, Olszyk D (2008) Effects of functionalized and nonfunctionalized single-walled carbon nanotubes on root elongation of select crop species. *Environmental Toxicology and Chemistry* **27**, 1922–1931. doi:10.1897/08-117.1

Chadegani M, Brink JJ, Shehata A, Ahmadjian V (1989) Optimization of protoplast formation, regeneration, and viability in *Microsporum gypseum*. *Mycopathologia* **107**, 33–50. doi:10.1007/BF00437588

Chang FP, Kuang LY, Huang CA, Jane WN, Hung Y, Yue-Ie CH, Mou CY (2013) A simple plant gene delivery system using mesoporous silica nanoparticles as carriers. *Journal of Materials Chemistry. B, Materials for Biology and Medicine* **1**, 5279–5287. doi:10.1039/c3tb20529k

Chen R, Ratnikova TA, Stone MB, Lin S, Lard M, Huang G, Hudson JS, Ke PC (2010) Differential uptake of carbon nanoparticles by plant and mammalian cells. *Small* **6**, 612–617. doi:10.1002/smll.200901911

Cherng JY, Talsma H, Verrijk R, Crommelin D, Hennink W (1999) The effect of formulation parameters on the size of poly ((2-dimethylamino) ethyl methacrylate)-plasmid complexes. *European Journal of Pharmaceutics and Biopharmaceutics* **47**, 215–224. doi:10.1016/S0939-6411(98)00103-9

Cifuentes Z, Custardoy L, De La Fuente JM, Marquina C, Ibarra MR, Rubiales D, Pérez-De-Luque A (2010) Absorption and translocation to the aerial part of magnetic carbon-coated nanoparticles through the root of different crop plants. *Journal of Nanobiotechnology* **26**, 1–8.

Collier C, Vossmeyer T, Heath AJ (1998) Nanocrystal superlattices. *Annual Review of Physical Chemistry* **49**, 371–404. doi:10.1146/annurev.physchem.49.1.371

Cunningham FJ, Goh NS, Demirer GS, Matos JL, Landry MP (2018) Nanoparticle-mediated delivery towards advancing plant genetic engineering. *Trends in Biotechnology* **36**, 882–897. doi:10.1016/j.tibtech.2018.03.009

Das A, Gupta P, Chakraborty D (2015) Physical methods of gene transfer: kinetics of gene delivery into cells – a review. *Agricultural Reviews (Karnal)* **36**, 61–66. doi:10.5958/0976-0741.2015.00007.0

Davis ME (2002) Non-viral gene delivery systems. *Current Opinion in Biotechnology* **13**, 128–131. doi:10.1016/S0958-1669(02)00294-X

De La Torre-Roche R, Hawthorne J, Deng Y, Xing B, Cai W, Newman LA, Wang Q, Ma X, Hamdi H, White JC (2013) Multiwalled carbon nanotubes and C60 fullerenes differentially impact the accumulation of weathered pesticides in four agricultural plants. *Environmental Science & Technology* **47**, 12539–12547. doi:10.1021/es4034809

Demirer GS, Landry MP (2017) Delivering genes to plants. *Chemical Engineering Progress* **113**, 40–45.

Demirer GS, Zhang H, Matos JL, Goh N, Cunningham F, Sung Y, Chang R, Aditham AJ, Chio L, Cho MJ (2019) High aspect ratio nanomaterials enable delivery of functional genetic material without DNA integration in mature plants. *Nature Nanotechnology* **14**, 456–464. doi:10.1038/s41565-019-0382-5

Dietz KJ, Herth S (2011) Plant nanotoxicology. *Trends in Plant Science* **16**, 582–589. doi:10.1016/j.tplants.2011.08.003

Dimitrov DS, Sowers AE (1990) Membrane electroporation: fast molecular exchange by electroosmosis. *Biochimica et Biophysica Acta (BBA) - Biomembranes* **1022**, 381–392. doi:10.1016/0005-2736(90)90289-Z

Dimkpa CO, Mclean JE, Latta DE, Manangón E, Britt DW, Johnson WP, Boyanov MI, Anderson AJ (2012) CuO and ZnO nanoparticles: phytotoxicity, metal speciation, and induction of oxidative stress in sand-grown wheat. *Journal of Nanoparticle Research* **14**, 1125–1140. doi:10.1007/s11051-012-1125-9

Dobson J (2006) Gene therapy progress and prospects: magnetic nanoparticle-based gene delivery. *Gene Therapy* **13**, 283–287. doi:10.1038/sj.gt.3302720

Eichert T, Goldbach HE (2008) Equivalent pore radii of hydrophilic foliar uptake routes in stomatous and astomatous leaf surfaces: further evidence for a stomatal pathway. *Physiologia Plantarum* **132**, 491–502. doi:10.1111/j.1399-3054.2007.01023.x

Eichert T, Kurtz A, Steiner U, Goldbach HE (2008) Size exclusion limits and lateral heterogeneity of the stomatal foliar uptake pathway for aqueous solutes and water-suspended nanoparticles. *Physiologia Plantarum* **134**, 151–160. doi:10.1111/j.1399-3054.2008.01135.x

El-Temsah YS, Joner EJ (2012) Impact of Fe and Ag nanoparticles on seed germination and differences in bioavailability during exposure in aqueous suspension and soil. *Environmental Toxicology* **27**, 42–49. doi:10.1002/tox.20610

Erathodiyil N, Ying JY (2011) Functionalization of inorganic nanoparticles for bioimaging applications. *Accounts of Chemical Research* **44**, 925–935. doi:10.1021/ar2000327

Faisal M, Saquib Q, Alatar AA, Al-Khedhairy AA, Hegazy AK, Musarrat J (2013) Phytotoxic hazards of NiO-nanoparticles in tomato: a study on mechanism of cell death. *Journal of Hazardous Materials* **250**, 318–332. doi:10.1016/j.jhazmat.2013.01.063

Feizi H, Kamali M, Jafari L, Moghaddam PR (2013) Phytotoxicity and stimulatory impacts of nanosized and bulk titanium dioxide on fennel (*Foeniculum vulgare* Mill). *Chemosphere* **91**, 506–511. doi:10.1016/j.chemosphere.2012.12.012

Feynman R (1960) There's plenty of room at the bottom [talk given on 29 December 1959]. *California Institute of Technology, Engineering and Science Magazine* **23**, 22–36.

Fisk HJ, Dandekar AM (2005) Electroporation: introduction and expression of transgenes in plant protoplasts. In *Transgenic Plants: Methods and Protocols.* (Ed. L Peña) pp. 79–80. Springer.

Forest V, Hochepied JF, Leclerc L, Trouvé A, Abdelkebir K, Sarry G, Augusto V, Pourchez J (2019) Towards an alternative to nano-QSAR for nanoparticle toxicity ranking in case of small datasets. *Journal of Nanoparticle Research* **21**, 95–109. doi:10.1007/s11051-019-4541-2

Fourches D, Pu D, Tassa C, Weissleder R, Shaw SY, Mumper RJ, Tropsha A (2010) Quantitative nanostructure–activity relationship (QNAR) modeling. *ACS Nano* **4**, 5703–5712. doi:10.1021/nn1013484

Fromm M, Taylor LP, Walbot V (1985) Expression of genes transferred into monocot and dicot plant cells by electroporation. *Proceedings of the National Academy of Sciences of the United States of America* **82**, 5824–5828. doi:10.1073/pnas.82.17.5824

Fu YQ, Li LH, Wang PW, Qu J, Fu YP, Wang H, Sun JR (2012) Delivering DNA into plant cell by gene carriers of ZnS nanoparticles. *Chemical Research in Chinese Universities* **28**, 672–676.

Gao J, Xu G, Qian H, Liu P, Zhao P, Hu Y (2013) Effects of nano-TiO2 on photosynthetic characteristics of *Ulmus elongata* seedlings. *Environmental Pollution* **176**, 63–70. doi:10.1016/j.envpol.2013.01.027

Gary DJ, Puri N, Won YY (2007) Polymer-based siRNA delivery: perspectives on the fundamental and phenomenological distinctions from polymer-based DNA delivery. *Journal of Controlled Release* **121**, 64–73. doi:10.1016/j.jconrel.2007.05.021

Geisler-Lee J, Wang Q, Yao Y, Zhang W, Geisler M, Li K, Huang Y, Chen Y, Kolmakov A, Ma X (2012) Phytotoxicity, accumulation and transport of silver nanoparticles by *Arabidopsis thaliana*. *Nanotoxicology* **7**, 323–337. doi:10.3109/17435390.2012.658094

Ghosh PS, Kim CK, Han G, Forbes NS, Rotello VM (2008) Efficient gene delivery vectors by tuning the surface charge density of amino acid-functionalized gold nanoparticles. *ACS Nano* **2**, 2213–2218. doi:10.1021/nn800507t

Giraldo JP, Landry MP, Faltermeier SM, Mcnicholas TP, Iverson NM, Boghossian AA, Reuel NF, Hilmer AJ, Sen F, Brew JA (2014) Plant nanobionics approach to augment photosynthesis and biochemical sensing. *Nature Materials* **13**, 400–408. doi:10.1038/nmat3890

Green M, Howman E (2005) Semiconductor quantum dots and free radical induced DNA nicking. *Chemical Communications* 121–123. doi:10.1039/b413175d

Grimsley N, Hohn B, Hohn T, Walden R (1986) Agroinfection: an alternative route for viral infection of plants by using the Ti plasmid. *Proceedings of the National Academy of Sciences of the United States of America* **83**, 3282–3286. doi:10.1073/pnas.83.10.3282

Gupta G, Arora S, Singh S, Singh N (2014) Advances in magnetofection magnetically guided nucleic acid delivery: a review. *Journal of Pharmaceutical Technology Research Management* **1**, 19–29.

Hajiahmadi Z, Shirzadian-Khorramabad R, Kazemzad M, Sohani MM (2019) Enhancement of tomato resistance to *Tuta absoluta* using a new efficient mesoporous silica nanoparticle-mediated plant transient gene expression approach. *Scientia Horticulturae* **243**, 367–375. doi:10.1016/j.scienta.2018.08.040

Hansen G, Wright MS (1999) Recent advances in the transformation of plants. *Trends in Plant Science* **4**, 226–231. doi:10.1016/S1360-1385(99)01412-0

Hao Y, Yang X, Shi Y, Song S, Xing J, Marowitch J, Chen J, Chen J (2013) Magnetic gold nanoparticles as a vehicle for fluorescein isothiocyanate and DNA delivery into plant cells. *Botany* **91**, 457–466. doi:10.1139/cjb-2012-0281

Hao Y, Xu B, Ma C, Shang J, Gu W, Li W, Hou T, Xiang Y, Cao W, Xing B (2019) Synthesis of novel mesoporous carbon nanoparticles and their phytotoxicity to rice (*Oryza sativa* L.). *Journal of Saudi Chemical Society* **23**(1), 75–82. doi:10.1016/j.jscs.2018.05.003

Herrero MJ, Sendra L, Miguel A, Aliño SF (2017) Physical methods of gene delivery. In *Safety and Efficacy of Gene-Based Therapeutics for Inherited Disorders*. (Ed. N Brunetti-Pierri) pp. 113–135. Springer.

Hischemöller A, Nordmann J, Ptacek P, Mummenhoff K, Haase M (2009) *In-vivo* imaging of the uptake of upconversion nanoparticles by plant roots. *Journal of Biomedical Nanotechnology* **5**, 278–284. doi:10.1166/jbn.2009.1032

Hong J, Peralta-Videa JR, Rico C, Sahi S, Viveros MN, Bartonjo J, Zhao L, Gardea-Torresdey JL (2014) Evidence of translocation and physiological impacts of foliar applied CeO2 nanoparticles on cucumber (*Cucumis sativus*) plants. *Environmental Science & Technology* **48**, 4376–4385. doi:10.1021/es404931g

Hoshino A, Fujioka K, Oku T, Suga M, Sasaki YF, Ohta T, Yasuhara M, Suzuki K, Yamamoto K (2004) Physicochemical properties and cellular toxicity of nanocrystal quantum dots depend on their surface modification. *Nano Letters* **4**, 2163–2169. doi:10.1021/nl048715d

Hotze EM, Phenrat T, Lowry GV (2010) Nanoparticle aggregation: challenges to understanding transport and reactivity in the environment. *Journal of Environmental Quality* **39**, 1909–1924. doi:10.2134/jeq2009.0462

Huang Y, Zhao L, Keller AA (2017) Interactions, transformations, and bioavailability of nano-copper exposed to root exudates. *Environmental Science & Technology* **51**, 9774–9783. doi:10.1021/acs.est.7b02523

Husaini AM, Abdin MZ, Parray GA, Sanghera GS, Murtaza I, Alam T, Srivastava DK, Farooqi H, Khan HN (2010) Vehicles and ways for efficient nuclear transformation in plants. *GM Crops* **1**, 276–287. doi:10.4161/gmcr.1.5.14660

Jaganathan D, Ramasamy K, Sellamuthu G, Jayabalan S, Venkataraman G (2018) CRISPR for crop improvement: an update review. *Frontiers of Plant Science* **9**, 985–1002. doi:10.3389/fpls.2018.00985

Jakutavičiūtė M, Ruzgys P, Tamošiūnas M, Maciulevičius M, Šatkauskas S (2017) Physical methods for drug and gene delivery through the cell plasma membrane. In *Transport across Natural and Modified Biological Membranes and its Implications in Physiology and Therapy.* (Eds J Kulbacka and S Satkauskas) pp. 73–92. Springer.

Jen CP, Chen YH, Fan CS, Yeh CS, Lin YC, Shieh DB, Wu CL, Chen DH, Chou CH (2004) A nonviral transfection approach *in vitro*: the design of a gold nanoparticle vector joint with microelectromechanical systems. *Langmuir* **20**, 1369–1374. doi:10.1021/la036154k

Jiang L, Ding L, He B, Shen J, Xu Z, Yin M, Zhang X (2014) Systemic gene silencing in plants triggered by fluorescent nanoparticle-delivered double-stranded RNA. *Nanoscale* **6**, 9965–9969. doi:10.1039/C4NR03481C

Jośko I, Oleszczuk P (2013) Manufactured nanomaterials: the connection between environmental fate and toxicity. *Critical Reviews in Environmental Science and Technology* **43**, 2581–2616. doi:10.1080/10643389.2012.694329

Kabata-Pendias A, Pendias H (1993) *Biogeochemia pierwiastkw ladowych.* Wydawnictwo Naukowe PWN, Warszawa.

Kanchiswamy CN, Malnoy M, Velasco R, Kim JS, Viola R (2015) Non-GMO genetically edited crop plants. *Trends in Biotechnology* **33**, 489–491. doi:10.1016/j.tibtech.2015.04.002

Karny A, Zinger A, Kajal A, Shainsky-Roitman J, Schroeder A (2018) Therapeutic nanoparticles penetrate leaves and deliver nutrients to agricultural crops. *Scientific Reports* **8**, 7589. doi:10.1038/s41598-018-25197-y

Khodakovskaya MV, De Silva K, Biris AS, Dervishi E, Villagarcia H (2012) Carbon nanotubes induce growth enhancement of tobacco cells. *ACS Nano* **6**, 2128–2135. doi:10.1021/nn204643g

Khodakovskaya MV, Kim BS, Kim JN, Alimohammadi M, Dervishi E, Mustafa T, Cernigla CE (2013) Carbon nanotubes as plant growth regulators: effects on tomato growth, reproductive system, and soil microbial community. *Small* **9**, 115–123. doi:10.1002/smll.201201225

Khromov AV, Gushchin VA, Timerbaev VI, Kalinina NO, Taliansky ME, Makarov VV (2018) Guide RNA design for CRISPR/Cas9-mediated potato genome editing. *Doklady. Biochemistry and Biophysics* **479**, 90–94. doi:10.1134/S1607672918020084

Kim SW, Kim T, Kim YS, Choi HS, Lim HJ, Yang SJ, Park CR (2012) Surface modifications for the effective dispersion of carbon nanotubes in solvents and polymers. *Carbon* **50**, 3–33. doi:10.1016/j.carbon.2011.08.011

Kirchner C, Liedl T, Kudera S, Pellegrino T, Muñoz Javier A, Gaub HE, Stölzle S, Fertig N, Parak WJ (2005) Cytotoxicity of colloidal CdSe and CdSe/ZnS nanoparticles. *Nano Letters* **5**, 331–338. doi:10.1021/nl047996m

Kole C, Kole P, Randunu KM, Choudhary P, Podila R, Ke PC, Rao AM, Marcus RK (2013) Nanobiotechnology can boost crop production and quality: first evidence from increased plant biomass, fruit yield and phytomedicine content in bitter melon (*Momordica charantia*). *BMC Biotechnology* **13**, 37. doi:10.1186/1472-6750-13-37

Kole C, Kumar DS, Khodakovskaya MV (2016) *Plant Nanotechnology: Principles and Practices.* Springer.

Kumar V, Guleria P, Kumar V, Yadav SK (2013) Gold nanoparticle exposure induces growth and yield enhancement in *Arabidopsis thaliana. Science of the Total Environment* **461**, 462–468. doi:10.1016/j.scitotenv.2013.05.018

Kwak SY, Lew TTS, Sweeney CJ, Koman VB, Wong MH, Bohmert-Tatarev K, Snell KD, Seo JS, Chua NH, Strano MS (2019) Chloroplast-selective gene delivery and expression in planta using chitosan-complexed single-walled carbon nanotube carriers. *Nature Nanotechnology* **14**, 447–455. doi:10.1038/s41565-019-0375-4

Lahiani MH, Dervishi E, Chen J, Nima Z, Gaume A, Biris AS, Khodakovskaya MV (2013) Impact of carbon nanotube exposure to seeds of valuable crops. *ACS Applied Materials & Interfaces* **5**, 7965–7973. doi:10.1021/am402052x

Landa P, Vankova R, Andrlova J, Hodek J, Marsik P, Storchova H, White JC, Vanek T (2012) Nanoparticle-specific changes in *Arabidopsis thaliana* gene expression after exposure to ZnO, TiO2, and fullerene soot. *Journal of Hazardous Materials* **241**, 55–62. doi:10.1016/j.jhazmat.2012.08.059

Larue C, Laurette J, Herlin-Boime N, Khodja H, Fayard B, Flank AM, Brisset F, Carriere M (2012a) Accumulation, translocation and impact of TiO2 nanoparticles in wheat (*Triticum aestivum*): influence of diameter and crystal phase. *Science of the Total Environment* **431**, 197–208. doi:10.1016/j.scitotenv.2012.04.073

Larue C, Veronesi G, Flank AM, Surble S, Herlin-Boime N, Carrière M (2012b) Comparative uptake and impact of TiO2 nanoparticles in wheat and rapeseed. *Journal of Toxicology and Environmental Health. Part A.* **75**, 722–734. doi:10.1080/15287394.2012.689800

Lee JY, Lu H (2011) Plasmodesmata: the battleground against intruders. *Trends in Plant Science* **16**, 201–210. doi:10.1016/j.tplants.2011.01.004

Lee S, Chung H, Kim S, Lee I (2013) The genotoxic effect of ZnO and CuO nanoparticles on early growth of buckwheat, *Fagopyrum esculentum*. *Water, Air, and Soil Pollution* **224**, 1668. doi:10.1007/s11270-013-1668-0

Lee K, Conboy M, Park HM, Jiang F, Kim HJ, Dewitt MA, Mackley VA, Chang K, Rao A, Skinner C (2017) Nanoparticle delivery of Cas9 ribonucleoprotein and donor DNA *in vivo* induces homology-directed DNA repair. *Nature Biomedical Engineering* **1**, 889. doi:10.1038/s41551-017-0137-2

Li YL, Zhu L, Liu Z, Cheng R, Meng F, Cui JH, Ji SJ, Zhong Z (2009) Reversibly stabilized multifunctional dextran nanoparticles efficiently deliver doxorubicin into the nuclei of cancer cells. *Angewandte Chemie* **121**, 10098–10102. doi:10.1002/ange.200904260

Li J, Qian J, Xu Y, Yan S, Shen J, Yin M (2019) A facile-synthesized star polycation is constructed as a highly efficient gene vector in pest management. *ACS Sustainable Chemistry & Engineering* **7**, 6316–6322. doi:10.1021/acssuschemeng.9b00004

Lin D, Xing B (2007) Phytotoxicity of nanoparticles: inhibition of seed germination and root growth. *Environmental Pollution* **150**, 243–250. doi:10.1016/j.envpol.2007.01.016

Lin D, Xing B (2008) Root uptake and phytotoxicity of ZnO nanoparticles. *Environmental Science & Technology* **42**, 5580–5585. doi:10.1021/es800422x

Lin S, Reppert J, Hu Q, Hudson JS, Reid ML, Ratnikova TA, Rao AM, Luo H, Ke PC (2009) Uptake, translocation, and transmission of carbon nanomaterials in rice plants. *Small* **5**, 1128–1132.

Liu Z, Sun X, Nakayama-Ratchford N, Dai H (2007) Supramolecular chemistry on water-soluble carbon nanotubes for drug loading and delivery. *ACS Nano* **1**, 50–56. doi:10.1021/nn700040t

Liu Q, Chen B, Wang Q, Shi X, Xiao Z, Lin J, Fang X (2009) Carbon nanotubes as molecular transporters for walled plant cells. *Nano Letters* **9**, 1007–1010. doi:10.1021/nl803083u

Liu Q, Zhao Y, Wan Y, Zheng J, Zhang X, Wang C, Fang X, Lin J (2010) Study of the inhibitory effect of water-soluble fullerenes on plant growth at the cellular level. *ACS Nano* **4**, 5743–5748. doi:10.1021/nn101430g

Liu Q, Zhang X, Zhao Y, Lin J, Shu C, Wang C, Fang X (2013) Fullerene-induced increase of glycosyl residue on living plant cell wall. *Environmental Science & Technology* **47**, 7490–7498. doi:10.1021/es4010224

Lodha A, Lodha M, Patel A, Chaudhuri J, Dalal J, Edwards M, Douroumis D (2012) Synthesis of mesoporous silica nanoparticles and drug loading of poorly water-soluble drug cyclosporin A. *Journal of Pharmacy & Bioallied Sciences* **4**, 92–94. doi:10.4103/0975-7406.94153

Loh XJ, Lee TC, Dou Q, Deen GR (2016) Utilising inorganic nanocarriers for gene delivery. *Biomaterials Science* **4**, 70–86. doi:10.1039/C5BM00277J

López-Saucedo F, Flores-Rojas GG, Meléndez-Ortiz HI, Morfín-Gutierrez A, Luna-Straffon MA, Bucio E (2019) Stimuli-responsive nanomaterials for drug delivery. In *Characterization and Biology of Nanomaterials for Drug Delivery: Nanoscience and Nanotechnology in Drug Delivery.* (Eds S Mohapatra, S Ranjan, N Dasgupta, R Kumar and S Thomas) pp. 375–424. Elsevier.

Lv J, Christie P, Zhang S (2019) Uptake, translocation, and transformation of metal-based nanoparticles in plants: recent advances and methodological challenges. *Environmental Science Nano* **6**, 41–59.

Lv J, Zhang S, Luo L, Zhang J, Yang K, Christie P (2015) Accumulation, speciation and uptake pathway of ZnO nanoparticles in maize. *Environmental Science. Nano* **2**, 68–77. doi:10.1039/C4EN00064A

Ma X, Geiser-Lee J, Deng Y, Kolmakov A (2010) Interactions between engineered nanoparticles (ENPs) and plants: phytotoxicity, uptake and accumulation. *Science of the Total Environment* **408**, 3053–3061. doi:10.1016/j.scitotenv.2010.03.031

Ma C, White JC, Dhankher OP, Xing B (2015) Metal-based nanotoxicity and detoxification pathways in higher plants. *Environmental Science & Technology* **49**, 7109–7122. doi:10.1021/acs.est.5b00685

Makhotenko AV, Snigir EA, Kalinina NO, Makarov VV, Taliansky ME (2018) Data on a delivery of biomolecules into *Nicotiana benthamiana* leaves using different nanoparticles. *Data in Brief* **16**, 1034–1037. doi:10.1016/j.dib.2017.12.031

Mao Y, Botella JR, Liu Y, Zhu JK (2019) Gene editing in plants: progress and challenges. *National Science Review* **6**, 421–437. doi:10.1093/nsr/nwz005

Martin-Ortigosa S, Wang K (2014) Proteolistics: a biolistic method for intracellular delivery of proteins. *Transgenic Research* **23**, 743–756. doi:10.1007/s11248-014-9807-y

Martin-Ortigosa S, Valenstein JS, Lin VSY, Trewyn BG, Wang K (2012a) Gold functionalized mesoporous silica nanoparticle mediated protein and DNA codelivery to plant cells via the biolistic method. *Advanced Functional Materials* **22**, 3576–3582. doi:10.1002/adfm.201200359

Martin-Ortigosa S, Valenstein JS, Sun W, Moeller L, Fang N, Trewyn BG, Lin VSY, Wang K (2012b) Parameters affecting the efficient delivery of mesoporous silica nanoparticle materials and gold nanorods into plant tissues by the biolistic method. *Small* **8**, 413–422. doi:10.1002/smll.201101294

Martin-Ortigosa S, Peterson DJ, Valenstein JS, Lin VSY, Trewyn BG, Lyznik LA, Wang K (2014) Mesoporous silica nanoparticle-mediated intracellular Cre protein delivery for maize genome editing via loxP site excision. *Plant Physiology* **164**, 537–547. doi:10.1104/pp.113.233650

Mikolajczyk A, Gajewicz A, Mulkiewicz E, Rasulev B, Marchelek M, Diak M, Hirano S, Zaleska-Medynska A, Puzyn T (2018) Nano-QSAR modeling for ecosafe design of heterogeneous TiO2-based nano-photocatalysts. *Environmental Science. Nano* **5**, 1150–1160. doi:10.1039/C8EN00085A

Miralles P, Church TL, Harris AT (2012) Toxicity, uptake, and translocation of engineered nanomaterials in vascular plants. *Environmental Science & Technology* **46**, 9224–9239. doi:10.1021/es202995d

Mitter N, Worrall EA, Robinson KE, Li P, Jain RG, Taochy C, Fletcher SJ, Carroll BJ, Lu GM, Xu ZP (2017) Clay nanosheets for topical delivery of RNAi for sustained protection against plant viruses. *Nature Plants* **3**, 16207. doi:10.1038/nplants.2016.207

Moorthy MS, Kim HB, Bae JH, Kim SH, Ha CS (2016) Design of core–shell magnetic mesoporous silica hybrids for pH and UV light stimuli-responsive cargo release. *RSC Advances* **6**, 29106–29115. doi:10.1039/C5RA28143A

Mortazavi S, Zohrabi Z (2018) Biolistic co-transformation of rice using gold nanoparticles. *Iran Agricultural Research* **37**, 75–82.

Navarro DA, Bisson MA, Aga DS (2012) Investigating uptake of water-dispersible CdSe/ZnS quantum dot nanoparticles by *Arabidopsis thaliana* plants. *Journal of Hazardous Materials* **211**, 427–435. doi:10.1016/j.jhazmat.2011.12.012

Nowack B, Bucheli TD (2007) Occurrence, behavior and effects of nanoparticles in the environment. *Environmental Pollution* **150**, 5–22. doi:10.1016/j.envpol.2007.06.006

Nuruzzaman M, Rahman M, Liu Y, Naidu R (2016) Nanoencapsulation, nano-guard for pesticides: a new window for safe application. *Journal of Agricultural and Food Chemistry* **64**, 1447–1483. doi:10.1021/acs.jafc.5b05214

O'Brien JA, Lummis SC (2011) Nano-biolistics: a method of biolistic transfection of cells and tissues using a gene gun with novel nanometer-sized projectiles. *BMC Biotechnology* **11**, 66. doi:10.1186/1472-6750-11-66

Pan Y, Li T, Cheng J, Telesca D, Zink JI, Jiang J (2016) Nano-QSAR modeling for predicting the cytotoxicity of metal oxide nanoparticles using novel descriptors. *RSC Advances* **6**, 25766–25775. doi:10.1039/C6RA01298A

Pantarotto D, Singh R, McCarthy D, Erhardt M, Briand JP, Prato M, Kostarelos K, Bianco A (2004) Functionalized carbon nanotubes for plasmid DNA gene delivery. *Angewandte Chemie International Edition* **43**, 5242–5246. doi:10.1002/anie.200460437

Peng C, Zhang H, Fang H, Xu C, Huang H, Wang Y, Sun L, Yuan X, Chen Y, Shi J (2015) Natural organic matter-induced alleviation of the phytotoxicity to rice (*Oryza sativa* L.) caused by copper oxide nanoparticles. *Environmental Toxicology and Chemistry* **34**, 1996–2003. doi:10.1002/etc.3016

Perreault F, Popovic R, Dewez D (2014) Different toxicity mechanisms between bare and polymer-coated copper oxide nanoparticles in *Lemna gibba*. *Environmental Pollution* **185**, 219–227. doi:10.1016/j.envpol.2013.10.027

Peteu SF, Oancea F, Sicuia OA, Constantinescu F, Dinu S (2010) Responsive polymers for crop protection. *Polymers* **2**, 229–251. doi:10.3390/polym2030229

Pokhrel LR, Dubey B (2013) Evaluation of developmental responses of two crop plants exposed to silver and zinc oxide nanoparticles. *Science of the Total Environment* **452**, 321–332. doi:10.1016/j.scitotenv.2013.02.059

Popat A, Hartono SB, Stahr F, Liu J, Qiao SZ, Lu GQM (2011) Mesoporous silica nanoparticles for bioadsorption, enzyme immobilisation, and delivery carriers. *Nanoscale* **3**, 2801–2818. doi:10.1039/c1nr10224a

Puzyn T, Bakhtiyor R, Gajewicz A, Hu X, Dasari T, Michalkova A, Hwang H, Toropov A, Leszczynska D, Leszczynski J (2011) Using nano-QSAR to predict the cytotoxicity of metal oxide nanoparticles. *Nature Nanotechnology* **6**, 175. doi:10.1038/nnano.2011.10

Qian H, Peng X, Han X, Ren J, Sun L, Fu Z (2013) Comparison of the toxicity of silver nanoparticles and silver ions on the growth of terrestrial plant model *Arabidopsis thaliana*. *Journal of Environmental Sciences (China)* **25**, 1947–1956. doi:10.1016/S1001-0742(12)60301-5

Que Q, Chilton MDM, Elumalai S, Zhong H, Dong S, Shi L (2019) Repurposing macromolecule delivery tools for plant genetic modification in the era of precision genome engineering. In *Transgenic Plants: Methods and Protocols* (rev. edn). (Eds S Kumar, P Barone and M Smith) pp. 3–18. Springer.

Răcuciu M, Miclăuş S, Creangă DE (2009) The response of plant tissues to magnetic fluid and electromagnetic exposure. *Romanian Journal of Biophysics* **19**, 73–83.

Rafsanjani M, Kiran U, Ali A, Abdin M (2016) Transformation efficiency of calcium phosphate nanoparticles for genetic manipulation of *Cichorium intybus* L. *Indian Journal of Biotechnology* **15**, 145–152.

Raliya R, Franke C, Chavalmane S, Nair R, Reed N, Biswas P (2016) Quantitative understanding of nanoparticle uptake in watermelon plants. *Frontiers of Plant Science* **7**, 1288. doi:10.3389/fpls.2016.01288

Ramkumar TR, Lenka SK, Arya SS, Bansal KC (2020) A short history and perspectives on plant genetic transformation. In *Biolistic DNA Delivery in Plants*. (Eds S Rustgi, H Luo) pp. 39–68. Humana, New York. doi:10.1007/978-1-0716-0356-7_3

Raven J (2003) Long-distance transport in non-vascular plants. *Plant, Cell & Environment* **26**, 73–85. doi:10.1046/j.1365-3040.2003.00920.x

Ren TB, Feng Y, Zhang Z-H, Li L, Li YY (2011) Shell-sheddable micelles based on star-shaped poly(ε-caprolactone)-SS-poly(ethyl glycol) copolymer for intracellular drug release. *Soft Matter* **7**, 2329–2331. doi:10.1039/c1sm05020f

Ruf S, Bock R (2017) Loopholes for smuggling DNA into pollen. *Nature Plants* **3**, 918. doi:10.1038/s41477-017-0072-y

Ryan JA, Overton KW, Speight ME, Oldenburg CN, Loo L, Robarge W, Franzen S, Feldheim DL (2007) Cellular uptake of gold nanoparticles passivated with BSA-SV40 large t antigen conjugates. *Analytical Chemistry* **79**, 9150–9159. doi:10.1021/ac0715524

Samboju NC, Yau KY, Burroughs F, Samuel JP, Webb SR (2014) Linear DNA molecule delivery using PEGylated quantum dots for stable transformation in plants. US Patent no. 8,653,327.

Samuel JP, Petolino JF, Samboju NC, Webb SR, Yau KY (2015) Nanoparticle mediated delivery of sequence specific nucleases. US Patent no. 9,187,755.

Sanford JC (1990) Biolistic plant transformation. *Physiologia Plantarum* **79**, 206–209. doi:10.1111/j.1399-3054.1990.tb05888.x

Santi L, Batchelor L, Huang Z, Hjelm B, Kilbourne J, Arntzen CJ, Chen Q, Mason HS (2008) An efficient plant viral expression system generating orally immunogenic Norwalk virus-like particles. *Vaccine* **26**, 1846–1854. doi:10.1016/j.vaccine.2008.01.053

Sattelmacher B (2001) The apoplast and its significance for plant mineral nutrition. *New Phytologist* **149**, 167–192. doi:10.1046/j.1469-8137.2001.00034.x

Sattelmacher B, Horst WJ (2007) *The Apoplast of Higher Plants: Compartment of Storage, Transport and Reactions*. Springer Science & Business Media, Netherlands.

Scherer F, Anton M, Schillinger U, Henke J, Bergemann C, Krüger A, Gänsbacher B, Plank C (2002) Magnetofection: enhancing and targeting gene delivery by magnetic force *in vitro* and *in vivo*. *Gene Therapy* **9**, 102–109. doi:10.1038/sj.gt.3301624

Schreiber L (2005) Polar paths of diffusion across plant cuticles: new evidence for an old hypothesis. *Annals of Botany* **95**, 1069–1073. doi:10.1093/aob/mci122

Schreiber L (2010) Transport barriers made of cutin, suberin and associated waxes. *Trends in Plant Science* **15**, 546–553. doi:10.1016/j.tplants.2010.06.004

Schwab F, Zhai G, Kern M, Turner A, Schnoor JL, Wiesner MR (2016) Barriers, pathways and processes for uptake, translocation and accumulation of nanomaterials in plants: critical review. *Nanotoxicology* **10**, 257–278.

Schymura S, Fricke T, Hildebrand H, Franke K (2017) Elucidating the role of dissolution in CeO2 nanoparticle plant uptake by smart radiolabeling. *Angewandte Chemie International Edition* **56**, 7411–7414. doi:10.1002/anie.201702421

Serag MF, Kaji N, Gaillard C, Okamoto Y, Terasaka K, Jabasini M, Tokeshi M, Mizukami H, Bianco A, Baba Y (2011) Trafficking and subcellular localization of multiwalled carbon nanotubes in plant cells. *ACS Nano* **5**, 493–499. doi:10.1021/nn102344t

Serag MF, Kaji N, Tokeshi M, Baba Y (2012) Introducing carbon nanotubes into living walled plant cells through cellulase-induced nanoholes. *RSC Advances* **2**, 398–400. doi:10.1039/C1RA00760B

Seshadri B, Bolan N, Naidu R (2015) Rhizosphere-induced heavy metal (loid) transformation in relation to bioavailability and remediation. *Journal of Soil Science and Plant Nutrition* **15**, 524–548.

Shen CX, Zhang QF, Li J, Bi FC, Yao N (2010) Induction of programmed cell death in *Arabidopsis* and rice by single-wall carbon nanotubes. *American Journal of Botany* **97**, 1602–1609. doi:10.3732/ajb.1000073

Silva AT, Nguyen A, Ye C, Verchot J, Moon JH (2010) Conjugated polymer nanoparticles for effective siRNA delivery to tobacco BY-2 protoplasts. *BMC Plant Biology* **10**, 291. doi:10.1186/1471-2229-10-291

Singh R, Pantarotto D, Mccarthy D, Chaloin O, Hoebeke J, Partidos CD, Briand JP, Prato M, Bianco A, Kostarelos K (2005) Binding and condensation of plasmid DNA onto functionalized carbon nanotubes: toward the construction of nanotube-based gene delivery vectors. *Journal of the American Chemical Society* **127**, 4388–4396. doi:10.1021/ja0441561

Smirnova E, Gusev A, Zaytseva O, Sheina O, Tkachev A, Kuznetsova E, Lazareva E, Onishchenko G, Feofanov A, Kirpichnikov M (2012) Uptake and accumulation of multiwalled carbon nanotubes change the morphometric and biochemical characteristics of *Onobrychis arenaria* seedlings. *Frontiers of Chemical Science and Engineering* **6**, 132–138. doi:10.1007/s11705-012-1290-5

Song U, Jun H, Waldman B, Roh J, Kim Y, Yi J, Lee EJ (2013a) Functional analyses of nanoparticle toxicity: a comparative study of the effects of TiO2 and Ag on tomatoes (*Lycopersicon esculentum*). *Ecotoxicology and Environmental Safety* **93**, 60–67. doi:10.1016/j.ecoenv.2013.03.033

Song U, Shin M, Lee G, Roh J, Kim Y, Lee EJ (2013b) Functional analysis of TiO2 nanoparticle toxicity in three plant species. *Biological Trace Element Research* **155**, 93–103. doi:10.1007/s12011-013-9765-x

Srinivasan C, Saraswathi R (2010) Nano-agriculture: carbon nanotubes enhance tomato seed germination and plant growth. *Current Science* **99**, 274–275.

Su Y, He Y, Lu H, Sai L, Li Q, Li W, Wang L, Shen P, Huang Q, Fan C (2009) The cytotoxicity of cadmium based, aqueous phase-synthesized, quantum dots and its modulation by surface coating. *Biomaterials* **30**, 19–25. doi:10.1016/j.biomaterials.2008.09.029

Sun D, Hussain HI, Yi Z, Siegele R, Cresswell T, Kong L, Cahill DM (2014) Uptake and cellular distribution, in four plant species, of fluorescently labeled mesoporous silica nanoparticles. *Plant Cell Reports* **33**, 1389–1402. doi:10.1007/s00299-014-1624-5

Tan XM, Lin C, Fugetsu B (2009) Studies on toxicity of multi-walled carbon nanotubes on suspension rice cells. *Carbon* **47**, 3479–3487. doi:10.1016/j.carbon.2009.08.018

Thornton PD, McConnell G, Ulijn RV (2005) Enzyme responsive polymer hydrogel beads. *Chemical Communications* **47**, 5913–5915. doi:10.1039/b511005j

Tiwari D, Dasgupta-Schubert N, Cendejas LV, Villegas J, Montoya LC, García SB (2014) Interfacing carbon nanotubes (CNT) with plants: enhancement of growth, water and ionic nutrient uptake in maize (*Zea mays*) and implications for nanoagriculture. *Applied Nanoscience* **4**, 577–591. doi:10.1007/s13204-013-0236-7

Torney F, Trewyn BG, Lin VSY, Wang K (2007) Mesoporous silica nanoparticles deliver DNA and chemicals into plants. *Nature Nanotechnology* **2**, 295. doi:10.1038/nnano.2007.108

Toropova AP, Toropov AA, Veselinović AM, Veselinović JB, Leszczynska D, Leszczynski J (2017) Quasi-SMILES as a novel tool for prediction of nanomaterials' endpoints. In *Multi-Scale Approaches in Drug Discovery: From Empirical Knowledge to in silico Experiments and Back.* (Ed A Speck-Blanche) pp. 191–221. Elsevier.

Tripathi DK, Ahmad P, Sharma S, Chauhan DK, Dubey NK (2017) *Nanomaterials in Plants, Algae, and Microorganisms: Concepts and Controversies*, **1.** Academic Press, UK.

Vaghchhipawala Z, Rojas CM, Senthil-Kumar M, Mysore KS (2011) Agroinoculation and agroinfiltration: simple tools for complex gene function analyses. In *Plant Reverse Genetics: Methods and Protocols.* (Ed A Pereira) pp. 65–76. Springer.

Vance ME, Kuiken T, Vejerano EP, Mcginnis SP, Hochella MF, Rejeski D, Hull MS (2015) Nanotechnology in the real world: redeveloping the nanomaterial consumer products inventory. *Beilstein Journal of Nanotechnology* **6**, 1769–1780. doi:10.3762/bjnano.6.181

Villagarcia H, Dervishi E, De Silva K, Biris AS, Khodakovskaya MV (2012) Surface chemistry of carbon nanotubes impacts the growth and expression of water channel protein in tomato plants. *Small* **8**, 2328–2334. doi:10.1002/smll.201102661

Villaverde JJ, Beatriz SM, Carmen LG, José Luis AP, Pilar SE (2018) Considerations of nano-QSAR/QSPR models for nanopesticide risk assessment within the European legislative framework. *Science of the Total Environment* **634**, 1530–1539. doi:10.1016/j.scitotenv.2018.04.033

Wang EC, Wang AZ (2014) Nanoparticles and their applications in cell and molecular biology. *Integrative Biology* **6**, 9–26. doi:10.1039/c3ib40165k

Wang X, Han H, Liu X, Gu X, Chen K, Lu D (2012a) Multi-walled carbon nanotubes can enhance root elongation of wheat (*Triticum aestivum*) plants. *Journal of Nanoparticle Research* **14**, 841. doi:10.1007/s11051-012-0841-5

Wang Z, Xie X, Zhao J, Liu X, Feng W, White JC, Xing B (2012b) Xylem-and phloem-based transport of CuO nanoparticles in maize (*Zea mays* L.). *Environmental Science & Technology* **46**, 4434–4441. doi:10.1021/es204212z

Wang S, Liu H, Zhang Y, Xin H (2015) The effect of CuO NPs on reactive oxygen species and cell cycle gene expression in roots of rice. *Environmental Toxicology and Chemistry* **34**, 554–561. doi:10.1002/etc.2826

Wang P, Lombi E, Zhao FJ, Kopittke PM (2016) Nanotechnology: a new opportunity in plant sciences. *Trends in Plant Science* **21**, 699–712. doi:10.1016/j.tplants.2016.04.005

Winkler DA, Mombelli E, Pietroiusti A, Tran L, Worth A, Fadeel B, McCall MJ (2013) Applying quantitative structure–activity relationship approaches to nanotoxicology: current status and future potential. *Toxicology* **313**, 15–23. doi:10.1016/j.tox.2012.11.005

Wong MH, Misra RP, Giraldo JP, Kwak SY, Son Y, Landry MP, Swan JW, Blankschtein D, Strano MS (2016) Lipid exchange envelope penetration (LEEP) of nanoparticles for plant engineering: a universal localization mechanism. *Nano Letters* **16**, 1161–1172. doi:10.1021/acs.nanolett.5b04467

Wroblewski T, Tomczak A, Michelmore R (2005) Optimization of *Agrobacterium*-mediated transient assays of gene expression in lettuce, tomato and Arabidopsis. *Plant Biotechnology Journal* **3**, 259–273. doi:10.1111/j.1467-7652.2005.00123.x

Wu FS, Feng TY (1999) Delivery of plasmid DNA into intact plant cells by electroporation of plasmolyzed cells. *Plant Cell Reports* **18**, 381–386. doi:10.1007/s002990050590

Wu P, Yan XP (2013) Doped quantum dots for chemo/biosensing and bioimaging. *Chemical Society Reviews* **42**, 5489–5521. doi:10.1039/c3cs60017c

Xu Q, Hu J, Xie K, Yang H, Du K, Shi G (2010) Accumulation and acute toxicity of silver in *Potamogeton crispus* L. *Journal of Hazardous Materials* **173**, 186–193. doi:10.1016/j.jhazmat.2009.08.067

Xu Z, Li B, Tang W, Chen T, Zhang H, Wang Q (2011) Glycopolypeptide-encapsulated Mn-doped ZnS quantum dots for drug delivery: fabrication, characterization, and *in vitro* assessment. *Colloids and Surfaces. B, Biointerfaces* **88**, 51–57. doi:10.1016/j.colsurfb.2011.05.055

Yang J, Cao W, Rui Y (2017) Interactions between nanoparticles and plants: phytotoxicity and defense mechanisms. *Journal of Plant Interactions* **12**, 158–169. doi:10.1080/17429145.2017.1310944

Yi Z, Hussain HI, Feng C, Sun D, She F, Rookes JE, Cahill DM, Kong L (2015) Functionalized mesoporous silica nanoparticles with redox-responsive short-chain gatekeepers for agrochemical delivery. *ACS Applied Materials & Interfaces* 7, 9937–9946. doi:10.1021/acsami.5b02131

Yu Z, Sun X, Song H, Wang W, Ye Z, Shi L, Ding K (2015) Glutathione-responsive carboxymethyl chitosan nanoparticles for controlled release of herbicides. *Materials Sciences and Applications* **6**, 591. doi:10.4236/msa.2015.66062

Zangi R, Filella M (2012) Transport routes of metalloids into and out of the cell: a review of the current knowledge. *Chemico-Biological Interactions* **197**, 47–57. doi:10.1016/j.cbi.2012.02.001

Zhai G, Walters KS, Peate DW, Alvarez PJ, Schnoor JL (2014) Transport of gold nanoparticles through plasmodesmata and precipitation of gold ions in woody poplar. *Environmental Science & Technology Letters* **1**, 146–151. doi:10.1021/ez400202b

Zhang M, Tao W, Pianetta PA (2007) Dynamics modelling of biolistic gene guns. *Physics in Medicine and Biology* **52**, 1485. doi:10.1088/0031-9155/52/5/017

Zhang Y, Liang Z, Zong Y, Wang Y, Liu J, Chen K, Qiu JL, Gao C (2016) Efficient and transgene-free genome editing in wheat through transient expression of CRISPR/Cas9 DNA or RNA. *Nature Communications* **7**, 12617. doi:10.1038/ncomms12617

Zhao L, Peralta-Videa JR, Ren M, Varela-Ramirez A, Li C, Hernandez-Viezcas JA, Aguilera RJ, Gardea-Torresdey JL (2012) Transport of Zn in a sandy loam soil treated with ZnO NPs and uptake by corn plants: electron microprobe and confocal microscopy studies. *Chemical Engineering Journal* **184**, 1–8. doi:10.1016/j.cej.2012.01.041

Zhao X, Meng Z, Wang Y, Chen W, Sun C, Cui B, Cui J, Yu M, Zeng Z, Guo S (2017) Pollen magnetofection for genetic modification with magnetic nanoparticles as gene carriers. *Nature Plants* **3**, 956. doi:10.1038/s41477-017-0063-z

Zhu ZJ, Wang H, Yan B, Zheng H, Jiang Y, Miranda OR, Rotello VM, Xing B, Vachet RW (2012) Effect of surface charge on the uptake and distribution of gold nanoparticles in four plant species. *Environmental Science & Technology* **46**, 12391–12398. doi:10.1021/es301977w

Zou X, Li P, Huang Q, Zhang H (2016) The different response mechanisms of *Wolffia globosa*: light-induced silver nanoparticle toxicity. *Aquatic Toxicology (Amsterdam, Netherlands)* **176**, 97–105. doi:10.1016/j.aquatox.2016.04.019

Index